Current Progress in Agronomy

Volume I

Current Progress in Agronomy
Volume I

Edited by **Jamie Hanks**

R CALLISTO
REFERENCE

New York

Published by Callisto Reference,
106 Park Avenue, Suite 200,
New York, NY 10016, USA
www.callistoreference.com

Current Progress in Agronomy: Volume I
Edited by Jamie Hanks

International Standard Book Number: 978-1-63239-138-4 (Hardback)

Printed in the United States of America.

Contents

Preface

The science of studying soil, water and crop management principles and practices is known as agronomy. It is concerned with improving the agricultural practices by drawing out methods that would increase the productivity by providing favourable conditions to the crops. Agronomic activities like restoring soil fertility, preparation of good seed bed, improving crop varieties, and managing soil resources help in getting a better yield and good quality of produce. Agronomy also takes care of crop rotation, irrigation, weed control, insecticides and pesticides and plant breeding. It also uses biotechnology to achieve its objectives, like introducing desired characteristics to a crop or eliminating undesired traits. It is an approach that will change the face of agriculture in the world.

Amendment of tephrosia improved fallows with inorganic fertilizers improves soil chemical properties. Moisture and salinity stress induced changes in biochemical constituents and water relations of different grape rootstock cultivars are also important techniques.

Plants confers drought and salt tolerance, along with an increase in crop biomass, impacting yield and its related traits in mycorrhizal and nonmycorrhizal inoculated wheat cultivars under different water regimes using multivariate statistics; Genetic transformation of common bean (phaseolus vulgaris l.) with the gus color marker, the bar herbicide resistance, and the barley (hordeum vulgare) hva1 drought tolerance genes are important subjects in crop management and agriculture.

The book is the collective effort of scientists and researchers from across the globe. I wish to thank them and the writers and researchers involved in its completion.

Editor

Comparison of Raw Dairy Manure Slurry and Anaerobically Digested Slurry as N Sources for Grass Forage Production

Olivia E. Saunders,[1] Ann-Marie Fortuna,[1, 2] Joe H. Harrison,[3] Elizabeth Whitefield,[3] Craig G. Cogger,[1] Ann C. Kennedy,[4] and Andy I. Bary[1]

[1] Department of Crop and Soil Sciences, Washington State University, Pullman, WA 99164, USA
[2] Department of Soil Science, North Dakota State University, 1301 12th Avenue North, Fargo, ND 58105, USA
[3] Department of Animal Sciences, Washington State University, Puyallup, WA 98371, USA
[4] Land Management and Water Conservation Research Unit, USDA-ARS, Washington State University, Pullman, WA 99164, USA

Correspondence should be addressed to Ann-Marie Fortuna, annmarie.fortuna@ndsu.edu

Academic Editor: H. Allen Torbert

We conducted a 3-year field study to determine how raw dairy slurry and anaerobically digested slurry (dairy slurry and food waste) applied via broadcast and subsurface deposition to reed canarygrass (*Phalaris arundinacea*) affected forage biomass, N uptake, apparent nitrogen recovery (ANR), and soil nitrate concentrations relative to urea. Annual N applications ranged from $600 \, kg \, N \, ha^{-1}$ in 2009 to $300 \, kg \, N \, ha^{-1}$ in 2011. Forage yield and N uptake were similar across slurry treatments. Soil nitrate concentrations were greatest at the beginning of the fall leaching season, and did not differ among slurry treatments or application methods. Urea-fertilized plots had the highest soil nitrate concentrations but did not consistently have greatest forage biomass. ANR for the slurry treatments ranged from 35 to 70% when calculations were based on ammonium-N concentration, compared with 31 to 65% for urea. Slurry ANR calculated on a total N basis was lower (15 to 40%) due to lower availability of the organic N in the slurries. No consistent differences in soil microbial biomass or other biological indicators were observed. Anaerobically digested slurry supported equal forage production and similar N use efficiency when compared to raw dairy slurry.

1. Introduction

There is a need for a set of best management practices that addresses how to utilize the growing quantity of reactive nitrogen (N) produced by livestock operations. Animal agriculture in the United States has become more specialized with farms consolidating and growing in size [1]. The number of dairy farms has decreased by 94% since 1960, but the number of animals has remained constant [2]. Animal consolidation has created challenges with respect to on-farm N surplus, waste management and nutrient loading in the environment [3, 4]. Annually in the United States, more than 5800 Mg of manure N is produced [5]. One approach to ameliorate negative environmental impacts associated with animal manures is through adoption of anaerobic digestion technologies to treat farm-generated manures and food processing wastes [6–9]. Digestion of wastes can provide a stable and consistent source of nutrients comparable to inorganic fertilizers such as urea.

Anaerobic digestion converts organic carbon into methane used to generate electricity, and it also converts organic N to plant available ammonium (NH_4^+), increasing the ratio of NH_4^+/total N in the effluent [10]. Carbon is removed during both the methane production and fiber removal processes, resulting in a smaller C : N ratio of the effluent [11]. Therefore, digested effluent can serve as low-cost source of readily available nutrients for crop production. Some studies [12] have found increased yield and N availability with application of anaerobically digested material as compared to nondigested material, possibly due to increased N availability and reduced carbon (C) content. Anaerobically digested manure can provide sufficient nutrients to support biomass and crop yields equivalent to synthetic fertilizers and raw manures [13, 14]. The apparent mineral nitrogen recovery

(ANR$_M$) of tall fescue (*Festuca* spp.) receiving raw dairy manure slurry was reported by Bittman et al. [13] as 55 and 51% at early and late applications, respectively, using the drag shoe method (band applied directly to soil, under plant canopy). When surface applied, the ANR$_M$ was 37% applied early and 40% when applied late. Similar results were presented by Cherney et al. [15] with orchardgrass (*Dactulis glomerata* L.) and tall fescue.

Perennial systems that contain living plants year round tend to remove more reactive N than annual systems. Mean reed canarygrass biomass measured in trials in Minnesota was 13 Mg ha^{-1} under modest N applications (168 kg N ha^{-1} yr^{-1}) [16]. Bermudagrass (*Cynodon* spp.) fertilized with 89–444 kg manure N ha^{-1} yielded a mean of 7.92 Mg dry matter ha^{-1} over four or five cuttings per year [17]. However, the forage crop recovered only 25% of the N applied over the four years included in the study. Reed canarygrass (*Phalaris arundinacea*) is an ideal candidate for N removal because of its ability to store any left-over N applied during the growing season in rhizomes overwinter, providing a significant advantage to the forage in early spring when soil-N may be limited [18].

As with any N source, application of manure N in excess of crop uptake can result in NO$_3^-$ leaching [19]. Up to 46% of applied manure N may persist in the soil, increasing the potential for loss of N after multiple applications in a growing season [20]. Some studies indicate that manure N poses less of a risk to leaching than the same amount of N in the form of synthetic fertilizer [21] due to immobilization of N that often occurs as humic materials build up in soil. Other researchers have determined manure increases NO$_3^-$ leaching [22]. Irrespective of the source and properties of the N fertilizer applied during winter months when plants are dormant, NO$_3^-$ leaching can be the main source of N loss [23].

Manure additions can enhance soil fertility and quality through their short and long-term contribution to soil C and N [24–27]. Current research demonstrates that long-term manure applications increase soil organic matter, basal respiration, microbial biomass, and enzymatic activity (measures of soil quality), while mineral fertilizers can decrease pH, enzymatic activity, and microbial biomass C [28]. Organic amendments such as manure also have an effect on microbial community structure in addition to enhancing the activity, C content, and size of soil microbial biomass [29, 30]. A study by Zhong et al. [31] demonstrated total phospholipid fatty acid (PLFA), gram-negative, and actinobacterial PLFA were highest in treatments of organic matter and organic matter + mineral NPK fertilizer. Functional diversity from organic manure and organic manure + mineral NPK fertilizers increased over time far more than with additions of synthetic fertilizers alone. We anticipate that long-term application of raw dairy slurry and digested slurry will enhance soil quality affecting microbial community structure and activity overtime.

The goal of this study was to determine the fate of applied N in anaerobically digested slurry (derived from mixed dairy slurry and food waste), raw dairy slurry, and urea during forage production. Specifically, we compare the biomass, ANR and N uptake of forages to determine which N source(s) has the potential to maintain forage biomass and reduce reactive N. In addition, we evaluated the effectiveness of subsurface deposition versus broadcast application of raw slurry and anaerobically digested slurry to improve forage biomass production, ANR and N uptake of forages, as well as reduce residual reactive N. Our hypotheses were (1) digested slurry would have more available N and generate a greater forage response than raw slurry, (2) subsurface deposition would conserve more N than surface application, resulting in greater forage response, particularly for the raw manure with higher solids content, and (3) application of digested slurry would reduce soil nitrate-N concentrations relative to urea.

2. Materials and Methods

2.1. Site Description. A field-based experiment, located on a commercial dairy in Monroe, Washington, was established in 2009. The field was mapped as 90% Puget silty clay loam (fine-silty, mixed, superactive, nonacid, mesic Fluvaquentic Endoaquepts) and 10% Sultan silt loam soils (fine-silty, isotic, mesic, Aquandic Dystrochrepts) [32] and had a history of manure applications. The site had climatic conditions typical of the Maritime Pacific Northwest with cool wet winters and dry summers. The 2009 growing season had a drier than normal summer, while 2011 had a cool spring and dry summer (Table 1). The 2010 season had the best growing conditions, with a warm spring and more summer rainfall than 2009 or 2011.

The experimental design included six treatments in a randomized complete block design with four replicates (3.6 m × 18 m). Treatments included two dairy manure slurries (raw and anaerobically digested), two slurry application methods (broadcast and subsurface deposition), inorganic N (pelletized urea), and a zero-fertilizer treatment that received 0 kg N ha^{-1}.

The raw dairy slurry was flushed from the barn floor and obtained fresh from a holding tank. Digested slurry was produced in an anaerobic digester with a plug-flow design, operating within mesophilic (23.5°C) conditions, with an approximate 17-day retention time, and storage capacity of ~6,100,000 liters. Liquid slurry from a single dairy consisting of 1,000 lactating cows was codigested with pre-consumer food-waste substrates. Food-waste consisted of no more than 30% of the total digester input and included whey, egg byproduct, processed fish, ruminant blood, biodiesel byproduct, and Daf grease (dissolved air flotation). After digestion, materials were passed through a rotating drum screen solid separator where solids were removed for composting and liquids pumped to a storage lagoon. The digested slurry applied to plots was obtained just after liquids-solids separation and prior to lagoon storage. A 250 mL subsample of each slurry was taken during each application (Table 2), cooled, and analyzed for total-nitrogen, nitrate-N, ammonium-N, total-phosphorus, and total solids [33] (Table 3).

TABLE 1: Average air temperature and total precipitation by month beginning at the start of plot implementation through the 3rd growing season.

Year	Month	Average air temp (°C)	Total precipitation (mm)
2009	April	9.2	61
	May	12.8	73
	June	**16.8**	**19**
	July	**20.0**	**6**
	August	**18.0**	**13**
	September	15.5	54
	October	9.9	100
	November	7.6	137
	December	1.4	35
2010	January	7.3	89
	February	7.0	44
	March	7.8	54
	April	9.4	55
	May	11.2	66
	June	**14.4**	**61**
	July	**17.0**	**7**
	August	**17.1**	**22**
	September	15.3	85
	October	10.6	85
	November	5.7	107
	December	5.3	184
2011	January	5.1	120
	February	3.4	80
	March	7.1	119
	April	7.2	79
	May	10.6	74
	June	**14.1**	**39**
	July	**15.7**	**18**
	August	**16.4**	**3**
	September	15.5	23
	October	10.2	71

Data from Washington State University AgWeatherNet, 21-Acres Station.

TABLE 2: Dates of forage harvest and fertilizer (slurry and urea) applications for field study in Monroe, WA for 2009–2011.

2009		2010		2011	
Forage Harvest	Fertilizer application[a]	Forage Harvest	Fertilizer application[a]	Forage harvest	Fertilizer application[a]
	17-Apr-09[b]		4-Mar-10		
7-May-09[c]	14-May-09	26-Apr-10	11-May-10	5-May-11	19-May-11
2-Jun-09	8-Jun-09	10-Jun-10	22-Jun-10	10-Jun-11	22-Jun-11
1-Jul-09	20-Jun-09[d]	7-Jul-10	15-Jul-10	14-Jul-10	4-Aug-11
28-Jul-09	11-Aug-09	12-Aug-10		22-Aug-10	31-Aug-11
31-Aug-09		15-Sep-10	30-Sep-10	20-Sep-10	
29-Sep-09		2-Dec-10		18-Oct-11	
30-Nov-09					

[a] Soil samples taken 1-day prior to fertilizer application.
[b] Early season manure application by grower, prior to plot establishment.
[c] Harvest prior to plot establishment, yield data from this harvest does not include replicates.
[d] Unintended slurry application from grower.

TABLE 3: Annual mean N and P concentrations of raw and anaerobically digested slurries applied to pasture plots.

	2009		2010		2011	
	Raw dairy Slurry	Digested slurry	Raw dairy slurry	Digested slurry	Raw dairy slurry	Digested slurry
Percent Total Solids (%)	2.8	1.9	3.4	2.0	3.4	1.4
Total N, mg kg^{-1}[a]	1441	1617	1653	2672	1475	2000
NH$_4$–N, mg kg^{-1}	707	1038	776	1253	760	930
Organic N, mg kg^{-1}[b]	734	578	877	1419	715	1070
Total P, mg kg^{-1}	350	300	331	292	330	210

[a]N and P concentrations reported as is.
[b]Organic nitrogen (N) = total N − NH$_4$–N.

A mix of reed canarygrass (*Phalaris arundinacea*) cv. "Palaton" and white clover (*Trifolium repens*) was over-seeded into the field at 62 kg ha^{-1} in May 2006, three years before the start of this experiment. Plots were sprayed with broad leaf herbicides on 18 June 2009, 10 July 2010, and 8 August 2011 with 1.17 L ha^{-1}, 2, 4-Dichlorophenoxy acetic acid, 73 mL ha^{-1} Carfentrazone-ethyl (*Aim*), and 410 mL ha^{-1}, dicamba (*Banvel*) to control the clover.

2.2. Slurry Application Method. Slurries were applied via two application methods, subsurface deposition and surface broadcast application. Subsurface deposition was accomplished with a 4169-liter capacity manure tank fitted to a National Volume Equipment pump (model MEC 4000/PALD) with a 3.05 meter Aerway Sub-Surface Deposition (Model AW1000-2B48-D) and custom Banderator attachment for application of manure through eight PVC pipes attached directly behind the Banderator tines. Tines were set 19 cm apart on the roller and allowed to drop 10 cm below the soil surface creating intermittent slices 12.5 cm in length at the surface. Visual observation of the plots suggested that the tines created slices at random locations throughout the growing season. Surface broadcast of raw and anaerobically digested slurries were accomplished using an Aerway system with the tines raised above the soil surface.

Application rates for the raw and anaerobically digested slurry were projected to be equal in total N and allowed to vary in ammonia-N, for a total yearly application of approximately 600 kg N ha^{-1} in 2009, 500 kg N ha^{-1} in 2010, and 300 kg N ha^{-1} in 2011 (Table 4). We reduced the amount of N applied on urea and slurry treatments each year of the study from 2009 to 2011 based on the fall soil nitrate concentrations. When soil nitrate-N is above 35 mg N kg^{-1} in the fall, it is recommended that applications be eliminated after August 1st, N application rates be reduced in the subsequent year by 25–40 percent and sidedress N at planting be eliminated [34]. An early season raw dairy slurry application (Table 2, application 1 in 2009) was applied by the grower to all plots prior to establishment of the field plots and is not included in the statistical analyses. An inadvertent application of 143 kg N ha^{-1} across all plots by the grower in June of 2009 (Table 2, application 4) is included in the analysis. This accounts for the higher annual application rate in 2009. Application rates were lowest in 2011 because wet

conditions prevented an early season application (Table 4), and the slurries had lower mean N concentrations. Plots were fertilized no more than five days after grass harvest. There were a total of five manure applications per year during 2009-2010 and four in 2011 (Table 2).

2.3. Field Management and Analysis. The aboveground biomass from grass swaths, 0.6 × 0.6 m, was harvested from the center of each plot every 28–35 d (Table 2) using hand-held hedge clippers. Three subsamples were taken within each of the four plot replicates for each treatment. The three subsamples were divided into grasses, clover, and weeds to adjust the aboveground biomass and ANR measurements. Due to herbicide applications, weeds were minimal all years. White clover biomass was significant in two of the cuttings in 2011. Samples were bagged, and weighed immediately. Forage was then dried at 55°C for 24 hrs, weighed, and ground in a Wiley Mill (Arthur H. Thomas Co., Philadelphia, PA) with a 1 mm screen. Ground samples were analyzed for forage nitrogen content with a Leco FP-528 Nitrogen Analyzer (Leco Corporation, St. Joseph, MI; AOAC, 2001) by Cumberland Valley Analytical Services Inc. (Hagerstown, MD).

Soil samples were collected from each plot and analyzed for Bray-1 P, exchangeable K, and pH at the beginning of the experiment 2009 and again at the end of 2010. Six soil cores per plot were taken to a 30-cm depth using a 2.54 cm diameter soil sampling probe and composited. Additional soil samples were collected for nitrate-N analysis monthly throughout the growing season using the same method, except for biweekly from mid-September through the end of November. Nitrate-N below the 30 cm depth was not measured. Soil chemical properties were analyzed by Soiltest Farm Consultants (Moses Lake, WA) using the methods of Gavlak et al. [33]. Ammonium-N was determined using a salicylate-nitroprusside method and nitrate-N using the cadmium reduction method. Soil samples for gravimetric water content were homogenized by mixing, and a subsample was dried at 38°C for 72 hrs [35].

Whole-soil phospholipid fatty acid (PLFA) procedures generally followed Bligh and Dyer [36] as described by Petersen and Klug [37] and modified by Ibekwe and Kennedy [38]. Fatty acid methyl esters were analyzed on a gas chromatograph (Agilent Technologies GC 6890, Palo Alto, CA)

TABLE 4: Application rate of fertilizer source at each application period, and seasonal total N and P inputs, 2009–2011.

Application	1	2	3	4	5	1	2	3	4	5	Seasonal total		
	Total N kg ha^{-1}					NH$_4^+$–N kg ha^{-1}					Total N kg ha^{-1}	NH$_4^+$–N kg ha^{-1}	Total P kg ha^{-1}
2009													
Control	111[a]	0	0	143[b]	0	51	0	0	83[b]	0	254	135	0
Urea	111[a]	112	112	143[b]	112	51	112	112	83[b]	112	590	471	0
Raw	111[a]	121	168	143[b]	47	51	64	78	83[b]	23	590	300	130
Digested	111[a]	176	115	143[b]	81	51	114	93	83[b]	47	626	389	121
2010													
Control	0	0	0	0	0	0	0	0	0	0	0	0	0
Urea	112	112	112	112	0	112	112	112	112	0	448	448	0
Raw	92	112	113	99	84	53	24	58	53	53	500	240	92
Digested	86	121	63	67	129	55	40	35	40	98	466	268	54
2011													
Control	0	0	0	0	0	0	0	0	0	0	0	0	0
Urea	0	112	112	112	0	0	112	112	112	0	336	336	0
Raw	0	52	60	84	63	0	34	24	29	47	258	135	51
Digested	0	33	136	69	70	0	23	52	33	27	308	134	26

[a]Early season slurry application by grower, prior to plot establishment.
[b]Application 4 in 2009 was an unintended application from the grower to all plots. Urea fertilizer considered equal to NH$_4^+$ in plant availability.

with a fused silica column and equipped with a flame ionizer detector and integrator. ChemStation (Agilent Technologies) operated the sampling, analysis, and integration of the samples. Extraction efficiencies were based on the nonadecanoic acid peak as an internal standard. Peak chromatographic responses were translated into mole responses using the internal standard and responses were recalculated as needed. Microbial groups were calculated based on the procedure of Pritchett et al. [39].

2.4. Slurry Analysis.
Slurries were analyzed for total N, ammonium-N, and total P (Table 3). Nitrogen was extracted via the Kjeldahl method [33]. Phosphorus was analyzed using a Thermo IRIS Advantage HX Inductively Coupled Plasma (ICP) Radial Spectrometer (Thermo Instrument Systems, Inc., Waltham, MA) by the Dairy One Forage Analysis Laboratory (Ithaca, NY).

2.5. Statistical Analyses and Calculations.
An analysis of variance (ANOVA) was run using SAS PROC MIXED on the aboveground forage biomass, nitrogen content in forage, soil nitrate-N, and soil biological groups for all treatments across the three years [40]. Data were analyzed as a randomized complete block design with each of the six treatments analyzed independently. Crop biomass and crop-nitrogen content from each year were analyzed separately using ANOVA with treatment and sample day as fixed effects. Significance is indicated with a $p < 0.05$ [40].

Forage apparent N recovery (ANR%) was calculated in 2010 and 2011 as a percentage of N (total and inorganic) applied during the season based on the work of Cogger et al. [41] and Bittman et al. [13]:

$$100 \times (\text{annual grass N uptake, treated}) \\ - (\text{annual grass N uptake, control})/\text{applied N.} \quad (1)$$

Estimates of N fixed in white clover were set to 80% of total N in clover biomass based on ^{15}N studies conducted in a pasture of similar forages and N fertilizer management [42]. Using the above correction, 80% of clover N was subtracted from the forage N uptake values used in the ANR calculations for the two cuttings in 2011 with significant amounts of clover.

3. Results

3.1. Baseline Soil Data.
Soil data sampled in May 2009 prior to the start of the field experiment (Table 5) indicated that fertility was not different across the field site. Organic matter, a source of inorganic-N, averaged 5.5 percent (55 g kg^{-1}). Bulk density of soils in the field ranged from 1.14 to 1.30 g cm^{-3}, with a mean average density of 1.21 g cm^{-3}.

3.2. Forage Biomass, N Uptake and ANR.
Analysis of variance results for cumulative forage biomass in 2009 to 2011 are presented in Table 6. Total yield was greatest in 2010 (14.1–18.0 Dry Mg ha^{-1}) and lowest in 2011 (9.2–11.1 Dry Mg ha^{-1}). The 2009 data (8.08–9.5 Dry Mg ha^{-1}) did not include the first cutting of the year (7.7 Mg ha^{-1}) because it was harvested before plots and treatments were established. The growing conditions in 2010 were the most favorable of the three seasons. Forage biomass in 2011 was reduced by cool spring temperatures and low summer rainfall (Table 1). Urea had the highest yield in 2009, (Table 6). In 2010, urea and digested broadcast slurry had higher yield than the digested slurry applied subsurface. Slurry type and application method did not affect yield in 2009 or 2011.

Similar trends occurred when comparing crop N uptake in the forage grasses (Table 6). Urea-treated plots accumulated the most plant N, ranging from 296 to 655 kg N ha^{-1} removed per year. Uptake of N in forage grasses was greatest in 2010 (Table 6). Slurry type and application method did not have a significant effect on N uptake any year.

TABLE 5: Soil pH, Bray-P, and exchangeable K at start of experiment and after two years of slurry applications.

Plot	pH	Bray P	NH$_4$OAc K
		mg kg^{-1}	
Baseline, 12 May 2009			
Control	6.0	173	591
Urea	6.0	176	608
Raw-subsurface	6.0	160	598
Raw-broadcast	6.0	165	632
Digested-subsurface	6.1	140	616
Digested-broadcast	6.0	168	612
2 December 2010			
Control	6.2a	173	379b
Urea	6.0b	176	286c
Raw-subsurface	6.3a	186	479a
Raw-broadcast	6.2a	185	465a
Digested-subsurface	6.2a	173	447ab
Digested-broadcast	6.2a	162	440ab

Letters in a column within a year indicate significant differences at $\rho = 0.05$, letters are not included when no significant differences were found. Samples from different dates were analyzed separately using an ANOVA.

TABLE 6: Annual forage yield and N uptake, 2009 to 2011.

Treatment	Forage yield Dry Mg ha^{-1}			Nitrogen uptake N kg ha^{-1}		
	2009[a]	2010	2011	2009[b]	2010	2011
Control	8.0^b	14.1^c	9.2^b	283^c	362^{cd}	192^c
Urea	9.5^a	18.0^a	11.1^a	389^a	655^a	296^a
Raw- subsurface	8.6^b	16.6^{ab}	10.5^a	330^b	507^b	263^b
Raw- broadcast	7.9^b	17.0^{ab}	10.8^a	308^{bc}	531^b	254^b
Digested-subsurface	8.6^b	16.1^b	10.9^a	332^b	501^b	239^b
Digested-broadcast	8.7^b	17.3^a	10.9^a	338^b	550^{ab}	255^b

Letters within a column indicate significant differences at $\rho = 0.05$.
[a]Values for forage yield from the first harvest prior to implementation of nitrogen fertilizer treatments and application method were 7.7 Mg ha^{-1}.
[b]The N content in forage yield from the first harvest prior to implementation of nitrogen fertilizer treatments and application method was 253 kg N ha^{-1}.

Nitrogen uptake was lowest in 2011, likely a result of lower N application rates (Table 4) and poorer weather during the spring and summer. Forages in 2011 also contained significant amounts of clover, an N fixer (27% of the dry mass of forage yield at harvest 1 and 34% at harvest 2). Less than 10% of the forage biomass was clover in 2009 and 2010.

In the first full season of the study (2010), the recovery of applied N in the forage (ANR) was higher than in 2011 (Table 7). More favorable weather patterns for growth in 2010 compared with 2011 probably increased ANR in 2010. Urea treatments had an ANR of 65% in 2010 and 31% in 2011. Calculations based on total N applied in slurries were lower, ranging from 29 to 40% in 2010 and 15 to 24% in 2011, and similar between the two types of slurry. ANR calculations based only on the amount of total NH$_4^+$–N applied in slurries were 52 to 70% in 2010 and 35 to 53% in 2011, similar to ANR observed for urea.

3.3. Soil Nitrate-N.
Plots receiving urea had the highest concentration of soil nitrate-N over the three seasons, while there were few differences among the slurry treatments (Table 8). Soil nitrate-N concentrations were highest in all fertilized treatments from July to the start of the fall rainy season, when the potential for leaching increases. Soil nitrate-N levels were greatest in 2009, likely because of the high rates of N applied that year. Lower soil nitrate-N in 2010 reflected the high N uptake during the favorable growing conditions that year. Soil nitrate-N increased again in 2011, particularly in the fall. This was despite a lower N application rate and may reflect the reduced yield and N uptake by the forages during the less favorable growing season in 2011.

3.4. Microbial Groups.
Microbial groups in general did not vary with treatment, but rather varied by year (Table 9). The control and urea treatments varied from the other treatments most consistently for most groups, while no consistent differences were observed among the slurry treatments. By 2011, the control treatment had significantly lower bacteria and anaerobic markers than the other treatments, but similar levels of overall microbial biomass and fungi.

TABLE 7: Apparent nitrogen recovery (ANR) in harvested forage as percentage of total and ammonium N applied, 2010 and 2011.

	ANR 2010		ANR 2011	
Treatment	% of Total N	% of NH$_4^+$–N	% of Total N	% of NH$_4^+$–N
Urea	65	65	31	31
Raw-subsurface	29	60	15	35
Raw-broadcast	34	70	24	47
Digested-subsurface	30	52	23	53
Digested-broadcast	40	70	20	46

Urea fertilizer considered equal to NH$_4^+$ in plant availability.

TABLE 8: Soil NO$_3^-$–N (mg kg^{-1}) at 0 to 30 cm depth, 2009–2011.

	Soil NO$_3^-$–N (mg kg^{-1})					
				Treatment		
Sample Date	Control	Urea	Raw subsurface	Raw broadcast	Digested subsurface	Digested broadcast
			2009			
12-May	20gh	21gh	18gh	19gh	19gh	20gh
4-Jun	18gh	28fg	24fg	23 g	30fg	24fg
6-Jul	35fe	80b	71bc	76bc	68c	65cd
3-Aug	34f	86ab	80b	76bc	86ab	71bc
9-Sep	20gh	91a	66cd	82ab	72bc	67c
21-Sep	20gh	81b	53de	62cd	78bc	55d
1-Oct	17gh	62cd	52de	50de	56d	45de
19-Oct	14gh	91a	35fe	54de	44e	45de
3-Nov	11h	54de	23g	30fg	29fg	23g
19-Nov	10h	22gh	9.8h	12h	11h	10h
30-Nov	11h	11gh	12h	12h	10h	12h
			2010			
26-Feb	12fg	11fg	15ef	15ef	13f	14ef
11-May	13f	23d	20de	20de	20de	18ef
16-Jun	6.1g	9.2fg	7.9g	7.0g	7.4g	7.3g
13-Jul	10fg	25cd	18e	18de	16ef	14ef
17-Aug	13fg	61a	23cd	22de	18ef	22de
30-Sep	18de	36b	23cd	28c	22de	19de
12-Oct	12fg	28bc	22de	22de	27cd	24cd
26-Oct	7.2g	12fg	15ef	17ef	21de	16ef
2-Dec	6.9g	8.2g	11fg	9.5fg	10fg	10fg
			2011			
4-Apr	6.1g	6.2g	7.0g	6.5g	7.0g	8.0g
21-Jun	7.9g	18ef	11fg	11fg	12fe	12fg
4-Aug	8.7g	21ef	15fg	17ef	18ef	18ef
30-Aug	12fg	43cd	22ef	16f	23ef	17ef
16-Sept	17ef	48bc	39cd	48bc	48bc	32d
29-Sept	17ef	46c	36d	42cd	55b	36d
13-Oct	19ef	66a	45c	44c	50bc	44c
4-Nov	8.4g	32d	28de	24e	30de	23ef

Letters within a year indicate significant differences at $\rho = 0.05$.

4. Discussion

4.1. Forage Biomass, N Uptake and ANR. Forage biomass, plant N-uptake, and nitrate concentrations during the 2009–2011 growing seasons were affected by seasonal and long-term N management (a history of manure applications) that resulted in high N uptake from the control treatments. Also, favorable growing conditions in 2010 allowed for a more productive field season in this year. For this study, total harvest yield during each season was within the range of other published work where animal manures were applied to forages harvested multiple times over a season [16, 17, 41].

While other studies have shown incorporation of manure to increase yield and crop N content by reducing gaseous losses [13], we did not see an improvement in crop N content from incorporation of slurries in this system. Forages grown

TABLE 9: Soil microbial analyses from field plots in the spring, 2009–2011.

	Biomass g kg^{-1}	Bacteria Mole percent[a]	Fungi Mole percent	Bacteria to fungi ratio	Anaerobe Mole percent	Mono-unsaturated Mole percent
			May 2009			
Control	535 ab	0.246	0.098	3.01	0.091	0.338
Urea	433 c	0.246	0.092	3.23	0.092	0.348
Raw-B	538 ab	0.243	0.093	3.18	0.094	0.335
Raw-SSD	454 bc	0.237	0.094	3.07	0.091	0.330
Digested-B	473 bc	0.242	0.092	3.18	0.093	0.324
Digested-SSD	623 a	0.238	0.083	3.45	0.091	0.322
			May 2010			
Control	610 a	0.243 b	0.071 abc	4.18 ab	0.115 ab	0.328 b
Urea	333 b	0.215 c	0.074 ab	3.48 c	0.101 b	0.322 b
Raw-B	401 a	0.266 ab	0.084 a	4.04 bc	0.116 ab	0.357 ab
Raw-SSD	297 b	0.268 a	0.066 bc	4.91 a	0.127 a	0.414 a
Digested-B	258 b	0.259 ab	0.071 abc	4.65 ab	0.123 a	0.398 a
Digested-SSD	279 b	0.267 ab	0.066 c	4.97 a	0.125 a	0.406 a
			April 2011			
Control	512	0.221 b	0.087 ab	3.11 d	0.082 c	0.341 b
Urea	447	0.250 a	0.078 c	3.96 a	0.094 ab	0.380 a
Raw-B	489	0.257 a	0.093 ab	3.35 bcd	0.092 b	0.335 b
Raw-SSD	428	0.253 a	0.095 a	3.25 cd	0.100 ab	0.345 b
Digested-B	441	0.258 a	0.085 bc	3.68 ab	0.101 ab	0.357 ab
Digested-SSD	491	0.255 a	0.085bc	3.61 abc	0.102 a	0.359 ab

Letters within a column within a year indicate significant differences at $\rho = 0.05$. No letters indicate no significant differences within that column.
[a]Mole percent = (mole substance in a mixture)/(mole mixture) %.

in plots with broadcast applied slurries took up the same amount of N or more N than with subsurface deposition, which may have been caused by plant-growth disturbance from the airway banderator when subsurface applying effluent. Additionally, the infiltration rate of the anaerobically digested slurry may have been rapid enough that gaseous losses in the field were not different among subsurface deposition and broadcast applications. From an agronomic perspective, the two slurry types performed equally well as urea over the three growing seasons. Anaerobically digested slurry was suitable for forage production when applied at rates equal to raw dairy slurry. Moller and Stinner [8] also reported no differences in N uptake between digested and undigested slurry. How the system will respond after many years of anaerobically digested slurry application is unclear as the quantity of organic N applied is less than that of raw dairy slurry, supplying less recalcitrant N to the pool of soil organic matter.

4.2. Soil Nitrate-N and Microbial Groups. We found few differences between slurry treatments in seasonal soil NO$_3^-$ concentrations. There was, however, significantly more nitrate-N in urea-treated plots on many dates, even though there was slightly less total N applied to the urea plots in some years. The spike in nitrate concentration in October on soils where urea was applied in place of slurries indicates a greater potential for N leaching from urea compared with

the slurries. All treatments declined in NO$_3^-$ concentrations to levels that were not significantly different from control treatments after the fall rains began. Lower soil nitrate-N during the growing season of 2010 compared with 2009 may be due in part to a lower amount of total nitrogen applied. Also, little rainfall during the 2009 growing season may have caused a buildup of soil nitrate in the surface layers. Higher late-season nitrate in 2011 compared with 2010 may have been the result of poorer growing conditions reducing N uptake.

Postharvest soil nitrate-N is a measure of residual plant-available N subject to leaching loss, and an indicator of excess applied N and/or poor yield. The recommended timing of postharvest soil nitrate testing in forage systems that utilize animal manure as a source of fertility in the Maritime Pacific Northwest is prior to October 15 [34]. Nitrate concentrations from soil samples collected from our site in mid-October showed that all treatments except the control exceeded 30 mg NO$_3$–N kg ha^{-1} in 2009 and 2011, with NO$_3$–N levels highest in the urea treatment. Fall nitrate-N levels above 30 mg kg^{-1} are considered excessive in manured pastures, and reduced rates and adjusted timing of applications are recommended [34].

While soil nitrate concentrations decreased during the fall 2009 months, it is likely that some of this nitrate was not entirely leached from the system, but stored in the canary grass rhizomes over winter as described by Partala et al. [18].

This is evident in the significantly higher yields and nitrogen content of forages during the early season harvest on 26 April 2010.

While the focus of this study is N, dairy manure also contains high levels of P. Runoff from high-P soils can lead to eutrophication in fresh water. Soil P levels were already excessive at the start of this study, because of the history of dairy manure applications at the site, and P tended to increase in the slurry-treated plots during the study (Table 5). The anaerobically digested slurry contained less P than the raw dairy slurry, probably because it had a lower solids content, which would lead to less P accumulation over time.

Microbial groups varied with year more than treatment in these field studies. Urea treatments varied from the other treatments to the greatest extent. The raw and anaerobically digested materials did not alter the soil microbial components as determined by PLFA. Our results may partially be the result of past manure applications.

5. Conclusions

Subsurface deposition did not increase yield or N uptake compared with surface broadcast application, possibly because the slurries were low enough in solids to infiltrate readily into the soil, and because the subsurface injectors could have disrupted plant growth. Anaerobically digested dairy slurry was shown to provide adequate soil fertility and N availability for crop uptake and forage production over the three field seasons. In the short term, anaerobically digested slurry did not significantly increase yield or N uptake compared with similar rates of raw slurry.

This study indicated that soil nitrates measured to a 30 cm depth were fairly consistent across slurry treatments and application methods during each of the field seasons. Soil nitrate-N was lower in 2010 due to favorable growing conditions and lower total applied N relative to 2009. Although urea treatments had the highest apparent N recovery value, the potential for nitrate leaching was also greatest under this management. Anaerobically digested slurry did not increase soil NO_3^- concentrations or alter the microbial composition and provided equal forage production and similar N use efficiency when compared to undigested dairy slurry.

Acknowledgments

Funding for this project was provided by WA Dairy Prod. Commission 113547-001, USDA-NRCS 68-3A 75-5-178 and Hatch Project 0711. The authors would like to thank our student aides: Jeni Maakad, Taylor Graves, Mike Pecharko, Jacob Turner, Clayton Waller, and Allison Wood. They would also like to welcome Vivienne Marie Whitefield to their work group born November 21, 2011.

References

[1] P. Gerber, H. A. Mooney, J. Dijkman, S. Tarawali, and C. DeHaan, *Livestock in a Changing Landscape: Experiences and Regional Perspectives*, vol. 2, Island Press, Washington, DC, USA, 2010.

[2] T. J. Centner, "Regulating concentrated animal feeding operations to enhance the environment," *Environmental Science and Policy*, vol. 6, no. 5, pp. 433–440, 2003.

[3] T. J. Centner, "Evolving policies to regulate pollution from animal feeding operations," *Environmental Management*, vol. 28, no. 5, pp. 599–609, 2001.

[4] A. N. Hristov, W. Hazen, and J. W. Ellsworth, "Efficiency of use of imported nitrogen, phosphorus, and potassium and potential for reducing phosphorus imports on Idaho dairy farms," *Journal of Dairy Science*, vol. 89, no. 9, pp. 3702–3712, 2006.

[5] R. L. Kellogg, C. H. Lander, D. C. Moffett, and N. Gollehon, *Manure Nutrients Relative to the Capacity of Cropland and Pastureland to Assimilate Nutrients: Spatial and Temporal Trends for the United States*, United States Department of Agriculture, Natural Resources Conservation Service, Washington, DC, USA, 2000.

[6] AgSTAR Program, "Managing manure with biogas recovery systems improved performance at competitive costs," Environmental Protection Agency, Office of Air and Radiation. 430-F02-004, 2002.

[7] H. Hou, S. Zhou, M. Hosomi et al., "Ammonia emissions from anaerobically-digested slurry and chemical fertilizer applied to flooded forage rice," *Water, Air, and Soil Pollution*, vol. 183, no. 1–4, pp. 37–48, 2007.

[8] K. Moller and W. Stinner, "Effects of organic wastes digestion for biogas production on mineral nutrient availability of biogas effluents," *Nutrient Cycling in Agroecosystems*, vol. 87, no. 3, pp. 395–413, 2010.

[9] M. A. Moser, "Anaerobic digesters control odors, reduce pathogens, improve nutrient manageability, can be cost competitive with lagoons, and provide energy too!," The AgSTAR Program, 2010, http://www.epa.gov/agstar/resources/man_.html.

[10] J. Michel, A. Weiske, and K. Moller, "The effect of biogas digestion on the environmental impact and energy balances in organic cropping systems using the life-cycle assessment methodology," *Renewable Agriculture and Food Systems*, vol. 25, no. 3, pp. 204–218, 2010.

[11] K. Moller, W. Stinner, A. Deuker, and G. Leithold, "Effects of different manuring systems with and without biogas digestion on nitrogen cycle and crop yield in mixed organic dairy farming systems," *Nutrient Cycling in Agroecosystems*, vol. 82, no. 3, pp. 209–232, 2008.

[12] H. C. de Boer, "Co-digestion of animal slurry can increase short-term nitrogen recovery by crops," *Journal of Environmental Quality*, vol. 37, no. 5, pp. 1968–1973, 2008.

[13] S. Bittman, C. G. Kowalenko, D. E. Hunt, and O. Schmidt, "Surface-banded and broadcast dairy manure effects on tall fescue yield and nitrogen uptake," *Agronomy Journal*, vol. 91, no. 5, pp. 826–833, 1999.

[14] E. R. Loria, J. E. Sawyer, D. W. Barker, J. P. Lundvall, and J. C. Lorimor, "Use of anaerobically digested swine manure as a nitrogen source in corn production," *Agronomy Journal*, vol. 99, no. 4, pp. 1119–1129, 2007.

[15] D. J. R. Cherney, J. H. Cherney, and E. A. Mikhailova, "Orchardgrass and tall fescue utilization of nitrogen from dairy manure and commercial fertilizer," *Agronomy Journal*, vol. 94, no. 3, pp. 405–412, 2002.

[16] C. C. Sheaffer, G. C. Marten, D. L. Rabas, N. P. Martin, and D. W. Miller, *Reed Canarygrass*, vol. 595, Minnesota Agricultural Experiment Station, 1990.

[17] M. A. Sanderson and R. M. Jones, "Forage yields, nutrient uptake, soil chemical changes, and nitrogen volatilization from bermudagrass treated with dairy manure," *Journal of Production Agriculture*, vol. 10, no. 2, pp. 266–271, 1997.

[18] A. Partala, T. Mela, M. Esala, and E. Ketoja, "Plant recovery of ^{15}N-labelled nitrogen applied to reed canary grass grown for biomass," *Nutrient Cycling in Agroecosystems*, vol. 61, no. 3, pp. 273–281, 2001.

[19] J. S. Angle, C. M. Gross, R. L. Hill, and M. S. McIntosh, "Soil nitrate concentrations under corn as affected by tillage, manure, and fertilizer applications," *Journal of Environmental Quality*, vol. 22, no. 1, pp. 141–147, 1993.

[20] G. R. Munoz, J. M. Powell, and K. A. Kelling, "Nitrogen budget and soil N dynamics after multiple applications of unlabeled or (15) nitrogen-enriched dairy manure," *Soil Science Society of America*, vol. 67, no. 3, pp. 817–825, 2003.

[21] H. Trindade, J. Coutinho, S. Jarvis, and N. Moreira, "Effects of different rates and timing of application of nitrogen as slurry and mineral fertilizer on yield of herbage and nitrate-leaching potential of a maize/Italian ryegrass cropping system in north-west Portugal," *Grass and Forage Science*, vol. 64, no. 1, pp. 2–11, 2009.

[22] J. M. Jemison and R. H. Fox, "Nitrate leaching from nitrogen-fertilized and manured corn measured with zero-tension pan lysimeters," *Journal of Environmental Quality*, vol. 23, no. 2, pp. 337–343, 1994.

[23] A. Bakhsh, R. S. Kanwar, C. Pederson, and T. B. Bailey, "N-source effects on temporal distribution of NO_3-N leaching losses to subsurface drainage water," *Water, Air, and Soil Pollution*, vol. 181, no. 1–4, pp. 35–50, 2007.

[24] T. G. Sommerfeldt, C. Chang, and T. Entz, "Long-term annual manure applications increase soil organic matter and nitrogen, and decrease carbon to nitrogen ratio," *Soil Science Society of America*, vol. 52, no. 6, pp. 1668–1672, 1988.

[25] W. E. Jokela, "Nitrogen fertilizer and dairy manure effects on corn yield and soil nitrate," *Soil Science Society of America*, vol. 56, no. 1, pp. 148–154, 1992.

[26] R. B. Ferguson, J. A. Nienaber, R. A. Eigenberg, and B. L. Woodbury, "Long-term effects of sustained beef feedlot manure application on soil nutrients, corn silage yield, and nutrient uptake," *Journal of Environmental Quality*, vol. 34, no. 5, pp. 1672–1681, 2005.

[27] J. Nyiraneza and S. Snapp, "Integrated management of inorganic and organic nitrogen and efficiency in potato systems," *Soil Science Society of America*, vol. 71, no. 5, pp. 1508–1515, 2007.

[28] G. F. Ge, Z. J. Li, F. L. Fan, G. X. Chu, Z. A. Hou, and Y. C. Liang, "Soil biological activity and their seasonal variations in response to long-term application of organic and inorganic fertilizers," *Plant and Soil*, vol. 326, no. 1, pp. 31–44, 2009.

[29] H. A. Ajwa and M. A. Tabatabai, "Decomposition of different organic materials in soils," *Biology and Fertility of Soils*, vol. 18, no. 3, pp. 175–182, 1994.

[30] H. Chu, X. J. Lin, T. Fujii et al., "Soil microbial biomass, dehydrogenase activity, bacterial community structure in response to long-term fertilizer management," *Soil Biology and Biochemistry*, vol. 39, no. 11, pp. 2971–2976, 2007.

[31] W. H. Zhong, T. Gu, W. Wang et al., "The effects of mineral fertilizer and organic manure on soil microbial community and diversity," *Plant and Soil*, vol. 326, no. 1-2, pp. 511–522, 2010.

[32] A. Debose and M. W. Klungland, *Soil Survey of Snohomish County Area*, USDA Soil Conservation Service, Washigton, DC, USA, 1983.

[33] R. G. Gavlak, D. A. Horneck, and R. O. Miller, *Soil, Plant, and Water Reference Methods for the Western Region*, vol. 125, Region Extension Publications, 3rd edition, 2005.

[34] D. M. Sullivan and C. G. Cogger, *Post-Harvest Soil Nitrate Testing for Manured Cropping Systems*, Oregon State University Extension Services, 2003.

[35] G. C. Topp and P. A. Ferré, "The soil solution phase," in *Methods of Soil Analysis. Part 4-Physical Methods*, J. H. Dane and G. C. Topp, Eds., vol. 5 of *Soil Science Society of America Book Series, Vol. 5*, pp. 422–423, Soil Science Society of America, 2nd edition, 2002.

[36] E. G. Bligh and W. J. Dyer, "A rapid method of total lipid extraction and purification," *Canadian Journal of Biochemistry and Physiology*, vol. 37, no. 8, pp. 911–917, 1959.

[37] S. O. Petersen and M. J. Klug, "Effects of sieving, storage, and incubation temperature on the phospholipid fatty acid profile of a soil microbial community," *Applied and Environmental Microbiology*, vol. 60, no. 7, pp. 2421–2430, 1994.

[38] A. M. Ibekwe and A. C. Kennedy, "Phospholipid fatty acid profiles and carbon utilization patterns for analysis of microbial community structure under field and greenhouse conditions," *FEMS Microbiology Ecology*, vol. 26, no. 2, pp. 151–163, 1998.

[39] K. A. Pritchett, A. C. Kennedy, and C. G. Cogger, "Management effects on soil quality in organic vegetable systems in western Washington," *Soil Science Society of America*, vol. 75, no. 2, pp. 605–615, 2011.

[40] SAS Institute Incorporation, *SAS User's Guide: Statistics. Version 9.2.*, SAS Institute Incorporation, Cary, NC, USA, 2008.

[41] C. G. Cogger, D. M. Sullivan, A. I. Bary, and S. C. Fransen, "Nitrogen recovery from heat-dried and dewatered biosolids applied to forage grasses," *Journal of Environmental Quality*, vol. 28, no. 3, pp. 754–759, 1999.

[42] F. V. Jørgensen and S. F. Ledgard, "Contribution from stolons and roots to estimates of the total amount of N_2 fixed by white clover (*Trifolium repens* L.)," *Annals of Botany*, vol. 80, no. 5, pp. 641–648, 1997.

2

Nitrogen Availability and Uptake by Sugarbeet in Years Following a Manure Application

Rodrick D. Lentz and Gary A. Lehrsch

North West Irrigation and Soils Research Laboratory, USDA-ARS, 3793 N 3600 W, Kimberly, ID, 83341-5076, USA

Correspondence should be addressed to Rodrick D. Lentz, rick.lentz@ars.usda.gov

Academic Editor: Mark Reiter

The use of solid dairy manure for sugarbeet production is problematic because beet yield and quality are sensitive to deficiencies or excesses in soil N, and soil N availability from manure varies substantially depending on the year of application. Experimental treatments included combinations of two manure rates (0.33 and 0.97 Mg total N ha^{-1}) and three application times, and non-manure treatments (control and urea fertilizer). We measured soil net N mineralization and biomass, N uptake, and yields for sprinkler-irrigated sugarbeet. On average, the 1-year-old, low-rate manure, and 1- and 2-year-old, high-rate manure treatments produced 1.2-fold greater yields, 1.1-fold greater estimated recoverable sugar, and 1.5-fold greater gross margins than that of fertilizer alone. As a group the 1-year-old, low-rate manure, and 2- and 3-year-old, high-rate-manure treatments produced similar cumulative net N mineralization as urea fertilizer; whereas the 1-year-old, high-rate manure treatment provided nearly 1.5-fold more N than either group. With appropriate manure application rates and attention to residual N and timing of sugarbeet planting, growers can best exploit the N mineralized from manure, while simultaneously maximizing sugar yields and profits.

1. Introduction

An estimated 20 million Mg manure is produced annually by the 9-million-cow US dairy herd. The regional dairy center in southern Idaho comprises 5.6% of the US total dairy herd and produces approximately 1.11 million Mg manure annually. In Idaho, much of the dairy manure is soil applied to supply crop nutrients and as a means of rebuilding soil organic carbon. The latter is particularly important for eroded soils, which are common in this historically furrow irrigated region [1]. To maximize their use of manure and minimize losses of nitrogen (N) to the environment, growers need to know how much N becomes available to crops from manure applications [2]. In addition, as competition increases for cropland in the region, farmers who rent acreage can expand the pool of land available to them if they are willing to utilize manured ground.

This is particularly important for sugarbeet (*Beta vulgaris* L.) growers because yield and beet quality parameters, sugar, and brei nitrate concentration are sensitive to both insufficient N [3, 4] and excess soil N [5, 6]. In addition, sugarbeet tends to incorporate soil residual N preferentially over fertilizer N, that is, sugarbeet will utilize more soil residual N and less applied fertilizer N than corn or tomato crops [7]. Applying excess N fertilizer early in the season or applying an optimal N application after June can divert photosynthate sugars normally used for beet root growth and sucrose accumulation to excess top growth [8]. By contrast, multiple small feedings of N to the sugarbeet from May through July can increase sucrose accumulation in roots [9]. Early research showed a positive influence of manure on beet yield and sucrose concentration [3, 5]. Still, planting sugarbeet in recently manured fields is not always recommended because N availability from the manure is not well quantified and is believed to occur too late in the season to improve yield and quality [10]. However, Lentz et al. [11] reported that (1) peak net N mineralization in manure-amended, irrigated soils coincided with maximum N uptake by beet and (2) first-year manure applications ≤20 Mg ha^{-1} (dry wt.) had no significant adverse effect on beet yield or quality.

Much of the N in dairy manure is in the organic form and only becomes available for uptake by crops via the time-dependent microbial-mediated process of mineralization.

Several studies have examined crop N uptake after multiple dairy manure applications, for barley (*Hordeum vulgare* L.) [12–14], corn (*Zea mays* L.) [15], wheat (*Triticum aestivum* L.) [16], and orchard grass (*Dactylis glomerata* L.) [17], or in the first year after a single manure application, for corn [18, 19], sugarbeet [11], and orchard grass (*Dactylis glomerata* L.) [17]. Relative to the total N applied in dairy manure, N recovery by corn, sugarbeet, and orchard grass in the first year after a single application ranged from −5 to 40% and averaged 21% [11, 17–19]. Crop N uptake from dairy manure amendments is influenced by type of crop [15], manure characteristics [20], organic amendment history [21], soil and location factors [18, 22, 23], application timing [12], and cropping management [24].

Far fewer studies have assessed N uptake by crops two and three or more years after a single manure application. For corn, the crop N recovery in the 2nd year after manure application was reported to be 9% by Klausner et al. [25], 8 to 15% by Ma et al. [15], and 15% by Eghball and Power [19], and N recovery for corn in years 3 through 4 was reported by Klausner et al. [25], being 2 or 3%. Similar studies for sugarbeet are lacking. One of the difficulties encountered when measuring crop N recovery from a manure application in successive years is the obfuscation caused by climatic variations between years [11]. Our objective was to (1) determine the effect of a single dairy manure application on sugarbeet yields and quality, N uptake, and N recovery for one, two, and three years after applying manure to a calcareous, irrigated, southern Idaho soil, and (2) employ an experimental approach that would reduce the confounding effects of climate between years.

2. Materials and Methods

We conducted the experiment at a site located 1.7 km southwest of Kimberly, ID (42E 31.12′N, 114E 22.47′W, elevation of 1190 m). The field plots were prepared in Portneuf silt loam soils (coarse silty, mixed, superactive, mesic Durinodic Xeric Haplocalcid). The experimental site had a history of maznure applications, receiving 40 to 75 Mg ha⁻¹ dry wt. every 3 yr between 1969 and 1986. In 1991 the uppermost 0.3 m (Ap horizon) of the Portneuf silt loam's profile was removed to expose the underlying Bk horizon and simulate an eroded profile [26]. For noneroded soil, the Ap horizon was left undisturbed. The site last received manure in 1994, 10 yrs before field plot preparations began for the current study. The eroded Portneuf soil profile is deep and calcareous, with textures ranging from silt loam to very fine sandy loam. Its surface soil (0 to 15 cm), that is, the Bk horizon, is a silt loam and contains on average 184 g kg⁻¹ clay, 609 g kg⁻¹ silt, 207 g kg⁻¹ sand, has a pH of 7.8 (H_2O saturated paste), electrical conductivity (EC) of 0.08 S m⁻¹, and includes 4.1 mg kg⁻¹ organic carbon, and 221 mg kg⁻¹ calcium carbonate equivalent. A silica and calcium carbonate-cemented horizon (20–60% cementation) occurs between depths of 33 to 130 cm in the eroded Portneuf. The soil has a mean cation exchange capacity of 190 mmol$_c$ kg⁻¹ and exchangeable sodium percentage of 1.5.

2.1. Experimental Design. Comparing sugarbeet yield and N uptake from a soil in years following a one-time manure application is problematic. Comparisons between annual sugarbeet measurements would be influenced not only by the treatment, but also by pest problems related to the continuous beet plantings and by climatic factors, which vary from year to year. To limit the effect of these confounding factors, we applied manure treatments at a 1x rate (average bulk application rate of 21.7 Mg ha⁻¹ dry wt. or 0.31 Mg total N ha⁻¹), and a nominal 3x rate (average bulk application of 68.9 Mg ha⁻¹ dry wt. or 0.97 Mg total N ha⁻¹) once only to a different set of field plots in the fall of each year 2004, 2005, and 2006. Thus, when sugarbeet was grown in 2007, the field plots included a set of two manure-rate treatments that were 1, 2, or 3 years old and were exposed to the same climatic conditions.

The experimental design was a randomized complete block with nine treatments and 4 replicates (Table 1). The experiment included the six manure treatments, with a manure-1x (m1) and manure-3x (m3) applied once to different plots of "eroded" Portneuf silt loam in 2004, 2005, and 2006. Three no-manure treatments were also included, urea fertilized (Fert) and control (no fertilizer or manure) treatments on eroded Portneuf soil, and a fertilized (urea) treatment on noneroded Portneuf soil (NE-Fert). No inorganic N fertilizer was applied to manure treatments. The Fert and NE-Fert treatments received 135 kg N ha⁻¹ as urea-N, based on a sugarbeet yield goal of 63 Mg ha⁻¹ [27] and a spring preplant soil test, which determined residual inorganic N present in the root zone (0–90 cm). The manure-1x application rate was a commonly applied rate in the region. At application time, we estimated the m1 manure would provide an average 107 kg N ha⁻¹ to crops in the first year after application, based upon earlier reports that 32% of total manure N was available to crops in the first year [28]. Since a soil test indicated that no P or K fertilizer was needed on our site, we applied none. Plots were 9.1 m wide × 21.3 m long and accommodated 16 rows of beets.

For each year that manure treatments were applied, we obtained solid dairy cattle (*Bos* species) manure that had been stockpiled at a local dairy through the summer. The manure's average total C concentration (standard error) was 217 g kg⁻¹ (58 g kg⁻¹), total N was 14.1 g kg⁻¹ (2.6 g kg⁻¹), and C : N ratio was 15.9 (1.5).

2.2. Field Operations. Manure was applied to designated plots on 18 Nov. 2004, 22 Dec. 2005, and 19 Oct. 2006 using a commercial spreader truck equipped with rooster-comb beaters. Two to four 0.15 m² trays were randomly placed in each plot prior to spreading to quantify application rate. The manure collected in each tray was weighed, mixed, subsampled for moisture, C, and N analyses, and then returned to the soil surface from which it had been collected. The field was disked to a depth of 0.1 m within 48 hours of manure application. Plots were not fertilized in 2005 prior to planting spring barley. Barley was harvested in mid-July 2005. In fall 2005 prior to manure application, surface residue was burned to destroy weedy growth that had occurred after harvest.

TABLE 1: Description of treatments.

Treatment name	Treatment ID	Soil type	Added N source	Bulk applic. rate, dry wt. Mg ha^{-1}	Year of application[†]	Treatment age (y) at time of measurement
Noneroded fertilizer	NE-Fert	Noneroded	Urea	0.29	Each year	1
Control	Control	Eroded	None	0	N/A	1
Fertilizer	Fert	Eroded	Urea	0.29	Each year	1
Manure-1x						
2006	m1-y1	Eroded	Dairy manure	17.4	2006	1
2005	m1-y2	Eroded	Dairy manure	32.5	2005	2
2004	m1-y3	Eroded	Dairy manure	23.0	2004	3
Manure-3x						
2006	m3-y1	Eroded	Dairy manure	56.7	2006	1
2005	m3-y2	Eroded	Dairy manure	78.4	2005	2
2004	m3-y3	Eroded	Dairy manure	71.7	2004	3

[†] All fertilizer was applied in spring 2007 while all manure was applied in fall of the year shown.

In March 2006 soil samples were taken from plots at 0-to-30 cm and 30-to-60 cm depths and analyzed for soil N, P, and K (described below). Levels of P and K in the soils were adequate for small grain. On 13 Apr. 2006 the Fert and NE-Fert treatments received 134 kg N ha^{-1} as urea via hand-held spreader, while the control and manure plots received none. The field was disked to 0.1 m depth and roller-harrowed prior to planting barley in late April 2006. Barley residue and volunteer growth was burned on 13 Oct. 2006 before manure was applied to the designated plots.

On 15 Mar. 2007 soil samples were collected from plots in 30 cm increments down to 90 cm. We applied urea to the Fert and NE-Fert treatments and immediately incorporated the material with a roller harrow. Sugarbeet seed was planted (cv. BETA 4023R) on 20 Apr. 2007 in rows 0.56 m apart, with an in-row spacing of 55 mm and later thinned (30 May 2007) to a population of 117,000 plants ha^{-1} (manufacturer or trade names are included for the readers' benefit. The USDA-ARS neither endorses nor recommends such products). Insect control was accomplished using a Poncho seed treatment (CropScience LP, Research Triangle Park, NC, USA). Standard commercial procedures were used to control weeds and diseases. A single cultivation was performed on 26 June 2007. Irrigation through the growing season was supplied via sprinkler to meet the crop's evapotranspiration requirements. The beet crop was harvested on 10 Oct. 2007. Meteorological data required to calculate crop evapotranspiration (ET) were acquired from a weather station located 5.6 km northeast of the experimental plots. A rain gauge located near the field plot measured growing season precipitation. Crop ET was estimated from the maximum reference ET calculated using the Kimberly-Penman ET model [29], adjusted with the appropriate daily crop coefficient. Mean monthly air temperature, and total monthly precipitation, and irrigation during the 36-month study (including the plot preparation period, Fall, 2004 through 2006, and 2007 growing season) are reported in Figure 1.

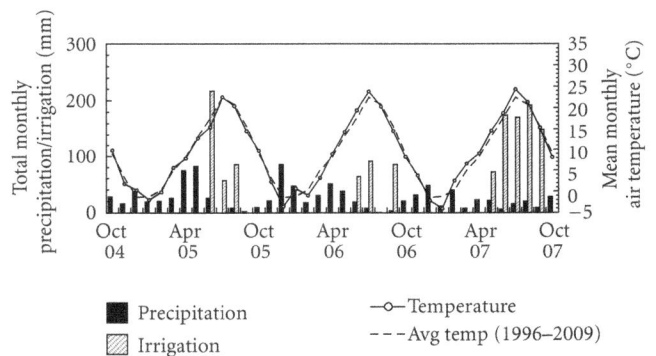

FIGURE 1: Total monthly precipitation and irrigation amounts, and mean monthly air temperature at the study site from fall 2004, when the first manure treatment was applied, through the 2007 growing season, when sugarbeet was planted on the experimental plots.

2.3. Sampling and Analyses. We measured N uptake in sugarbeet four times (1 June, 13 July, 20 Aug., and 27 Sep.) during the 2007 growing season by sampling total biomass of plant tops (aboveground tissue) and roots from 2 m of two adjacent rows (4 m total). The shredded sugarbeet roots and other aboveground plant tissue were dried at 65°C for dry matter determination. After grinding the dried tissue in a Thomas Wiley mill (Swedesboro, NJ) to pass an 865 μm screen, its total-N concentration was determined on a Thermo-Finnigan FlashEA1112 CNS analyzer (CE Elantech Inc., Lakewood, NJ, USA).

Sugarbeet yields were determined on 10 Oct. 2007 from two samples in each plot, each consisting of two adjacent 7.6 m long rows. Beet root subsamples collected for each of the two plot samples were analyzed for soil tare, as well as quality factors such as brei nitrate, brei conductivity, and sugar concentration, by the Amalgamated Sugar Company laboratory (Paul, ID, USA). Plot values were computed as the arithmetic mean of the two samples. The projected gross margin for each plot was computed as gross revenue minus operating costs. The gross revenue was calculated as the product of beet yield (tons, wet wt.) and the 2007 grower beet payment, which varied from $33.47 to $39.23 per ton of beets (wet wt.) depending on beet sugar percentage. Operating costs [30] were assumed equal for all treatments except for differences related to fertilizer and manure application. Manure costs included only transport and spreading fees based on a 3.2 km one-way haul distance. Manure application costs were amortized across two years, hence the manure cost in the 3rd year after application was zero. Costs associated with annual fertilizer treatments included the price of product and its application.

A buried bag technique [31] was used to measure net N mineralization in plot soils during the 2007 growing season. Briefly, three 5.7 cm diameter soil cores, 0-to-30 cm deep were collected on 25 Apr. 2007 in each plot (one from three of the plot's four quadrants), composited, and passed through a 0.4 cm screen. A subsample of the composited soil was collected to determine baseline (or initial) inorganic N and soil water content. The remaining soil was placed in 10 μm thick, 5 cm diam. polyethylene tubes sealed on one end. After being filled, the tubes were sealed on the remaining end, resulting in three 30 cm long soil columns that were inserted into the sample holes created previously. The bag's polyethylene film was only slightly permeable to water vapor but allowed good gas exchange between the enclosed soil and that surrounding it [31, 32]. A single bag was pulled from each plot on 15 June, 1 Aug., and 2 Oct. 2007. The net N mineralization during the period between burial and retrieval was calculated by subtracting the baseline inorganic N concentration (NO_3-N + NH_4-N) of the initial soil (collected 25 Apr.) from that of the retrieved bag. A positive difference indicated net N mineralization, while a negative value indicated net N immobilization during the period. We measured net N mineralization using buried bags for 25 Apr. to 15 June, 15 June to 1 Aug., and 1 Aug. to 2 Oct. The latter two period values were computed by difference relative to the previously retrieved buried bag sample. In addition, we estimated the net N mineralization in the not-yet-planted plots from 15 Mar. to 25 Apr. as the difference in soil inorganic N concentration (0–30 cm) between the two dates. We reported the net N mineralized as mg N kg^{-1} soil. Cumulative available soil N (0-to-30 cm depth) during the growing season was computed as the sum of the initial soil inorganic N present on 15 Mar., added fertilizer N (if any), and net N mineralized across the four periods.

The March 2006 and 2007 field soil samples and buried bag soil samples were air dried at 35°C and ground to pass a 2 mm screen. Soil N was extracted using a 2 M KCl solution. Within 6 h of extraction, the NO_3-N concentration in each extract was determined using an automated flow injection analyzer (Lachat Instruments, Loveland, CO, USA) after cadmium reduction (Method 12-107-04-1-B) while NH_4-N concentration was determined using a salicylate-hypochlorite method (Method 12-107-06-2-A). The soil's inorganic N concentration was calculated as the sum of the NO_3-N and NH_4-N concentrations (mg N kg^{-1} of dry soil). Bicarbonate extractable P [33] and exchangeable K [34] (except without

the addition of charcoal) were determined on field soil samples using ICP-OES. Manure C and N concentrations were determined on a freeze-dried sample with the CNS analyzer described above.

2.4. Statistical Analysis. Crop yield, biomass, N uptake, and quality factors for sugarbeet (brei nitrate, brei conductivity) were examined separately for each reporting interval via analysis of variance (ANOVA), PROC Mixed [35]. The statistical model included treatment as the fixed effect and block as the random effect. Treatment means were separated using the Tukey option [35]. We also included several single-degree-of-freedom orthogonal contrasts in the analysis. These included up to five class comparisons, where a class represents a combination of treatments: (1) no-manure versus manure treatments, where the no-manure class is control + Fert + NE-Fert and manure is m1 + m3; (2) manure-1x versus manure-3x treatments, averaged across all years; (3) manure-1x versus manure-3x treatments for years 2 and 3 only; (4) manure only treatments (m1 + m3) in year-1 versus years 2 and 3; (5) m1-y1 + m3-y1 + m3-y2 versus NE-Fert + Fert. The last contrast (number 5) tested the hypothesis that, relative to fertilizer applications, the effect of manure on the crop was influenced by the interaction between the factors, manure rate and age of application. Since the manure-1x added less C and N to the soil, its influence on the crop would diminish more rapidly than the manure-3x applications. All analyses were conducted using a $P = 0.05$ significance level. An identical statistical approach was used to analyze treatment effects on cumulative available soil N.

Since the experiment was conducted at a single location, findings pertain principally to that location. With judicious foresight, however, inferences made and conclusions drawn may apply to other locations with similar climatic conditions and crop management practices.

3. Results and Discussion

Meteorological data presented in Figure 1 portray the climatic conditions that prevailed during the years when the experimental plots were being developed and for 2007, the year that sugarbeet was grown on the site. The 2007 growing season was warmer than average, specifically during the February–July period, which was on average 1.5°C warmer than the 1996-to-2009 mean. The plots received 175 mm of annual rainfall in 2007, or 70% of the 1996-to-2009 mean value. The increased early-summer heat units coupled with abundant irrigation water supplies and the delay of hard frost until after October (instead of late September) contributed to near optimal 2007 sugarbeet yields in southern Idaho [36].

3.1. Sugarbeet Biomass and N Uptake. Several treatment effects were significant for the sugarbeet cumulative biomass (Table 2) and N uptake (Table 3) within each measurement period. The contrast tests identified several relationships with respect to treatment classes. First, the no-manure and manure treatments on the whole produced similar cumulative biomass and N uptake in sugarbeet tops, roots, and whole

TABLE 2: The influence of treatment on the total cumulative biomass for 2007 sugarbeet plant components. Table gives P values for treatment effects, and single-degree-of-freedom orthogonal comparisons derived from an analysis of variance.

	Accumulated sugarbeet biomass											
Source of variation	Tops				Roots				Whole plant			
	1 June	13 July	20 Aug	27 Sept	1 June	13 July	20 Aug	27 Sept	1 June	13 July	20 Aug	27 Sept
					P values							
Treatment (TRT)	**	***	***	**	0.36	**	***	**	0.02	***	***	***
Orthogonal contrasts[†]												
No manure versus manure	0.64	0.26	0.7	0.5	0.9	0.6	0.8	0.25	0.7	0.4	0.9	0.23
Man: m1 versus m3	**	**	***	0.06	**	0.07	***	0.13	**	*	***	*
Man: y1 versus y2 & y3	0.9	***	**	0.29	0.82	**	**	*	0.9	***	**	*
Man y2 & y3: m1 versus m3	**	**	***	*	**	*	***	*	**	**	***	**

[†] No manure: NEFert + control + Fert; Man: manure = m1 + m3 where m1: manure-1x; m3: manure-3x; y1, y2, y3: fall manure applied 1, 2, and 3 years in the past, respectively.
*$P < 0.05$, **$P < 0.01$, ***$P < 0.001$.

TABLE 3: The influence of treatment on N uptake in 2007 sugarbeet plant components. Table gives P values for treatment effects, and single-degree-of-freedom orthogonal comparisons derived from an analysis of variance.

	N uptake by beet biomass components											
Source of variation	Tops				Roots				Whole plant			
	1 June	13 July	20 Aug	27 Sept	1 June	13 July	20 Aug	27 Sept	1 June	13 July	20 Aug	27 Sept
					P values							
Treatment (TRT)	*	***	***	***	*	**	***	***	*	***	***	***
Orthogonal contrasts[†]												
No manure versus manure	0.26	0.25	0.78	0.52	0.38	0.35	0.75	0.7	0.65	0.26	0.9	0.6
Man-1x versus Man-3x	***	*	***	*	**	0.08	***	**	***	*	***	*
Man: y1 versus y2 & y3	0.9	***	*	0.15	0.87	***	**	0.11	0.9	***	*	0.09
Man y2 & y3: m1 versus m3	**	**	***	*	*	0.06	***	**	**	**	***	**

[†] No manure: NEFert + control + Fert; Man: manure = m1 + m3 where m1: manure-1x; m3: manure-3x; y1, y2, y3: fall manure applied 1, 2, and 3 years in the past, respectively.
*$P < 0.05$, **$P < 0.01$, ***$P < 0.001$.

plants. This result was partly due to substantial variability among treatment responses within each class, for example, the control treatment values were about half those of other no-manure treatments, and the m1-y3 values were about half that of other manure treatments (Table 4). However, the result shown in Table 3 suggests that no-manure treatments provided similar quantities of soil N to sugarbeet on average as did the manure treatments. As a consequence, season-long, total biomass production, and N uptake were similar between the groups (Table 4). See the discussion later in this section.

Second, compared to the manure-1x treatments, the manure-3x in general resulted in 1.12x greater season-long cumulative biomass production and 1.37x greater N uptake (Table 4). In addition, the relative difference between manure-3x and manure-1x responses was greater in the 2nd and 3rd years after manure application (i.e., comparing results of the m1 versus m3 contrast in y2, y3 with that of the m1 versus m3 contrast averaged across all years in Table 4). The disproportionately smaller increase in both

biomass and N uptake in response to a tripling of the manure rate indicated that the manure-3x treatment supplied excess N, and/or crop utilization of manure N decreased with increasing manure application [37]. As time since application increased (comparing the m1 versus m3 contrast in y2, y3), the N supplied by manure-1x apparently was less able to support beet growth than the manure-3x, causing a greater difference in biomass production and N uptake between the manure rate classes.

Third, the year of manure application affected total sugarbeet biomass and N uptake more during the early-June to mid-July period than at season's end (Tables 2 and 3). By 13 July 2007 the y1 manure treatments produced 1.4x greater sugarbeet biomass with 1.7x greater N uptake than the average for y2 and y3 manure treatments (Table 4). These disparities declined from that date onward. Thus by season's end, the sugarbeet in 2-year-old and 3-year-old manure plots had largely caught up to those of the 1-year-old manure treatments, such that differences were no longer significant. Thus y1 manure treatments generally provided greater

TABLE 4: Accumulated total biomass and N uptake in sugarbeet (tops and roots) at four times during the 2007 growing season.

Treatment[†]	Biomass				N uptake			
	1 June	13 July	20 Aug	27 Sept	1 June	13 July	20 Aug	27 Sept
		Mg ha^{-1} (dry wt.)				Kg ha^{-1}		
No manure								
NE-Fert	0.15a[‡]	7.3ab	17.5a	25.0bc	6.57a	124.6ab	183.8ab	257.4ab
Control	0.06b	4.2b	10.4c	17.1c	2.81b	52.7b	103.1b	152.5b
Fert	0.10ab	8.4a	20.9a	30.6a	4.85ab	164.6a	288.4a	383.3a
Manure-1x								
m1-y1	0.07ab	8.0a	18.2ab	25.3ab	3.18ab	140.3a	201.6ab	268.4ab
m1-y2	0.08ab	4.8b	12.1bc	21.8ab	3.6ab	62.6b	119.4b	211.0b
m1-y3	0.07b	4.1b	11.7bc	17.3c	2.98ab	56.4ab	123.2b	151.4b
Manure-3x								
m3-y1	0.13ab	7.6a	19.6a	24.8abc	5.89ab	140.5a	251.8a	300.2ab
m3-y2	0.13ab	7.0ab	18.4a	25.3ab	5.86ab	118.6ab	236.8a	315.4ab
m3-y3	0.13ab	6.0ab	17.3a	22.7b	5.44ab	92.4ab	211.4a	252.6ab
Treatment classes for orthogonal contrasts								
No manure	0.10	6.6	16.2	24.1	4.7	114.0	189.4	262.9
Manure	0.10	6.3	16.2	22.8	4.5	101.8	191.1	246.2
Manure-1x	0.07b	5.6b	14.0b	21.5b	3.3b	86.4b	146.9b	207.6b
Manure-3x	0.13a	6.9a	18.4a	24.1a	5.7a	117.2a	235.2a	284.8a
Manure Year-1	0.10	7.8a	18.9a	24.8	4.5	140.4a	227.8a	273.4
Manure Year 2 & 3	0.10	5.5b	14.9b	21.8	4.5	82.5b	172.7b	232.6
Year 2 & 3: m1	0.08b	4.5b	11.9b	19.6b	3.3b	59.5b	121.3b	181.2b
Year 2 & 3: m3	0.13a	6.5a	17.85a	24.0a	5.65a	105.5a	224.1a	284.0a

[†]NE-Fert: noneroded fertilizer (all other treatments on eroded soil); m1: manure-1x; m3: Manure-3x; y1, y2, y3: fall manure applied 1, 2, and 3 years in the past, respectively; manure: m1 + m3; no manure: NEFert + Control + Fert.
[‡]Within a given plant component and sample date, treatment means followed by the same lower case letter are not significantly different ($P < 0.05$). Not displayed if effect was not significant in the ANOVA (Table 5).

available soil N than y2 and y3 manure during the June-July sugarbeet growth period, but in later months, either soil N availability declined or some factor interfered with the growth and N uptake in y1 manure beets. We hypothesize that the former was the case, resulting from increased N immobilization for y1 beets during the June and July. The release of abundant, readily metabolized C from manure in y1 may have stimulated microbial growth [38, 39]. Lentz et al. [11] showed that immobilization in manure-amended soils was greater in y1 after application compared to y2 and y3 (see later discussion).

Fourth, when y1 manure treatments as a class were compared with y2 and y3 manure treatments, y1 had 1.14x greater season-long total biomass production and 1.18x greater total N uptake (Table 4). Within a manure treatment and measurement period, however, the magnitude and significance of the differences between y1 manure treatments and y2 or y3 manure treatments were greater and more common for manure-1x than for manure-3x treatments (Table 4). This suggests that manure-3x treatments, regardless of age, provided adequate N for the crop. Furthermore, the m1-y1 treatment resulted in similar sugarbeet biomass production and N uptake as any manure-3x treatment no

matter the year applied. This indicates that the m1-y1 treatment also provided adequate N for the sugarbeet.

The Fert and NE-Fert treatments consistently produced greater season-long crop biomass and N uptake than the control, although the difference was significant only for Fert after 1 June (Table 4, Figure 2), reflecting the greater N availability in the two fertilized treatments compared to the control. The NE-Fert produced greater sugarbeet biomass and N uptake than Fert on 1 June, day of year (DOY) 152, whereas the opposite tendency was observed at later dates. This likely resulted because seedlings emerged later and stand counts were 15% smaller (after thinning) in Fert plots relative to NE-Fert (data not presented). Later in the season, the lesser plant density for Fert compared to NE-Fert and other treatments (after thinning) may have rendered it less susceptible to a powdery mildew outbreak [40], which was identified in the field in midsummer and subsequently treated with fungicide and sulfur.

3.2. Sugarbeet Yield, Quality, and Profitability. Clean beet yields for all treatments ranged from 56.4 to 101.1 Mg ha^{-1} and averaged 83.0 Mg ha^{-1} (Table 5). These yields compare favorably with the average 2007 sugarbeet yield for southern

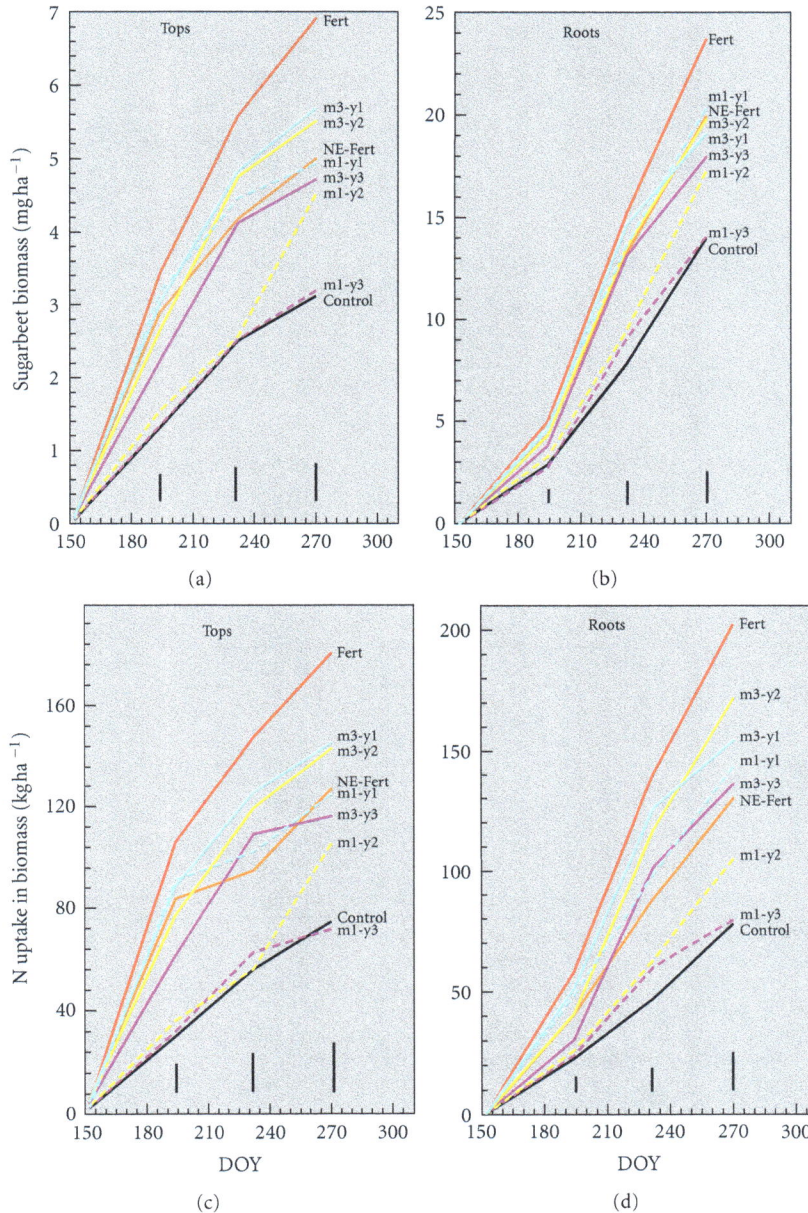

FIGURE 2: The effect of treatments on biomass accumulation in sugarbeet tops, that is, aboveground tissue (a) and root (b) components, and on N uptake in sugarbeet tops (c) and roots (d) in 2007. (Measured on dates (DOY) 1 June (152), 13 July (194), 20 Aug. (232), and 27 Sep. (270). Bar length represents the mean standard error ($n = 4$) for the 9 treatments at the given measurement date.

Idaho growers, 76.6 Mg ha^{-1} [36]. Sugarbeet yield and quality were affected by treatments, whether considered individually or when compared as classes (contrasts). The m1-y1, m3-y1, and m3-y2 treatments produced 1.3 to 1.8 times greater root yields than NE-Fert, control, m1-y2, and m1-y3 treatments (Table 5). Contrast tests showed that yields increased about 1.2-fold when (1) manure instead of fertilizer or no amendment was added to soil; (2) the manure amendment rate was increased from 1x to 3x; (3) sugarbeet was planted in the first year after fall manure application instead of waiting until the 2nd or 3rd year after application (Table 5).

Sugar concentration in beets ranged from 15.6 to 17.7% and averaged 16.7% (Table 5) with concentrations being generally greater in lower-yielding treatments, as expected. Our study's mean sugar concentration was nearly equivalent to the average sugar concentration obtained by southern Idaho growers in 2007, that is, 16.8% [36]. The NE-Fert and control treatments produced greater beet sugar concentrations than Fert, m3-y2, and m3-y3 treatments (mean 17.6 versus 16.0). Beet sugar concentrations decreased slightly (3–6% on average) when (1) manure was applied instead of fertilizer or no amendment; (2) manure application was

TABLE 5: Treatment and orthogonal contrast mean values for sugarbeet yield, quality, and gross margin parameters.

Treatment[†]	Clean beet root yield[‡]	Sugar	Est. Recov. sugar[‡]	Brei nitrate	Brei conductivity	Gross margin[§]
	Mg ha^{-1}	%	Mg ha^{-1}	mg kg^{-1}	dS m^{-1}	\$US ha^{-1}
No manure						
NE-Fert	75.2c[§]	17.7a	11.6ab	59.8c	0.58d	979bc
Control	56.4c	17.6a	8.5b	106.8c	0.68cd	292 d
Fert	90.2ab	16.4b	12.3ab	187.3bc	0.88bc	1272ab
Manure-1x						
m1-y1	101.0a	16.7ab	14.0a	147.1bc	0.91b	1884a
m1-y2	72.8bc	16.8ab	10.4ab	143.6bc	0.77bcd	731bcd
m1-y3	64.5c	16.8ab	9.1b	185.4bc	0.81bc	484cd
Manure-3x						
m3-y1	97.7a	16.8ab	13.6a	149.1bc	0.95b	1676a
m3-y2	101.1a	16.0b	13.2a	259.6ab	0.96b	1510ab
m3-y3	88.5ab	15.6b	11.2ab	308.5a	1.02a	1138b
Treatment classes for contrasts						
No manure	73.9b	17.2a	10.8	118.0b	0.71a	848b
Manure	87.6a	16.5b	11.9	198.9a	0.90b	1237a
Manure-1x	79.4b	16.8a	11.2b	158.7b	0.83b	1033b
Manure-3x	95.8a	16.1b	12.7a	239.1a	0.98a	1441a
Manure y1	99.4a	16.8a	13.8a	148.1b	0.93	1780a
Manure y2 & y3	81.7b	16.3b	11.0b	224.3a	0.89	966b
Year 2 & 3: m1	68.7b	16.8a	9.8b	164.5b	0.79b	607b
Year 2 & 3: m3	94.8a	15.8b	12.2a	284.1a	0.99a	1324a
m1-y1, m3-y1, m3-y2	99.9a	16.5b	13.6a	185.3	0.9a	1690a
NE-Fert, Fert	82.7b	17.1a	12.0b	123.6	0.7b	1126b

[†]NE-Fert: noneroded fertilizer (all other treatments on eroded soil); m1: manure-1x; m3: manure-3x; y1, y2, y3: fall manure applied 1, 2, and 3 years in the past, respectively; manure: m1 + m3; no manure: NEFert + Control + Fert.
[‡]Clean yield: yield minus soil tare; Est. Recov. Sugar: estimated amount of sugar extractable from beets per unit area.
[§]Gross margin: gross revenue minus operating costs.
[¶]For a given yield or quality parameter, treatment means or means for individual orthogonal contrasts followed by the same lower case letter are not significantly different ($P < 0.05$). Not displayed if effect was not significant in the ANOVA.

increased from 1x to 3x; or (3) sugarbeet was planted in the first year after fall manure application instead of waiting until the 2nd or 3rd year after application. These results are consistent with the concept that increasing N availability decreases beet root sugar concentration [8, 9].

Increased nitrate and soluble impurity (conductivity) concentrations in sugarbeet brei (fresh macerated beet root) are associated with a decrease in the quantity of sugar recovered from the sugarbeet and increased sugar extraction costs [4, 27]. When the manure application rate increased from 1x to 3x, brei nitrate increased an average 1.6-fold (from 158.7 to 239 mg kg^{-1}) and conductivity increased 1.2-fold on average. Brei conductivity of manure treatments in year 1 did not differ from the mean value for year 2 and year 3. The m3-y3 treatment produced the greatest brei nitrate concentrations in beet roots, 309 mg kg^{-1}. While this value exceeded the 250 mg kg^{-1} target level recommended for southern Idaho [27], it was still well below the mean value for the 2007, southcentral Idaho sugarbeet crop, 351 mg kg^{-1} (S. Camp, Amalgamated Sugar Co., personal communication, 2010).

The control produced the least estimated recoverable sugar, 8.5 Mg ha^{-1} (Table 5). The treatments m1-y1, m3-y1, and m3-y2 produced the greatest estimated recoverable sugar values (mean 13.6 Mg ha^{-1}), which were 1.5x greater than that of the two least performing treatments, m1-y3 and control (mean 8.8 Mg ha^{-1}), and 1.1x that of the two fertilizer treatments (mean 12.0 Mg ha^{-1}). In addition, the estimated recoverable sugar in beets increased 1.22-fold, on average, when manure application was increased from the 1x to 3x rate or sugarbeet was planted in the first year after fall manure application instead of waiting until the 2nd or 3rd year after application.

The gross margins listed in Table 5 integrate treatment effects on beet yield and quality and fertilizer or manure costs, and provide a measure of treatment effects on profitability. An examination of individual manure treatments revealed that all except m1-y3 produced similar or greater gross margins than either the Fert or NE-Fert. Contrast tests showed that 1) the average gross margin for m1-y1, m3-y1, and m3-y2 manure treatments was 1.5-fold greater

TABLE 6: Treatment and contrast class mean values for soil (0–30 cm) inorganic N concentrations in spring (before and after planting), cumulative net N mineralization, and cumulative available N during the growing season.

Treatment[†]	Soil N 15 Mar.	Soil N 25 Apr.	Cum. net N mineralized 25 Apr.–27 Sept.	Cum. available N 15 Mar.–27 Sept.
			mg kg^{-1}	
No manure				
NE-Fert	8.3b[‡]	39.0a	18.0bc	57.0bc
Control	10.9b	12.6b	14.6c	27.1d
Fert	8.9b	33.6a	19.6bc	62.8bc
Manure-1x				
m1-y1	13.5b	31.8a	32.8ab	64.6b
m1-y2	10.0b	17.9b	23.7bc	41.6c
m1-y3	8.3b	14.1b	22.5bc	36.7cd
Manure-3x				
m3-y1	24.1a	45.9a	41.4a	87.3a
m3-y2	12.7b	27.8ab	28.5bc	56.3bc
m3-y3	13.7b	28.7ab	29.1b	57.8bc
Treatment classes for contrasts				
No manure	9.4	28.4	17.1b	49.0b
Manure	13.7	27.7	29.7a	57.4a
Manure-1x	10.6	21.3b	26.3b	47.6b
Manure-3x	16.8	34.1a	32.9a	67.1a
Manure y1	18.8	38.9a	37.1a	76.0a
Manure y2 & y3	11.2	22.1b	26.0b	48.1b
Year 2 & 3: m1	9.2b	16.0b	23.1b	39.2b
Year 2 & 3: m3	13.2a	28.3a	28.8a	57.1a

[†] NE-Fert: noneroded fertilizer (all other treatments on eroded soil); m1: manure-1x; m3: manure-3x; y1, y2, y3: fall manure applied 1, 2, and 3 years in the past, respectively; manure: m1 + m3; No manure: NEFert + Control + Fert.
[‡] For a given yield or quality parameter, treatment means or means for individual orthogonal contrasts followed by the same lower case letter are not significantly different ($P < 0.05$). Not displayed if effect was not significant in the ANOVA.

than that for NE-fert and Fert treatments; 2) manure-3x treatments as a whole produced 1.4-fold greater gross margin than manure-1x treatments; 3) the mean gross margin for y1 manure treatment class was 1.8-fold greater than the y2 and y3 manure mean value; and 4) in the 2nd and 3rd year after manure application the manure-3x treatments on average resulted in a 2.2-fold greater gross margin than the manure-1x treatments.

Thus, fall manure applications 1 or 2 years prior to growing sugarbeet can potentially widen profit margins relative to conventional fertilizers applied preplant in the spring. While the m1-y1 manure treatment produced the greatest mean gross margin, use of greater manure application rates might be advisable. Manure quality often varies and the 1x rate leaves less margin for error. A greater application rate every two years, rather than one, halves application costs. Moreover, a high N-demand crop such as corn could be grown the year before sugarbeet to efficiently and profitably use the N mineralized in the first 12 months after the manure was applied [11]. On the other hand, mineralized N (as NO_3^-) could be leached below the sugarbeet's root zone before or during the beet growing season. Note that our margin analysis does not account for extra costs that may arise due to manure use, for example, additional management costs associated with increased weed pressure.

The influence of increasing manure applications on sugarbeet yields and estimated recoverable sugar in year 1 were also investigated by Lentz et al. [11] in 2003 for similar soils in southern Idaho. Lentz et al. [11] reported that, in contrast to the results of this study, sugarbeet root yields and recoverable sugar decreased as manure application rates increased. This difference was likely due to less initial residual soil N and less C and total N in the manure used in the current study relative to those in 2003. In sum, these factors decreased the N available in the 2007 soils which in turn reduced the possibility that excessive N mineralized from manure amendments would limit beet yields and recoverable sugar values [2, 8].

3.3. N Mineralization and Availability. The contrast tests for the season-long (25 Apr. to 27 Sept.) cumulative net N mineralization (Table 6) established that (1) manure treatments taken as a class produced 1.7x greater N than no-manure treatments; (2) manure-3x treatments on average produced 1.3x greater N than manure-1x treatments; (3) N

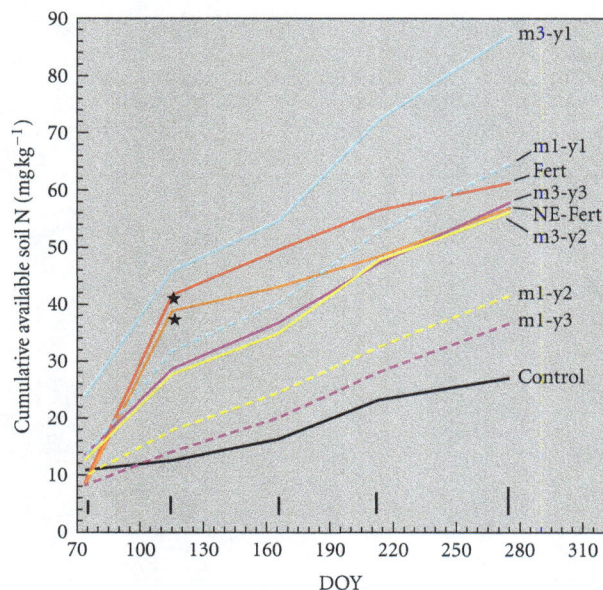

★ Includes added fertilizer N

FIGURE 3: Cumulative net soil N available through mineralization and any fertilizer addition in the 0-to-30 cm soil during the 2007 sugarbeet growing season (measured on dates (DOY) 15 Mar. (74), 25 Apr. (115), 15 June (166), 1 Aug. (213), and 2 Oct. (275). Bar length represents the mean standard error ($n = 4$) for the 9 treatments at the given measurement date.

mineralized from y1 manure treatments as a class was 1.4x greater than that for the y2 and y3 manure treatment class mean. Hence, the cumulative available soil N from manure amendments generally declined as application rate decreased and time since application increased (Table 6, Figure 3).

The net N mineralized in the uppermost 0.3 m soil profile during the growing season (25 Apr. to 27 Sept.) for year-1 manure treatments was 32.8 mg kg^{-1} for manure-1x, or 2.2 times the control value, and 41.4 mg kg^{-1} for manure-3x, or 2.8 times that of the control (Table 6). These net N mineralization values for the year-1 treatments were similar to those reported by Lentz et al. [11] for comparable treatments in 2003, that is, 32.6 (manure-1x) and 48.7 mg kg^{-1} (manure-3x). Net N mineralized during the growing season for year-2 and year-3 manure treatments was reduced an average 30% in comparison to year-1 manure (Table 6). Findings from the 2007 growing season showed that fertilizer and the m1-y1, m3-y2, and m3-y3 treatments supplied similar amounts of cumulative N. In contrast m3-y1 provided nearly 1.5x more N ($P < 0.0001$), and m1-y2 and m1-y3 provided 37% less N ($P < 0.0001$) than the mean fertilizer treatment value (Table 6, Figure 3). The control and m1-y3 treatments provided the least cumulative available soil N, produced the least biomass, and led to the least N being incorporated into crop tissue (Figures 2 and 3).

For manure treatments, mineralized N accumulated at a slower rate in the interval from 25 Apr. to 15 June (DOY 115 to 166) than for other intervals (revealed as a decrease

in slope in Figure 3). This slowing of the rate was most pronounced for the larger and more recent manure applications. This corroborates observations made by Lentz et al. [11], who described an identical phenomenon in their experiment conducted on similar soils at Kimberly, ID. The slowing rate of net N mineralization was likely due to immobilization of manure N that occurred after soils warmed during this early summer period. Mean soil temperatures at the 10 cm depth exceeded 21°C by mid-June (data not shown). Seasonal N mineralization data from an Ontario, Canada, experiment also showed a subtle dip in mineralization rate during this period, but the researchers described a more substantial decrease in N mineralization rate after DOY 227 [15]. The researchers attributed the substantial decreases to the release of carbonaceous root exudates and subsequent N immobilization [15]. Similar declines in N mineralization during the early summer period were reported for coastal Alabama soils amended with composted dairy manure [41].

The pattern of crop biomass accumulation and N-uptake in sugarbeet tops and roots (Figure 2) generally followed that of soil N availability (Figure 3). There were two exceptions. First, while Fert and NE-Fert treatments provided similar soil N, Fert produced substantially greater season-long crop biomass and N uptake than the NE-Fert (Figure 2, Table 4). This may be related to the differences in stand density and mildew pathology, as discussed previously. Second, though the net N mineralized for the m3 treatment was greater in y1 than for y2 or y3 (Figure 3), the extra N mineralized in y1 did not result in greater season-long crop N uptake (Figure 2, Table 4). This reveals that the N derived from the 3x manure (applied in the previous fall) was not utilized efficiently, presumably because it exceeded crop needs. Moreover, the excess soil mineral N in the 3x treatments was subject to leaching losses.

4. Conclusions

This study quantifies the effects of stock-piled dairy manure applications made 1, 2, or 3 years previously on sugarbeet. Results of this and a previous, related study [11] on calcareous, southern Idaho soil indicate that the influence of manure N applications on soil N availability, N uptake, and sugarbeet yield and quality was a function of residual inorganic soil N at the start of the growing season, the amount of Fall-applied manure added, and the year in which the manure was applied. A Fall manure application alone, when applied at an appropriate rate and planted to sugarbeet in either the first or second year after application provided adequate N nutrition for the production of a high quality sugarbeet crop. Furthermore, these manure treatments (m1-y1, m3-y1, and m3-y2) increased estimated recoverable sugar yields an average of 1.1-fold and increased gross margins an average of 1.5-fold relative to conventional fertilizer treatments. The increases in recoverable sugar and gross margins documented in this study are likely to vary from one site to another as a function of soil type, climate, and growing conditions. Our results illustrate nonetheless how proper manure management can increase sugarbeet yields and producer profit margins.

Abbreviations

Fert: Fertilizer on eroded soil
NE-Fert: Fertilizer on noneroded soils
EC: Electrical conductivity
y1, y2, y3: Fall manure applied 1, 2, and 3 years in the past, respectively.

References

[1] C. W. Robbins, B. E. Mackey, and L. L. Freeborn, "Improving exposed subsoils with fertilizers and crop rotations," *Soil Science Society of America Journal*, vol. 61, no. 4, pp. 1221–1225, 1997.

[2] J. N. Carter, D. T. Westermann, and M. E. Jensen, "Sugarbeet yield and quality as affected by nitrogen level," *Agronomy Journal*, vol. 68, no. 1, pp. 49–55, 1976.

[3] R. Gardner and D. W. Robertson, "Comparison of the effects of manures and commercial fertilizers on the yield of sugar beets," *Proceedings American Society of Sugar Beet Technology*, vol. 4, pp. 27–32, 1946.

[4] A. P. Draycott and D. R. Christenson, *Nutrients for Sugarbeet Production: Soil Plant Relationships*, CABI Publishing, Cambridge, Mass, USA, 2003.

[5] A. D. Halvorson and G. P. Hartman, "Long-term nitrogen rates and sources influence sugarbeet yield and quality," *Agronomy Journal*, vol. 67, pp. 389–393, 1975.

[6] J. N. Carter, "Effect of nitrogen and irrigation levels, location and year on sucrose concentration of sugarbeets in Southern Idaho," *Journal of the American Society of Sugar Beet Technologists*, vol. 21, no. 3, pp. 286–306, 1982.

[7] F. J. Hills, F. E. Broadbent, and O. A. Lorenz, "Fertilizer nitrogen utilization by corn, tomato, and sugarbeet," *Agronomy Journal*, vol. 75, pp. 423–426, 1983.

[8] J. N. Carter and D. J. Traveller, "Effect of time and amount of nitrogen uptake on sugarbeet growth and yield," *Agronomy Journal*, vol. 73, pp. 665–671, 1981.

[9] F. N. Anderson and G. A. Peterson, "Effect of incrementing nitrogen application on sucrose yield of sugarbeet," *Agronomy Journal*, vol. 80, pp. 709–712, 1988.

[10] J. G. Davis and D. G. Westfall, "Fertilizing sugar beets," Colorado State University Cooperative Extension Fact Sheet No. 0.542, Colorado State University, Fort Collins, Colo, USA, 2010.

[11] R. D. Lentz, G. A. Lehrsch, B. Brown, J. Johnson-Maynard, and A. B. Leytem, "Dairy manure nitrogen availability in eroded and noneroded soil for sugarbeet followed by small grains," *Agronomy Journal*, vol. 103, pp. 628–643, 2011.

[12] E. M. Hansen, I. K. Thomsen, and M. N. Hansen, "Optimizing farmyard manure utilization by varying the application time and tillage strategy," *Soil Use and Management*, vol. 20, no. 2, pp. 173–177, 2004.

[13] J. J. Miller, B. W. Beasley, C. F. Drury, and B. J. Zebarth, "Barley yield and nutrient uptake for soil amended with fresh and composted cattle manure," *Agronomy Journal*, vol. 101, no. 5, pp. 1047–1059, 2009.

[14] E. B. Mallory, T. S. Griffin, and G. A. Porter, "Seasonal nitrogen availability from current and past applications of manure," *Nutrient Cycling in Agroecosystems*, vol. 88, no. 3, pp. 351–360, 2010.

[15] B. L. Ma, L. M. Dwyer, and E. G. Gregorich, "Soil nitrogen amendment effects on nitrogen uptake and grain yield of maize," *Agronomy Journal*, vol. 91, no. 4, pp. 650–656, 1999.

[16] B. Gagnon, R. R. Simard, R. Robitaille, M. Goulet, and R. Rioux, "Effect of composts and inorganic fertilizers on spring wheat growth and N uptake," *Canadian Journal of Soil Science*, vol. 77, no. 3, pp. 487–495, 1997.

[17] V. R. Kanneganti and S. D. Klausner, "Nitrogen recovery by orchard grass from dairy manure applied with or without fertilizer nitrogen," *Communications in Soil Science and Plant Analysis*, vol. 25, no. 15-16, pp. 2771–2781, 1994.

[18] P. P. Motavalli, K. A. Kelling, and J. C. Converse, "First-year nutrient availability from injected dairy manure," *Journal of Environmental Quality*, vol. 18, no. 2, pp. 180–185, 1989.

[19] B. Eghball and J. F. Power, "Phosphorus- and nitrogen-based manure and compost applications: corn production and soil phosphorus," *Soil Science Society of America Journal*, vol. 63, no. 4, pp. 895–901, 1999.

[20] T. S. Griffin, Z. He, and C. W. Honeycutt, "Manure composition affects net transformation of nitrogen from dairy manures," *Plant and Soil*, vol. 273, no. 1-2, pp. 29–38, 2005.

[21] E. B. Mallory and T. S. Griffin, "Impacts of soil amendment history on nitrogen availability from manure and fertilizer," *Soil Science Society of America Journal*, vol. 71, no. 3, pp. 964–973, 2007.

[22] R. Xie and A. F. Mackenzie, "Urea and manure effects on soil nitrogen and corn dry matter yields," *Soil Science Society of America Journal*, vol. 50, no. 6, pp. 1504–1509, 1986.

[23] G. Wen, J. J. Schoenau, J. L. Charles, and S. Inanaga, "Efficiency parameters of nitrogen in hog and cattle manure in the second year following application," *Journal of Plant Nutrition and Soil Science*, vol. 166, no. 4, pp. 490–498, 2003.

[24] T. B. Parkin, T. C. Kaspar, and C. Cambardella, "Oat plant effects on net nitrogen mineralization," *Plant and Soil*, vol. 243, no. 2, pp. 187–195, 2002.

[25] S. D. Klausner, V. R. Kanneganti, and D. R. Bouldin, "An approach for estimating a decay series for organic nitrogen in animal manure," *Agronomy Journal*, vol. 86, no. 5, pp. 897–903, 1994.

[26] G. A. Lehrsch and D. C. Kincaid, "Sprinkler irrigation effects on infiltration and near-surface unsaturated hydraulic conductivity," *Transactions of the ASABE*, vol. 53, no. 2, pp. 397–404, 2010.

[27] A. Moore, J. Stark, B. Brown, and B. Hopkins, "Southern Idaho fertilizer guide: sugar beets," CIS 1174, University of Idaho Extension, Idaho, Moscow, 2009.

[28] B. Eghball, B. J. Wienhold, J. E. Gilley, and R. A. Eigenberg, "Mineralization of manure nutrients," *Journal of Soil and Water Conservation*, vol. 57, no. 6, pp. 470–473, 2002.

[29] J. L. Wright, D. T. Westermann, and G. A. Lehrsch, "Studying nitrate-N leaching with a bromide tracer in an irrigated silt loam soil," in *Best Management Practices for Irrigated Agriculture and the Environment*, J. Schaack, A. W. Freitag, and S. S. Anderson, Eds., pp. 229–242, U.S. Committee on Irrigation and Drainage, Denver, Colo, USA, 1998, Proceedings of the Irrigation and Drainage Water Management Conference, Fargo, ND, USA, 1997.

[30] P. Patterson, The economics of growing sugarbeets in southern Idaho: a short run gross margin analysis. AEES No 09-01, US Department of Agricultural Economics and Rural Sociology, University of Idaho Extension, 2009, http://www.amalgamatedsugar.com/articles/SugarbeetPacket.pdf.

[31] K. S. Balkcom, A. M. Blackmer, and D. J. Hansen, "Measuring soil nitrogen mineralization under field conditions," *Communications in Soil Science and Plant Analysis*, vol. 40, no. 7-8, pp. 1073–1086, 2009.

[32] C. F. Eno, "Nitrate production in the field by incubating the soil in polyethylene bags," *Soil Science Society Of America Proceedings*, vol. 24, no. 4, pp. 277–279, 1960.

[33] S. R. Olsen, C. V. Cole, F. S. Watatanabe, and L. A. Dean, "Estimation of available phosphorus in soils by extraction with sodium bicarbonate," USDA Circ. 939, U.S. Government Printing Office, Washington, DC, USA, 1954.

[34] J. J. Schoenau and R. E. Karamanos, "Sodium bicarbonate-extractable P, K, and N," in *Soil Sampling and Methods of Analysis*, M. R. Carter, Ed., pp. 51–58, Lewis Publishers, Boca Raton, Fla, USA, 1993.

[35] SAS Institute Inc., *SAS Online Documentation, Version 9.2 [CD-ROM]*, SAS Institute, Inc., Cary, NC, USA, 2009.

[36] S. Kraus, "Beet growers enjoy record yields," Ag Weekly, p. 29, 2007.

[37] R. D. Lentz and G. A. Lehrsch, "Net nitrogen mineralization from past year's manureand fertilizer applications," *Soil Science Society of America Journal*. In press, https://www.soils.org/publications/sssaj/view/76-3/s11-0282-6-4-23-2012.pdf.

[38] B. C. Liang, E. G. Gregorich, M. Schnitzer, and H.-R. Schulten, "Characterization of water extracts of two manures and their adsorption on soils," *Soil Science Society of America Journal*, vol. 60, no. 6, pp. 1758–1763, 1996.

[39] B. Marschner and K. Kalbitz, "Controls of bioavailability and biodegradability of dissolved organic matter in soils," *Geoderma*, vol. 113, no. 3-4, pp. 211–235, 2003.

[40] J. J. Burdon and G. A. Chilvers, "Host density as a factor in plant disease ecology," *Annual Review of Phytopathology*, vol. 20, pp. 143–166, 1982.

[41] D. B. Watts, H. A. Torbert, and S. A. Prior, "Soil property and landscape position effects on seasonal nitrogen mineralization of composted dairy manure," *Soil Science*, vol. 175, no. 1, pp. 27–35, 2010.

3

Determining Critical Soil pH for Grain Sorghum Production

Katy Butchee,[1] Daryl B. Arnall,[2] Apurba Sutradhar,[2] Chad Godsey,[2] Hailin Zhang,[2] and Chad Penn[2]

[1] *Department of Agriculture, Western Oklahoma State College, 2801 N. Main, Altus, OK 73521, USA*
[2] *Department of Plant and Soil Sciences, Oklahoma State University, Stillwater, OK 74078, USA*

Correspondence should be addressed to Daryl B. Arnall, b.arnall@okstate.edu

Academic Editor: David Clay

Grain sorghum (*Sorghum bicolor* L.) has become a popular rotation crop in the Great Plains. The transition from conventional tillage to no-tillage production systems has led to an increase in the need for crop rotations. Some of the soils of the Great Plains are acidic, and there is concern that grain sorghum production may be limited when grown on acidic soils. The objective of this study was to evaluate the effect of soil pH for grain sorghum production. Potassium chloride-exchangeable aluminum was also analyzed to determine grain sorghum's sensitivity to soil aluminum (Al) concentration. The relationship between relative yield and soil pH was investigated at Lahoma, Perkins, and Haskell, Oklahoma, USA with soil pH treatments ranging from 4.0–7.0. Soil pH was altered using aluminum sulfate or hydrated lime. Soil acidity reduced grain sorghum yield, resulting in a 10% reduction in yield at soil pH 5.42. Potassium chloride-exchangeable aluminum levels above 18 mg kg^{-1} resulted in yield reductions of 10% or greater. Liming should be considered to increase soil pH if it is below these critical levels where grain sorghum will be produced.

1. Introduction

Producers throughout the Great Plains are converting areas of conventional tillage to no-tillage systems [1]. A key component of successful no-tillage production systems is the integration of crop rotations, which help break weed, disease, and insect cycles [1]. Due to its ability to tolerate warm and relatively dry climates, grain sorghum (*Sorghum bicolor* L.) is well suited for crop rotations in the Central Great Plains.

Grain sorghum has traditionally been grown on soils with a pH of >6.5 [2]; however, a review of soil test results in 2005 by the Potash and Phosphate Institute [3] observed that 46% of the tested samples in Oklahoma had a soil pH of <6.0. The use of aluminum (Al) tolerant wheat varieties and banding of phosphorus (P) fertilizers has allowed producers to grow winter wheat in unfavorable pH conditions. Because of this, many producers are not accustomed to considering liming in their management decisions. With the integration of grain sorghum as a rotation crop, acidic soils may need to be limed; whereas, this may not have been necessary when continuous winter wheat was produced.

This study focused on evaluating relative yields of grain sorghum with respect to soil pH, which are useful for determining if there is a yield reduction associated with soil acidity. This information will be helpful for producers when determining if liming an acidic soil where grain sorghum will be produced is economical and necessary.

The exact quantitative effect of soil pH on grain sorghum yield has not previously been established. The majority of research relating to soil acidity in the central Great Plains has focused on winter wheat, while some studies have focused on determining the most acid-tolerant varieties of grain sorghum [4, 5]. Determining the behavior of grain sorghum grown on soil varying in pH will be a useful tool for educating producers and agronomists about the importance of liming acidic soils.

Previous research concerning grain sorghum and soil pH determined that as reactive Al concentration increased, the symptoms of Al toxicity also increased [6]. Ohki studied the relationship between root Al concentration and growth and found the Al critical toxicity level for grain sorghum was 54 mmol kg^{-1} tissue dry tissue matter.

TABLE 1: Description of soil series at Perkins, Lahoma, and Haskell, Oklahoma.

Location	Soil series
Perkins, OK	Teller series (fine loamy, mixed, active, thermic Udic Argiustolls) and konawa series (fine loamy, mixed, active, thermic Ultic Haplustalfs)
Lahoma, OK	Grant series (fine silty, mixed, superactive, thermic Udic Argiustolls)
Haskell, OK	Taloka series (fine, mixed, active, thermic Mollic Albaqualfs)

Duncan et al. [7] observed different grain sorghum genotypes to determine their acid tolerance. Grain yields dropped from 2069 kg ha^{-1} at soil pH 4.8 to 163 kg ha^{-1} at soil pH 4.4. There was also a significant yield decrease from 4279 kg ha^{-1} to 3,557 kg ha^{-1} at soil pH 5.5 to 5.1, respectively. This decrease was attributed to Al or Mn toxicity. The study indicated that the majority of plants grown at soil pH of 4.4 did not reach the reproductive growth stage with some of the plants dying. A 35% decrease in yield was observed from soil pH 5.1 to 4.8 and a 92% decrease from soil pH 4.8 to 4.4.

Tan and Keltjens [8] determined that Al toxicity was evident as damage to the roots and through the reduction of magnesium (Mg) availability. Grain sorghum plants grown in acid soils may express water stress due to root damage, which can limit their ability to extract water from the soil. Liming a soil with pH of 4.3 and raising it to pH 4.7 alleviated the Al toxicity.

Flores et al. [9] conducted an experiment to determine the variations in growth and yield associated with Al saturation of the soil. They studied both susceptible and tolerant genotypes of grain sorghum grown in both 40% (pH 4.6) and 60% (pH 4.1) Al saturation. The study determined that the acid-tolerant genotypes grown at 60% Al saturation had lower root mass scores and delayed flowering. There were, however, no differences in yield and growth traits for the acid-tolerant genotypes grown at 40% or 60% Al saturation. The susceptible genotypes showed an improvement in yield and growth traits in the lower Al saturation than the higher Al saturation. Flores et al. [9] concluded that all sorghum genotypes grown at Al saturation above 70% performed poorly.

2. Materials and Methods

The field experiment was established in 2009 at the Cimarron Valley Research Station near Perkins (Teller series), Oklahoma, the North Central Research Station near Lahoma (Grant Series), Oklahoma, and the Eastern Research Station near Haskell (Taloka series), Oklahoma (Table 1).

The experimental design was Randomized Complete Block (RCBD) with four replications. Plot size was 6 m long × 3 m wide with 4.6 m alleys between each replication at Lahoma and Perkins and 3 m alleys between each replication at Haskell. Soil was amended in each location to obtain six target soil pH ranging from 4.0 to 7.0.

For each growing season, soil samples were taken from each plot prior to planting to determine actual soil pH. Soil probes were used to obtain 15–20 cores from each plot to a depth of 15 cm. The soil samples were dried at 60°C overnight and ground to pass through a 2 mm sieve. A 1 : 1 soil : water suspension and glass electrode were used to measure soil pH and buffer index [10, 11]. 1 M KCl solution was used to extract soil NO$_3$–N and NH$_4$–N and quantified using a Flow Injection Autoanalyzer (Lachat Instruments, Milwaukee, WI, USA). Mehlich III solution was used to extract plant-available P and K [12], and the amount of P and K were quantified using a Spectro Ciros inductively coupled plasma (ICP) spectrophotometer [13]. Soil sample results were used to generate N, P, and K rates that were applied as a blanket application over each trial in 2009 and 2010.

A previous laboratory experiment determined the rates of aluminum sulfate (Al$_2$(SO$_4$)$_3$) and hydrated lime (Ca(OH)$_2$) needed to achieve a specific change in soil pH at each location. In this laboratory experiment, composite soil samples were collected from each of the experimental sites. Five incremental rates of Al$_2$(SO$_4$)$_3$ and 5 incremental rates of Ca(OH)$_2$ were each added to 1/2 kg subsamples from each of the locations to develop a response curve which could be used to determine the amount of material needed to reach a desired soil pH. The Al$_2$(SO$_4$)$_3$ and Ca(OH)$_2$ were mixed with the soil and wetted. The soil pH of each of the subsamples was measured at 2 weeks, 3 weeks, and 4 weeks from mixing. The change in pH associated with the different rates of Al$_2$(SO$_4$)$_3$ and Ca(OH)$_2$ was used when determining the Al$_2$(SO$_4$)$_3$ and Ca(OH)$_2$ rates needed to reach target pH in this study. Ca(OH)$_2$ was applied to raise the actual pH to the target pH. Al$_2$(SO$_4$)$_3$ was applied to lower the actual pH to the target pH. Based on the average total soil Al concentrations for this soil ($\sim$12,000 mg kg^{-1}), the highest alum amendment only added Al at 2.3% of the total background soil Al concentration. Table 2 lists the initial pH of each location and the amount of Ca(OH)$_2$ and Al$_2$(SO$_4$)$_3$ needed to change soil ph by 1.0. The plots were cultivated to incorporate the Ca(OH)$_2$ and Al$_2$(SO$_4$)$_3$ several months prior to planting. Grain sorghum was planted at a seeding rate of 123,500 seeds ha^{-1}. The middle two rows of each plot were harvested by hand or with a small plot combine. Grain was dried and weighed. Grain yield was corrected to 14% moisture content.

Additional soil samples were taken midseason and postharvest in each growing season to determine actual soil pH during growth, as well as nutrient levels. The final set of soil samples in 2010 were analyzed for the concentration of extractable Al in the soil. A 2.0 gram subsample from each plot was extracted with 20 mL of 1 M potassium chloride (KCl). Samples were placed on a shaker for 30 minutes and filtered. The amount of Al extracted with 1 M KCl (Al$_{KCl}$) was quantified using inductively coupled plasma spectrometry (ICP) [13].

TABLE 2: Initial soil pH values of each trial location and the amount of hydrated lime and alum (Mg ha^{-1}) required to change soil pH by 1.0 unit.

Location	Initial soil pH	Mg ha^{-1} hydrated lime 1.0 pH change	Mg ha^{-1} alum 1.0 pH change
Perkins, OK	4.86	1.69	1.52
Lahoma, OK	5.5	1.31	2.78
Haskell, OK	5.2	4.1	2.17

After harvest in 2010, deep soil cores were taken to 91 cm using a Giddings probe. Samples were taken from 3 plots with target soil pH 4.0, 6.0, and 7.0 at Perkins and Lahoma, OK, USA. These samples were analyzed for soil pH to determine the variation in soil pH within the profile. Samples were not taken at Haskell, OK, USA due to equipment and travel constraints.

Plant counts were taken from the two middle rows of each treatment 1–3 weeks after emergence in 2009 and 2010. Plant height measurements were taken from 5 random plants within the two middle rows of each treatment at the 7 leaf stage at Lahoma and Perkins and at the 8-leaf stage at Haskell in 2010. Also in 2010, the Greenseeker was used to collect Normalized Difference Vegetative Index (NDVI) readings from the middle two rows of each treatment at the 2–5 leaf stage at Lahoma and Perkins and at the 8 leaf stage at Haskell. NDVI is calculated as:

$$NDVI = \left[\frac{(NIR - Red)}{(NIR + Red)} \right]. \tag{1}$$

NIR and Red are near-infrared (780 nm) wavelengths, and red (671 nm) wavelengths respectively [14]. These readings provide a measurement of biomass, plant health, and plant vigor. The red light emitted from the Greenseeker is absorbed by plant chlorophyll. Healthier plants have a higher NDVI value because they absorb more red light and reflect more near-infrared light [15]. The number of heads in the middle two rows of each plot was counted prior to being harvested by combine or by hand in 2010. Grain was collected and weighed to calculate yield.

The use of relative yield, rather than absolute yield, allows the removal of some bias associated with multiple locations and varying growing conditions in this study. Relative yield in this study was expressed as a percentage of maximum yield potential for that location. Relative yield was calculated as:

$$\text{Relative yield}_{avg} = \left[\frac{(\text{Actual yield})}{(\text{Average of 3 highest yields for that site})} \right]. \tag{2}$$

Data analysis was generated using SAS software, Version 9.2 of the SAS system (Copyright 2008) SAS Institute Inc. Data was analyzed using quadratic least squares regression (PROC GLM) and nonlinear regression (PROC NLIN).

3. Results and Discussion

3.1. Soil Profile pH. Soil profile pH results indicate that soil pH was altered to a depth of approximately 31 cm at Perkins

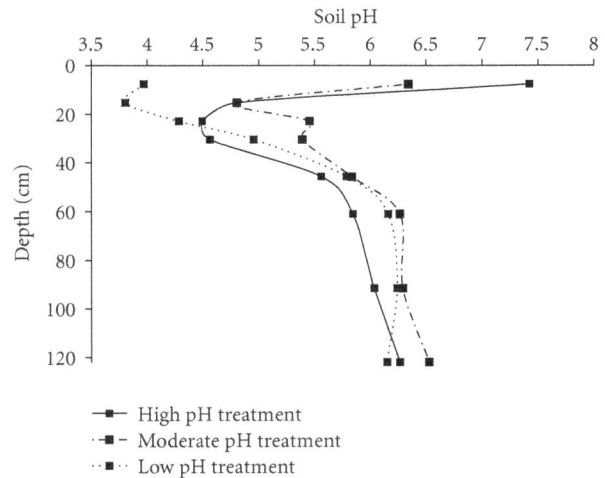

FIGURE 1: Postharvest soil profile pH at Perkins, Oklahoma for target pH treatments of 4.0, 6.0, and 7.0 (2010).

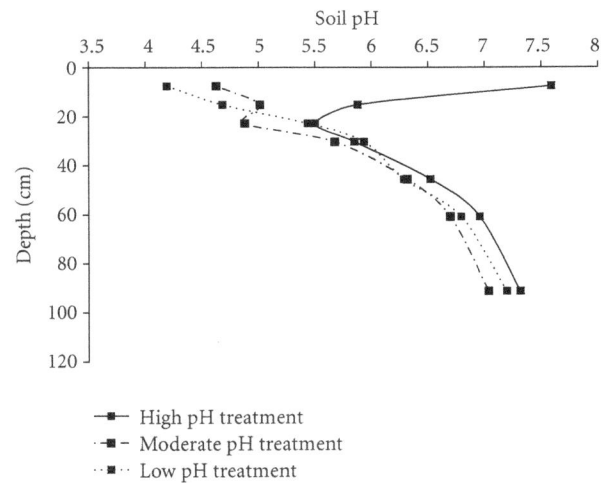

FIGURE 2: Postharvest soil profile pH at Lahoma, OK, USA for target pH treatments of 4.0, 6.0, and 7.0 (2010).

and Lahoma (Figures 1 and 2); however, soil pH varied from target pH throughout the profile. This variability could have masked the effect of high and low pH treatments as roots penetrated below the altered depth of the soil. However, this scenario is indicative of many Oklahoma acidic soils that are typically only acidic in the top 15 cm due to production practices [16]. The Lahoma location has a slight slope, and sheet erosion likely caused the treatment with target pH of

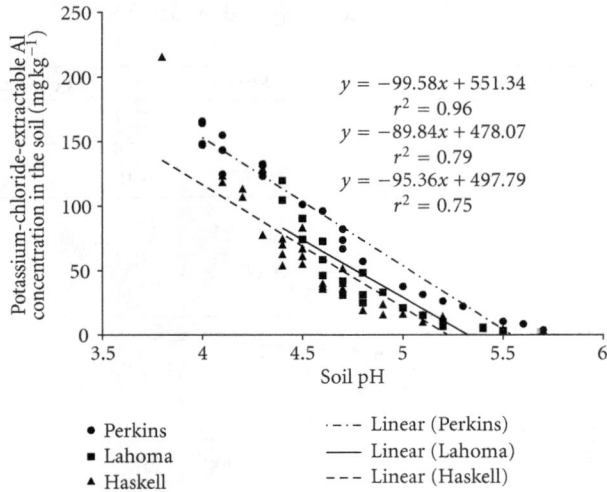

FIGURE 3: Relationship between potassium-chloride-extractable Al concentration in the soil and soil pH from a 15 cm composite soil sample for each treatment at Perkins, Lahoma, and Haskell, Oklahoma (2010).

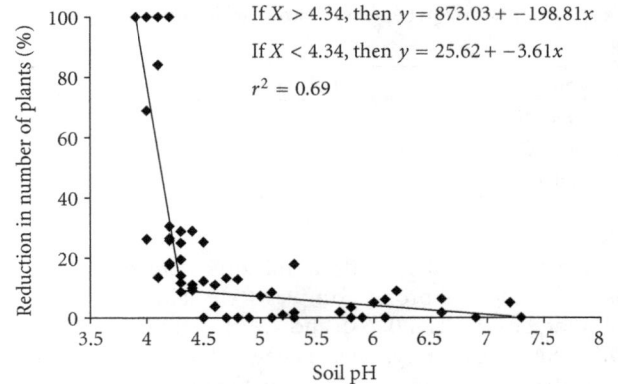

FIGURE 4: Percent reduction in number of grain sorghum plants from emergence to immediately prior to harvest at Perkins, Lahoma, and Haskell, Oklahoma (2010).

6 being much lower in the top 15 cm (Figure 2) as the target pH of the plot upslope was 4.0. After emergence a heavy rain drove sediment down slope into the plot.

3.2. Extractable Aluminum Concentration in Soil. Aluminum toxicity is one of the primary concerns when addressing soil acidity; therefore, potassium-chloride-extractable Al was analyzed in all plots in 2010. Soil pH and potassium-chloride-extractable Al concentrations in the soil were highly correlated at all sites with r^2 of 0.98, 0.93, and 0.95 at Perkins, Lahoma, and Haskell, respectively. Potassium-chloride-extractable Al concentrations increased as soil pH decreased (Figure 3). Regardless of the Al added with the alum, the increase in extractable Al is due to the decrease in pH. Soil soluble Al is controlled by pH conditions, not by total soil Al concentrations. For example, Warren et al. [17] applied normal and alum-treated poultry litter to two different Virginia soils. Even though the alum-treated litter added more Al to the soil compared to normal litter, the relationship between soil exchangeable Al and pH was unaffected by litter type (i.e., amount of Al added). Similarly, Moore and Edwards [18] found that application of N in the form of ammonium nitrate increased soil exchangeable Al due to decreasing pH, while the addition of Al from alum-treated litter had no impact on soil exchangeable Al. Soil Al mostly resides in the octahedral sheets of 1:1 and 2:1 clay minerals and also amorphous and crystalline Al oxides/hydroxides. Any Al added to the soil will immediately precipitate into an amorphous Al hydroxide mineral as a function of the soil pH. Similarly, the solubility of Al in soil minerals containing Al will be mostly controlled by pH; as pH decreases below 7, Al solubility increases.

3.3. Growth Factors. Plant emergence and plant height exhibited a negative response ($\alpha = 0.05$) to soil acidity when analyzed across all locations in 2009 and 2010 (Table 3). A negative response to soil acidity was also observed in NDVI measurements ($\alpha = 0.05$) at all locations, which demonstrates the reduction in biomass and plant vigor of plants in acidic treatments (Table 3). The number of heads plot^{-1} at harvest was reduced in low pH treatments at all locations in 2009 and 2010 suggesting that plant mortality increased with decreasing soil pH. Plant counts at emergence were higher than the number of heads at harvest. This suggests that soil acidity had an impact on stand establishment but even more of an effect on plant mortality through the growing season in 2010. The reduction in plant counts as the season progressed was correlated with soil pH (Figure 4). Plant mortality was not reduced when soil pH > 5.5, but the number of plants significantly decreased when soil pH < 4.43 (Figure 4). The plants located in treatments with lower soil pH likely had increased root pruning as a result of soil acidity, which prevented the roots from penetrating into the more neutral subsoil. Since these plants were not able to explore less acidic soil for nutrients, the plants did not survive. In contrast, plants located in treatments with moderate soil pH likely had less root pruning and were able to penetrate into more neutral subsoil and reach additional nutrients, thus allowing them to survive.

3.4. Relative Yield. Relative yield exhibited a negative response to soil acidity and was significant ($\alpha = 0.05$) when evaluated in quadratic least squares regression at two of the five grain sorghum site years (Table 4). Nonlinear regression generated a yield plateau at 0.71 relative yield and soil pH 4.54. Assuming that most producers would not be willing to sustain yield losses of greater than 10%, relative yield 0.90 was chosen as the yield plateau level. The regression equation generated from PROC NLIN ($y = 0.3513 \times -1.0051$) was used to determine that the soil pH at relative yield 0.90 was 5.42 (Figure 5).

There was a considerable amount of variation in the yield response to soil pH among locations and years, which was likely due to environmental impacts other than soil pH. For example, results from the 2009 season show less

TABLE 3: Results from quadratic least squares regression when evaluating the effect of soil pH on grain sorghum emergence, plant height, NDVI, and mortality (2009 and 2010).

	DF	Mean square	F	Prob F	r^2	R
			2009			
Emergence	2	0.21	5.37	0.0077	0.17	0.48
			2010			
Emergence	2	0.19	19.60	<0.0001	0.44	0.54
Plant height	2	1.01	43.08	<0.0001	0.63	0.62
NDVI	2	0.92	122.18	<0.0001	0.83	0.78
Mortality	2	20351.59	35.42	<0.0001	0.59	0.53

TABLE 4: Results from quadratic least squares regression when evaluating the effect of soil pH on grain sorghum relative yield (2009 and 2010).

	DF	Mean square	F	Prob F	r^2
			2009		
Lahoma	2	0.06	2.31	0.133	0.24
Perkins	2	0.04	0.35	0.7082	0.04
All locations	2	0.18	3.23	0.0542	0.18
			2010		
Lahoma	2	0.06	1.59	0.2371	0.17
Perkins	2	1.24	37.67	<0.0001	0.83
Haskell	2	0.25	15.65	0.0002	0.68
All locations	2	1.12	22.82	<0.0001	0.47

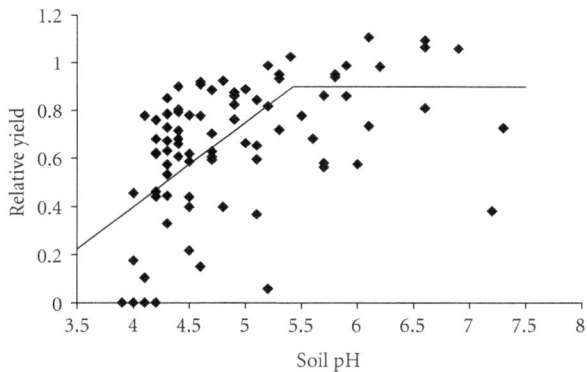

FIGURE 5: Relationship of grain sorghum relative yield and soil pH at Lahoma, Perkins, and Haskell, Oklahoma with yield plateau occurring at 0.90 with critical soil pH 5.42 (2009 and 2010).

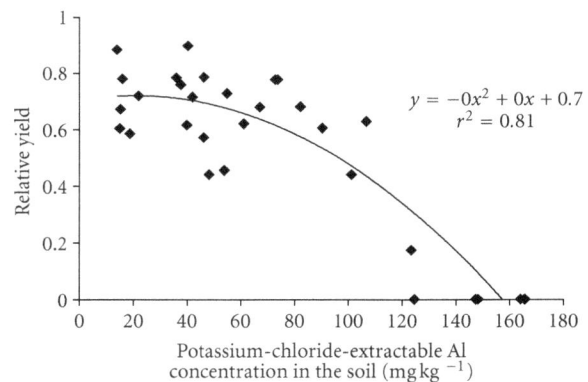

FIGURE 6: Quadratic relationship of potassium-chloride-extractable Al concentration in soil and grain sorghum relative yield at Perkins, Lahoma, and Haskell, Oklahoma (2010).

significance overall when compared to 2010 results. One possible explanation for this inconsistency among years could be soil moisture levels. Oklahoma Mesonet soil moisture graphs indicate that on average the period from planting to 30 days after planting in 2009 had higher fractional water index when compared to 2010 at all locations. The improved soil moisture conditions of 2009 could have masked the effect of soil pH as compared to 2010 by allowing roots to penetrate below the acidic surface soil earlier in the season. The concentration of extractable Al in the soil was analyzed in 2010 and was found to be highly correlated with soil pH ($r^2 = 0.90$) and relative yield ($r^2 = 0.81$) (Figures 3 and 6).

3.5. Environmental Impacts. The environment played a significant role in the severity of soil acidity stresses observed in this study. Higher soil moisture in 2009 compared to 2010 could have masked the effect of soil pH and reduced the negative effects on yield. Damage incurred from birds was also an outside environmental impact that could not be controlled. Also in 2010, a compaction layer was observed at Perkins that could have prevented roots from penetrating into more neutral subsoil, thereby emphasizing the effects of soil acidity in the top 15 cm seen at that location.

4. Conclusions

The results from this study varied from location to location and year to year; however, a trend was detected that confirms that soil acidity reduced grain sorghum yield. This study demonstrated that the environment played a significant role in the degree of soil acidity stresses observed in grain sorghum production. The critical levels and relative yield models developed in this study will be helpful when making liming decisions. Depending on environmental factors, these estimated yield reductions may not hold true in all situations.

The yield reductions associated with soil acidity can be substantial. However, when producers consider liming, all factors should be taken into account. For example, if commodity prices are low, land is rented, or there is not high yield potential, the cost of liming could outweigh the reward. The estimates developed in this study will provide producers with an additional tool to determine if liming a field is necessary and economical.

Potassium-chloride-extractable Al concentration in the soil, which is related to parent material and soil CEC, negatively affected crop response to soil acidity. Differences in potassium-chloride-extractable Al concentration can cause soil acidity symptoms associated with Al toxicities to occur at higher or lower soil pH than expected. For this reason, it could be beneficial when developing liming recommendations to consider Al concentration in the soil in addition to soil pH and buffer index.

In this study, at grain sorghum relative yield 0.90, the critical soil pH was 5.42. The models developed in this study will provide producers with a tool to estimate yield reductions at a given soil pH (Figure 5). As producers incorporate grain sorghum into rotations, it is recommended that soil pH be tested and limed if soil pH is 5.42 or below to ensure that significant yield reductions associated with soil acidity are avoided. Future research concerning crop response to soil pH may need to include additional locations and deep tillage so that soil pH is altered deeper than 15 cm.

References

[1] J. Edwards, F. Epplin, B. Hunger et al., "No-till wheat production in Oklahoma," Oklahoma Cooperative Extension Service Fact Sheet 2132, Oklahoma State University, Stillwater, Okla, USA, 2006.

[2] P. L. Mask, A. Hagan, and C. C. Mitchell Jr., "Production Guide for Grain Sorghum," Bulletin ANR-0502, Alabama Cooperative Extension System, Alabama A&M University and Auburn Universities, Auburn, Ala, USA, 1988.

[3] Potash and Phosphate Institute, "Soil test levels in North America: summary update," PPI/PPIC/FAR Technical Bulletin 2005-1, Norcross, Ga, USA, 2005.

[4] R. R. Duncan, "Sorghum genotype comparisons under variable acid soil stress," Journal of Plant Nutrition, vol. 10, no. 9–16, pp. 1079–1088, 1987.

[5] S. K. Kariuki, H. Zhang, J. L. Schroder et al., "Hard red winter wheat cultivar responses to a pH and aluminum concentration gradient," Agronomy Journal, vol. 99, no. 1, pp. 88–98, 2007.

[6] K. Ohki, "Aluminum stress on sorghum growth and nutrient relationships," Plant and Soil, vol. 98, no. 2, pp. 195–202, 1987.

[7] R. R. Duncan, W. Dobson Jr., and C. D. Fisher, "Leaf elemental concentrations and grain yield of sorghum grown on an acid soil," Communications in Soil Science and Plant Analysis, vol. 11, no. 7, pp. 699–707, 1980.

[8] K. Tan and W. G. Keltjens, "Analysis of acid-soil stress in sorghum genotypes with emphasis on aluminium and magnesium interactions," Plant and Soil, vol. 171, no. 1, pp. 147–150, 1995.

[9] C. I. Flores, R. B. Clark, and L. M. Gourley, "Growth and yield traits of sorghum grown on acid soil at varied aluminum saturations," Plant and Soil, vol. 106, no. 1, pp. 49–57, 1988.

[10] J. T. Sims, "Lime requirement," in Methods of Soil Analysis, Part 3-Chemical Methods, D. L. Sparks, Ed., pp. 491–515, American Society of Agronomy, Soil Science Society of America, Madison, Wis, USA, 1st edition, 1996.

[11] F. J. Sikora, "A buffer that mimics the smp buffer for determining lime requirement of soil," Soil Science Society of America Journal, vol. 70, no. 2, pp. 474–486, 2006.

[12] A. Mehlich, "Mehlich 3 soil test extractant: a modification of Mehlich 2 extractant," Communications in Soil Science & Plant Analysis, vol. 15, no. 12, pp. 1409–1416, 1984.

[13] P. N. Soltanpour, G. W. Johnson, S. M. Workman, J. B. Jones Jr., and R. O. Miller, "Inductively coupled plasma emission spectrometry and inductively coupled plasma-mass spectrometry," in Methods of Soil Analysis, Part 3-Chemical Methods, D. L. Sparks, Ed., pp. 91–139, American Society of Agronomy, Soil Science Society of America, Madison, Wis, USA, 1st edition, 1996.

[14] R. W. Mullen, K. W. Freeman, W. R. Raun, G. V. Johnson, M. L. Stone, and J. B. Solie, "Identifying an in-season response index and the potential to increase wheat yield with nitrogen," Agronomy Journal, vol. 95, no. 2, pp. 347–351, 2003.

[15] Y. Lan, H. Zhang, R. Lacey, C. Hoffman, and W. Wu, "Development of an integrated sensor and instrumentation system for measuring crop conditions," Agricultural Engineering International, vol. 11, Manuscript IT 08 1115, 2009.

[16] F. Gray and M. H. Roozitalab, Benchmark and Key Soils of Oklahoma: A Modern Classification System, Oklahoma State University, Agricultural Experiment Station, Stillwater, Okla, USA, 1976.

[17] J. G. Warren, S. B. Phillips, G. L. Mullins, D. Keahey, and C. J. Penn, "Environmental and production consequences of using alum-amended poultry litter as a nutrient source for corn," Journal of Environmental Quality, vol. 35, no. 1, pp. 172–182, 2006.

[18] P. A. Moore and D. R. Edwards, "Long-term effects of poultry litter, alum-treated litter, and ammonium nitrate on phosphorus availability in soils," Journal of Environmental Quality, vol. 36, no. 1, pp. 163–174, 2007.

4

Required Lateral Inlet Pressure Head for Automated Subsurface Drip Irrigation Management

Moncef Hammami,[1] Khemaies Zayani,[2] and Hédi Ben Ali[3]

[1] *University of Carthage, High School of Agriculture of Mateur, 7030 Mateur, Tunisia*
[2] *University of Carthage/High Institute of Environmental Sciences and Technologies of Borj Cedria,*
  B.P. 1003, 2050 Hammam Lif, Tunisia
[3] *Agricultural Investment Promotion Agency, 6000 Gabès, Tunisia*

Correspondence should be addressed to Moncef Hammami; hammami.moncef@ymail.com

Academic Editor: Vicente Gimeno

Subsurface drip irrigation (SDI) is one of the most promising irrigation systems. It is based on small and frequent water supplies. Because SDI emitters are buried, their discharges are dependent on the water status at the vicinity of the outlets. This paper was targeted to design the SDI laterals accounting for the soil water-retention characteristics and the roots water extraction. The proposed approach provides systematic triggering and cut-off of irrigation events based on fixed water suctions in the vadose zone. In doing so, the soil water content is maintained at an optimal threshold ascertaining the best plant growth. Knowing the soil water-retention curve, the appropriate water suction for the plant growth, and the emitter discharge-pressure head relationship, the developed method allows the computation of the required hydraulics of the lateral (e.g., inlet pressure head, inside diameter, etc.). The proposed approach is a helpful tool for best SDI systems design and appropriate water management. An illustrative example is presented for SDI laterals' design on tomato crop.

1. Introduction

In subsurface drip irrigation (SDI), water seeps from the buried emitters into the soil and spreads out in the vadose zone under the conjugate effect of capillary and gravity forces [1, 2]. Thus, SDI system allows the direct application of water to the rhizosphere maintaining dry the nonrooted topsoil. This pattern generates numerous advantages such as minimizing soil evaporation and then evapoconcentration phenomenon. The rationale is that SDI improves the water application uniformity, increases the laterals and emitters longevity, reduces the occurrence of soil-borne diseases, and allows the control of weeds infestation. Several field trials revealed relevant profits on managing SDI for crop production. Nevertheless, the appropriate depth of buried laterals remains debatable [3–6]. Comparing evaporation from surface and subsurface drip irrigation systems, Evett et al. [7] reported that 51 mm and 81 mm were saved with drip laterals buried at 15 cm and 30 cm, respectively. Neelam and

Rajput [1] recorded maximum onion yield (25.7 t ha^{-1}) with drip laterals buried at 10 cm. According to these authors, the maximum drainage occurred when drip laterals are laid at 30 cm depth. On the other hand, numerous studies were devoted to the analysis of the effect of the soil properties on the SDI emitters discharge and water distribution uniformity [8–10]. The analytical approach proposed by Sinobas et al. [2] predicts reasonably well the soil water suction and the pressure head distributions in the laterals and SDI units [11].

The water oozes from the buried emitters under the conjugate effect of the inlet lateral pressure head and the water suction in the surrounding soil. Therefore, the emitter discharge is high at the beginning of watering because the root zone is yet relatively dry. Gradually, as the pore space at the dripper outlet is filled with water, a positive pressure head develops, which may cause a decrease in dripper discharge [12]. If the discharge is larger than the infiltration capacity, the resulting overpressure near the nozzle tends to lessen the flow rate [9, 13]. It should be stressed that the overpressure phenomenon is more likely to occur near the emitter outlet in fine textured soils than in coarse ones. Indeed under field conditions, Shani et al. [12] highlighted that the pressure head increases rapidly up to 8 m after 10 to 15 minutes of irrigation in fine soil. For similar soil and flow rate conditions, Gil et al. [14] obtained lower overpressures in pots. They observed that the greater the emitter flow rate is, the higher the overpressure will be.

The emitter flow rate is a power function of the pressure head difference between the nozzle's inlet and outlet. The increase of the pressure head at the emitter outlet induces a flow-rate decrease. Reciprocally, plant roots water uptake generates soil drying. Thus, the resulting decrease of the soil pressure head fosters an increase of the emitter flow rate and/or the energy saving [15]. Accordingly, the emitter discharge variations are governed by the soil moisture variations and the roots water uptake. To reap the best from this opportunity, the SDI systems design should account for the influence of soil water status on the buried emitters discharge. Lazarovitch [16] and Clothier and Green [17] recommended the use of sufficiently low flow rates to match the roots water uptake. This paper is devoted to the development of a new approach to design SDI systems for a systematic irrigation management, energy saving, and least labor cost.

2. Basics of the Approach

The pressurized irrigation systems are customarily designed so that the mean pressure head throughout the pipe is equal to the nominal pressure head. On the other hand, irrigation management is based on the replenishment of the soil holding capacity. Hence, the soil moisture should range between predetermined and minimum allowable water contents. Hereinafter, we assume that the average pressure head is equal to the emitter operating pressure head. The emitter discharge equation may be expressed by [18]

$$q = KH^x, \qquad (1)$$

where $q[L^3T^{-1}]$ and $H[L]$ are the emitter discharge and the emitter pressure head, whereas $K[L^{3-x}T^{-1}]$ and $x[-]$ are fitting parameters.

Equation (1) is valid for a pressure head higher than or equal to 5.0 m. It is worth pointing out that most long-path turbulent flow emitters and pressure-compensating outlets require an operating pressure head fulfilling this condition. For buried emitters, the emitter pressure head is lumped with the water suction near the outlets:

$$H = h_e - h_i, \qquad (2)$$

where h_e and h_i refer to the pressure heads [L] at the inner and outer tips of the emitter, respectively. For emitters laid on the ground, h_i is the atmospheric pressure. Conversely, for buried emitters, h_i is a spatial-temporal variable dependent on the prevailing soil water content. Hereinafter, we will consider the sigmoid retention curve of Van Genuchten [19] as follows:

$$\theta = \theta_r + \frac{(\theta_s - \theta_r)}{[1 + (\alpha h)^n]^m}, \qquad (3)$$

where $\theta[L^3L^{-3}]$ and $h[L]$ refer to the volumetric water content and to the soil suction head, respectively. The residual water contents θ_r, $\alpha[L^{-1}]$, n, and m are inferred by fitting scattered data (θ, h) according to (3), and θ_s refers to the saturated soil water content. The dimensionless parameters n and m are expressed by the Mualem [20] relationship as follows:

$$m = 1 - \frac{1}{n}. \qquad (4)$$

The soil capillary capacity $C[L^{-1}]$ is derived straightforwardly by differentiating (3) with respect to the suction head as follows:

$$C = \frac{d\theta}{dh} = \frac{-mn\alpha(\theta_s - \theta_r)(\alpha h^{n-1})}{[1 + (\alpha h)^n]^{m+1}}. \qquad (5)$$

Equation (5) shows that additional increase of the suction head produces an additional water release from the soil. Besides, the value of C is the highest if the second derivative of the soil moisture content with respect to the suction head is zero. Under these conditions, the crops absorb the maximum water from the root zone for the same additional energy increment. Further analysis provides the coordinates of the inflexion point of the retention curve as well as the maximum capillary capacity as follows:

$$h_{op} = \frac{-m^{1/n}}{\alpha},$$

$$\theta_{op} = \theta_r + \frac{(\theta_s - \theta_r)}{(1 + m)^m}, \qquad (6)$$

$$C_{max} = \frac{nm^{m+1}\alpha(\theta_s - \theta_r)}{(1 + m)^{m+1}},$$

where h_{op}, θ_{op}, and C_{max} refer to the optimal water suction, optimal soil water content, and maximum capillary capacity, respectively. Therefore, the design of SDI systems should ascertain a suction head at the emitter outlet that matches the optimal water status within the root zone. Combining (1) and (2) yields the following:

$$q = K(h_e - h_{op})^x. \qquad (7)$$

Equations (6) and (7) highlight the dependence of the emitter discharge on the pressure heads at the inner and outer tips of the nozzle. Inasmuch as the soil is more or less dry at the beginning of the irrigation, the discharge decreases with the elapsed time. Incidentally as the soil becomes wetter, the soil pressure head increases and the emitter discharge stabilizes to a minimum value. Experimental results of Gil et al. [14] showed that the decrease of the flow rate is steeper in loamy than in sandy soils. Yao et al. [13] recorded that the wetted soil volume in medium loam and sandy loam is virtually invariant as the inlet pressure head increased from 60 to 150 cm. This increase of pressure head may lead to the back-pressure development. Yao et al. [13] recommended that the emitter discharge should be matched to the soil conditions, so that back-pressure occurrence is avoided. According to Ben-Gal et al. [21] and Lazarovitch et al. [9], one of the main issues with SDI systems is the soil saturation. This phenomenon induces temporary asphyxia of crops and may stop the emitter discharge even though the moistened bulb is not yet spatially well extended.

According to (1) and (2), the emitter discharge is null whenever the outlet pressure head (h_i) matches the pre-determined inlet one (h_e). Afterwards, the redistribution process provides drier rooted soil profiles. Subsequently, the pressure near the emitter (h_i) decreases until the pressure differential between the outlet tips overtakes a minimum value Δh_{min} required for the emitter operation. This threshold Δh_{min} is dependent on the structural form, dimension, and material of the emitter pathway. For any emitter model, Δh_{min} may be inferred from the emitter discharge-pressure head relationship provided by the manufacturer.

Thus, the next irrigation is automatically triggered once the following inequality is fulfilled:

$$h_e - h_i = h_e - h_{op} \geq \Delta h_{min}. \tag{8}$$

Therefore, the required minimum pressure head at the emitter inlet h_{min}^* should comply with

$$h_{min}^* \geq h_{op} + \Delta h_{min}. \tag{9}$$

It is worth emphasizing that the suction head at the vicinity of the emitter cannot be lastingly maintained constant and equal to h_{op}. Unavoidable fluctuations of the suction head are expected owing to evapotranspiration and water redistribution processes. For the sake of convenience, the suction head in the root zone should be circumscribed within a prescribed interval $[h_{op} + \Delta h_{op}, h_{op} - \Delta h_{op}]$. Therefore, the minimum required emitter inlet pressure head h_{req}^{min} is given by

$$h_{req}^{min} = h_{op} - \Delta h_{op} + \Delta h_{min}, \tag{10a}$$

whereas the maximum required emitter inlet pressure head h_{req}^{max} is given by

$$h_{req}^{max} = h_{op} + \Delta h_{op} + \Delta h_{min}. \tag{10b}$$

The magnitude of the interval $[h_{op} \pm \Delta h_{op}]$ should account for the sensitivity of the crop to the water stress. As a matter of fact, for tomato crop, the reduction of the water requirement by 20% produces 20% yield decrease [22]. Contrariwise, the decrease of the onion water requirement by 20% induces only 2% of yield decrease [5]. It should be highlighted that these yield reductions are more or less significant according to the physiological stages.

3. The Required Lateral Pressure Head

For a buried lateral equipped with N identical emitters, the inlet discharge Q will vary within the following limits:

$$Nq_{min} \leq Q \leq Nq_{max}, \tag{11}$$

TABLE 1: Tolerable soil pressure head variations for some crops.

Crop	Upper pressure (cm)	Lower pressure (cm)	References
Spring wheat	−25	−1000	Li et al. [23]
Tomato	−2	−800	Gärdenäs et al. [24] Hanson et al. [25]
Soybeans	−25	−800	Clemente et al. [26]
Grape	−2	−1000	Hanson et al. [25]
Grass	−25	−800	Clemente et al. [26]

where q_{max} and q_{min} are the maximum and minimum emitters average discharges, respectively. For design purpose, only the maximum average emitters discharge is considered. Therefore, the lateral inner diameter is designed to allow the conveyance of the upper bound of the discharge. Consequently, the minimum pressure head required at the upstream end of nontapered flat lateral is

$$h_{Lm} = Z_d + J_L + \Delta h_{min} + h_{op} - \Delta h_{op}, \tag{12a}$$

whereas the maximum pressure head required at the upstream end of the lateral is

$$h_{LM} = Z_d + J_L + \Delta h_{min} + h_{op} + \Delta h_{op}, \tag{12b}$$

with $Z_d[L]$ and $J_L[L]$ being the emitters burial depth and the head loss along the lateral, respectively. By convention, the gravitational potential Z_d is computed negatively downwards.

According to the aforementioned basics, the design procedure of SDI systems should lead to the automation of water supplies. Indeed, the irrigation events are triggered whenever the mean pressure head within the root zone is reduced to the minimum prescribed value ($h_{op} - \Delta h_{op}$). They would be automatically ended once the pressure head within the root zone exceeds the maximum value ($h_{op} + \Delta h_{op}$). From theoretical standpoint, a self-regulation of the flow rate by soil water properties and moisture conditions should prevail. Moreover, the emitters discharge variations due to the head losses are offset by soil pressure head gradients. Accordingly, the irrigation events as well as the uniformity of the flow rates are controlled by the soil suction head at the emitters depth of burial. These results corroborate those obtained by Gil et al. [14] who recorded higher flow rates' variability with surface emitters than with buried ones.

Tolerable soil pressure head variations for some crops are summarized in Table 1.

It is worth pointing out that the abovementioned approach remains valid regardless of the used soil water-retention relationship. The following steps summarize the proposed design procedure of SDI laterals.

4. The Design Steps

(1) Carry out simultaneous in situ field measurements of moisture contents and suction heads.

(2) Fit the experimental dataset (θ, h) in accordance with the appropriate soil water-retention curve (for instance, (5)).

(3) Derive twice the moisture content with respect to the suction head and infer h_{op}.

(4) Select the proper interval of the soil suction head Δh_{op} for the considered crop (for instance, data provided in Table 1).

(5) For the used emitters type, calculate the minimum inlet pressure head h_{min}^* using (9).

(6) Calculate the minimum and maximum required emitter inlet pressure heads using (10a) and (10b), respectively.

(7) Using (11), calculate the required lateral inlet discharge.

(8) Deduce the minimum and maximum required lateral inlet pressure heads, using (12a) and (12b), respectively.

5. Study Case

To illustrate the proposed procedure, let us consider a polyethylene nontapered flat pipe 100 m long with equidistant in line-emitters spaced 0.40 m. The laterals irrigate tomato crop on homogeneous sandy soil. Following Patel and Rajput [5], the emitters depth is 15 cm. Therefore, the design steps are the following.

(1) Simultaneous in situ measurements of the soil water contents and suction heads were performed [27] on three randomized points during water redistribution. In each soil profile, suction heads were measured using three tensiometers installed at 10, 30, and 50 cm depth. Soil cores sampled at the same depths were used to determine gravimetrically the correspondent soil water contents. For each depth, the average of the three measurements was considered.

(2) Experimental data were fitted in accordance to the Van Genuchten [19] model [27]. Scattered and fitted data are depicted in Figure 1. The inferred fitting parameters $(\theta_r, m, n,$ and $\alpha)$ are summarized in Table 2.

(3) Using (6), the optimum suction head h_{op} is approximately −12 cm. This value is within the optimal range of the suction head for tomato crop [24, 25]. To prevent asphyxia risk or relative water stress at upper (−2 cm) and lower (−800 cm) tolerable pressure heads, Δh_{op} being equal to 400 cm is acceptable.

(4) Thus, the prescribed soil pressure head limits for tomato crop are the following:

$$h_{op} - \Delta h_{op} \approx -12 - 400 \text{ cm} = -412 \text{ cm},$$
$$h_{op} + \Delta h_{op} \approx -12 + 400 = 388 \text{ cm}. \tag{13}$$

In order to avoid eventual back-pressure development, the suction head should be maintained within [−412 cm, 0.0 cm].

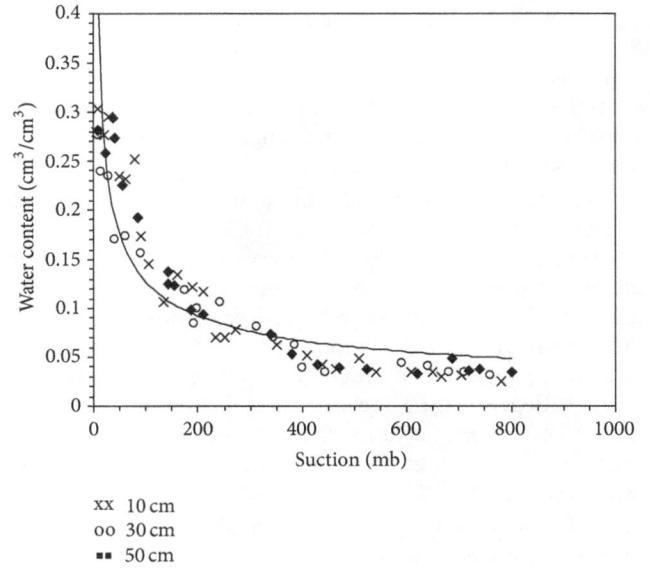

FIGURE 1: Water retention curve: measured data at different soil depths.

TABLE 2: Fitting parameters of Van Genuchten's equation for the considered sandy soil.

θ_s (cm^3/cm^3)	θ_r (cm^3/cm^3)	α (cm^{-1})	n	R^2
0.38	0.02	0.05	1.70	0.991

(5) A trapezoidal labyrinth long-path emitter with a minimal differential operating pressure head of $\Delta h_{min} = 500$ cm is used. The discharge-pressure head relationship of these emitters is [28]

$$q = 0.752(h_e - h_i)^{0.478}, \tag{14}$$

where q = emitter discharge (l/h), h_e = the emitter inlet pressure head (m), and h_i = the emitter outlet pressure head (m).

(6) Using (10a) and (10b), the required emitter inlet pressure h_{req} should comply with

$$(-12 - 400 + 500) = 88 \le h_{req} \text{ (cm)} \le (0 + 500) = 500. \tag{15}$$

To maintain an optimal suction head within the root zone (−12 cm) and to compensate the minimum differential operating pressure head ($\Delta h_{min} = 500$ cm), the optimal required emitter inlet pressure would be $h_{oreq} = (-12 + 500) = 488$ cm. Compared with the pressure heads customarily required for on-surface drippers (approximately 1000 cm), the obtained value underlines an outstanding energy saving with SDI systems. Thus, according to (14), the corresponding emitter discharge q is given as

$$0.752[0.88 - 0.00]^{0.478} = 0.707 \le q \text{ (l/h)}$$
$$\le 0.752[5.00 - (-4.12)]^{0.478} \tag{16}$$
$$= 2.163.$$

As long as the lowest differential pressure head (0.88 m) is less than the minimum differential operating pressure head ($\Delta h_{min} = 500$ cm), the emitter discharge should lay within the interval [0.00, 2.163]. Nevertheless, the optimal required emitter discharge matching the optimal soil suction head q_{op} would be

$$q_{op} = 0.752[4.88 - (-0.12)]^{0.478} \approx 1.623 \text{ l/h}. \quad (17)$$

(7) The number of emitters along the lateral equals 100 m/0.4 m = 250. According to (11), the optimal required discharge at the lateral inlet tip is

$$Q_{op} = 250 \times 1.623 = 405.75 \text{ l/h}. \quad (18)$$

The head loss gradient j may be estimated by Watters and Keller's formula [29] as

$$j = \alpha Q^{\beta} D^{-\gamma}, \quad (19)$$

where Q and D are the discharge and the lateral inner diameter, respectively. For j (m/m), Q (l/h), and D (mm), the parameters of (19) are $\beta = 1.75$, $\gamma = 4.75$, and $\alpha = 14.709598\upsilon^{0.25}$ where υ (m^2 s^{-1}) refers to the kinematic viscosity of water. At 20°C, α is equal to 0.4655. Considering (19) and an inner diameter of 16 mm, the head loss throughout the lateral J_L is given by [29]

$$J_L = \frac{\alpha Q_{max}^{\beta} D^{-\gamma}}{1 + \beta} L = \frac{0.4655(405.75)^{1.75}(16.0)^{-4.75}}{1 + 1.75} 100$$

$$= 1.184 \text{ m}. \quad (20)$$

This value would be doubled if we account for the emitters connection head losses as computed by the Juana et al. [30] approach.

(8) Using (12a) and (12b) and accounting for emitters connection head losses, the required pressure head at the inlet tip of the lateral would be

$$(-15 + 2 * 118.4 + 500 - 12 - 400)$$

$$= 309.8 \leq h_L \text{ (cm)} \quad (21)$$

$$\leq (-15 + 2 * 118.4 + 500 + 0) = 721.8.$$

In the same way, the optimal required pressure head h_{Lo} at the lateral inlet would be

$$h_{Lo} = (-15 + 2 * 118.4 - 12 + 500) = 709.8 \text{ cm}. \quad (22)$$

Therefore, it is possible to ensure a complete automation of the SDI system via the installation of an overhead basin whose water level is constant.

6. Conclusion

Apart from labor's reduction, the water, and energy savings, the SDI system offers the opportunity of a completely automated irrigation management. In fact, the wise control of the soil water content variation at the vicinity of the outlets is a milestone in managing subdrip irrigation. The rationale is that the flow rate of buried drippers is function of the pressure head at the soil depth of burial. Consequently, the temporal variation of the flow rates is dependent on soil water redistribution and roots water uptake. The design procedure developed in this paper provides the appropriate emitters discharge and the inlet lateral pressure head that fit the plant roots water uptake. Knowing the soil retention curve and the roots water uptake, the procedure provides guidelines to design SDI laterals. The main objective of the design is to ascertain optimal suction head within the emitters depth of burial so that irrigation events are automatically controlled by the soil moisture variations. The illustrative study case showed that soil water content could be circumscribed within an interval suitable for plant growth. This approach could be a helpful tool for the best SDI systems design and the best water supplies management. Nevertheless, it is worth to underline that the current approach completely overlooks the effect of burial on drippers' clogging.

References

[1] P. Neelam and T. B. S. Rajput, "Dynamics and modeling of soil water under subsurface drip irrigated onion," *Agricultural Water Management*, vol. 95, no. 12, pp. 1335–1349, 2008.

[2] L. R. Sinobas, M. Gil, L. Juana, and R. Sánchez, "Water distribution in laterals and units of subsurface drip irrigation. I: simulation," *Journal of Irrigation and Drainage Engineering*, vol. 135, no. 6, pp. 721–728, 2009.

[3] G. L. Grabow, R. L. Huffman, R. O. Evans, D. L. Jordan, and R. C. Nuti, "Water distribution from a subsurface drip irrigation system and dripline spacing effect on cotton yield and water use efficiency in a coastal plain soil," *Transactions of the ASABE*, vol. 49, no. 6, pp. 1823–1835, 2006.

[4] M. M. Kandelous and J. Šimůnek, "Numerical simulations of water movement in a subsurface drip irrigation system under field and laboratory conditions using HYDRUS-2D," *Agricultural Water Management*, vol. 97, no. 7, pp. 1070–1076, 2010.

[5] N. Patel and T. B. S. Rajput, "Effect of subsurface drip irrigation on onion yield," *Irrigation Science*, vol. 27, no. 2, pp. 97–108, 2009.

[6] E. D. Vories, P. L. Tacker, S. W. Lancaster, and R. E. Glover, "Subsurface drip irrigation of corn in the United States Mid-South," *Agricultural Water Management*, vol. 96, no. 6, pp. 912–916, 2009.

[7] S. R. Evett, T. A. Howell, and A. D. Schneider, "Energy and water balances for surface and subsurface drip irrigated corn," in *Proceedings of the 5th International Micro irrigation Congress*, pp. 135–140, Orlando, Fla, USA, April 1995.

[8] J. E. Ayars, C. J. Phene, R. B. Hutmacher et al., "Subsurface drip irrigation of row crops: a review of 15 years of research at the Water Management Research Laboratory," *Agricultural Water Management*, vol. 42, no. 1, pp. 1–27, 1999.

[9] N. Lazarovitch, J. Šimůnek, and U. Shani, "System-dependent boundary condition for water flow from subsurface source," *Soil Science Society of America Journal*, vol. 69, no. 1, pp. 46–50, 2005.

[10] B. Safi, M. R. Neyshabouri, A. H. Nazemi, S. Massiha, and S. M. Mirlatifi, "Water application uniformity of a subsurface

drip irrigation system at various operating pressures and tape lengths," *Turkish Journal of Agriculture and Forestry*, vol. 31, no. 5, pp. 275–285, 2007.

[11] L. R. Sinobas, M. Gil, L. Juana, and R. Sánchez, "Water distribution in laterals and units of subsurface drip irrigation. II: field evaluation," *Journal of Irrigation and Drainage Engineering*, vol. 135, no. 6, pp. 729–738, 2009.

[12] U. Shani, S. Xue, R. Gordin-Katz, and A. W. Warrick, "Soil-limiting flow from subsurface emitters. I: pressure measurements," *Journal of Irrigation and Drainage Engineering*, vol. 122, no. 5, pp. 291–295, 1996.

[13] W. W. Yao, X. Y. Ma, J. Li, and M. Parkes, "Simulation of point source wetting pattern of subsurface drip irrigation," *Irrigation Science*, vol. 29, no. 4, pp. 331–339, 2011.

[14] M. Gil, L. R. Sinobas, L. Juana, R. Sánchez, and A. Losada, "Emitter discharge variability of subsurface drip irrigation in uniform soils: effect on water-application uniformity," *Irrigation Science*, vol. 26, no. 6, pp. 451–458, 2008.

[15] P. D. Colaizzi, A. D. Schneider, S. R. Evett, and T. A. Howell, "Comparison of SDI, LEPA, and spray irrigation performance for grain sorghum," *Transactions of the American Society of Agricultural Engineers*, vol. 47, no. 5, pp. 1477–1492, 2004.

[16] N. Lazarovitch, *The effect of soil water potential, hydraulic properties and source characteristic on the discharge of a subsurface source [Ph.D. thesis]*, Faculty of Agriculture of the Hebrew University of Jerusalem, 2008.

[17] B. E. Clothier and S. R. Green, "Roots: the big movers of water and chemical in soil," *Soil Science*, vol. 162, no. 8, pp. 534–543, 1997.

[18] Z. Khemaies and H. Moncef, "Design of level ground laterals in trickle irrigation systems," *Journal of Irrigation and Drainage Engineering*, vol. 135, no. 5, pp. 620–625, 2009.

[19] M. T. Van Genuchten, "A closed-form equation for predicting the hydraulic conductivity of unsaturated soils," *Soil Science Society of America Journal*, vol. 44, no. 5, pp. 892–898, 1980.

[20] Y. Mualem, "A new model for predicting the hydraulic conductivity of unsaturated porous media," *Water Resources Research*, vol. 12, no. 3, pp. 513–522, 1976.

[21] A. Ben-Gal, N. Lazorovitch, and U. Shani, "Subsurface drip irrigation in gravel-filled cavities," *Vadose Zone Journal*, vol. 3, no. 4, pp. 1407–1413, 2004.

[22] J. Doorenbos, A. H. Kassem, C. L. M. Bentverlsen et al., "Yield Response to water," FAO Paper 33, 1987.

[23] K. Y. Li, R. De Jong, and J. B. Boisvert, "An exponential root-water-uptake model with water stress compensation," *Journal of Hydrology*, vol. 252, no. 1–4, pp. 189–204, 2001.

[24] A. I. Gärdenäs, J. W. Hopmans, B. R. Hanson, and J. Šimůnek, "Two-dimensional modeling of nitrate leaching for various fertigation scenarios under micro-irrigation," *Agricultural Water Management*, vol. 74, no. 3, pp. 219–242, 2005.

[25] B. R. Hanson, J. Šimůnek, and J. W. Hopmans, "Evaluation of urea-ammonium-nitrate fertigation with drip irrigation using numerical modeling," *Agricultural Water Management*, vol. 86, no. 1–2, pp. 102–113, 2006.

[26] R. S. Clemente, R. De Jong, H. N. Hayhoe, W. D. Reynolds, and M. Hares, "Testing and comparison of three unsaturated soil water flow models," *Agricultural Water Management*, vol. 25, no. 2, pp. 135–152, 1994.

[27] M. Hammami and K. Zayani, "Effect of trickle irrigation strategies on tomato yield and roots' distribution," *World Journal of Agricultural Sciences*, vol. 5, pp. 847–855, 2009.

[28] W. Qingsong, S. Yusheng, D. Wenchu, L. Gang, and H. Shuhuai, "Study on hydraulic performance of drip emitters by computational fluid dynamics," *Agricultural Water Management*, vol. 84, no. 1-2, pp. 130–136, 2006.

[29] G. Z. Watters and J. Keller, "Trickle irrigation tubing hydraulics," Tech. Rep. 78-2015, ASCE, Reston, Va, USA, 1978.

[30] L. Juana, L. R. Sinobas, A. Losada, and M. Asce, "Determining minor head losses in drip irrigation laterals. I: methodology," *Journal of Irrigation and Drainage Engineering*, vol. 128, no. 6, pp. 376–384, 2002.

5

Changes in Soluble-N in Forest and Pasture Soils after Repeated Applications of Tannins and Related Phenolic Compounds

Jonathan J. Halvorson,[1] **Javier M. Gonzalez,**[1] **and Ann E. Hagerman**[2]

[1] *USDA-ARS Appalachian Farming Systems Research Center, 1224 Airport Road, Beaver, WV 25813-9423, USA*
[2] *Department of Chemistry and Biochemistry, Miami University, 160 Hughes Laboratories, 701 East High Street, Oxford, OH 45056, USA*

Correspondence should be addressed to Jonathan J. Halvorson, jonathan.halvorson@ars.usda.gov

Academic Editor: Dexter B. Watts

Tannins (produced by plants) can reduce the solubility of soil-N. However, comparisons of tannins to related non-tannins on different land uses are limited. We extracted soluble-N from forest and pasture soils (0–5 cm) with repeated applications of water (Control) or solutions containing procyanidin from sorghum, catechin, tannic acid, β-1,2,3,4,6-penta-O-galloyl-D-glucose (PGG), gallic acid, or methyl gallate (10 mg g^{-1} soil). After eight treatments, samples were rinsed with cool water (23°C) and incubated in hot water (16 hrs, 80°C). After each step, the quantity of soluble-N and extraction efficiency compared to the Control was determined. Tannins produced the greatest reductions of soluble-N with stronger effects on pasture soil. Little soluble-N was extracted with cool water but hot water released large amounts in patterns influenced by the previous treatments. The results of this study indicate hydrolyzable tannins like PGG reduce the solubility of labile soil-N more than condensed tannins like sorghum procyanidin (SOR) and suggest tannin effects will vary with land management. Because they rapidly reduce solubility of soil-N and can also affect soil microorganisms, tannins may have a role in managing nitrogen availability and retention in agricultural soils.

1. Introduction

Tannins are reactive secondary metabolites produced by plants that affect important biological, chemical, and physical processes in soil and couple primary productivity to biogeochemical cycles [1–4]. Tannin effects on decomposition and nitrogen availability in soil have been a subject of research for more than fifty years [5, 6]. However, development of strategies for use of tannins as soil management tools has lagged, in part because few studies have specifically related them to improving plant productivity or soil fertility. Early tannin research was conducted on temperate agricultural soils [7–9], while recent work has concentrated more on their role in forest ecosystems [10–12] and tropical soils [13–15]. These studies, however, have tended to emphasize the impacts of tannins on microbially mediated processes rather than on the more immediate abiotic interactions between tannins and soil and have made little attempt to frame their findings into the context of landscape effects.

Tannins are believed to affect the nitrogen cycle through several direct and indirect mechanisms that reduce rates of net mineralization or nitrification. Some tannins are directly toxic to plants or microorganisms [16, 17] but their effects vary with particular tannin chemistry or among taxonomic groups [18]. Some tannins or related phenolic compounds are used by soil microorganisms as substrates increasing microbial demand for nitrogen and immobilization in microbial biomass [2, 12, 19]. Tannins can also reduce rates of mineralization or decomposition by affecting the activity of enzymes [20, 21] or by forming complexes with other proteins or organic nitrogen compounds via reversible noncovalent processes such as hydrogen bonding and hydrophobic interactions (cf. [2, 22, 23]). The availability of the nitrogen sequestered in tannin-protein complexes varies among species of plants, taxa of microorganisms, or even among strains of mycorrhizae [11, 24–28]. Tannins and related phenolics may also affect soil-N through interactions with inorganic soil fractions [3, 23, 29]. For example, tannin-related phenolic compounds can interact with nitrite, produced during nitrification, to form more recalcitrant organic forms in a process termed nitrosation [30–32].

Our earlier studies revealed some tannins were rapidly sorbed by soil and reduced the solubility of labile soil-N [33, 34]. Significant amounts of retained tannin-C remained in soil even after repeated rinses with hot water [33]. A single application of a gallotannin (β-1,2,3,4,6-penta-O-galloyl-D-glucose) produced a persistent reduction in the solubility of organic-N not observed with gallic acid, its simple monomeric constituent, suggesting the rapid formation of stable complexes with soil [33]. Tannic acid also influenced the recovery and composition of Bradford-reactive soil protein, associated with glomalin, produced by arbuscular mycorrhizae [35]. These observations suggested plant tannins are capable of affecting critical soil ecosystem processes such as formation of soil organic matter and rates of nutrient cycles and thus may have a role in managing nitrogen availability and retention in soil.

This report is a portion of a two-part study designed to expand the body of basic information about the effects of tannins and related non-tannin phenolics on soil organic matter and nutrient cycling. In part 1, we reported patterns of sorption of phenolic-C and showed Appalachian forest and pasture soils had a high affinity and a fixed capacity for tannins while related phenolic compounds were retained less [34]. This work summarizes the effects of repeated applications of chemically well-defined hydrolyzable and condensed tannins (polymers) and related non-tannin phenolic substances (monomers) on the solubility of soil-N. We compared surface soil from pasture to soil from the surrounding woodlands to gain insight into the magnitude of change associated with conversion from woodlands to silvopasture and assessed their potential significance on a landscape basis. Our ultimate goal is to gather and develop information needed to devise new management strategies that use the phenolic compounds added by plant residues, leachates, livestock manure or, from intentional amendments, to achieve desired agronomic or environmental goals.

2. Materials and Methods

2.1. Sample Collection and Preparation. Surface soil (0–5 cm) was collected from four farm units in Southern West Virginia, each with areas in forest (mixed deciduous or pine) and pasture use as described in greater detail by Halvorson and Gonzalez [36] and Halvorson et al. [34]. Each sample consisted of a composite of 10 soil cores (6.35 cm in diameter) collected along transects. In the lab, composite samples were sieved (2 mm), dried at 55°C, and stored until further analysis.

2.2. Soil Properties. Bulk density (BD) was determined gravimetrically from intact soil cores [36]. Soil chemical properties were determined for each composite sample (Table 1). Soil pH and electrical conductance (EC) were measured by electrode (1 : 1 soil : water). Total inorganic-N (TIN) was estimated as the sum of water extractable inorganic-N plus KCl extractable-N remaining in soil after the first extraction

reported in previous work [33]. Total soil-C content was determined by dry combustion [37] with a FlashEA 1112 NC Analyzer (CE Elantech, Lakewood, NJ). The texture of the soil samples was determined by hydrometer (Midwest labs, Omaha, NE, http://www.midwestlabs.com/). Cation exchange capacity (CEC) was measured at the inherent soil pH by exchange with cobalt hexamine trichloride [38–40].

2.3. Effects of Test Compounds on Extraction of Soil-N. As part of a larger study [34], soluble-N extracted from soil was determined by difference after each of eight repeated applications of aqueous treatment solutions followed by a sequential extraction with cool (23°C) and hot (80°C) water [41, 42].

2.3.1. Test Compounds. Soil samples were treated with deionized water (Control) or with solutions containing model tannins or non-tannin phenolic compounds (organic acids and flavonoids), selected to represent a range of phenolic compounds of varying complexity present in the plant-soil continuum [43]. Our representative condensed tannin was a polymeric flavonoid-based procyanidin isolated from sorghum grain [*Sorghum bicolor* (L.) Moench] (SOR) [44]. We also evaluated tannic acid (TA), a common but imprecisely defined mixture of galloyl esters, and β-1,2,3,4,6-penta-O-galloyl-D-glucose (PGG), a well-defined gallotannin purified from the tannic acid. Non-tannin phenolics included the flavonoid catechin (CAT), the phenolic acid gallic acid (GA), and its ester, methyl gallate, (MG) (Figure 1, Table 2).

2.3.2. Procedure. Samples of soil (2.5 g) were weighed into *Oak Ridge* centrifuge tubes and treated with 25 mL of deionized water (Control) or with 25 mL of test solution to yield a final amendment of 10 mg test compound g^{-1} soil. After reciprocal shaking at 200 rpm for 1 hour at room temperature, samples were centrifuged for 3 min at 11,952 g and decanted. Supernatants were analyzed for soluble-N with a Shimadzu TOC-VCPN analyzer equipped with a TNM-1 module (Shimadzu Scientific Instruments, Columbia, MD). The treatment application step was repeated seven more times by adding an additional 25 mL of Control or treatment solution to the soil pellet, shaking, and extracting as above. After the final treatment, all samples were extracted with 25 mL of cool (23°C) water and assayed again. Finally, more water (25 mL) was added to soil samples, which were then vortexed, incubated overnight in a hot water bath (16 hrs, 80°C), and assayed for soluble-N as before.

Data were corrected to account for any nitrogen added from the treatments, or carryover from the previous treatment step. Treatment effects were determined for absolute values (mg kg^{-1} soil) but the amount of net soluble-N extracted by treatment solutions was also expressed relative to the Control and used to determine if treatments decreased (net Treatment < net Control) or increased (net Treatment > net Control) extraction of soluble-N

TABLE 1: Selected soil properties adapted from Halvorson et al. [34]. Mean (SEM), $n = 4$.

Land use	BD[†]	pH(1:1)	EC (μS cm soil^{-1})	TIN[‡] (mg N kg soil^{-1})	Total C (mg C g soil^{-1})	Total N (mg N g soil^{-1})	CEC (cmolc kg soil^{-1})	Sand %	Silt %	Clay %
Forest	0.95 (0.03)	4.47 (0.26)	103 (13)	17.3 (3.4)	56.0 (3.5)	4.0 (0.2)	9.8 (1.9)	73.0 (3.1)	20.5 (2.2)	6.5 (1.0)
Pasture	1.14 (0.09)	5.27 (0.08)	151 (14)	26.5 (4.4)	42.9 (2.4)	4.5 (0.4)	10.1 (1.0)	69.0 (6.0)	25.0 (5.3)	6.0 (0.8)

[†] Bulk density (BD) for the study was determined gravimetrically.
[‡] Total inorganic-N (TIN) was estimated as the sum of water extractable inorganic-N plus KCl extractable-N remaining in soil [33].

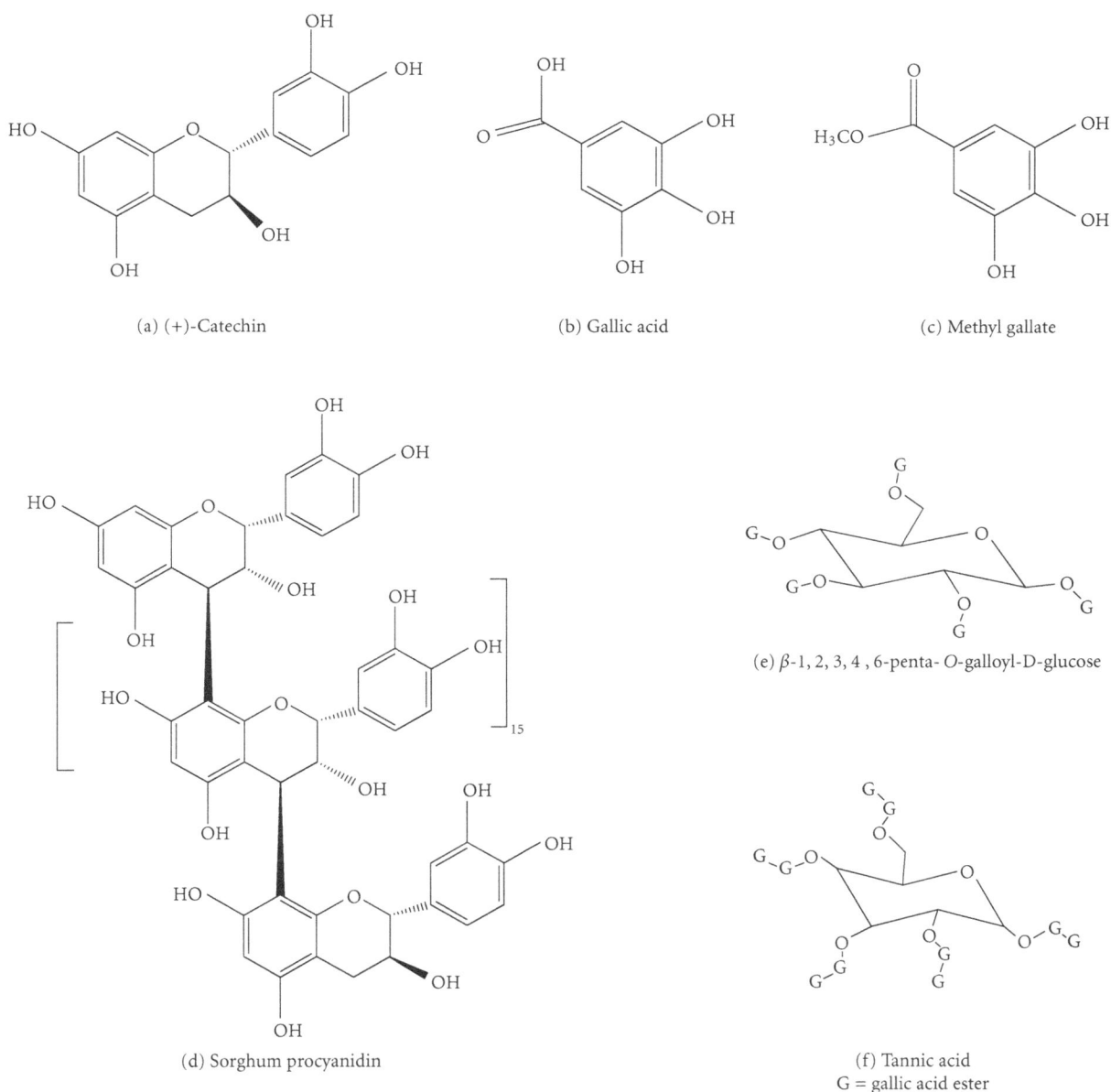

(a) (+)-Catechin

(b) Gallic acid

(c) Methyl gallate

(d) Sorghum procyanidin

(e) β-1, 2, 3, 4 , 6-penta- O-galloyl-D-glucose

(f) Tannic acid
G = gallic acid ester

FIGURE 1: Chemical structures for (a) (+)-catechin (CAT), (b) gallic acid (GA), (c) methyl gallate (MG), (d) sorghum procyanidin (SOR), (e) β-1,2,3,4,6-penta-o-galloyl-d-glucose (PGG), and (f) tannic acid (TA). The structure shown for tannic acid is a representative molecule for tannic acid, an imprecisely defined mixture of galloyl esters [44]. Figure redrawn from [34].

TABLE 2: Details about treatment compounds [34].

Treatment	Class	Source	Compound characteristics					Solution characteristics¶			
			MW†	C‡ (%)	N‡ (%)	C (g mol⁻¹)	$K_{ow}^§$	Soluble-C (mg kg⁻¹ soil)	Soluble-N (mg kg⁻¹ soil)	Phenolics# (mmol GA equiv kg⁻¹ soil)	pH (1mg/mL)
Methyl 3,4,5-trihydroxybenzoate, 98% (methyl gallate)(MG)	Phenolic organic ester	Indofine Chemical Co., Hillsborough, NJ	184	51.7	0.084	96	6.3	5141	0.1	37.2	4.4
Gallic acid, certified (GA)	Phenolic organic acid	Fisher Scientific, Pittsburgh, PA	170	47.7	0.106	85	0.3	4739	0	52.6	3.3
Tannic acid, certified (TA)	Mixture of gallotannins	Fisher Scientific, Pittsburgh, PA	902	49.4	0.142	474	ND	5042	2.5	30.7	3.5
β-1,2,3,4,6 penta-O-galloyl-D-glucose (PGG)	Gallotannin	Purified from Tannic Acid (Fisher)	941	49.7	0.099	492	129	5037	0.7	26.1	5.1
(+)-Catechin hydrate, >98% (CAT)	Flavonoid	Sigma, St Louis, MO	290	61.6	0	180	2.4	6067	1.0	32.1	5.6
[(4β- > 8)-epicatechin₁₅-(4β- > 8)-catechin] (Sorghum procyanidin) (SOR)	Polymeric flavonoid	[Sorghum bicolor (L.) Moench]	4624	48.6	0.094	2880	0.002	5189	8.2	13.7	6.0

† Number average molecular weight for TA estimated by RP-HPLC by Hagerman (Personal Communication) and used to calculate weighed average $g\,C\,mol^{-1}$.

‡ Total C and N were determined in triplicate with a FlashEA 1112 NC Analyzer (CE Elantech, Lakewood, NJ).

§ Octanol-water partition coefficients (K_{ow}) for PGG CAT and SOR from [45] GA and MG from Lu et al. [46]. Low values correspond to hydrophilic compounds while higher values are indicative of hydrophobic ones.

¶ Supplied g^{-1} soil application⁻¹.

Determined by the Modified Prussian Blue assay.

from soil. We determined percentage change in soluble-N attributable to treatments, ΔSol-N, as

$$\Delta\text{Sol-N} = 100 * \frac{(\text{Sol-N}_{trt} - \text{Sol-N}_{control})}{\text{Sol-N}_{control}}, \qquad (1)$$

where Sol-N_{trt} and $\text{Sol-N}_{control}$ indicate the amount of net soluble-N extracted from soil samples treated with phenolic compounds or water alone, respectively.

2.4. Statistical Analysis. Significant effects of test compounds and land use were identified by analysis of variance (ANOVA) with SAS 9.1 and PROC MIXED using a model that accounted for both fixed (land use, treatment) and random (sample location) effects [47, 48]. We used the KR (Kenward-Roger) option to calculate degrees of freedom and selected covariance structures to minimize Akaike's Information Criterion. Assumptions of data normality were evaluated and appropriate data transformations identified with SAS/ASSIST. We assumed a value of 5% as the minimum criterion for significance. Significant main effects were separated by pairwise comparisons among means, adjusted by the Tukey-Kramer method. The SLICE option in PROC MIXED was used to test significant Treatment × Use interaction. Significant deviation of ΔSol-N from zero, indicative of a meaningful change in soluble-N due to the treatment, was determined by the LSMEANS statement in PROC MIXED. Values indicated in text and graphs are the arithmetic mean, ± the standard error of the mean, expressed on air-dry soil basis.

3. Results

3.1. General Patterns. Multiple applications of phenolic solutions produced overall extraction pattern for soluble-N shown in Figure 2. The first of the eight cycles had the greatest effect, extracting 50–60% of the cumulative total amount, while each of the seven subsequent applications removed incrementally less N. The rinse with cool water after the treatment cycles extracted little additional soluble-N but the final incubation in hot water released a relatively large pulse of soluble-N from all treatments and the control. The distinct patterns of ΔSol-N, established with the first treatment application, generally varied little with subsequent phenolic treatments or cool water rinse but showed some differences between forest and pasture soils (Figures 3(a) and 3(b)).

3.2. Treatment with Phenolic Compounds. The amount of soluble-N extracted by the first treatment cycle varied with simple main effects of Treatment and Use. Soluble-N was comparable for most treatments with amounts from samples treated with MG slightly greater than the Control (Table 3). In contrast, both TA and PGG reduced amounts of soluble-N. Pasture soil produced more soluble-N than forest soil. Extraction efficiency of treatments compared to the water Control, ΔSol-N, varied as a Treatment × Use interaction. Treatment with MG significantly increased the amount of soluble-N from forest soil by about 19% compared to 9% for pasture soil (Figure 4). Treatment with CAT did not affect

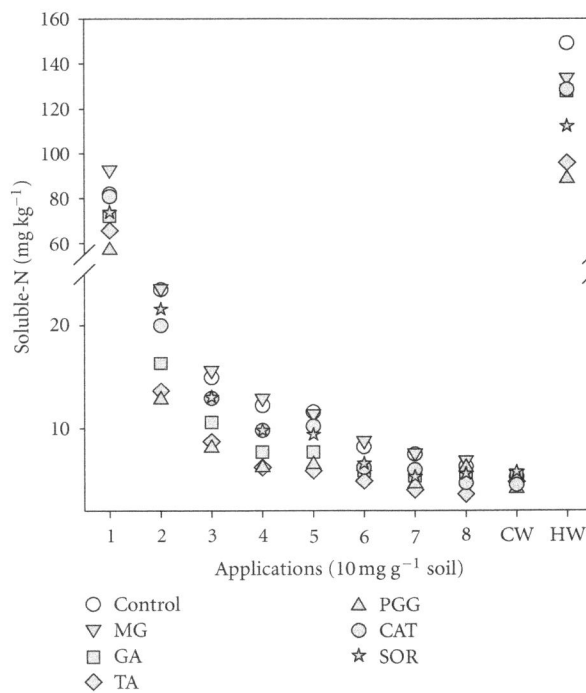

FIGURE 2: Soluble-N extracted from 0–5 cm soil with eight treatment cycles of water (Control) or treatment solutions (10 mg g^{-1} soil) followed by extractions with cool (CW) and hot (HW) water. Values are the mean of 4 farms × 2 uses (forest and pasture), $n = 8$. Treatment abbreviations are defined in Figure 1.

soluble-N from pasture or forest soil. Gallic acid produced no effect on forest soil, but reduced soluble-N from pasture soil by 19%. Soluble-N was reduced from both soil types by SOR, TA, and PGG, but decreases were stronger in pasture soil. Greatest reductions resulted from the PGG treatment, which lowered soluble-N from forest and pasture by 24 and 34%, respectively.

Patterns established with the first treatment cycle persisted throughout the seven subsequent applications of phenolics and thus cumulative extractions, after all eight treatment cycles, varied with main effects of Treatment and Use (Table 3). While MG slightly increased soluble-N, amounts were significantly reduced by GA, TA, and PGG treatments. Pasture soil yielded more soluble-N than forest soil. Cumulative treatment ΔSol-N varied as a Treatment × Use interaction. Eight treatment cycles with MG extracted 13% more soluble-N from forest soil than water alone but had little effect on pasture soil. By contrast, treatment with CAT reduced soluble-N by about 9% for both soils. Soluble-N was also reduced by the other phenolic compounds but treatment effects were discernibly stronger for pasture soil. Greatest reductions, with PGG, resulted in 28 and 40% less soluble-N, than the water control, from forest and pasture, respectively.

3.3. Cool and Hot Water Extractions. Only small amounts, 5.1 ± 0.2 mg kg^{-1}, of soluble-N, were extracted by cool water, after the final treatment cycle, limiting the interpretive value

TABLE 3: Soluble-N (mg kg⁻¹ soil) extracted with treatment solutions[†].

Trt	Single treatment			Cumulative after treatment 8			Cool water			Hot water			Final total		
	Forest	Pasture	Avg. (n = 8)	Forest	Pasture	Avg. (n = 8)	Forest	Pasture	Avg. (n = 8)	Forest	Pasture	Avg. (n − 8)	Forest	Pasture	Avg. (n − 8)
H₂O	70 (1)	94 (10)	82 (7)AB	147 (4)	186 (19)	166 (11)AB	4.9 (0.3)	5.8 (0.8)	5.3 (0.4)AB	107 (4)	191 (27)	149 (20)A	259 (6)	383 (45)	321 (31)A
MG	83 (3)	102 (10)	93 (6)A	166 (6)	193 (19)	179 (11)A	5.5 (0.5)	5.7 (0.5)	5.6 (0.3)A	98 (5)	170 (26)	134 (18)AB	269 (10)	368 (44)	319 (28)A
CAT	70 (1)	91 (8)	81 (6)AB	137 (3)	165 (14)	151 (9)BC	4.2 (0.4)	4.8 (1.0)	4.5 (0.5)BC	91 (3)	166 (24)	129 (18)AB	231 (6)	336 (37)	284 (26)A
SOR	65 (1)	82 (8)	74 (5)BC	136 (3)	154 (13)	145 (7)BC	6.2 (0.3)	5.4 (0.5)	5.8 (0.3)A	95 94	129 (8)	112 (8)BC	237 (6)	289 (20)	263 (14)A
GA	69 (1)	75 (7)	72 (3)BC	131 (3)	131 (12)	131 (6)CD	5.5 (0.4)	5.0 (0.4)	5.3 (0.3)AB	107 (3)	148 (19)	128 (12)AB	244 (5)	284 (31)	264 (16)A
TA	61 (1)	70 (5)	66 (3)C	110 (2)	116 (9)	113 (5)DE	5.0 (0.7)	4.5 (0.7)	4.8 (0.5)ABC	86 (3)	106 (9)	96 (6)CD	201 (4)	226 (17)	213 (10)B
PGG	53 (2)	61 (6)	57 (3)D	105 (3)	110 (9)	108 (4)E	4.2 (0.2)	4.0 (0.5)	4.1 (0.2)C	78 (4)	99 (8)	89 (6)D	188 (4)	213 (17)	200 (9)B
Avg. (n = 28)	67 (2)Y	82 (4)X		133 (4)Y	151 (8)X		5.1 (0.2)	5.0 (0.2)		95 (2)Y	144 (9)X		233 (6)Y	300 (16)X	

[†] Treatment solutions consisted of a water control or supplied 10 mg of methyl gallate (MG), gallic acid (GA), catechin (CAT) condensed tannin from sorghum (SOR), tannic acid (TA), or β-1,2,3,4,6-penta-O-galloyl-D-glucose (PGG) per g soil. Data are arithmetic average (standard error) (n = 4). Main effects of land use (column averages, n = 28) or treatments (row averages, n = 8) are denoted by capital letters (Tukey's HSD, $P < 0.05$).

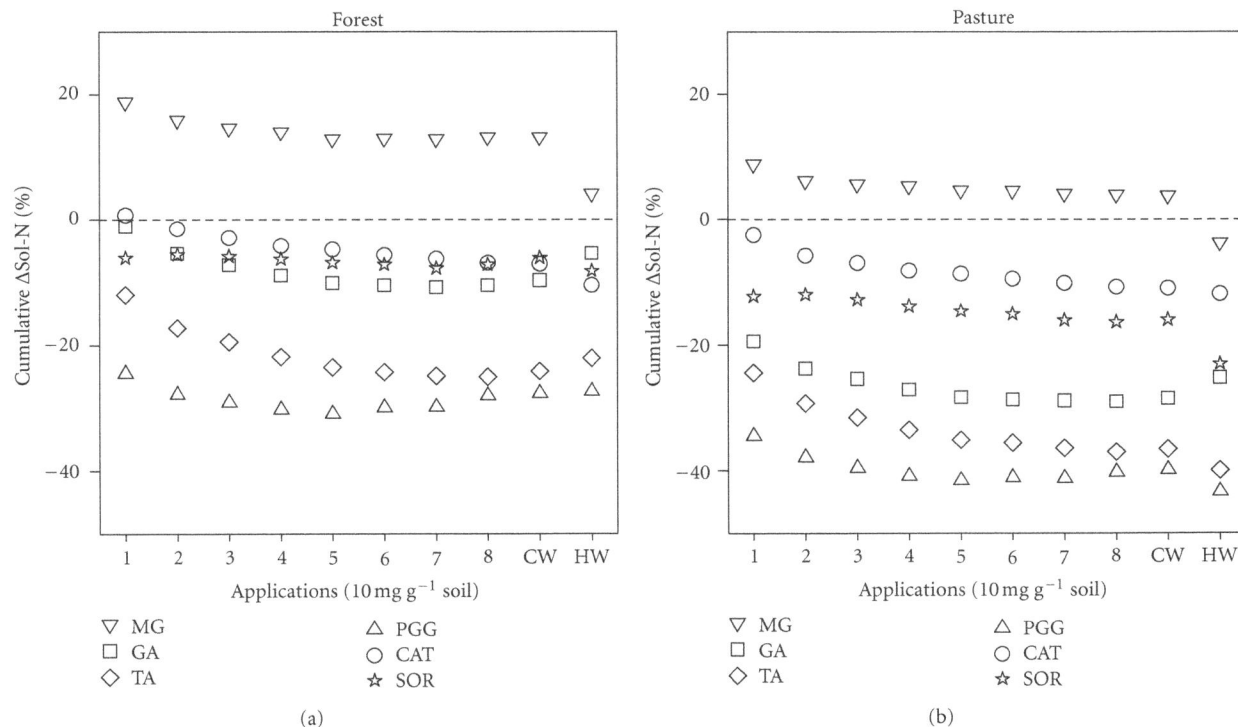

FIGURE 3: Percent change in soluble-N extractions attributable to treatments, ΔSol-N for (a) forest and (b) pasture soils calculated with (1) ($n = 4$). Treatment abbreviations are defined in Figure 1.

of intertreatment ΔSol-N comparisons (Table 3). Variation among treatment effects was less than 2 mg kg^{-1} soil but samples treated with PGG yielded less cool water soluble-N than others.

In contrast, hot water extracted large amounts of soluble-N that varied with simple main effects of Treatment and Use (Table 3). The greatest amount of soluble-N, extracted from the Control treatment, was comparable to the amount extracted by the preceding eight treatment cycles (Table 3). Significantly less soluble-N was extracted with hot water from TA- and PGG-treated samples than other treatments. Pasture produced about 50% more soluble-N than forest soil.

Hot water ΔSol-N varied as an interaction between Treatment and Use (Figure 5). Gallic acid had no effect on hot water extractions from forest soil, compared to the Control, but reduced soluble-N from pasture samples by 22%. Previous treatments with MG and CAT reduced soluble-N similarly from forest and pasture soils, by 10 and 14%, respectively. Reductions in soluble-N, observed for samples treated with SOR, TA, or PGG, were greater in pasture than forest soil. Hot water soluble-N was reduced from samples previously treated with PGG, by 27% for forest soil and by 47% for pasture.

3.4. Cumulative Extraction of Soluble-N. Final cumulative soluble-N extracted by eight treatment cycles and subsequent cool and hot rinses differed by Treatment and Use (Table 3). Treatment effects segregated into two groups with less total soluble-N extracted from soils treated with TA or PGG.

Pasture soil yielded an average of 300 mg kg^{-1} soluble-N or about 29% more than forest soil.

Final cumulative ΔSol-N varied as an interaction between Treatment and Use (Figure 6). In both forest and pasture soils, ΔSol-N was highest from samples treated with MG and lowest from samples treated with TA or PGG. The repeated treatments with MG did not significantly influence cumulative extraction of soluble-N from forest or pasture soil. Forest soil was also unaffected by GA but soluble-N was reduced from pasture soil by 25% (nearly 100 mg kg^{-1}). Treatments with CAT reduced soluble-N similarly from both land uses by an average of 11% (about 37 mg kg^{-1}). Tannins produced the greatest reductions of soluble-N with significantly stronger effects on pasture soil. The condensed tannin, SOR, reduced cumulative extractions of soluble-N from forest and pasture soils by 8 and 23%, respectively, (21 and 93 mg kg^{-1}). Tannic acid, reduced soluble-N from forest soil by 22% (58 mg kg^{-1}) and by 40% (156 mg kg^{-1}) from pasture soil. The hydrolyzable tannin, PGG, produced greatest reductions, 27 and 43% from forest and pasture soils, respectively, (71 and 176 mg kg^{-1}).

4. Discussion

When expressed on a landscape basis, the size of the pool of soil N affected by tannins and other phenolic compounds appears significant. Water alone (Control) extracted the equivalent of 33 and 54 kg of soluble-N ha^{-1} from 0–5 cm of forest and pasture soils with the first treatment cycle

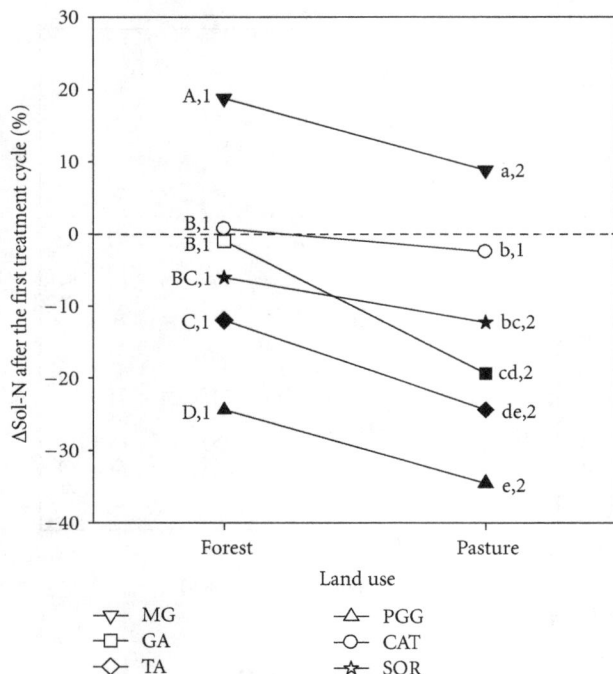

FIGURE 4: Treatment × Use interactions for ΔSol-N after the first application of treatment solutions (10 mg g^{-1} soil). Within each land use, letters denote differences among treatments. Differences between uses are denoted by numbers. Filled symbols denote significant deviations from the control. Treatment abbreviations are defined in Figure 1.

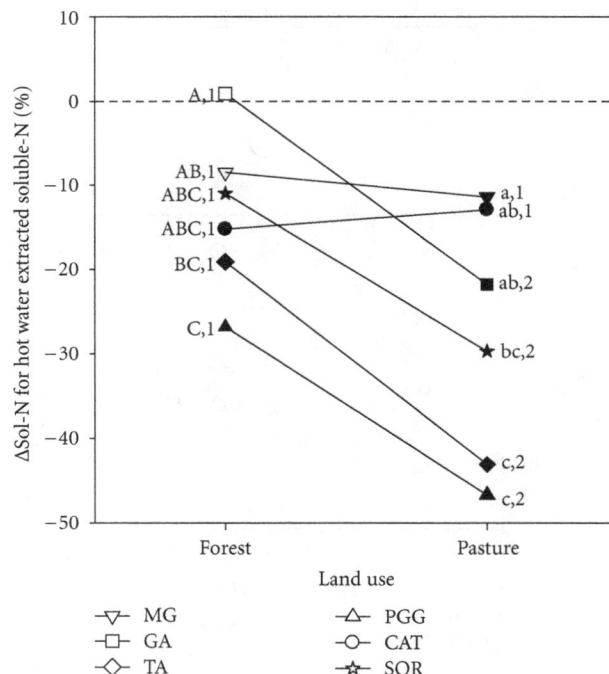

FIGURE 5: Treatment × Use interactions for ΔSol-N calculated for hot water extractions (not cumulative). Within each land use, letters denote differences among treatments. Differences between uses are denoted by numbers. Filled symbols denote significant deviations from the control. Treatment abbreviations are defined in Figure 1.

(Tables 1 and 3). Compared against these baseline values, the first treatment with MG increased losses of soluble-N from forest and pasture soils by 5-6 kg N ha^{-1} while PGG conserved 8 and 19 kg N ha^{-1} in forest and pasture samples, respectively. These reductions of soluble-N were strongly correlated with the amount of phenolic treatment-C sorbed by the soil [34] (Figure 7).

Incremental extractions of nitrogen after eight successive treatment applications were small suggesting a majority of the most labile pool of soil-N had been removed (Figure 2). The difference between Control extractions, equivalent to 70 and 106 kg ha^{-1} from forest and pasture soils, and treatment extractions infer the magnitude of the soil-N pool most responsive to the phenolic inputs. These suggest the eight treatments with MG mobilized an additional 19 and 7 mg N kg^{-1} forest or pasture soil, compared to water, or 9 and 4 kg N ha^{-1}. Conversely, PGG reduced the solubility of labile soil-N by 41 and 76 mg kg^{-1} in forest and pasture soils, respectively, equivalent to retention of 21 and 43 kg N ha^{-1}.

Negligible extractions of soluble-N with cool water, exhibiting only small differences among treatments, were unexpected (Table 3). We had, instead, anticipated relatively large releases of soluble-N from samples that had previously retained N once the treatments were omitted.

In contrast to cool water, hot water released large amounts of soluble-N, accounting for between 40 and 50% of the cumulative extraction, suggesting it originated from a

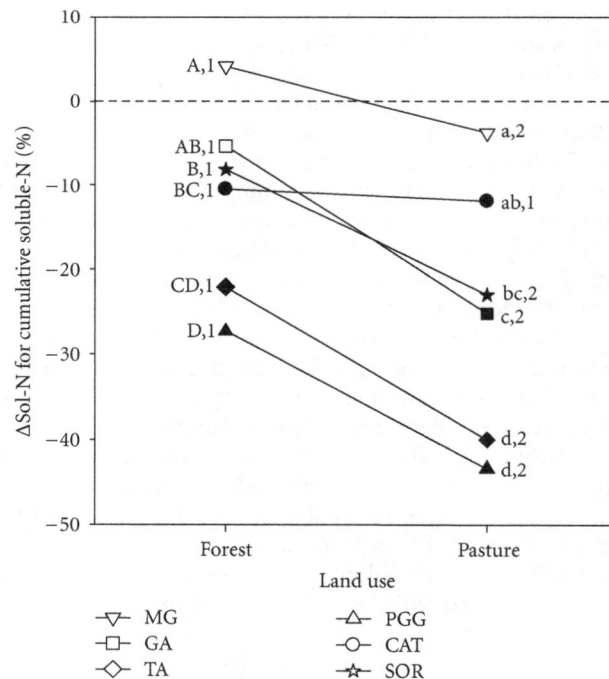

FIGURE 6: Treatment × Use interactions for ΔSol-N calculated for cumulative extractions. Within each land use, letters denote differences among treatments. Differences between uses are denoted by numbers. Filled symbols denote significant deviations from the control. Treatment abbreviations are defined in Figure 1.

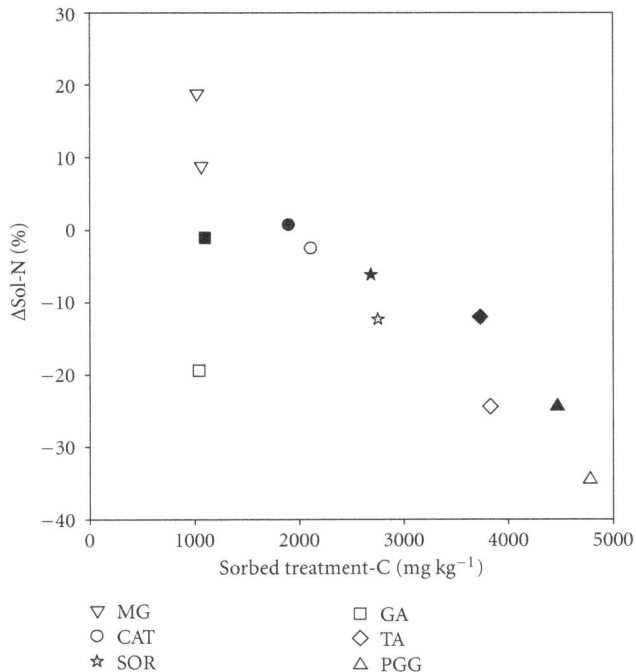

FIGURE 7: Relationship between ΔSol-N and sorbed treatment-C after the first application of treatments. Data from forest and pasture soils are denoted by filled and open symbols, respectively. Values for treatment-C are from [34]. Treatment abbreviations are defined in Figure 1.

different pool of soil organic matter. The hot water-soluble pool from forest soil, 51 kg N ha^{-1}, was less than half that from pasture soil, 110 kg N ha^{-1}. Hot water-extractable soil-N is thought to be primarily composed of unspecified forms of organic-N, particularly amino-N and amides, associated with soil microbial biomass and organic matter, with the remainder consisting of NH_4-N, generated by hydrolysis of heat-labile organic N [41, 42, 49]. In a previous experiment, Halvorson et al. [33] found organic-N comprised more than 80% of soluble-N extracted by hot water after a single treatment with GA, TA or PGG.

The patterns of treatment effects observed for hot water soluble-N, along with those observed for the preceding extractions, suggest phenolic compounds, like MG, may simply affect the efficacy of the extraction process while others including the tannins and CAT somehow increase the ability for organic-N to resist hydrolysis or physically restrain it in the soil matrix. While eight treatments with MG increased soluble-N in supernatants by an average 7 kg N ha^{-1} (Table 3), hot water soluble-N from samples treated with MG was reduced by 8 kg N ha^{-1} (Figure 5). Thus, final cumulative extractions from MG-treated forest and pasture soils, though distinct from each other (129 compared to 211 kg N ha^{-1}), were not appreciably different from cumulative extractions with water (Figure 6). In contrast, the PGG treatment reduced soluble-N in treatment supernatants, the cool water rinse, and the final hot water incubation with reductions of hot water soluble-N equivalent

to 14 and 53 kg N ha^{-1} from forest and pasture (Figures 4 and 5, Table 3). All together, PGG reduced soluble-N from the forest soil by 71 mg kg^{-1} soil or 35 kg ha^{-1} and from pasture soil by 170 mg kg^{-1} soil or 97 kg ha^{-1}.

Reductions in N availability by condensed tannins are attributed to formation of protein-tannin complexes with soil protein that are more recalcitrant than those formed with hydrolyzable tannins in part because they are less available to microorganisms as substrate [10, 50–52]. However, the strong reductions in solubility of labile soil-N, observed for TA and PGG, suggest hydrolyzable tannins may play a more prominent role in abiotic immobilization of organic soil-N than previously thought. A study by Hagerman et al. [45] concluded PGG, a nonpolar tannin (Table 2), precipitated with bovine serum albumin (BSA) by forming a hydrophobic coat around the protein while the more polar, SOR, formed hydrogen-bond cross-links between BSA molecules.

In addition to precipitation, polymerization resulting from condensation reactions between phenolic and amino-containing compounds in solution can occur. Oxidation of phenolics, forming semiquinones and quinones, can be carried out biotically by the polyphenol oxidase enzymes and/or abiotically by redox reactions with manganese and iron oxides [53, 54]. When PGG is oxidized, it forms covalently linked complexes with protein [55].

Additions of some tannins/phenolics to the soil resulted in the dissolution of Mn oxides, evinced by increased Mn content in the supernatant [56]. Mobilization of other soil metals such as Fe, Al and Si by tannins has also been observed [29, 57]. During the dissolution of Mn oxides, the Mn may be reduced from insoluble Mn(III and IV) to soluble Mn(II), which is available for plants. Phenolic compounds could be oxidized, forming quinones and semiquinones, the latter a highly reactive radical that readily self-polymerizes or copolymerizes with other compounds [58]. Tannic acid, gallic acid, and other polyhydroxy phenols with OH–OH in the orthopositions are known to be highly effective in the dissolution of Mn in soils with high Mn oxides content [59] or in synthetic Mn oxide [60]. Thus, the lower soluble-N extracted with PGG, GA, or TA solutions compared to other compounds might be due in part to redox reactions with soil Mn together with oxidation of these organic compounds into quinones or semiquinones and formation of "humic-like" polymers with amino-containing compounds that were retained in the soil matrix.

The significant interactions between treatment and land use, observed for ΔSol-N, indicated the effects of phenolic plant inputs were of greater consequence in land managed as pasture than the surrounding woodlands (Figures 4–6). Variations in soil microbial community composition, related to land use, have been suggested to explain variations in the mineralization of tannin-protein complexes [13]; however this study and our related work [33] also suggest land use can affect the initial reactions between tannins and soil.

The effects of tannins on soil organic matter and nutrient cycling have important implications for livestock production in mixed systems such as silvopastures that include a mixture of forages together with browse and overstory tree species [61–64]. Appalachian silvopasture soil typically differs from the surrounding unmanaged woodland because it is limed to increase soil pH, receives additional N-inputs from fertilizers and manure, and can develop greater bulk density due to compaction by livestock (Table 1). In addition, transitioning from either forest or pasture to a silvopastoral mixed stand, containing both forages and overstory, may affect soil nitrogen pools by redistributing the patterns of nutrients in soils and biomass. Decreases in soil C and N have been associated with afforestation, especially in the case of pines [65]. Two years after conversion from a mixed hardwood woodland, Staley et al. [66] reported losses of organic-C and -N from West Virginia silvopasture soil of 17 and 9%, respectively, which they attributed to litter decomposition.

Along with their effects on soil organic matter and the availability of nitrogen, tannins and related phenolic compounds can interact with important metals in soil, such as Ca, Mn, Al, and Fe, probably through chelation and oxidation/reduction reactions [67, 68]. Interactions between metals and phenolic compounds may inhibit or promote plant growth in forest soil. For example, tannic acid has been reported to reduce the rate of root growth by itself but has also been shown to mitigate the toxic effects of Al on roots [69] and in soil [10, 26]. In addition tannins affect nutrient value of forages (e.g., [70, 71]) and animal health [72, 73]. Thus, tannins can link plant productivity, ruminant physiology, pathogen survival, and environmental quality in agroecosystems.

The results of this study indicate hydrolyzable tannins like PGG can quickly reduce the solubility of labile soil-N more than condensed tannins like SOR and suggest tannin effects will vary with land management. However information about short-term reactions that incorporate tannin-C onto the soil matrix and immobilize soil-N must be considered together with their potential for chemical and biological degradation [33]. Further work is required to determine the persistence of tannin effects as they are degraded by soil microorganisms or other soil biota [74], physically lost by leaching, or chemically oxidized after interacting with soil metals [67, 68].

Tannins and other phenolic compounds affect a number of important biological, chemical, and physical processes in plants animals and soil. Studies such as this improve our understanding of the effects of natural phenolic inputs on soil organic matter and nutrient cycling and will ultimately lead to new management strategies. Future research on tannins and other plant polyphenolics in soil ecosystems should be focused towards understanding their effects on plant productivity and soil function. Functional definitions linking specific tannin chemistry to soil processes are required that can also serve as a rationale for comparing tannins. Research remains hampered by a lack of standardized methods that simplify the extraction, identification, and quantification of tannins from plants and soil and that can be adapted to field measurements. Experimental field work will remain difficult until suitable model tannins can be identified that are available in reasonable quantities and expense.

Because they are chemically and biologically active, tannins appear to have the potential to be used to improve nitrogen retention by soils but additional work is needed to determine how long their effects can last and whether the retained N is readily available to crop or forage plants. The effects of tannins on nutrient cycling will likely be influenced by specific tannin chemistry [13, 75, 76], vary with tannin concentrations, and the quantity and quality of soil N [77].

Disclaimer

USDA is an equal opportunity provider and employer. Mention of trade names or commercial products in this publication is solely for the purpose of providing specific information and does not imply recommendation or endorsement by the US Department of Agriculture.

Acknowledgments

The authors thank J. Harrah and T. Robertson for excellent analytical assistance. This work is dedicated to P. J. A. Halvorson.

References

[1] S. Hättenschwiler and P. M. Vitousek, "The role of polyphenols in terrestrial ecosystem nutrient cycling," *Trends in Ecology and Evolution*, vol. 15, no. 6, pp. 238–242, 2000.

[2] T. E. C. Kraus, R. A. Dahlgren, and R. J. Zasoski, "Tannins in nutrient dynamics of forest ecosystems—a review," *Plant and Soil*, vol. 256, no. 1, pp. 41–66, 2003.

[3] J. D. Horner, J. R. Gosz, and R. G. Cates, "The role of carbon-based plant secondary metabolites in decomposition in terrestrial ecosystems," *American Naturalist*, vol. 132, no. 6, pp. 869–883, 1988.

[4] M. von Lützow, I. Kögel-Knabner, K. Ekschmitt et al., "Stabilization of organic matter in temperate soils: mechanisms and their relevance under different soil conditions—a review," *European Journal of Soil Science*, vol. 57, no. 4, pp. 426–445, 2006.

[5] C. B. Coulson, R. I. Davies, and D. A. Lewis, "Polyphenols in plant, humus and soil. 1. Polyphenols of leaves, litter and superficial humus from mull and mor sites," *Journal of Soil Science*, vol. 11, no. 1, pp. 20–29, 1960.

[6] W. R. C. Handley, "Further evidence for the importance of residual leaf protein complexes in litter decomposition and the supply of nitrogen for plant growth," *Plant and Soil*, vol. 15, no. 1, pp. 37–73, 1961.

[7] J. Lewis and R. Starkey, "Vegetable tannins, their decomposition and effects of decomposition of some organic compounds," *Soil Science*, vol. 106, pp. 241–247, 1968.

[8] J. Basaraba, "Influence of vegetable tannins on nitrification in soil," *Plant and Soil*, vol. 21, no. 1, pp. 8–16, 1964.

[9] J. Basaraba and R. L. Starkey, "Effect of plant tannins on decomposition of organic substances," *Soil Science*, vol. 101, no. 1, pp. 17–23, 1966.

[10] P. Kraal, K. G. J. Nierop, J. Kaal, and A. Tietema, "Carbon respiration and nitrogen dynamics in Corsican pine litter amended with aluminium and tannins," *Soil Biology and Biochemistry*, vol. 41, no. 11, pp. 2318–2327, 2009.

[11] G. D. Joanisse, R. L. Bradley, C. M. Preston, and G. D. Bending, "Sequestration of soil nitrogen as tannin-protein complexes may improve the competitive ability of sheep laurel (*Kalmia angustifolia*) relative to black spruce (*Picea mariana*)," *New Phytologist*, vol. 181, no. 1, pp. 187–198, 2009.

[12] T. E. C. Kraus, R. J. Zasoski, R. A. Dahlgren, W. R. Horwath, and C. M. Preston, "Carbon and nitrogen dynamics in a forest soil amended with purified tannins from different species," *Soil Biology and Biochemistry*, vol. 36, no. 2, pp. 309–321, 2004.

[13] R. Mutabaruka, K. Hairiah, and G. Cadisch, "Microbial degradation of hydrolysable and condensed tannin polyphenol-protein complexes in soils from different land-use histories," *Soil Biology and Biochemistry*, vol. 39, no. 7, pp. 1479–1492, 2007.

[14] K. Sivapalan, V. Fernando, and M. W. Thenabadu, "Humified phenol-rich plant residues and soil urease activity," *Plant and Soil*, vol. 70, no. 1, pp. 143–146, 1983.

[15] K. Sivapalan, V. Fernando, and M. W. Thenabadu, "N-mineralization in polyphenol-rich plant residues and their effect on nitrification of applied ammonium sulphate," *Soil Biology and Biochemistry*, vol. 17, no. 4, pp. 547–551, 1985.

[16] A. Scalbert, "Antimicrobial properties of tannins," *Phytochemistry*, vol. 30, no. 12, pp. 3875–3883, 1991.

[17] F. A. Einhellig, "Mode of allelochemical action of phenolic compounds," in *Allelopathy: Chemistry and Mode of Action*, F. A. Macías et al., Ed., pp. 217–238, CRC Press, Boca Raton, Fla, USA, 2004.

[18] T. K. Bhat, B. Singh, and O. P. Sharma, "Microbial degradation of tannins—a current perspective," *Biodegradation*, vol. 9, no. 5, pp. 343–357, 1998.

[19] J. P. Schimel, R. G. Cates, and R. Ruess, "The role of balsam poplar secondary chemicals in controlling soil nutrient dynamics through succession in the Alaskan taiga," *Biogeochemistry*, vol. 42, no. 1-2, pp. 221–234, 1998.

[20] R. E. Benoit and R. L. Starkey, "Enzyme inactivation as a factor in the inhibition of decomposition of organic matter by tannins," *Soil Science*, vol. 105, no. 4, pp. 203–208, 1968.

[21] G. D. Joanisse, R. L. Bradley, C. M. Preston, and A. D. Munson, "Soil enzyme inhibition by condensed litter tannins may drive ecosystem structure and processes: the case of *Kalmia angustifolia*," *New Phytologist*, vol. 175, no. 3, pp. 535–546, 2007.

[22] M. C. Rillig, B. A. Caldwell, H. A. B. Wösten, and P. Sollins, "Role of proteins in soil carbon and nitrogen storage: controls on persistence," *Biogeochemistry*, vol. 85, no. 1, pp. 25–44, 2007.

[23] B. Adamczyk, S. Adamczyk, A. Smolander, and V. Kitunen, "Tannic acid and Norway spruce condensed tannins can precipitate various organic nitrogen compounds," *Soil Biology and Biochemistry*, vol. 43, no. 3, pp. 628–637, 2011.

[24] E. Castells, "Indirect effects of phenolics on plant performance by altering nitrogen cycling: another mechanism of plant–plant negative interactions," in *Allelopathy in Sustainable Agriculture and Forestry*, R. S. Zeng, A. U. Mallik, and S. M. Luo, Eds., pp. 137–156, Springer, New York., NY, USA, 2008.

[25] N. Wurzburger and R. L. Hendrick, "Plant litter chemistry and mycorrhizal roots promote a nitrogen feedback in a temperate forest," *Journal of Ecology*, vol. 97, no. 3, pp. 528–536, 2009.

[26] R. R. Northup, R. A. Dahlgren, and J. G. McColl, "Polyphenols as regulators of plant-litter-soil interactions in northern California's pygmy forest: a positive feedback?" *Biogeochemistry*, vol. 42, no. 1-2, pp. 189–220, 1998.

[27] G. D. Bending and D. J. Read, "Effects of the soluble polyphenol tannic acid on the activities of ericoid and ectomycorrhizal fungi," *Soil Biology and Biochemistry*, vol. 28, no. 12, pp. 1595–1602, 1996.

[28] G. D. Bending and D. J. Read, "Nitrogen mobilization from protein-polyphenol complex by ericoid and ectomycorrhizal fungi," *Soil Biology and Biochemistry*, vol. 28, no. 12, pp. 1603–1612, 1996.

[29] J. Kaal, K. G. J. Nierop, and J. M. Verstraten, "Retention of tannic acid and condensed tannin by Fe-oxide-coated quartz sand," *Journal of Colloid and Interface Science*, vol. 287, no. 1, pp. 72–79, 2005.

[30] S. El Azhar, R. Verhe, M. Proot, P. Sandra, and W. Verstraete, "Binding of nitrite-N on polyphenols during nitrification," *Plant and Soil*, vol. 94, no. 3, pp. 369–382, 1986.

[31] S. El Azhar, J. Vandenabeele, and W. Verstraete, "Nitrification and organic nitrogen formation in soils," *Plant and Soil*, vol. 94, no. 3, pp. 383–399, 1986.

[32] R. D. Fitzhugh, G. M. Lovett, and R. T. Venterea, "Biotic and abiotic immobilization of ammonium, nitrite, and nitrate in soils developed under different tree species in the Catskill Mountains, New York, USA," *Global Change Biology*, vol. 9, no. 11, pp. 1591–1601, 2003.

[33] J. J. Halvorson, J. M. Gonzalez, A. E. Hagerman, and J. L. Smith, "Sorption of tannin and related phenolic compounds and effects on soluble-N in soil," *Soil Biology and Biochemistry*, vol. 41, no. 9, pp. 2002–2010, 2009.

[34] J. J. Halvorson, J. M. Gonzalez, and A. E. Hagerman, "Repeated applications of tannins and related phenolic compounds are retained by soil and affect cation exchange capacity," *Soil Biology and Biochemistry*, vol. 43, no. 6, pp. 1139–1147, 2011.

[35] J. J. Halvorson and J. M. Gonzalez, "Tannic acid reduces recovery of water-soluble carbon and nitrogen from soil and affects the composition of Bradford-reactive soil protein," *Soil Biology and Biochemistry*, vol. 40, no. 1, pp. 186–197, 2008.

[36] J. J. Halvorson and J. M. Gonzalez, "Bradford reactive soil protein in Appalachian soils: distribution and response to incubation, extraction reagent and tannins," *Plant and Soil*, vol. 286, no. 1-2, pp. 339–356, 2006.

[37] D. W. Nelson and L. E. Sommers, "Total carbon, organic carbon and organic matter," in *Methods of Soil Analysis Part 3: Chemical Methods*, D. L. Sparks et al., Ed., Soil Science Society of America Books Series No. 5, pp. 961–1010, Soil Science Society of America, Inc., Madison, Wis, USA, 1996.

[38] H. Ciesielski and T. Sterckeman, "Determination of cation exchange capacity and exchangeable cations in soils by means of cobalt hexamine trichloride. Effects of experimental conditions," *Agronomie*, vol. 17, no. 1, pp. 1–7, 1997.

[39] H. Ciesielski and T. Sterckeman, "A comparison between three methods for the determination of cation exchange capacity and exchangeable cations in soils," *Agronomie*, vol. 17, no. 1, pp. 9–16, 1997.

[40] ISO 23470:2007, "Soil quality—determination of effective cation exchange capacity (CEC) and exchangeable cations using a hexamminecobalt trichloride solution," ISO/TC 190, Soil quality Subcommittee SC3, Chemical methods and soil characteristics, 2007.

[41] A. Ghani, M. Dexter, and K. W. Perrott, "Hot-water extractable carbon in soils: a sensitive measurement for determining impacts of fertilisation, grazing and cultivation," *Soil Biology and Biochemistry*, vol. 35, no. 9, pp. 1231–1243, 2003.

[42] D. Curtin, C. E. Wright, M. H. Beare, and F. M. McCallum, "Hot water-extractable nitrogen as an indicator of soil nitrogen availability," *Soil Science Society of America Journal*, vol. 70, no. 5, pp. 1512–1521, 2006.

[43] C. Gallet and P. Lebreton, "Evolution of phenolic patterns in plants and associated litters and humus of a mountain forest ecosystem," *Soil Biology and Biochemistry*, vol. 27, no. 2, pp. 157–165, 1995.

[44] A. E. Hagerman, *The Tannin Handbook*, Miami University, Oxford, Ohio, USA, 2002.

[45] A. E. Hagerman, M. E. Rice, and N. T. Ritchard, "Mechanisms of Protein Precipitation for Two Tannins, Pentagalloyl Glucose and Epicatechin16 (4 → 8) Catechin (Procyanidin)," *Journal of Agricultural and Food Chemistry*, vol. 46, no. 7, pp. 2590–2595, 1998.

[46] Z. Lu, G. Nie, P. S. Belton, H. Tang, and B. Zhao, "Structure-activity relationship analysis of antioxidant ability and neuroprotective effect of gallic acid derivatives," *Neurochemistry International*, vol. 48, no. 4, pp. 263–274, 2006.

[47] R. C. Littell, G. A. Milliken, W. W. Stroup, and R. D. Wolfinger, *SAS System for Mixed Models*, SAS Institute Inc., Cary, NC, USA, 1996.

[48] SAS, "SAS OnlineDoc, Version 8," SAS Institute, Inc., Cary, NC, USA, 1999.

[49] P. Leinweber, H. R. Schulten, and M. Korschens, "Hot water extracted organic matter: chemical composition and temporal variations in a long-term field experiment," *Biology and Fertility of Soils*, vol. 20, no. 1, pp. 17–23, 1995.

[50] R. L. Bradley, B. D. Titus, and C. P. Preston, "Changes to mineral N cycling and microbial communities in black spruce humus after additions of $(NH_4)_2SO_4$ and condensed tannins extracted from *Kalmia angustifolia* and balsam fir," *Soil Biology and Biochemistry*, vol. 32, no. 8-9, pp. 1227–1240, 2000.

[51] K. G. J. Nierop and J. M. Verstraten, "Fate of tannins in Corsican pine litter," *Journal of Chemical Ecology*, vol. 32, no. 12, pp. 2709–2719, 2006.

[52] B. Adamczyk, V. Kitunen, and A. Smolander, "Polyphenol oxidase, tannase and proteolytic activity in relation to tannin concentration in the soil organic horizon under silver birch and Norway spruce," *Soil Biology and Biochemistry*, vol. 41, no. 10, pp. 2085–2093, 2009.

[53] F. J. Stevenson, *Humus Chemistry: Genesis, Composition, Reactions*, John Wiley & Sons, 2nd edition, 1994.

[54] P. J. Hernes, R. Benner, G. L. Cowie, M. A. Goi, B. A. Bergamaschi, and J. I. Hedges, "Tannin diagenesis in mangrove leaves from a tropical estuary: a novel molecular approach," *Geochimica et Cosmochimica Acta*, vol. 65, no. 18, pp. 3109–3122, 2001.

[55] Y. Chen and A. E. Hagerman, "Reaction pH and protein affect the oxidation products of β-pentagalloyl glucose," *Free Radical Research*, vol. 39, no. 2, pp. 117–124, 2005.

[56] J. J. Halvorson, J. M. Gonzalez, and A. E. Hagerman, "Effects of tannins on soil carbon, cation exchange capacity, and metal solubility," in *Proceedings of the 94th Ecological Society of America Annual Meeting*, Albuquerque, NM, USA, August 2009.

[57] J. Kaal, K. G. J. Nierop, and J. M. Verstraten, "Interactions between tannins and goethite- or ferrihydrite-coated quartz sand: influence of pH and evaporation," *Geoderma*, vol. 139, no. 3-4, pp. 379–387, 2007.

[58] R. Liu, "Comment on "surface complexation of catechol to metal oxides: an ATR-FTIR, adsorption, and dissolution study"," *Environmental Science and Technology*, vol. 44, no. 16, pp. 6517–6518, 2010.

[59] N. V. Hue, S. Vega, and J. A. Silva, "Manganese toxicity in a Hawaiian Oxisol affected by soil pH and organic amendments," *Soil Science Society of America Journal*, vol. 65, no. 1, pp. 153–160, 2001.

[60] A. T. Stone and J. J. Morgan, "Reduction and dissolution of manganese(III) and manganese(IV) oxides by organics: 2. Survey of the reactivity of organics," *Environmental Science and Technology*, vol. 18, no. 8, pp. 617–624, 1984.

[61] R. P. Udawatta and L. D. Godsey, "Agroforestry comes of age: putting science into practice," *Agroforestry Systems*, vol. 79, no. 1, pp. 1–4, 2010.

[62] C. M. Feldhake, J. P. S. Neel, and D. P. Belesky, "Establishment and production from thinned mature deciduous-forest silvopastures in Appalachia," *Agroforestry Systems*, vol. 79, no. 1, pp. 31–37, 2010.

[63] J. P. S. Neel, C. M. Feldhake, and D. P. Belesky, "Influence of solar radiation on the productivity and nutritive value of herbage of cool-season species of an understorey sward in a mature conifer woodland," *Grass and Forage Science*, vol. 63, no. 1, pp. 38–47, 2008.

[64] V. D. Nair, S. G. Haile, G. A. Michel, and P. K. R. Nair, "Environmental quality improvement of agricultural lands through silvopasture in Southeastern United States," *Scientia Agricola*, vol. 64, no. 5, pp. 513–519, 2007.

[65] S. T. Berthrong, E. G. Jobbágy, and R. B. Jackson, "A global meta-analysis of soil exchangeable cations, pH, carbon, and nitrogen with afforestation," *Ecological Applications*, vol. 19, no. 8, pp. 2228–2241, 2009.

[66] T. E. Staley, J. M. Gonzalez, and J. P. S. Neel, "Conversion of deciduous forest to silvopasture produces soil properties indicative of rapid transition to improved pasture," *Agroforestry Systems*, vol. 74, no. 3, pp. 267–277, 2008.

[67] E. H. Majcher, J. Chorover, J. M. Bollag, and P. M. Huang, "Evolution of CO_2 during birnessite-induced oxidation of ^{14}C-labeled catechol," *Soil Science Society of America Journal*, vol. 64, no. 1, pp. 157–163, 2000.

[68] H. M. Appel, "Phenolics in ecological interactions: the importance of oxidation," *Journal of Chemical Ecology*, vol. 19, no. 7, pp. 1521–1552, 1993.

[69] T. B. Kinraide and A. E. Hagermann, "Interactive intoxicating and ameliorating effects of tannic acid, aluminum (Al^{3+}), copper (Cu^{2+}), and selenate (SeO_4^{2-}) in wheat roots: a descriptive and mathematical assessment," *Physiologia plantarum*, vol. 139, no. 1, pp. 68–79, 2010.

[70] T. N. Barry and W. C. McNabb, "The implications of condensed tannins on the nutritive value of temperate forages fed to ruminants," *British Journal of Nutrition*, vol. 81, no. 4, pp. 263–272, 1999.

[71] B. R. Min, T. N. Barry, G. T. Attwood, and W. C. McNabb, "The effect of condensed tannins on the nutrition and health of ruminants fed fresh temperate forages: a review," *Animal Feed Science and Technology*, vol. 106, no. 1–4, pp. 3–19, 2003.

[72] H. Hoste, F. Jackson, S. Athanasiadou, S. M. Thamsborg, and S. O. Hoskin, "The effects of tannin-rich plants on parasitic nematodes in ruminants," *Trends in Parasitology*, vol. 22, no. 6, pp. 253–261, 2006.

[73] I. Mueller-Harvey, "Unravelling the conundrum of tannins in animal nutrition and health," *Journal of the Science of Food and Agriculture*, vol. 86, no. 13, pp. 2010–2037, 2006.

[74] M. Coulis, S. Hättenschwiler, S. Rapior, and S. Coq, "The fate of condensed tannins during litter consumption by soil animals," *Soil Biology and Biochemistry*, vol. 41, no. 12, pp. 2573–2578, 2009.

[75] T. E. C. Kraus, Z. Yu, C. M. Preston, R. A. Dahlgren, and R. J. Zasoski, "Linking chemical reactivity and protein precipitation to structural characteristics of foliar tannins," *Journal of Chemical Ecology*, vol. 29, no. 3, pp. 703–730, 2003.

[76] S. Coq, J. M. Souquet, E. Meudec, V. Cheynier, and S. Hättenschwiler, "Interspecific variation in leaf litter tannins drives decomposition in a tropical rain forest of French Guiana," *Ecology*, vol. 91, no. 7, pp. 2080–2091, 2010.

[77] J. M. Talbot and A. C. Finzi, "Differential effects of sugar maple, red oak, and hemlock tannins on carbon and nitrogen cycling in temperate forest soils," *Oecologia*, vol. 155, no. 3, pp. 583–592, 2008.

6

Interference of Selected Palmer Amaranth (*Amaranthus palmeri*) Biotypes in Soybean (*Glycine max*)

Aman Chandi,[1] **David L. Jordan,**[1] **Alan C. York,**[1] **Susana R. Milla-Lewis,**[1]
James D. Burton,[2] **A. Stanley Culpepper,**[3] **and Jared R. Whitaker**[4]

[1] *Department of Crop Science, North Carolina State University, P.O. Box 7620, Raleigh, NC 27695-7620, USA*
[2] *Department of Horticulture Science, North Carolina State University, P.O. Box 7609, Raleigh, NC 27695, USA*
[3] *Department of Crop and Soil Sciences, University of Georgia, P.O. Box 478, Tifton, GA 31794, USA*
[4] *Department of Crop and Soil Sciences, University of Georgia-Southeast District, P.O. Box 8112, Statesboro, GA 30460, USA*

Correspondence should be addressed to David L. Jordan, david_jordan@ncsu.edu

Academic Editor: Kassim Al-Khatib

Palmer amaranth (*Amaranthus palmeri* S. Wats.) has become difficult to control in row crops due to selection for biotypes that are no longer controlled by acetolactate synthase inhibiting herbicides and/or glyphosate. Early season interference in soybean [*Glycine max* (L.) Merr.] for 40 days after emergence by three glyphosate-resistant (GR) and three glyphosate-susceptible (GS) Palmer amaranth biotypes from Georgia and North Carolina was compared in the greenhouse. A field experiment over 2 years compared season-long interference of these biotypes in soybean. The six Palmer amaranth biotypes reduced soybean height similarly in the greenhouse but did not affect soybean height in the field. Reduction in soybean fresh weight and dry weight in the greenhouse; and soybean yield in the field varied by Palmer amaranth biotypes. Soybean yield was reduced 21% by Palmer amaranth at the established field density of 0.37 plant m^{-2}. When Palmer amaranth biotypes were grouped by response to glyphosate, the GS group reduced fresh weight, dry weight, and yield of soybean more than the GR group. The results indicate a possible small competitive disadvantage associated with glyphosate resistance, but observed differences among biotypes might also be associated with characteristics within and among biotypes other than glyphosate resistance.

1. Introduction

Palmer amaranth is one of the most troublesome weeds of agronomic crops in the southeastern United States [1–3] because of its competitive ability, C$_4$ photosynthesis, higher water use efficiency, and rapid growth rate [4, 5]. This weed also possesses drought tolerance mechanisms which allow survival under limited water availability [6–8] and it adapts readily to shading [9]. Several biotypes of Palmer amaranth have evolved resistance to herbicides representing different modes of action, including 5-enolpyruvylshikimate-3-phosphate synthase (EPSPS) inhibitors, mitotic inhibitors, acetolactate synthase (ALS) inhibitors, and photosynthetic inhibitors [10], which make it challenging to control in cropping systems.

Species, density, and time of emergence with respect to the crop determine the relative competitiveness of pigweed species [11–14]. Interference of pigweed species including common waterhemp (*Amaranthus rudis* Sauer) [15–17], Palmer amaranth [11, 17, 18], and redroot pigweed (*Amaranthus retroflexus* L.) [14, 17, 19–22] has been evaluated in soybean. Soybean yield reduction as a result of interference increased from 17% to 68% with an increase in Palmer amaranth density from 0.33 to 10 plants m^{-1} of row length [11]. Furthermore, in the same study, correlation between soybean yield reduction and Palmer amaranth density was linear up to two Palmer amaranth plants m^{-1} of row, indicating that intraspecific interference between adjacent Palmer amaranth plants began at greater densities. Monks and Oliver [18] studied the competitive influence of common cocklebur (*Xanthium strumarium* L.), johnsongrass [*Sorghum helpens* (L.) Pers.], Palmer amaranth, sicklepod [*Cassia obtusifolia* (L.) H. S. Irwin and Barneby], and tall morningglory [*Ipomea purpurea* (L.) Roth] on biomass and yield of two soybean cultivars in Arkansas. Their results indicated a reduction in biomass of both soybean cultivars when growing within 50 cm of Palmer amaranth and reduction in soybean seed yield within a distance of 25 cm of Palmer amaranth. The distance of influence of Palmer amaranth was among the greatest for the weeds evaluated in this study. Among the three pigweed species (Palmer amaranth, redroot pigweed, and common waterhemp) interfering with soybean, Palmer amaranth accumulated the greatest biomass, followed by common waterhemp and then redroot pigweed [17]. Further, Palmer amaranth planted along with soybean at a density of 8 plants m^{-1} of row resulted in the greatest (79%) reduction in soybean yield, followed by common waterhemp (56%) and then redroot pigweed (38%).

Interference of Palmer amaranth has also been studied in other crops [23–29]. Palmer amaranth growing at a density of 0.9 plants m^{-2} resulted in up to 92% reduction in cotton (Gossypium hirsutum L.) lint yield [23]. While a linear decrease in cotton yield was observed from 13% to 54% with an increase in Palmer amaranth density from 1 to 10 plants/m², volume and biomass of Palmer amaranth remained unaffected by intraspecific competition at all densities [24]. Palmer amaranth reduced corn leaf area index (LAI) and corn grain yield from 11% to 91% as density increased from 0.5 to 8 plants/m² [25, 26]. There was a negative linear relationship between grain sorghum [Sorghum bicolor (L.) Moench] yield and density of Palmer amaranth. Increasing weed density decreased grain sorghum yield by reducing the numbers of grains produced in panicles [27]. Season-long Palmer amaranth interference in peanut (Arachis hypogaea L.) reduced peanut canopy diameter and one plant m^{-1} of row resulted in a predicted yield loss of up to 28% [28]. Yield reductions from 30 to 94% were reported in sweet potato [Ipomoea batatas (L.) Lam.] by Palmer amaranth densities ranging from 0.5 to 6.5 plants m^{-1} row [29].

Many biotypes of Palmer amaranth have developed confirmed resistance to glyphosate in the southern United States, making it difficult to manage [30–33]. Herbicide-resistant weed biotypes sometimes have a fitness penalty compared with nonresistant wild types [34–45]. Several components of fitness of maternally inherited triazine-resistant smooth pigweed (Amaranthus hybridus L.) were reduced, including early seedling emergence, early growth, mid-season leaf number, and total above-ground biomass, but differences varied among years and populations [36, 37]. Evolved resistance in Powell's amaranth (Amaranthus powellii S. Wats.) to ALS-inhibiting herbicides resulted in thinner roots and stems and reduced leaf area, resulting in a 67% reduction in aboveground vegetative mass and a 30% reduction in seed biomass [38]. A mutant of blackgrass (Alopecurus myosuroides Huds.) resistant to herbicides that inhibit acetyl coenzyme A carboxylase (ACCase) when grown in competition with wheat (Triticum aestivum L.) under limited water supply had a 6%, 42%, and 26% reduction in height, vegetative, and reproductive biomass, respectively, as compared to the wild biotype [39]. The proportion of resistant individuals in segregating F_2 populations of rigid ryegrass (Lolium rigidum Gaud.) decreased as compared to susceptible individuals

over a period of 4 years [40, 41]. Baucom and Mauricio [42] reported a high fitness cost of glyphosate resistance in tall morningglory. Glyphosate-resistant genotypes produced fewer seeds as compared to susceptible genotypes in the absence of selection pressure from glyphosate.

Determining relative differences in interference of GR and GS Palmer amaranth biotypes in soybean could be of benefit to evaluate possible competitive disadvantage associated with GR trait. Therefore, greenhouse experiment was conducted to compare early season interference by selected GR and GS biotypes of Palmer amaranth grown with soybean. A field experiment was conducted to study the effect of season-long interference by these biotypes on soybean.

2. Materials and Methods

2.1. Greenhouse Experiment. Seeds from six Palmer amaranth biotypes [31] collected from fields in Georgia and North Carolina during the fall of 2005 were grown along with Roundup Ready soybean cultivar AG6301 (Monsanto Company, St. Louis, MO 63167, USA) in 15 cm round plastic pots containing commercial potting soil (Fafard 4P potting mix, Conrad Fafard Inc. Agawam, MA 01001). Three Palmer amaranth biotypes were GR (one from North Carolina and two from Georgia) and three were GS (one from North Carolina and two from Georgia) (Figure 1). Approximately six soybean seeds and 25 Palmer amaranth seeds were planted in two parallel rows spaced 2.5 cm apart in each pot. Seedlings were thinned to one soybean and one Palmer amaranth plant pot^{-1} 10 days after emergence (DAE). Controls included a single soybean or Palmer amaranth plant pot^{-1}. Plants were fertilized (Scotts Starter Fertilizer, The Scotts Company LLC, Marysville, OH 43041, USA) with 25 mL of a 4.6 g L^{-1} fertilizer solution per pot every 10 days to ensure optimum plant growth. Pots were spaced sufficiently enough to avoid shading from adjacent pots during the entire duration of experiment. Plants were irrigated daily using an overhead sprinkler system. The greenhouse was maintained at $35 \pm 5°C$, and natural illumination was supplemented for 14 hours each day with metal halide lighting ($400 \mu mol\, m^{-2}\, s^{-1}$) (Hubbell Lighting, Inc., Greenville, SC 29607). The experimental design was a randomized complete block with treatments replicated 10 times and the experiment was repeated.

Height of the Palmer amaranth and soybean was determined every 5 days beginning 1 week after pots were thinned, corresponding to 15, 20, 25, 30, 35, and 40 DAE. Plant height was measured from the soil surface to the base of the upper most fully expanded leaf for both soybean and Palmer amaranth plants. At 40 DAE, Palmer amaranth and soybean plants were severed at the soil surface to determine shoot fresh weight and dry weight. The samples were dried in paper bags in oven at 60°C for 72 hours for dry weight measurements.

Data for percent reduction in plant height and percent reduction in fresh and dry weight relative to controls without interference were subjected to ANOVA using Proc. GLM (Statistical Analysis Systems, version 9.1, SAS Institute Inc.,

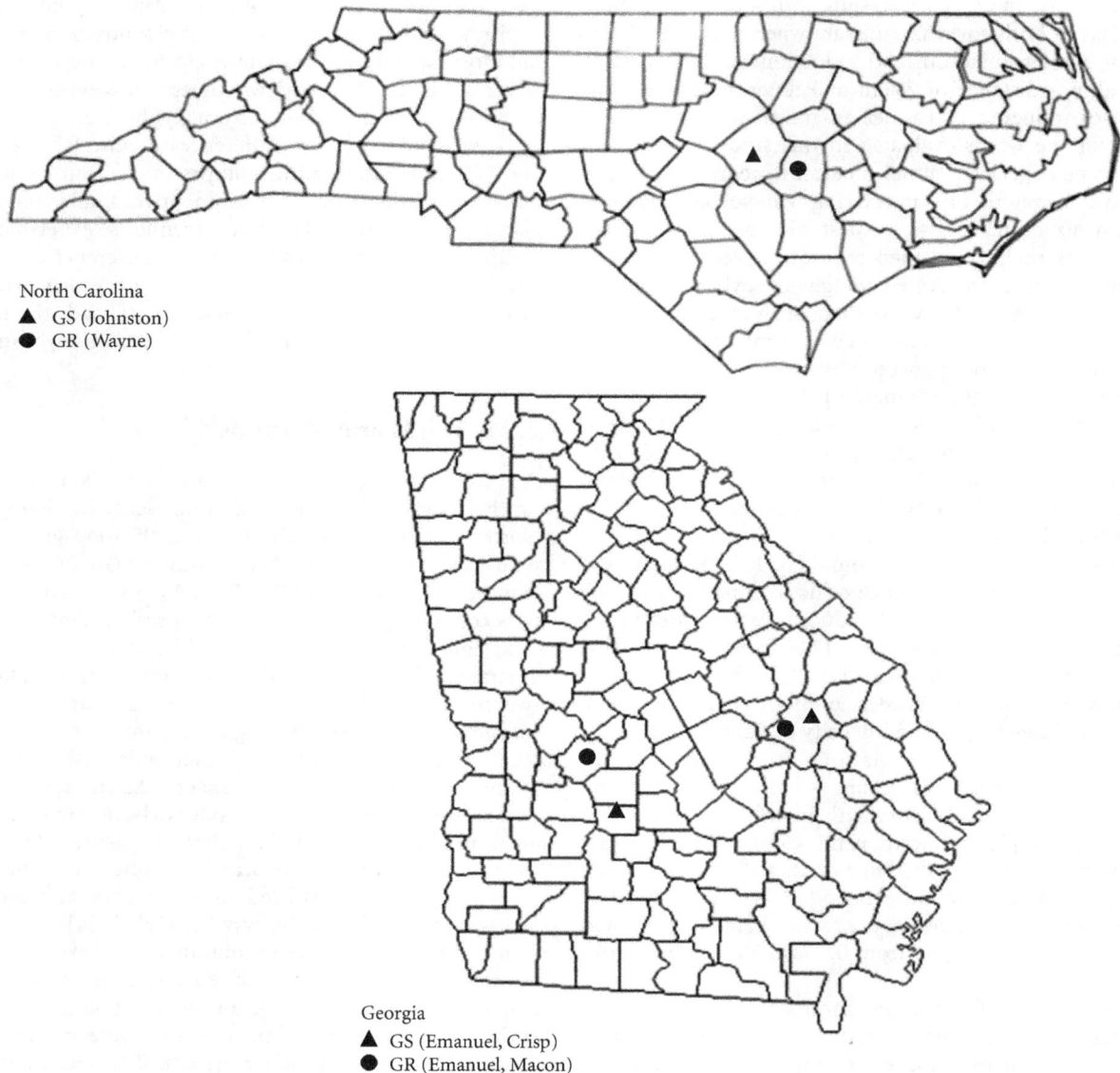

FIGURE 1: Locations of North Carolina and Georgia biotypes used in the study.

SAS Campus Drive, Cary, NC 2751, USA). Due to lack of interaction data were pooled over the two runs. In a separate analysis, data were grouped for biotypes expressing resistance or susceptibility to glyphosate (GR biotype group and GS biotype group) and subjected to ANOVA. Means of significant effects were separated using Fisher's Protected LSD test.

2.2. Field Experiment. The experiment was conducted in conventionally planted (row to row distance = 91 cm) Roundup Ready soybean cultivar AG6301 during 2008 and 2009 at the Cunningham Research Station near Kinston, NC on a Norfolk loamy sand (fine-loamy, kaolinitic, thermic Typic Kandiudults). Plot size was one row by 6 m, and two border rows were included between experimental units. Approximately 10 seeds of each Palmer amaranth biotype

were planted 3 cm apart and 4 cm to the side of soybean row immediately after planting soybean in the middle of the plot. Both GR and GS biotypes were thinned to one plant per plot (one per row) about 20 DAE. At 35 DAE, Palmer amaranth plants were covered with large plastic bags and potassium salt of glyphosate (Roundup Weathermax, Monsanto Company, St. Louis, MO 63167, USA) at 1.1 kg ae ha^{-1} was applied over the entire test area to control other weeds. For the reminder of the season, weeds other than the one desired Palmer amaranth per plot were removed by hand. A control was included without Palmer amaranth. The experimental design was a randomized complete block with eight replications.

Soybean height was recorded 30, 60, 90, and 120 DAE at a distance of 30, 60, 90, and 120 cm on either side of Palmer amaranth within the soybean row. Mature soybean plants were harvested manually in row sections of 0 to 30,

31 to 60, 61 to 90, 91 to 120, and 121 to 150 cm on either side of Palmer amaranth. Soybean plants were threshed using stationary thresher. Observations on soybean were converted to percent reduction in height and yield relative to the control in absence of Palmer amaranth.

Data for percent reduction in soybean height and yield were subjected to ANOVA as explained earlier when considering Palmer amaranth biotypes individually. In a separate analysis, data were grouped for biotypes expressing resistance or susceptibility to glyphosate (GR and GS biotype groups) and subjected to ANOVA. Means of significant main effects and interactions were separated using Fisher's Protected LSD test. Due to lack of interaction data were pooled over the two runs. Percent reduction in soybean yield was linearly regressed against distance from Palmer amaranth using Sigmaplot 12.0 (Systat Software Inc. 1735 Technology Drive, Suite 430, San Jose, CA 95110, USA). The regression expression used was $y = ax + b$, where, y = percent reduction in yield, x = distance from Palmer amaranth plant, and a and b are constants.

3. Results and Discussion

3.1. Greenhouse Experiment. Differences among individual Palmer amaranth biotypes were not observed for soybean height reduction at any of the six times that height was recorded (data not shown). Averaged over biotypes, soybean height was reduced 10, 10, 8, 11, 10, and 5% at 15, 20, 25, 30, 35, and 40 DAE (data not shown). Differences among individual Palmer amaranth biotypes were observed for soybean fresh weight ($P \leq 0.05$) and dry weight ($P \leq 0.10$) reduction. Both fresh weight and dry weight reduction was similar with the Emanuel (GR), Macon, (GR), Crisp (GS), Emanuel (GS), and Johnston (GS) biotypes (Table 1). The Wayne (GR) biotype reduced fresh and dry weight less than the Macon (GR), Crisp (GS), and Emanuel (GS) biotypes.

When Palmer amaranth biotypes were grouped with respect to response to glyphosate, no differences in soybean height reduction were noted between the two groups (data not shown). However, differences among Palmer amaranth biotype groups were observed for soybean fresh weight ($P \leq 0.05$) and dry weight ($P \leq 0.10$) reduction. The GS biotype group reduced fresh weight and dry weight of soybean more than the GR biotype group (Table 1). Soybean fresh weight was reduced 31% and 23% as result of interference from the GS and GR Palmer amaranth biotype groups, respectively. Interference from GS and GR biotype groups reduced soybean dry weight by 27% and 21%, respectively.

3.2. Field Experiment. Interference of Palmer amaranth biotypes and biotype groups did not affect soybean height at 30, 60, 90, or 120 DAE and at distance of 30, 60, 90, or 120 cm from Palmer amaranth (data not shown). These results were unexpected given the competitive ability of Palmer amaranth [1–9, 11, 17, 18]. Chivinge and Schweppenhauser [46] reported that competition with smooth pigweed reduced branching, shoot dry weight, leaf area index, number of pods per plant, and grain yield of soybean, but plant height,

TABLE 1: Percent reduction in soybean fresh weight and dry weight at harvest (40 days after emergence) caused by early season interference by Palmer amaranth in greenhouse experiment.[a,b]

Palmer amaranth biotype	Fresh weight reduction	Dry weight reduction
	%	
Individual Palmer amaranth biotypes[c]		
Emanuel (GR)	24ab	23ab
Macon (GR)	31a	28a
Wayne (GR)	13b	11b
Crisp (GS)	31a	26a
Emanuel (GS)	37a	31a
Johnston (GS)	26ab	24ab
Palmer amaranth biotypes grouped by response to glyphosate[d]		
GR group	23z	21z
GS group	31y	27y

[a]Data are pooled over runs of the experiment. Abbreviations: GR, glyphosate resistant; GS, glyphosate susceptible.
[b]Means within a parameter and analysis followed by the same letter are not significantly different according to Fisher's Protected LSD test at $P \leq 0.05$ for fresh weight reduction and $P \leq 0.10$ for dry weight reduction.
[c]Consists of six Palmer amaranth biotypes.
[d]Consists of a group of three glyphosate-resistant (GR) and a group of three glyphosate-susceptible (GS) Palmer amaranth biotypes.

number of seeds per pod, and 1000-seed weight were not affected. The main effects of year, biotype, distance, and their interactions were not significant for percent reduction in soybean height when Palmer amaranth biotypes were considered individually or as biotype groups at 30, 60, 90, and 120 DAE and at distance of 30, 60, 90, and 120 cm from Palmer amaranth (data not shown). When Palmer amaranth biotypes were considered individually, the main effects of year, biotype, and distance from Palmer amaranth were significant for percent reduction in soybean yield (Table 2). However, the interactions of these factors were not significant. Similar results were obtained when Palmer amaranth biotypes were grouped based on response to glyphosate (GR and GS biotype groups).

There were differences in soybean yield reduction as a result of interference from individual Palmer amaranth biotypes averaged over years and five 30 cm distance increments from Palmer amaranth (Table 3). Interference from all GS biotypes reduced soybean yield similarly. Among GR biotypes, interference from the Emanuel biotype reduced soybean yield more than Macon and Wayne biotypes. Soybean yield reduction by the GR Emanuel biotype was similar to GS biotypes. When biotype groups were compared, interference from the GS biotype group reduced soybean yield more than the GR biotype group. The GS biotype group reduced soybean yield 23% compared with 19% reduction by the GR biotype group.

A significant effect of distance from Palmer amaranth was reflected in increasing yield of soybean as distance from Palmer amaranth increased (Figure 2). The greatest yield reduction, 34%, was noted at 15 cm from Palmer amaranth

TABLE 2: $P > F$ for percent reduction in soybean yield caused by season-long interference by Palmer amaranth in field experiment.[a]

Source of variation	Individual Palmer amaranth biotypes[b]	Palmer amaranth biotypes grouped by response to glyphosate[c]
	$P > F$ value	
Year	<0.0001	<0.0001
Biotype	0.0013	0.0010
Year × biotype	0.9135	0.4459
Distance	<0.0001	<0.0001
Year × distance	0.1296	0.1457
Biotype × distance	0.1980	0.8564
Year × biotype × distance	0.9150	0.6201
Coefficient of variation (%)	90.4	96.3

[a]Data are pooled over runs of the experiment.
[b]Consists of six Palmer amaranth biotypes.
[c]Consists of a group of three glyphosate-resistant (GR) and a group of three glyphosate-susceptible (GS) Palmer amaranth biotypes.

TABLE 3: Percent reduction in soybean yield caused by season-long interference by Palmer amaranth in field experiment.[a,b]

Palmer amaranth biotype	Yield reduction
	%
Individual Palmer amaranth biotypes[c]	
Emanuel (GR)	24a
Macon (GR)	17b
Wayne (GR)	16b
Crisp (GS)	24a
Emanuel (GS)	22a
Johnston (GS)	24a
Palmer amaranth biotypes grouped by response to glyphosate[d]	
GR	19z
GS	23y

[a]Data are pooled over five 30 cm distance increments from Palmer amaranth and runs (years) of the experiment. Abbreviations: GR, glyphosate resistant; GS, glyphosate susceptible.
[b]Means within a parameter and analysis followed by the same letter are not significantly different according to Fisher's Protected LSD test at $F \leq 0.05$.
[c]Consists of six Palmer amaranth biotypes.
[d]Consists of a group of three glyphosate-resistant and a group of three glyphosate-susceptible Palmer amaranth biotypes.

while yield was reduced 28%, 22%, 16%, and 11% at 45 cm, 75 cm, 105 cm, and 135 cm from Palmer amaranth, respectively. Soybean yield and biomass reduction when growing within a distance of 25 cm and 50 cm of Palmer amaranth, respectively, were reported by Monks and Oliver [18].

Palmer amaranth density established in this experiment was 0.37 Palmer amaranth plants m^{-2} or 0.33 Palmer amaranth plants m^{-1} of row length (calculated based on effective harvested plot size of 3 m by 0.91 m). Soybean yield loss corresponding to this density was 22% (averaged over five 30 cm distance increments from Palmer amaranth). Soybean

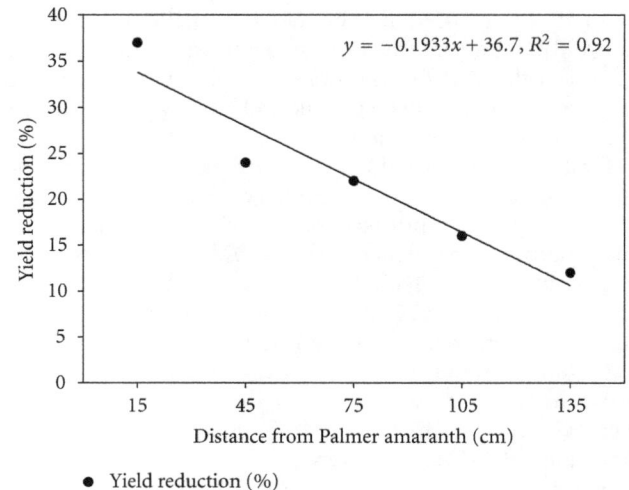

FIGURE 2: Percent reduction in soybean yield as influenced by distance from Palmer amaranth in the field experiment.

yield loss of 17% at a Palmer amaranth density of 0.33 m^{-1} of row was reported by Klingaman and Oliver [11]. Herbicide Application Decision Support System for field crops (WebHADSS version 2004.0.3) predicted soybean yield loss of 11.2% at density of 0.37 Palmer amaranth plants m^{-2}.

Results from both the greenhouse and field studies indicate a possible small competitive disadvantage associated with the glyphosate resistance trait in the biotypes of Palmer amaranth examined. A fitness penalty associated with glyphosate resistance [40–42, 45] as well as other herbicides [36–39] has been reported previously. A fitness penalty associated with GR Palmer amaranth has not been reported. On the other hand, the observed differences may be due to reasons unrelated to glyphosate resistance. A wide range of phenotypic variation has been reported in Palmer amaranth accessions [47–50]. Genetic variability in Palmer amaranth biotypes used in this study was assessed in another experiment using Amplified Fragment Length Polymorphisms [51]. Pair-wise genetic similarity values were found to be relatively low, averaging 0.34. The variation among GR and GS biotype groups used in this study was also found to be less than the overall genetic variability present within all the individual biotypes. It is possible that a high degree of phenotypic and genetic variability present among and within Palmer amaranth biotypes used in the study was responsible for the observed differences in interference.

4. Conclusions

Collectively, results from these experiments indicate that interference in soybean can vary among Palmer amaranth biotypes. Although data suggest that there may be a small competitive disadvantage due to the GR trait, a larger pool of biotypes is needed to conclusively define a fitness cost to glyphosate resistance in Palmer amaranth. The observed differences in interference may have been associated with inherent diversity existing within and among Palmer amaranth biotypes.

Acknowledgments

Syngenta Crop Protection and Monsanto Company provided partial funding for this research. Appreciation is expressed to Rick Seagroves, Jamie Hinton, Peter Eure, and Gurinderbir Chahal for technical assistance and to Navinderpal Singh and Chuck Foresman for providing advice on this project.

References

[1] T. M. Webster, "Weed survey-southern states," *Proceedings Southern Weed Science Society*, vol. 57, pp. 404–426, 2004.

[2] T. M. Webster, "Weed survey-southern states," *Proceedings Southern Weed Science Society*, vol. 58, pp. 291–306, 2005.

[3] T. M. Webster and H. D. Coble, "Changes in the weed species composition of the southern United States: 1974 to 1995," *Weed Technology*, vol. 11, no. 2, pp. 308–317, 1997.

[4] C. C. Black, T. M. Chen, and R. H. Brown, "Biochemical basis for plant competition," *Weed Science*, vol. 17, no. 3, pp. 338–344, 1969.

[5] M. J. Horak and T. M. Loughin, "Growth analysis of four *Amaranthus* species," *Weed Science*, vol. 48, no. 3, pp. 347–355, 2000.

[6] J. Ehleringer, "Ecophysiology of *Amaranthus palmeri*, a sonoran desert summer annual," *Oecologia*, vol. 57, no. 1-2, pp. 107–112, 1983.

[7] G. Place, D. Bowman, M. Burton, and T. Rufty, "Root penetration through a high bulk density soil layer: differential response of a crop and weed species," *Plant and Soil*, vol. 307, no. 1-2, pp. 179–190, 2008.

[8] S. R. Wright, M. W. Jennette, H. D. Coble, and T. W. Rufty, "Root morphology of young *Glycine max*, Senna obtusifolia, and *Amaranthus palmeri*," *Weed Science*, vol. 47, no. 6, pp. 706–711, 1999.

[9] P. Jha, J. K. Norsworthy, M. B. Riley, D. G. Bielenberg, and W. Bridges Jr., "Acclimation of palmer amaranth (*Amaranthus palmeri*) to shading," *Weed Science*, vol. 56, no. 5, pp. 729–734, 2008.

[10] I. Heap, "The International Survey of Herbicide Resistant Weeds," 2012, http://www.weedscience.org.

[11] T. E. Klingaman and L. R. Oliver, "Palmer amaranth (*Amaranthus palmeri*) interference in soybeans (*Glycine max*)," *Weed Science*, vol. 42, no. 4, pp. 523–527, 1994.

[12] R. J. Aldrich, "Predicting crop yield reductions from weeds," *Weed Technology*, vol. 1, no. 3, pp. 199–206, 1987.

[13] S. Z. Knezevic, M. J. Horak, and R. L. Vanderlip, "Relative time of redroot pigweed (*Amaranthus retroflexus* L.) emergence is critical in pigweed-sorghum [*Sorghum bicolor* (L.) Moench] competition," *Weed Science*, vol. 45, no. 4, pp. 502–508, 1997.

[14] P. Cowan, S. E. Weaver, and C. J. Swanton, "Interference between pigweed (Amaranthus spp.), barnyardgrass (*Echinochloa crus-galli*), and soybean (*Glycine max*)," *Weed Science*, vol. 46, no. 5, pp. 533–539, 1998.

[15] R. G. Hartzler, B. A. Battles, and D. Nordby, "Effect of common waterhemp (*Amaranthus rudis*) emergence date on growth and fecundity in soybean," *Weed Science*, vol. 52, no. 2, pp. 242–245, 2004.

[16] S. M. Jones, R. J. Smeda, G. S. Smith, and W. G. Johnson, "The effect of waterhemp competition on soybean yield," *Proceedings of North Central Weed Science Society*, vol. 53, article 146, 1998.

[17] C. N. Bensch, M. J. Horak, and D. Peterson, "Interference of redroot pigweed (*Amaranthus retroflexus*), Palmer amaranth (*A. palmeri*), and common waterhemp (*A. rudis*) in soybean," *Weed Science*, vol. 51, no. 1, pp. 37–43, 2003.

[18] D. W. Monks and L. R. Oliver, "Interactions between soybean (*Glycine max*) cultivars and selected weeds," *Weed Science*, vol. 36, no. 6, pp. 770–774, 1988.

[19] A. Dieleman, A. S. Hamill, S. F. Weise, and C. J. Swanton, "Empirical models of pigweed (*Amaranthus* spp.) interference in soybean (*Glycine max*)," *Weed Science*, vol. 43, no. 4, pp. 612–618, 1995.

[20] A. Légère and M. M. Schreiber, "Competition and canopy architecture as affected by soybean (*Glycine max*) row width and density of redroot pigweed (*Amaranthus retroflexus*)," *Weed Science*, vol. 37, no. 1, pp. 84–92, 1989.

[21] P. L. Orwick and M. M. Schreiber, "Interference of redroot pigweed (*Amaranthus retroflexus*) and robust foxtail (*Setaria viridis* var. *robusta-alba* or var. *robusta-purpurea*) in soybeans (*Glycine max*)," *Weed Science*, vol. 27, no. 6, pp. 665–674, 1979.

[22] J. L. Shurtleff and H. D. Coble, "Interference of certain broadleaf weed species in soybean (*Glycine max*)," *Weed Science*, vol. 33, pp. 654–657, 1985.

[23] M. W. Rowland, D. S. Murray, and L. M. Verhalen, "Full-season Palmer amaranth (*Amaranthus palmeri*) interference with cotton (*Gossypium hirsutum*)," *Weed Science*, vol. 47, no. 3, pp. 305–309, 1999.

[24] G. D. Morgan, P. A. Baumann, and J. M. Chandler, "Competitive impact of Palmer amaranth (*Amaranthus palmeri*) on cotton (*Gossypium hirsutum*) development and yield," *Weed Technology*, vol. 15, no. 3, pp. 108–112, 2001.

[25] R. A. Massinga and R. S. Curie, "Impact of Palmer amaranth (*Amaranthus palmeri*) on corn (*Zea mays*) grain yield and quality of forage," *Weed Technology*, vol. 16, no. 3, pp. 532–536, 2002.

[26] R. A. Massinga, R. S. Currie, M. J. Horak, and J. Boyer, "Interference of Palmer amaranth in corn," *Weed Science*, vol. 49, no. 2, pp. 202–208, 2001.

[27] J. W. Moore, D. S. Murray, and R. B. Westerman, "Palmer amaranth (*Amaranthus palmeri*) effects on the harvest and yield of grain sorghum (*Sorghum bicolor*)," *Weed Technology*, vol. 18, no. 1, pp. 23–29, 2004.

[28] I. C. Burke, M. Schroeder, W. E. Thomas, and J. W. Wilcut, "Palmer amaranth interference and seed production in peanut," *Weed Technology*, vol. 21, no. 2, pp. 367–371, 2007.

[29] S. L. Meyers, K. M. Jennings, J. R. Schultheis, and D. W. Monks, "Interference of palmer amaranth (*Amaranthus palmeri*) in sweetpotato," *Weed Science*, vol. 58, no. 3, pp. 199–203, 2010.

[30] A. S. Culpepper, T. L. Grey, W. K. Vencill et al., "Glyphosate-resistant Palmer amaranth (*Amaranthus palmeri*) confirmed in Georgia," *Weed Science*, vol. 54, no. 4, pp. 620–626, 2006.

[31] A. S. Culpepper, J. R. Whitaker, A. W. MacRae, and A. C. York, "Weed science: distribution of glyphosate-resistant palmer amaranth (*Amaranthus palmeri*) in Georgia and North Carolina during 2005 and 2006," *Journal of Cotton Science*, vol. 12, no. 3, pp. 306–310, 2008.

[32] J. K. Norsworthy, G. M. Griffith, R. C. Scott, K. L. Smith, and L. R. Oliver, "Confirmation and control of glyphosate-resistant Palmer amaranth (*Amaranthus palmeri*) in Arkansas," *Weed Technology*, vol. 22, no. 1, pp. 108–113, 2008.

[33] L. E. Steckel, C. L. Main, A. T. Ellis, and T. C. Mueller, "Palmer amaranth (*Amaranthus palmeri*) in Tennessee has low level glyphosate resistance," *Weed Technology*, vol. 22, no. 1, pp. 119–123, 2008.

[34] J. S. Holt, "Ecological fitness of herbicide-resistant weeds," *Proceedings of Second International Weed Control Congress*, pp. 387–392, 1996.

[35] M. A. Jasieniuk, A. L. Brûlè-Babel, and I. N. Morrison, "The evolution and genetics of herbicide resistance in weeds," *Weed Science*, vol. 44, no. 1, pp. 176–193, 1996.

[36] N. Jordan, "Effects of the triazine-resistance mutation on fitness in *Amaranthus hybridus* (smooth pigweed)," *Journal of Applied Ecology*, vol. 33, no. 1, pp. 141–150, 1996.

[37] N. Jordan, "Fitness effects of the triazine resistance mutation in *Amaranthus hybridus*: relative fitness in maize and soyabean crops," *Weed Research*, vol. 39, no. 6, pp. 493–505, 1999.

[38] F. J. Tardif, I. Rajcan, and M. Costea, "A mutation in the herbicide target site acetohydroxyacid synthase produces morphological and structural alterations and reduces fitness in Amaranthus powellii," *New Phytologist*, vol. 169, no. 2, pp. 251–264, 2006.

[39] Y. Menchari, B. Chauvel, H. Darmency, and C. Délye, "Fitness costs associated with three mutant acetyl-coenzyme A carboxylase alleles endowing herbicide resistance in black-grass *Alopecurus myosuroides*," *Journal of Applied Ecology*, vol. 45, no. 3, pp. 939–947, 2008.

[40] C. Preston and A. M. Wakelin, "Resistance to glyphosate from altered herbicide translocation patterns," *Pest Management Science*, vol. 64, no. 4, pp. 372–376, 2008.

[41] A. M. Wakelin and C. Preston, "The cost of glyphosate resistance: is there a fitness penalty associated with glyphosate resistance in annual ryegrass?" in *Proceedings of 15th Australian Weeds Conference*, pp. 515–518, 2006.

[42] R. S. Baucom and R. Mauricio, "Fitness costs and benefits of novel herbicide tolerance in a noxious weed," *Proceedings of the National Academy of Sciences of the United States of America*, vol. 101, no. 36, pp. 13386–13390, 2004.

[43] S. I. Warwick, "Herbicide resistance in weedy plants: physiology and population biology," *Annual Review of Ecology and Systematics*, vol. 22, no. 1, pp. 95–114, 1991.

[44] M. M. Vila-Aiub, P. Neve, and S. B. Powles, "Fitness costs associated with evolved herbicide resistance alleles in plants," *New Phytologist*, vol. 184, no. 4, pp. 751–767, 2009.

[45] B. P. Pedersen, P. Neve, C. Andreasen, and S. B. Powles, "Ecological fitness of a glyphosate-resistant *Lolium rigidum* population: growth and seed production along a competition gradient," *Basic and Applied Ecology*, vol. 8, no. 3, pp. 258–268, 2007.

[46] O. A. Chivinge and M. A. Schweppenhauser, "Competition of soybean with blackjack (*Bidens pilosa* L.) and pigweed (*Amaranthus hybridus* L.)," *African Crop Science Journal*, vol. 3, no. 1, pp. 73–82, 1995.

[47] N. R. Burgos, Y. I. Kuk, and R. E. Talbert, "*Amaranthus palmeri* resistance and differential tolerance of *Amaranthus palmeri* and *Amaranthus hybridus* to ALS-inhibitor herbicides," *Pest Management Science*, vol. 57, no. 5, pp. 449–457, 2001.

[48] B. J. Gossett and J. E. Toler, "Differential control of Palmer amaranth (*Amaranthus palmeri*) and smooth pigweed (*Amaranthus hybridus*) by postemergence herbicides in soybean (*Glycine max*)," *Weed Technology*, vol. 13, no. 1, pp. 165–168, 1999.

[49] M. J. Horak and D. E. Peterson, "Biotypes of palmar amaranth (*Amaranthus palmeri*) and common waterhemp (*Amaranthus*

[50] J. A. Bond and L. R. Oliver, "Comparative growth of Palmer amaranth (*Amaranthus palmeri*) accessions," *Weed Science*, vol. 54, no. 1, pp. 121–126, 2006.

[51] A. Chandi, *Characterization and management of selected herbicide resistant weed populations [Ph.D. dissertation]*, North Carolina State University, Raleigh, NC, USA, 2011.

rudis) are resistant to imazethapyr and thifensulfuron," *Weed Technology*, vol. 9, no. 1, pp. 192–195, 1995.

7

Inheritance of Evolved Glyphosate Resistance in a North Carolina Palmer Amaranth (*Amaranthus palmeri*) Biotype

Aman Chandi,[1] Susana R. Milla-Lewis,[1] Darci Giacomini,[2]
Philip Westra,[2] Christopher Preston,[3] David L. Jordan,[1] Alan C. York,[1]
James D. Burton,[4] and Jared R. Whitaker[5]

[1] *Department of Crop Science, North Carolina State University, Box 7620, Raleigh, NC 27695, USA*
[2] *Department of Bioagricultural Sciences and Pest Management, Colorado State University,*
  1179 Campus Delivery, Fort Collins, CO 80523, USA
[3] *School of Agriculture, Food & Wine, University of Adelaide, Adelaide, SA 5005, Australia*
[4] *Department of Horticulture Science, North Carolina State University, Box 7609, Raleigh, NC 27695, USA*
[5] *Crop and Soil Sciences, University of Georgia Southeast District, P.O. Box 8112, Statesboro, GA 30460, USA*

Correspondence should be addressed to David L. Jordan, david_jordan@ncsu.edu

Academic Editor: Kent Burkey

Inheritance of glyphosate resistance in a Palmer amaranth biotype from North Carolina was studied. Glyphosate rates for 50% survival of glyphosate-resistant (GR) and glyphosate-susceptible (GS) biotypes were 1288 and 58 g ha^{-1}, respectively. These values for F1 progenies obtained from reciprocal crosses (GR × GS and GS × GR) were 794 and 501 g ha^{-1}, respectively. Dose response of F1 progenies indicated that resistance was not fully dominant over susceptibility. Lack of significant differences between dose responses for reciprocal F1 families suggested that genetic control of glyphosate resistance was governed by nuclear genome. Analysis of F1 backcross (BC1F1) families showed that 10 and 8 BC1F1 families out of 15 fitted monogenic inheritance at 2000 and 3000 g ha^{-1} glyphosate, respectively. These results indicate that inheritance of glyphosate resistance in this biotype is incompletely dominant, nuclear inherited, and might not be consistent with a single gene mechanism of inheritance. Relative 5-enolpyruvylshikimate-3-phosphate synthase (*EPSPS*) copy number varied from 22 to 63 across 10 individuals from resistant biotype. This suggested that variable *EPSPS* copy number in the parents might be influential in determining if inheritance of glyphosate resistance is monogenic or polygenic in this biotype.

1. Introduction

Glyphosate has become the world's most widely used herbicide since its commercialization in 1974, because it is effective, economical, and comparatively safe to the environment [1, 2]. Being nonselective, glyphosate is used to control a wide array of weed species including both grasses and broadleaf weeds [3, 4]. When glyphosate-resistant (GR) crops {including canola (*Brassica napus* L.), corn (*Zea mays* L.), cotton (*Gossypium hirsutum* L.), and soybean (*Glycine max* (L.) Merr.)} were commercialized beginning from 1996 to 1998, glyphosate revolutionized production of these crops by enabling growers to use this herbicide for weed control in standing crops [1]. With widespread adoption of glyphosate-resistant corn, cotton, and soybean, glyphosate replaced

many previously used selective herbicides and intensified pressure of glyphosate on weeds.

Currently, glyphosate resistance has been confirmed in 13 weed species in the United States, including Palmer amaranth (*Amaranthus palmeri* S. Wats.) [5]. Palmer amaranth is among the most competitive weeds of southern cropping systems [6] and populations of Palmer amaranth have evolved resistance to glyphosate in recent years because of repeated applications of this herbicide [7]. Evolution of glyphosate resistance in weed species poses a great risk to the continued success of GR crops [4]. Although the extent of GR Palmer amaranth biotypes has been well documented [8–10], information is limited about the genetic control of resistance in this species. Mode of inheritance, among other factors, is an important component affecting the evolution of resistance [11, 12]. The number of genes involved in governing resistance and their interactions influence not only the enrichment of resistance in a population, but also gene flow among populations [13–15].

Resistance to herbicides has been studied in many weed species. In most studies, resistance is the result of a single gene [16]; however, resistance as the result of more than one gene has been reported [17, 18]. Resistance in all situations except target site resistance to the triazine herbicides is encoded on the nuclear genome [16]. Target site triazine resistance is encoded on the chloroplastic genome [19], accounting for its cytoplasmic inheritance. Typically, resistance is partially dominant over susceptibility [16]; however, both completely dominant and, less commonly, recessive inheritance patterns of resistance have also been reported. Herbicide resistance was found to be inherited as a single, partially dominant, nuclear trait in common sow thistle (*Sonchus oleraceus* L.), Italian ryegrass (*Lolium multiflorum* Lam.), prickly lettuce (*Lactuca serriola* L.), rigid ryegrass (*Lolium rigidum* Gaud.), and wild oats (*Avena fatua* L.) [20–24]. Inheritance of resistance to dinitroaniline herbicides in green foxtail (*Setaria viridis* (L.) P. Beauv.), and goosegrass (*Eleusine indica* (L.) Gaertn.) was found to be controlled by a single nuclear recessive gene [12, 25].

The mode of inheritance of glyphosate resistance has been studied in a few weed species. In rigid ryegrass [26, 27], horseweed (*Conyza canadensis* L.) [28], and goosegrass [29], resistance was inherited as a single dominant or partially dominant nuclear gene with no influence from maternal effects. Other studies have pointed to more complex mechanisms of inheritance. For example, genetic control of glyphosate resistance in a population of rigid ryegrass from California was reported to be incompletely dominant and multigenic, involving at least two nuclear genes [17]. Inheritance of glyphosate resistance has also been suggested to follow a polygenic additive pattern in biotypes of Palmer amaranth from Georgia [30]. Probable involvement of one or more minor genes in conferring resistance to glyphosate at lower doses has also been reported in a GR rigid ryegrass population from Australia [26]. Furthermore, earlier studies on glyphosate resistance and the inheritance of the *EPSPS* gene in Palmer amaranth have shown that increased copy numbers of this gene confer resistance to glyphosate [31]. Preliminary results of an experiment carried out by Giacomini et al. [32] indicated a wide range, from 1 to 80, in *EPSPS* copy number in the majority of the F1 populations studied indicating that inheritance of those additional copies from parents to progeny can be highly unpredictable. Research on the inheritance of evolved glyphosate resistance in Palmer amaranth has been somewhat limited. The present study was conducted with the objective to further investigate the mechanism of inheritance of evolved glyphosate resistance in a Palmer amaranth biotype from North Carolina.

2. Materials and Methods

2.1. Generation of F1 Families. Seeds from a GR biotype from Wayne county (glyphosate rate for 50% visible control = 1770 g ha^{-1}) and a GS biotype from Johnston county

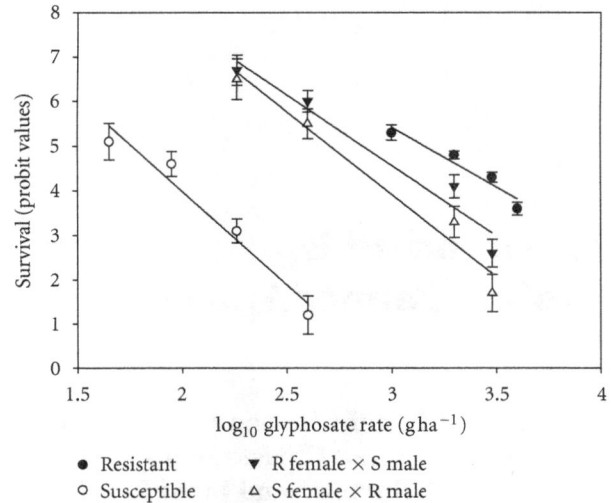

FIGURE 1: Glyphosate dose responses of glyphosate-resistant (GR), glyphosate-susceptible (GS), F1 R × S (GR female × GS male), and F1 S × R (GS female × GR male) Palmer amaranth populations. Points are mean values ± S.E. Best fit curves for these respective populations are: $y = 13.4 − 2.7x$, $P = 0.0390$, $r^2 = 0.92$; $y = 12.4 − 4.2x$, $P = 0.0221$, $r^2 = 0.96$; $y = 14 − 3.1x$, $P = 0.0246$, $r^2 = 0.95$; and $y = 15 − 3.7x$, $P = 0.0170$, $r^2 = 0.97$.

TABLE 1: Percent survival of parent Palmer amaranth biotypes (GR and GS) and F1 families (R × S and S × R) after treatment with glyphosate*.

Biotype	Herbicide rate (g/ha)			
	180	400	2000	3000
		%		
GR Parent	—	—	44a	32a
GS Parent	10b	3c	—	—
F1 R × S	86a	81a	22b	10b
F1 S × R	83a	68a	22b	4b

*Abbreviations used: GR: glyphosate-resistant; GS: glyphosate-susceptible; R × S and S × R: reciprocal crosses where, the first alphabet denotes female parent. Means within a glyphosate rate followed by the same letter are not significantly different according to Fisher's Protected LSD test at $P \leq 0.05$.

(glyphosate rate for 50% visible control = 89 g ha^{-1}) in North Carolina [33] were subjected to two cycles of recurrent selection with glyphosate at 840 g ha^{-1} in the greenhouse in order to obtain the individuals used as parents for all experiments. Glyphosate rates required for 50% visible control of GR and GS biotypes in recurrent generations were similar (data not shown) suggesting that the biotypes used for this study were homozygous for the glyphosate resistance trait. After collection, seeds were kept at −20°C for a month and subsequently planted in excess in 10-cm square pots containing commercial potting mix (Fafard 4P potting mix, Conrad Fafard Inc., Agawam, MA 01001) in the greenhouse. Seedlings were thinned to one plant per pot 8 days after emergence. The greenhouse was maintained at 35 ± 5°C and natural lighting was supplemented for 14 h each day with metal halide lighting (Hubbell Lighting, Inc.,701 Millennium Blvd, Greenville, SC 29607) delivering

TABLE 2: Chi-square analysis of segregation for glyphosate resistance at 2000 g/ha in BC1F1 families assuming monogenic and two gene additive inheritance*.

Backcross family	Observed			One gene model				Two gene additive model			
				Expected		χ^2	P value	Expected		χ^2	P value
	Alive	Dead	Total	Alive	Dead			Alive	Dead		
BC1F1 S × R1	8	38	46	6.01	39.99	0.759	0.384	9.01	36.99	0.142	0.707
BC1F1 S × R2	7	45	52	6.79	45.21	0.007	0.932	10.19	41.81	1.241	0.265
BC1F1 S × R3	12	38	50	6.53	43.47	5.267	0.022	9.80	40.20	0.616	0.432
BC1F1 S × R4	20	32	52	6.79	45.21	29.539	<0.0001	10.19	41.81	11.750	0.001
BC1F1 S × R5	3	49	52	6.79	45.21	2.436	0.119	10.19	41.81	6.308	0.012
BC1F1 S × R6	3	37	40	5.22	34.77	1.090	0.296	7.84	32.16	3.713	0.054
BC1F1 S × R7	7	23	30	3.92	26.08	2.787	0.095	5.88	24.12	0.266	0.606
BC1F1 R × S1	19	33	52	6.79	45.21	25.235	<0.0001	10.19	41.81	9.477	0.002
BC1F1 R × S2	15	37	52	6.79	45.21	11.407	0.001	10.19	41.81	2.825	0.093
BC1F1 R × S3	4	48	52	6.79	45.21	1.321	0.250	10.19	41.81	4.675	0.031
BC1F1 R × S4	33	19	52	6.79	45.21	116.308	<0.0001	10.19	41.81	63.516	<0.0001
BC1F1 R × S5	6	46	52	6.79	45.21	0.106	0.744	10.19	41.81	2.142	0.143
BC1F1 R × S6	6	46	52	6.79	45.21	0.106	0.744	10.19	41.81	2.142	0.143
BC1F1 R × S7	7	45	52	6.79	45.21	0.007	0.932	10.19	41.81	1.241	0.265
BC1F1 R × S8	4	23	27	3.53	23.47	0.073	0.787	5.29	21.71	0.391	0.532
Total						196.449	<0.0001			110.446	<0.0001
Pooled	154	559	713	93.14	619.86	45.750	<0.0001	139.70	573.30	1.819	0.177
Homogeneity						150.699	<0.0001			108.627	<0.0001
Response of parental controls											
GR Parent	39	13	52								
GS Parent	0	52	52								
F1 R × S	15	37	52								
F1 S × R	11	36	47								

*Abbreviations used: GR: glyphosate-resistant; GS: glyphosate-susceptible; R × S and S × R: reciprocal crosses where, the first alphabet denotes female parent; BC1F1: backcross family; F1: first filial.

$400 \, \mu\text{mol} \, \text{m}^{-2} \, \text{s}^{-1}$. Palmer amaranth is dioecious, therefore female and male plants were selected and paired for crosses. Plants were enclosed in dialysis tubing (Carolina Biological Supply Co., P.O. Box 6010, Burlington, NC 27216) to prevent any unintentional crossing until they were ready to be paired for crosses. When plants began to form inflorescences, reciprocal crosses were set up between GR and GS individuals to generate 16 F1 families. There were two sets of F1 families: one originating from GR females (8 R × S families) and another from GS females (8 S × R families). One plant each of the GR and GS biotypes were paired together and encased in dialysis tubing well before pollen shed in order to prevent entry of foreign pollen. Pollination was ensured by tapping the tubing twice every day. Single female plants were also enclosed in dialysis tubing as controls to check for the efficacy of the tubing in preventing entry of foreign pollen. Seeds from each female plant were harvested upon maturation, kept in separate envelopes to constitute individual F1 families, and stored at −20°C for 50 days to break seed dormancy.

2.2. F1 Dose Response Experiments.

A dose response study for glyphosate was conducted on GS and GR biotypes and F1 families. Fifteen seeds each for GS, GR and the 16 F1 families

were planted in 15-cm round pots containing commercial potting mix in the greenhouse. The potassium salt of glyphosate (Roundup Weathermax, Monsanto Company, St. Louis, Mo 63167) was applied to Palmer amaranth plants at the 3- to 4-leaf stage. Glyphosate was applied at 45, 90, 180, and $400 \, \text{g} \, \text{ha}^{-1}$ to the susceptible biotype, at 1000, 2000, 3000, and $4000 \, \text{g} \, \text{ha}^{-1}$ to the resistant biotype, and at 180, 400, 2000, and $3000 \, \text{g} \, \text{ha}^{-1}$ to the F1 families. Glyphosate was applied in $140 \, \text{L} \, \text{ha}^{-1}$ at 207 kPa using a CO_2-pressurized backpack sprayer with a flat-fan nozzle (8002 Spray nozzles, Spraying Systems Company, Wheaton, IL 60189). There were six replicate pots for each herbicide rate for the GR and GS biotypes, and one pot for each of the 16 F1 families. Thus, the different families were used as replicates for each herbicide rate. Visible estimates of percent survival were taken 2 wk after treatment. Data from each group of maternal and paternal F1 families were pooled for analysis.

2.3. Data Analysis for F1 Dose Response Experiments.

Probit values for percent survival rates were plotted versus $\log_{10}$ herbicide doses to develop a dose response curve using SIGMAPLOT 11.2 (SigmaPlot, version 11.0, Systat Software, Inc., 1735 Technology Drive, Suite 430, San Jose, CA, 95110). The values for 50% percent survival were calculated

TABLE 3: Chi-square analysis of segregation for glyphosate resistance at 3000 g/ha in BC1F1 families assuming monogenic and two gene additive inheritance*.

Backcross family	Observed			One gene model Expected		χ^2	P value	Two gene additive model Expected		χ^2	P value
	Alive	Dead	Total	Alive	Dead			Alive	Dead		
BC1F1 S × R1	1	38	39	3.12	35.88	1.566	0.211	4.68	34.32	3.288	0.070
BC1F1 S × R2	2	50	52	4.16	47.84	1.219	0.270	6.24	45.76	3.274	0.070
BC1F1 S × R3	3	47	50	4	46	0.272	0.602	6	44	1.705	0.192
BC1F1 S × R4	4	48	52	4.16	47.84	0.007	0.935	6.24	45.76	0.914	0.339
BC1F1 S × R5	0	52	52	4.16	47.84	4.522	0.033	6.24	45.76	7.091	0.008
BC1F1 S × R6	2	41	43	3.44	39.56	0.655	0.418	5.16	37.84	2.199	0.138
BC1F1 S × R7	1	32	33	2.64	30.36	1.107	0.293	3.96	29.04	2.514	0.113
BC1F1 R × S1	21	31	52	4.16	47.84	74.097	<0.0001	6.24	45.76	39.674	<0.0001
BC1F1 R × S2	11	41	52	4.16	47.84	12.224	<0.0001	6.24	45.76	4.126	0.042
BC1F1 R × S3	2	50	52	4.16	47.84	1.219	0.270	6.24	45.76	3.274	0.070
BC1F1 R × S4	24	28	52	4.16	47.84	102.849	<0.0001	6.24	45.76	57.441	<0.0001
BC1F1 R × S5	8	44	52	4.16	47.84	3.853	0.050	6.24	45.76	0.564	0.453
BC1F1 R × S6	9	43	52	4.16	47.84	6.121	0.013	6.24	45.76	1.387	0.239
BC1F1 R × S7	14	36	50	4	46	27.174	<0.0001	6	44	12.121	<0.0001
BC1F1 R × S8	1	24	25	2	23	0.543	0.461	3	22	1.515	0.218
Total						237.429	<0.0001			141.087	<0.0001
Pooled	103	605	708	56.64	651.36	41.245	<0.0001	84.96	623.04	4.353	0.037
Homogeneity						196.184	<0.0001			136.734	<0.0001
Response of parental controls											
GR Parent	35	17	52								
GS Parent	0	52	52								
F1 R × S	6	44	50								
F1 S × R	10	40	50								

*Abbreviations used: R: glyphosate-resistant; S: glyphosate-susceptible; R × S and S × R: reciprocal crosses where, the first alphabet denotes female parent; BC1F1: backcross family; F1: first filial.

from the regression equations. Data for percent survival for glyphosate rates common across F1 populations and GR and GS parents were subjected to ANOVA using the GLM procedure in SAS (Statistical Analysis Systems, version 9.2, SAS Institute Inc., SAS Campus Drive, Cary, NC 27513). Means were separated using Fisher's Protected LSD test at $P \leq 0.05$.

2.4. Generation of Backcross Families. Sixteen selected surviving males after $400 \, g \, ha^{-1}$ glyphosate application from each of the F1 families were transplanted into round pots (10 cm diameter by 12 cm deep) and each paired with a female from the original GS biotype already in the pot in the greenhouse. Both inflorescences were encased together in the same manner as outlined in the generation of F1 families. Fertilization was ensured by tapping the tubing twice every day. Single female plants were also enclosed in dialysis tubing as controls. Seeds were harvested individually from each female and kept in separate envelopes forming 16 BC1F1 families. However, seed yield from one of the backcrosses was low and therefore this family was not included for further evaluation of glyphosate resistance. Seeds were stored at −20°C for 50 days to break dormancy.

2.5. Treating BC1F1 Families with Glyphosate. All fifteen BC1F1 families were assessed for resistance to glyphosate along with F1 families (both R × S and S × R), GR, and GS biotypes as controls in the greenhouse. Seeds were planted in excess in 15-cm round pots and 10 day old seedlings were transplanted into nursery trays. About 50 seedlings were transplanted per tray and each treated with 2000 and $3000 \, g \, ha^{-1}$ of glyphosate at the 4- to 5-leaf stage. Glyphosate was applied as described previously. Numbers of surviving and dead individuals were recorded 3 wk after treatment.

2.6. Data Analysis for BC1F1 Dose-response Experiments. Data for the observed number of surviving and dead plants were compared with predicted values by subjecting data to chi-square tests in order to understand if genetic control of glyphosate resistance in this biotype was governed by single or multiple genes. The proportion of surviving susceptible and F1 individuals at each dose tested was used to calculate expected survival of BC1F1 families assuming monogenic inheritance [34]. The following equation was used for calculating expected survival:

$$YX = 0.5 \, (WRS + WSS), \qquad (1)$$

where, YX = expected proportion alive, WRS = proportion of individuals observed alive in the F1 biotype (averaged over R × S and S × R families), WSS = proportion of individuals alive in GS biotype.

A similar calculation was used to test for fit to a two gene additive model. A homogeneity test was conducted in order to test whether data could be pooled over backcross families.

2.7. EPSPS Gene Copy Number Determination. Copy number of the *EPSPS* gene was determined in 2 and 10 plants of the original GS and GR biotypes, respectively, in order to elucidate whether or not gene copy number played a role in the inheritance of resistance in derived F1 and BC1F1 populations. The following procedure is a modification of the quantitative real-time polymerase chain reaction (qPCR) method followed by Gaines et al. [31]. Genomic DNA was extracted from fresh tissue using DNEasy Plant Mini Kits (Qiagen, 27220 Turnberry Lane, Suite 200, Valencia, CA 91355) and checked for quality by gel electrophoresis. Amount and purity of the samples was determined with a ND-1000 Nanodrop instrument (Thermo Scientific, 28W092 Commercial Avenue, Barrington, IL 60010). The DNA was diluted to a 2-ng/μL concentration in highly purified 18 mΩ water. Using PerfeCTa SYBR Green Supermix with ROX (Quanta Biosciences, 202 Perry Parkway, Gaithersburg, MD 20877), the SYBR Green Supermix, upstream and downstream primers, and water were combined in a 1.5-mL tube to create a SYBR Green master mix. The genomic DNA templates (10 ng) were run with each primer set in triplicate in 12.5-μL reaction volumes using the SYBR Green master mix on a Polymerase Chain Reaction (PCR) plate. The plate was covered by Microseal "B" film (Bio-Rad, 2000 Alfred Nobel Drive, Hercules, CA 94547), which is optically clear for real-time PCR detection, and fluorescence data were captured in real time during each amplification cycle. The ABI Prism 7000 Real-Time PCR Detection System (Applied Biosystems, Foster City, CA 94547) was run with the following thermoprofile: 15 min at 95 C, 40 cycles of 95 C for 30 sec, and 60 C for 1 min, and finally a melt-curve analysis to check for primer-dimers. No-template controls, consisting of 10 μL of Master Mix and 2.5 μL of water, served as the negative controls for this procedure. No primer-dimers and no amplification products were seen in the melt-curve analysis and the controls, respectively. The melting peaks for both primer sets were 81C.

Primer efficiency curves were created for each primer set by using a 1/10x dilution series of genomic DNA from a resistant plant. The *EPSPS* primers EPSF1 (5′-ATG-TTGGACGCTCTCAGAACTCTTGGT-3′) × EPSR8 (5′-TGAATTTCCTCCAGCAACGGCAA-3′) (195-bp product) had an efficiency of 95.16% and the *ALS* primers ALSF2 (5′-GCTGCTGAAGGCTACGCT-3′) × ALSR2 (5′-GCG-GGACTGAGTCAAGAAGTG-3′) (118-bp product) had an efficiency of 95.62%. These efficiencies are very similar and thus directly comparable in later calculations.

Threshold cycles (Ct) were calculated by the ABI Prism 7000 program, and relative copy number was determined by using a modified version of the $2^{-\Delta\Delta Ct}$ method from [35].

The *ALS* gene was used as a reference gene present in the genome at a copy number of one. Quantification of *EPSPS* was calculated by finding ΔCt = (Ct, *ALS*-Ct, *EPSPS*) and calculating 2ΔCt to get a relative *EPSPS* copy number count.

3. Results and Discussion

Percent survival data indicated much reduced control of the GR Palmer amaranth biotype compared with the GS biotype (Figure 1). Glyphosate rate for 50% percent survival of GR and GS biotypes were 1288 and 58 g ha^{-1}, respectively. Whitaker et al. [33] using these same biotypes of Palmer amaranth reported the values for 50% percent visible control to be 1769 and 89 g ha^{-1} for GR and GS biotypes, respectively. Differences in values between their results and those observed in our study most likely reflect differences in methodology (i.e., percent survival versus percent visible control).

The values for 50% percent survival for combined GR × GS and GS × GR F1 families were 794 and 501 g ha^{-1}, respectively. Both sets of F1 families were treated with 180, 400, 2000, and 3000 g ha^{-1} of glyphosate (Table 1). Glyphosate rates common for the GR parent biotype and F1 families were 2000 and 3000 g ha^{-1}, while glyphosate rates common for the GS parent biotype and F1 families were 180 and 400 g ha^{-1}. Therefore, comparisons between parent biotypes and both sets of F1 families for percent survival were made at glyphosate rates that were common among them. Although GR parents were not exposed to 180 and 400 g ha^{-1} of glyphosate, it was expected that all individuals would have survived at these rates because the glyphosate rate for 50% survival for this biotype was 1288 g ha^{-1}. Similarly, although GS parents were not exposed to 2000 and 3000 g ha^{-1} of glyphosate, all individuals would have died at these rates given that the rate for 50% survival for this biotype was 58 g ha^{-1}. At any given dose, percent survival of F1 progenies showed lower levels of resistance as compared to the GR parent biotype and higher levels of resistance than the GS parent biotype (Figure 1 and Table 1). Response of both sets of maternal and paternal resistant F1 families was closer to that of the resistant parent (Figure 1). Dose-response behavior of F1 families indicated that resistance was not fully dominant over susceptibility. Inheritance of glyphosate resistance as an incompletely dominant trait controlled by nuclear genes has been reported in other weed species [17, 27–29]. Values for percent survival of both sets of F1 families at a given glyphosate dose were not significantly different from each other (Figure 1 and Table 1), indicating that genetic control of glyphosate resistance is governed by the nuclear genome and that there is no maternal or cytoplasmic inheritance involved. These results are also similar to those of Gaines [30], who reported that resistance in GR Palmer amaranth was due to an incompletely dominant, nuclear-inherited gene.

To determine if genetic control of resistance involves single or multiple genes, BC1F1 families were developed and treated with glyphosate at 2000 and 3000 g ha^{-1}. The homogeneity chi-square was significant ($P < 0.0001$); therefore, data could not be pooled over backcross families.

Thus, tests on individual families were considered (Tables 2 and 3). Segregation of resistance in 10 out of 15 individual backcross families tested at $2000\,\mathrm{g\,ha^{-1}}$ glyphosate and 8 out of 15 individual backcross families tested at $3000\,\mathrm{g\,ha^{-1}}$ glyphosate conformed to a monogenic inheritance (Tables 2 and 3). When tested against a 2-gene additive model, the homogeneity test was also significant ($P < 0.0001$). When individual families were considered, 10 out of 15 families fitted the model at $2000\,\mathrm{g\,ha^{-1}}$ and $3000\,\mathrm{g\,ha^{-1}}$ glyphosate. The inheritance analysis suggested that the monogenic model fitted the data better than the two-gene additive model.

While in the majority of the BC1F1 families inheritance of glyphosate resistance was consistent with a single gene hypothesis, this was not the case in a few families. An examination of the data showed that there was an excess of resistant individuals in these families (Tables 2 and 3). Moreover, in three families (BC1F1 S × R4, BC1F1 R × S1, and BC1F1 R × S4) the excess of resistant individuals at both glyphosate rates used was large. This suggests some other form of inheritance is occurring in these families. In an earlier study on the inheritance of glyphosate resistance in Palmer amaranth biotypes from Georgia, an overabundance of resistant progeny was found in some but not all of the families evaluated [30]. Further studies on the inheritance of EPSPS gene in these populations showed that inheritance of additional copies of the gene from parents to progeny was highly unpredictable [31, 32]. Because the copy number determination technique was not available to us at the time dose response experiments were performed, the EPSPS gene copy number in the F1 parents and the BC1F1 individuals could not be determined. This information could have aided in providing a better interpretation of the results. However, plants from the GR and GS biotypes used to generate our F1 and BC1F1 families were analyzed for copy number. While the GS parent possessed only one copy of EPSPS relative to als, the number of copies in the GR parent ranged from 22 to 63 (Figure 2). These results confirm that increased EPSPS copy number is closely associated with glyphosate resistance in the resistant Palmer amaranth biotype used in this study. Moreover, EPSPS copy number in this biotype is highly variable. Given that the inheritance of glyphosate resistance in individuals with increased number of copies of EPSPS has been demonstrated to be unpredictable in previous studies [30, 32], it is likely that this is also the cause for the variability in the inheritance of glyphosate resistance observed in this study.

4. Conclusions

Collectively, these data suggest that glyphosate resistance in the studied biotype of Palmer amaranth is incompletely dominant, and nuclear-inherited with no maternal or cytoplasmic effects involved. While resistance is consistent with a single gene mechanism of inheritance for many backcross families, a number of individual backcross families seem to follow polygenic inheritance. Differences in copy number of the EPSPS gene in the resistant biotype of Palmer amaranth studied and the unpredictable behavior in the inheritance

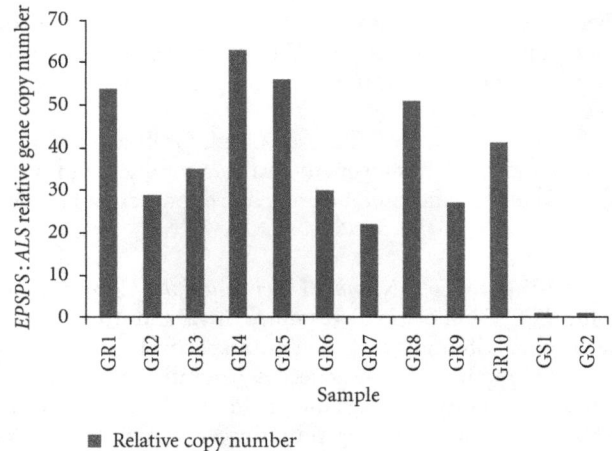

FIGURE 2: *EPSPS* gene copy number relative to *ALS* from 10 glyphosate-resistant (GR) and 2 glyphosate-susceptible (GS) plants.

of these copies might be responsible for the variable results among families regarding the number of genes involved in inheritance of glyphosate resistance. Further analysis looking at EPSPS copy numbers in each generation of GR Palmer amaranth progenies might help elucidate these issues.

Acknowledgments

Monsanto Company and Syngenta Crop Protection provided partial funding. Appreciation is expressed to C. Foresman and Dr. V. Shivrain for discussions relative to research and to R. Seagroves, P. Eure, and J. Hinton for technical assistance.

References

[1] G. M. Dill, C. A. CaJacob, and S. R. Padgette, "Glyphosate-resistant crops: adoption, use and future considerations," *Pest Management Science*, vol. 64, no. 4, pp. 326–331, 2008.

[2] S. O. Duke and S. B. Powles, "Glyphosate: a once-in-a-century herbicide," *Pest Management Science*, vol. 64, no. 4, pp. 319–325, 2008.

[3] A. D. Baylis, "Why glyphosate is a global herbicide: strengths, weaknesses and prospects," *Pest Management Science*, vol. 56, no. 4, pp. 299–308, 2000.

[4] S. B. Powles, "Evolved glyphosate-resistant weeds around the world: lessons to be learnt," *Pest Management Science*, vol. 64, no. 4, pp. 360–365, 2008.

[5] I. Heap, "The International Survey of Herbicide Resistant Weeds," 2012, http://www.weedscience.org/.

[6] T. M. Webster, "Weed survey—Southern states: broadleaf crops subsection," *Proceedings Southern Weed Science Society*, vol. 58, pp. 291–294, 2005.

[7] A. S. Culpepper, J. R. Whitaker, A. W. MacRae, and A. C. York, "Weed science: distribution of glyphosate-resistant palmer amaranth (*Amaranthus palmeri*) in Georgia and North Carolina during 2005 and 2006," *Journal of Cotton Science*, vol. 12, no. 3, pp. 306–310, 2008.

[8] A. S. Culpepper, T. L. Grey, W. K. Vencill et al., "Glyphosate-resistant Palmer amaranth (*Amaranthus palmeri*) confirmed in Georgia," *Weed Science*, vol. 54, no. 4, pp. 620–626, 2006.

[9] J. K. Norsworthy, G. M. Griffith, R. C. Scott, K. L. Smith, and L. R. Oliver, "Confirmation and control of glyphosate-resistant Palmer amaranth (*Amaranthus palmeri*) in Arkansas," *Weed Technology*, vol. 22, no. 1, pp. 108–113, 2008.

[10] L. E. Steckel, C. L. Main, A. T. Ellis, and T. C. Mueller, "Palmer amaranth (*Amaranthus palmeri*) in Tennessee has low level glyphosate resistance," *Weed Technology*, vol. 22, no. 1, pp. 119–123, 2008.

[11] A. J. Diggle and P. Neve, "The population dynamics and genetics of herbicide resistance—a modeling approach," in *Herbicide Resistance and World Grains*, pp. 61–100, CRC Press, Boca Raton, Fla, USA, 2001.

[12] M. Jasieniuk, A. L. Brule-Babel, and I. N. Morrison, "Inheritance of trifluralin resistance in green foxtail (*Setaria viridis*)," *Weed Science*, vol. 42, no. 1, pp. 123–127, 1994.

[13] J. Gressel and L. A. Segel, "Interrelating factors controlling the rate of appearance of resistance: the outlook for the future," in *Herbicide Resistance In Plants*, pp. 325–347, John Wiley & Sons, New York, NY, USA, 1982.

[14] B. D. Maxwell, M. L. Roush, and S. R. Radosevich, "Predicting the evolution and dynamics of herbicide resistance in weed populations," *Weed Technology*, vol. 4, no. 1, pp. 2–13, 1990.

[15] M. L. Roush, S. R. Radosevich, and B. Maxwell, "Future outlook for herbicide-resistance research," *Weed Technology*, vol. 4, no. 1, pp. 208–214, 1990.

[16] C. Preston and C. A. Mallory-Smith, "Biochemical mechanisms, inheritance, and molecular genetics of herbicide resistance in weeds," in *Herbicide Resistance and World Grains*, pp. 23–60, CRC Press, Boca Raton, Fla, USA, 2001.

[17] M. Simarmata, S. Bughrara, and D. Penner, "Inheritance of glyphosate resistance in rigid ryegrass (*Lolium rigidum*) from California," *Weed Science*, vol. 53, no. 5, pp. 615–619, 2005.

[18] R. Busi, M. M. Vila-Aiub, and S. B. Powles, "Genetic control of a cytochrome P450 metabolism-based herbicide resistance mechanism in *Lolium rigidum*," *Heredity*, vol. 106, no. 5, pp. 817–824, 2011.

[19] J. Hirschberg and L. McIntosh, "Molecular basis of herbicide resistance in *Amaranthus hybridus*," *Science*, vol. 222, no. 4630, pp. 1346–1349, 1983.

[20] P. Boutsalis and S. B. Powles, "Inheritance and mechanism of resistance to herbicides inhibiting acetolactate synthase in *Sonchus oleraceus* L.," *Theoretical and Applied Genetics*, vol. 91, no. 2, pp. 242–247, 1995.

[21] K. J. Betts, N. J. Ehlke, D. L. Wyse, J. W. Gronwald, and D. A. Somers, "Mechanism of inheritance of diclofop resistance in Italian ryegrass (*Lolium multiflorum*)," *Weed Science*, vol. 40, no. 2, pp. 184–189, 1992.

[22] C. A. Mallory-Smith, D. C. Thill, M. J. Dial, and R. S. Zemetra, "Inheritance of sulfonylurea herbicide resistance in *Lactuca* spp.," *Weed Technology*, vol. 4, no. 4, pp. 787–790, 1990.

[23] F. J. Tardif, C. Preston, J. A. M. Holtum, and S. B. Powles, "Resistance to acetyl-coenzyme a carboxylase-inhibiting herbicides endowed by a single major gene encoding a resistant target site in a biotype of *Lolium rigidum*," *Australian Journal of Plant Physiology*, vol. 23, no. 1, pp. 15–23, 1996.

[24] B. G. Murray, I. N. Morrison, and A. L. Brule-Babel, "Inheritance of acetyl-CoA carboxylase inhibitor resistance in wild oat (*Avena fatua*)," *Weed Science*, vol. 43, no. 2, pp. 233–238, 1995.

[25] L. Zeng and W. V. Baird, "Genetic basis of dinitroaniline herbicide resistance in a highly resistant biotype of goosegrass (*Eleusine indica*)," *Journal of Heredity*, vol. 88, no. 5, pp. 427–432, 1997.

[26] D. F. Lorraine-Colwill, S. B. Powles, T. R. Hawkes, and C. Preston, "Inheritance of evolved glyphosate resistance in *Lolium rigidum* (Gaud.)," *Theoretical and Applied Genetics*, vol. 102, no. 4, pp. 545–550, 2001.

[27] A. M. Wakelin and C. Preston, "Inheritance of glyphosate resistance in several populations of rigid ryegrass (*Lolium rigidum*) from Australia," *Weed Science*, vol. 54, no. 2, pp. 212–219, 2006.

[28] I. A. Zelaya, M. D. K. Owen, and M. J. VanGessel, "Inheritance of evolved glyphosate resistance in *Conyza canadensis* (L.) Cronq," *Theoretical and Applied Genetics*, vol. 110, no. 1, pp. 58–70, 2004.

[29] C. H. Ng, W. Ratnam, S. Surif, and B. S. Ismail, "Inheritance of glyphosate resistance in goosegrass (*Eleusine indica*)," *Weed Science*, vol. 52, no. 4, pp. 564–570, 2004.

[30] T. A. Gaines, *Molecular genetics of glyphosate resistance in palmer amaranth (Amaranthus palmeri L.) [Ph.D. dissertation]*, Colorado State University, Fort Collins, Colo, USA, 2009.

[31] T. A. Gaines, W. Zhang, D. Wang et al., "Gene amplification confers glyphosate resistance in *Amaranthus palmeri*," *Proceedings of the National Academy of Sciences of the United States of America*, vol. 107, no. 3, pp. 1029–1034, 2010.

[32] D. A. Giacomini, S. Ward, T. A. Gaines, and P. Westra, "Inheritance of EPSPS gene amplification in Palmer amaranth," *Proceedings Weed Science Society of America*, 2011, Abstract no. 85.

[33] J. R. Whitaker, *Distribution, biology, and management of glyphosate-resistant Palmer amaranth in North Carolina [Ph.D. dissertation]*, North Carolina State University, Raleigh, NC, USA, 2009.

[34] B. E. Tabashnik, "Determining the mode of inheritance of pesticide resistance with backcross experiments," *Journal of Economic Entomology*, vol. 84, no. 3, pp. 703–712, 1991.

[35] K. J. Livak and T. D. Schmittgen, "Analysis of relative gene expression data using real-time quantitative PCR and the $2^{-\Delta\Delta C}$T method," *Methods*, vol. 25, no. 4, pp. 402–408, 2001.

8

Genetic Transformation of Common Bean (*Phaseolus vulgaris* L.) with the *Gus* Color Marker, the *Bar* Herbicide Resistance, and the Barley (*Hordeum vulgare*) *HVA1* Drought Tolerance Genes

Kingdom Kwapata, Thang Nguyen, and Mariam Sticklen

Department of Plant, Soil and Microbial Sciences, Michigan State University, East Lansing, MI 48824, USA

Correspondence should be addressed to Mariam Sticklen, stickle1@msu.edu

Academic Editor: Antonio M. De Ron

Five common bean (*Phaseolus vulgaris* L.) varieties including "Condor," "Matterhorn," "Sedona," "Olathe," and "Montcalm" were genetically transformed via the Biolistic bombardment of the apical shoot meristem primordium. Transgenes included *gus* color marker which visually confirmed transgenic events, the *bar* herbicide resistance selectable marker used for *in vitro* selection of transgenic cultures and which confirmed Liberty herbicide resistant plants, and the barley (*Hordeum vulgare*) late embryogenesis abundant protein (*HVA1*) which conferred drought tolerance with a corresponding increase in root length of transgenic plants. Research presented here might assist in production of better *P. vulgaris* germplasm.

1. Introduction

The common bean (*Phaseolus vulgaris* L.) is a very important source of vegetable protein, especially in those regions of the world in which animal proteins are scarce. Common bean provides 22% of the total protein requirement worldwide [1]. Conventional breeding has contributed significantly to the trait improvement of *P. vulgaris*. However, breeding cannot add certain genes that do not exist naturally in the *P. vulgaris* gene pool. Due to this limitation of plant breeding, new trait improvement approaches such as interspecific horizontal gene transfer via genetic engineering need to be utilized in order to complement the limitations encountered by conventional breeding of this crop [2, 3].

Mostly, *Agrobacterium*-mediated transformation and the gene gun microprojectiles bombardment method have been used for genetic transformation of *P. vulgaris*. However, neither system has shown as high as those seen in genetic transformation of cereals [4]. Researchers have unsuccessfully attempted to transform *P. vulgaris* protoplast, either via polyethylene glycol or electroporation [5]. A relatively advanced *Agrobacterium*-mediated transformation of *P. vulgaris* has been reported on the use of sonication and

vacuum infiltration for transfer of a group of 3 LEA (late embryogenesis abundant protein) genes from *Brassica napus* [6]. Although the transformation efficiency using this system was low, transgenic plants exhibited a high growth rate under salt and water stress. A recent report [7] on transformation of *P. vulgaris* varieties Mwitemania and Rose coco using the *gus* color marker gene reveals the importance of specificity of *Agrobacterium* strains in expression of *gus* gene in *P. vulgaris*. For example, infecting of *P. vulgaris* explants with EHA 105 (pCAMBIA 1201) or EHA 105 (pCAMBIA 1301) resulted in blue GUS coloration; however, it did not show the GUS expression when the explants were infected with LBA 4404 (pBI 121) *Agrobacterium* strain.

Using Biolistic bombardment of a construct containing the *bar* gene, Aragão et al. [3] developed transgenic *P. vulgaris* which conferred resistance to glufosinate ammonium, the active ingredient of Liberty herbicide (Aventis, Strasbourg, France), at concentrations of 500 g ha^{-1} in greenhouses and 400 g ha^{-1} in the field. *P. vulgaris* was also genetically engineered by Bonfim et al. [8] using RNAi-hairpin construct to silence the AC1 region of the viral genome of Bean Golden Mosaic Gemini Virus (BGMGV). However, out of 2,706 plants, only 18 putative transgenic lines were obtained.

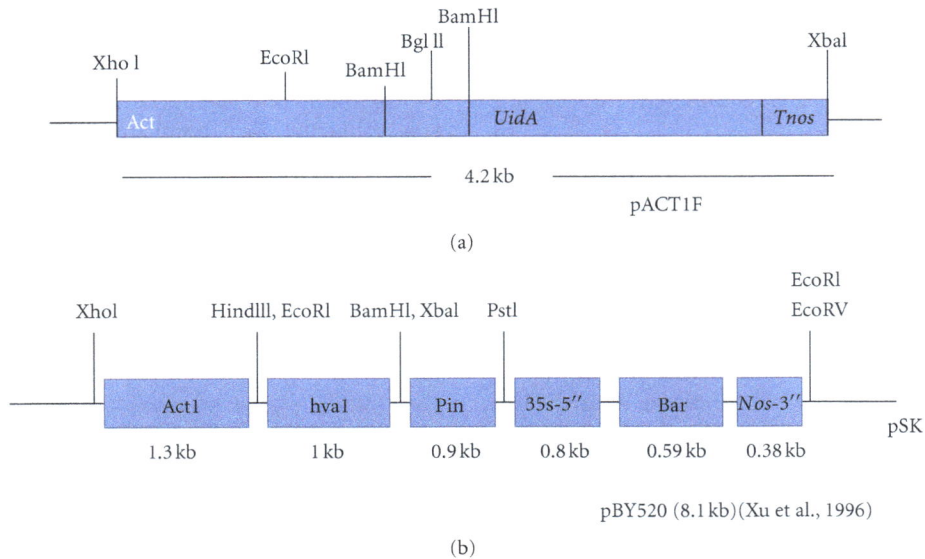

FIGURE 1: Plasmid constructs. (a) Linear map of pACT1F plasmid vector. Rice actin promoter (Act1), *gus* gene (*uidA*), and nopaline synthase terminator (*Tnos*); (b) linear map of pBY520 plasmid vector. Rice actin promoter (Act1) and potato protease inhibitor II (Pin II-3′) terminator, Barley or *Hordeum vulgare* (*HVA1*) LEA 3 gene, Cauliflower Mosaic Virus 35S promoter, *bar* gene and nopaline synthase terminator (*Nos-3′*).

Of the 18 putative transgenic plants, only one plant exhibited resistance to the virus. Field trials of the progenies of the single transgenic plant showed resistance to this virus [9]. Vianna et al. [10] developed an approach of transferring the transgene assembly as fragment pieces of DNA, as opposed to the entire plasmid into *P. vulgaris*. A protocol was published [11] on a relatively efficient genetic transformation of *P. vulgaris*. Due to the "troublesome" nature of *P. vulgaris* genetic transformation, an article describes a method called "transgenic composite" of *P. vulgaris* via the use of *Agrobacterium rhizogenes* transformation of derooted seedlings [12].

The efficiency of genetic engineering of *P. vulgaris* has remained a challenge. A relatively recent report explains the effect of *in vitro* conditions on indirect organogenesis (multiple shoots from meristem and cotyledon-derived callus regeneration) for production of an average of 0.5 shoot per callus clump. Indirect regeneration of different genotypes of *P. vulgaris* was also reported [13]. Kwapata et al. [14] cite that an *in vitro* culture of a single apical shoot meristem primordium could produce as many as 20 multiple shoots, which is a relatively higher number as compared to the work previously presented. However, this *in vitro* regeneration efficiency is still very low when compared to the desired 100s regenerated from the *in vitro* cultures of each apical shoot meristem primordia of cereal crops [4].

Genetic transformation of *P. vulgaris* can improve the biotic and abiotic stress tolerance. Biotic stress factors such as diseases result in *P. vulgaris* yield loss. Brazil just announced [15] the commercial use of golden mosaic virus resistant *P. vulgaris* that was developed via RNA interference by blocking the replication of the virus gene [8]. This is indeed a major step in the acceptance of biosafety of transgenic *P. vulgaris*.

Also, researchers from Denmark recently reported cloning of the bean common mosaic virus (BCMV) gene and its application for development of BCMV resistance [16].

Biotic stresses, including drought cause plants to lose cellular turgidity, followed by the aggregation and misfolding of proteins and yield losses [17]. A major group of abiotic stress tolerance genes coding for the late embryogenesis proteins include a class of heat shock proteins (Hsp) that are extremely hydrophilic and resilient towards heat, such that they do not coagulate at boiling temperatures. The LEA proteins play a role in water binding, ion sequestration, and macromolecule and membrane stabilization [18]. In the research presented here, the barley *HVA1* [19] gene was transferred into *P. vulgaris*, as this gene encodes a type III LEA protein. The Barley *HVA1* gene has previously been transferred to rice [20], wheat [21, 22], sugarcane [23], creeping bentgrass [24], mulberry [25], and oat [26, 27]. In all cases, plants developed tolerance to abiotic stresses such as drought and/or salt. Here we report the transfer of Barley *HVA1* gene to different varieties of *P. vulgaris* and report the development of drought tolerance of transgenic plants at greenhouse level.

2. Materials and Methods

2.1. Plasmids and Explant. Two different plasmid vectors were used in this research (Figure 1). Plasmids used included (a) pACT1F harboring the *gus* gene and (b) pBY520 harboring the *HVA1* and the *bar* gene, which confers drought tolerance and Liberty herbicide (glufosinate ammonium) resistance, respectively.

Explant preparation: the explant used to standardize the genetic transformation was *P. vulgaris* var. "Sedona."

Dry seeds were rinsed in tap water for 1 min, then rinsed three times with distilled water, soaked in 75% ethanol for 4 min, and again rinsed three times with distilled water. Then, the seeds were soaked in 20% commercial Clorox while steering for 15 min.

Seed coats of the surface-sterilized seeds were removed, and meristems were dissected under a light microscope under a laminar flow hood. The meristem dissection took place by removal of the cotyledons and the hypocotyls, leaving the meristem as an intact explant.

The meristem explants were cultured in Murashige and Skoog (MS) [28] medium containing 2.5 mg^{-1} benzyl adenine (BA; Sigma-Aldrich, Inc. Steinheim, Germany) and 0.1 mg^{-1} indole acetic acid (IAA; Sigma-Aldrich, Inc. Steinheim, Germany). Cultures were maintained under *in vitro* conditions and in a dark chamber for 5–7 days or until the explants grew to about 5–7 mm long. Then, 10 of the elongated apical meristems were placed in a circle in a Petri dish on top of MS medium, bombarded with gene constructs using the Biolistic gene via the helium particle delivery model PDS-1000 (DuPont, Wilmington, DE).

The pACT1-F construct containing the *gus* gene was coated onto 50 μg L^{-1} of 10 μm tungsten particles with 2.5 M calcium chloride and 0.1 M spermidine suspended in a solution of 1:1 (v/v) of 75% ethanol and 50% glycerol. The coated plasmid DNA was bombarded into the explants using three levels of pressure (500, 1000, or 1100 psi), plasmid concentrations of 1.5 μg or 3.0 μg, and with three levels of bombardment frequencies (1, 2 or 3 time). A total of 10 apical meristems were used for each bombardment condition.

The bombarded shoot meristems were transferred to regeneration medium [14] and kept under *in vitro* condition at room temperature with 16 h photoperiod and light intensity of 45–70 umol m^{-2}s^{-1}.

The bombarded shoot meristems were histologically stained to visualize the *gus* gene expression, and three longitudinal hand-cross-sections of each bombarded shoot meristem were made to identify the bombardment criteria that lead to expression of *gus* gene in relative location of *P. vulgaris* meristem subepidermal layer. Mean of transient transformation efficiencies (number of meristems showing blue spots) was used as preliminary data to identify the most acceptable criteria of bombardment (Table 1).

The most effective criteria were then used for stable transformation of the five varieties of *P. vulgaris*. The GUS histological assay bombarded versus control wild-type meristems included histochemical staining with 5-bromo-4-chloro-3-indoyl-β-D-glucuronicacid salt (X-gluc). Samples were dipped into GUS substrate buffer, according to published records [29], and incubated at 37°C for 24 hours. The tissue samples were washed with 100 percent ethanol to remove other colorations.

The statistical design used in this portion of research was a completely randomized design (CRD). An Analysis of Variance (ANOVA) was used to test the statistical significance at an alpha level of 0.001. Standard deviations were used to compare variability.

2.2. Stable Genetic Transformation.

Stable genetic transformation of *P. vulgaris* was performed using the Biolistic delivery for bombardment of a 1:1 ratio mixture of the two plasmids into the apical shoot subepidermal cell layer area using the ideal bombardment criteria (Table 1). The bombarded explants were cultured in regeneration media [14] without the use of any chemical selections for 24 hours.

The selection of stable transgenic plants was based on the use of *gus* color marker gene and 4 mg L^{-1} of glufosinate ammonium selection for the *bar* herbicide resistance marker gene. The *in vitro* regeneration of putatively transgenic *P. vulgaris* explants followed a previous report [14].

2.3. Confirmation of Transgene Integration and Expression

2.3.1. Polymerase Chain Reaction (PCR). Polymerase chain reaction (PCR) was used for detection of integration of *bar* and *HVA1* transgenes in four generations (T$_0$–T$_3$) of plants that were putatively transformed with Biolistic gun. The primers used were (1) *bar* F, 5′-ATG AGC CCA GAA CGA CG-3′ (forward primer); *bar* R, 5′-TCA CCT CCA ACC AGA ACC AG-3′ (reverse primer); (2) *HVA1* F, 5′-TGG CCT CCA ACC AGA ACC AG-3′ (forward primer); *HVA1* R, 5′-ACG ACT AAA GGA ACG GAA AT-3′ (reverse primer).

2.3.2. Southern Blot Hybridization. The Southern blot hybridization analysis was conducted to determine the stability of transformation and to determine the copy numbers of the *bar* and *HVA1* transgenes. The DIG High Prime DNA Labeling and Detection Starter Kit (Roche Co., Cat. No. 1 585 614) was used as per manufacturer's instructions. Transgenic and control wild-type nontransgenic genomic DNA was isolated using methods described [30]. The DIG-labeled probes for *bar* and *HVA1* were synthesized using primers for specific genes as described previously. Those transgenic plants that integrated 1-2 copies of transgenes were kept for further studies.

2.3.3. Northern Blot Hybridization. Northern blot analysis was conducted using the DIG-labeled Northern Starter Kit (Roche Co., Cat. No. 12039672910). Total RNA from the leaves of transgenic and the control wild-type nontransgenic plants was isolated using TRI reagent (Sigma-Aldrich, St. Louis, MO) as per manufacturer's instructions. A total of 30 μg of RNA per sample was loaded onto a 1.2% (m/v) agarose-formaldehyde denaturing gel as described [31] and transferred to a Hybond-N+ membrane (Amersham-Pharmacia Biotech) and fixed with a UV crosslinker (Stratalinker UV Crosslinker 1800, Stratagene, CA). The RNA or DNA DIG-labeled probe, containing the coding region of the gene of interest, was used for detection of transcripts.

2.4. Biological Activity Tests

2.4.1. Herbicide Resistance Assay. Following a glufosinate ammonium *in vitro* culture kill curve studies (data not shown), an optimum 4 mg L^{-1} of glufosinate ammonium

Genetic Transformation of Common Bean (Phaseolus vulgaris L.) with the Gus Color Marker, the Bar Herbicide Resistance, and the Barley (Hordeum vulgare) HVA1 Drought Tolerance Genes

65

was used in the *in vitro* culture of putatively transgenic shoot regeneration and rooting media.

Different concentrations of Liberty herbicide (50, 100, 150, 250, or 350 mg L^{-1}) were used to find the ideal foliar spray concentration of trifoliate transgenic plants.

In vitro germination of progeny seeds in MS medium [28] containing 4 mg L^{-1} of glufosinate ammonium was used to indentify segregation ratio of the *bar* transgene in transgenic progenies.

2.4.2. Drought Tolerance Test.
The *HVA1* transgenic and wild-type control seeds were collected, and seedlings were grown in 15 cm clay pots containing BACCTO High Porosity Professional Planting Mix (Michigan Peat Company, Houston, TX) in a growth chamber for three weeks or until trifoliate leaves appeared. Plants were watered daily for 21 days, after which moisture was withheld for 21 days. Then, water was applied to plants continuously for up to 14 days, and the percentage of plants recovered was recorded. Also, percent plant leaf abscission was used as an indirect measure of degree of plant wilting. In reality, the number of green leaves on plants after 21 days of moisture withdrawal was used to find percent plant leaf abscission.

3. Results and Discussions

3.1. Explant.
Our results show that the apical shoot meristem primordium might be a good explant for genetic transformation of common beans. The apical shoot meristem in *P. vulgaris* is an undifferentiated meristematic tissue in a small and relatively round shape, which is composed of different cell layers. The top layer or the "Epidermal Cell Layer" divides horizontally and will not differentiate. The layer beneath the Epidermal Cell layer is the "Subepidermal Cell Layer" (also called the primordial cell layer or stem cell layer) normally divides indefinitely and differentiates into gametes resulting into fertile plants. Therefore, it is the Subepidermal Cell Layer that needs to be targeted via the Biolistic gun for genetic transformation. Using the gus color marker gene, the researchers of this report tried to standardize the Biolistic delivery bombardment to hit this layer.

3.2. Transient Expression of the Gus Marker Gene.
Bombarding the explants twice at the approximate distance of 4 cm between the gun barrel and target explants, using a pressure setting of 1100 psi, with a concentration of 1.5 µg of plasmid DNA per bombardment yielded the highest GUS activity efficiency of 8.4% (Table 1). Mean transient transformation was calculated by counting the mean of number of bombarded meristems that showed blue spots.

The transient transformation frequency of the GUS expression is shown in Figure 2. The number of clear blue spots was seen 15 days after bombardment.

3.3. Stable Transformation.
PCR was performed for all *bar* and HVA1 transgenes used, among which results are only shown for integration of HVA1 transgene in all four *P. vulgaris* cultivars (Figure 3(a)). Southern blot hybridizations were performed in multiple samples of PCR-positive plants,

TABLE 1: Transient expression of GUS using different gene gun pressure (psi), DNA plasmid concentration and bombardment frequency for optimizing the Biolistic bombardment conditions for the pCATIF containing the *gus* gene.

Bombardment pressure (psi)	Concentration of plasmid DNA (µg)	Bombardment frequency	Mean transformation percent
500	1.5	1	0.1 ± 0.04
500	1.5	2	0.2 ± 0.10
500	1.5	3	0.4 ± 0.30
500	3	1	0.1 ± 0.04
500	3	2	0.6 ± 0.32
500	3	3	0.7 ± 0.32
1000	1.5	1	2.9 ± 0.67
1000	1.5	2	3.9 ± 1.4
1000	1.5	3	5.1 ± 1.2
1000	3	1	5.6 ± 1.0
1000	3	2	8.1 ± 0.3
1000	3	3	7.4 ± 1.0
1100	1.5	1	7.2 ± 0.70
1100	1.5	2	8.4 ± 0.74
1100	1.5	3	8.2 ± 0.50
1100	3	1	7.5 ± 0.69
1100	3	2	4.8 ± 0.93
1100	3	3	3.3 ± 0.92

among which data are only shown for integration of HVA1 gene in different *P. vulgaris* cultivars. After Southern blot hybridization analysis, transgenic plants that showed the integration of at most two copies of transgenes (e.g., see Figure 3(b)) were kept for transcription analysis. Transcription analysis via RT-PCR showed that HVA1 has transcribed in all transgenic plants. However RNA blotting confirmed that only certain transgenic plants sufficiently transcribed their transgenes (e.g., see Figure 3(c)). This is because RT-PCR is much more sensitive than the RNA blotting.

The GUS bioassay was a method of selecting the transgenic shootlets. All Southern blot-positive progenies of *P. vulgaris* varieties ("Matterhorn," "Condor," "Sedona," "Olathe," and "Montcalm") showed GUS expression. Figure 4 represents expression of GUS protein in seeds and pods of T3 of "Matterhorn."

Because glufosinate ammonium was included in the *in vitro* cultures of all putatively transgenic shoots, roots and plantlets, all transgenic plant progenies were resistant to 150 mg L^{-1} of Liberty herbicide (Figure 5). Lower concentrations did not kill wild-type control nontransgenic plants, and higher concentrations killed transgenic plants as well as their wild-type control non-transgenic counterparts.

Most drought tolerant HVA1 transgenic plants were "Sedona" and "Matterhorn" which persisted for 21 days without irrigation. They showed symptoms of drought stress but recovered only after three days when moisture application resumed. The wild-type control plants died or showed severe symptoms of drought stress, with most of their leaves

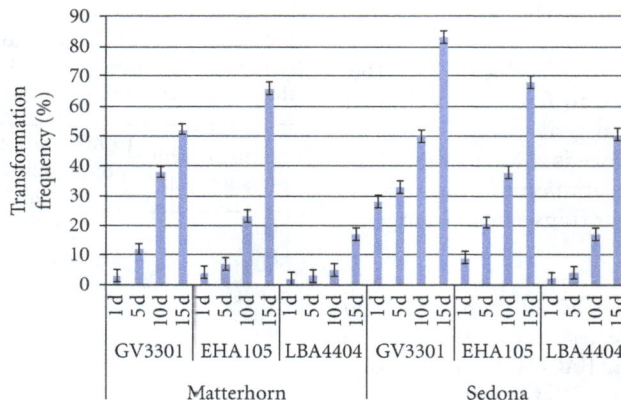

FIGURE 2: Percent transient expression of GUS at different number of days after bombardment.

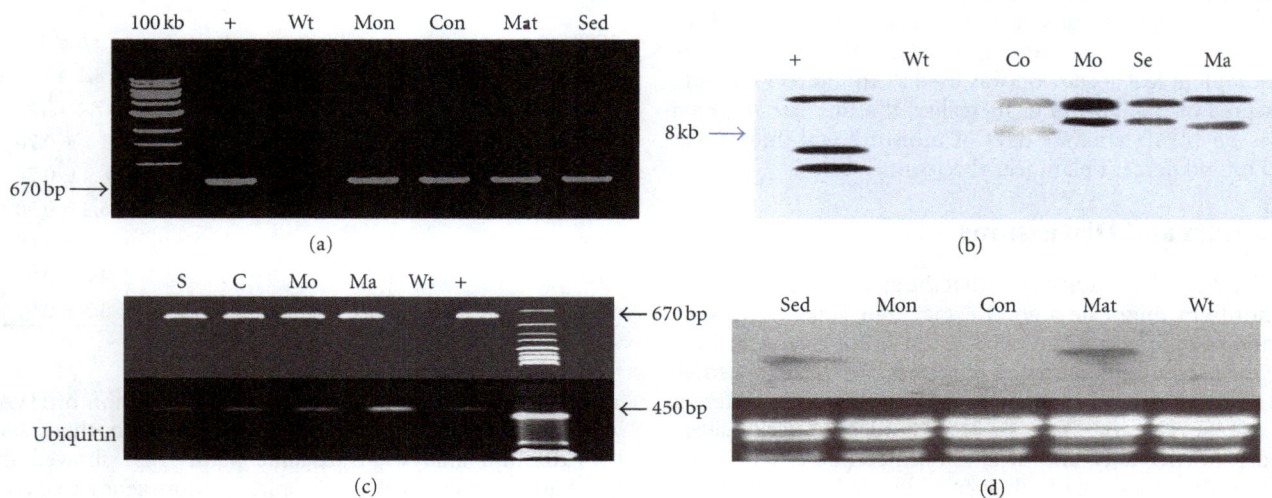

(a)

(b)

(c)

(d)

FIGURE 3: Molecular analysis confirming the integration and transcription of HVA1 transgene in plants. (a) PCR results of T3 transgenic plants. The expected band size is 670 bp; (b) Southern blot hybridization showing integration of *HVA1* gene digested with BamH1. The results indicate that there are two copies of transgene in all varieties. (c) RT-PCR of *HVA1* expression for T2 transgenic plants. Like in PCR, here the expected band size is 670 bp for *HVA1*. Below the RT-PCR is the cDNA loading control showing the expression of ubiquitin with the expected band size of 450 bp. (d) Northern blot analysis also confirmed the transcription of *HVA1* gene in Sedona and Matterhorn. Wt: wild type shows no transgene integration; C. RNA transcription analysis of *HVA1* gene in T3 transgenic plants. Mat: "Matterhorn" and Sed: "Sedona" showed some expression. The remaining lanes, Wt: wild type, Mon: "Montcalm" and Con: "Condor" showed no transcriptions.

(a)

(b)

FIGURE 4: GUS biological activity shown after histochemical assays in pods and seeds of T3 "Matterhorn."

Genetic Transformation of Common Bean (Phaseolus vulgaris L.) with the Gus Color Marker, the Bar Herbicide Resistance, and the Barley (Hordeum vulgare) HVA1 Drought Tolerance Genes

67

(a)

(b)

(c)

(d)

FIGURE 5: Although not completely resistant, the trifoliate stage of the third generation transgenic (T2) plants that had transcribed *bar* gene showed more resistance to foliar spray of 150 mg L^{-1} Liberty herbicide than their wild-type control non-transgenic counterpart plants. "Condor" (a), "Matterhorn" (b), "Montcalm" (c), and "Sedona" (d). "Matterhorn" seems to be more resistant to the herbicide.

(a)

(b)

(c)

(d)

FIGURE 6: Drought tolerance assays. (a) "Matterhorn" plants before drought induction; (b) after 21 days of continuous water withholding; (c) "Matterhorn" drought recovered plants after water reapplication; 1: control non-transgenic plant that was watered throughout the experiment; 2: "Matterhorn" transgenic plant after 21 days of no-irrigation, 3: wild-type non-transgenic plant after 21 days of no-irrigation; (d) root growth in plants after 21 days of drought stress. 1: Control non-transgenic plant roots; these were watered daily, 2: transgenic plant roots after 21 days of no-irrigation, and 3: wild-type non-transgenic plant roots after 21 days of no-irrigation.

being wilted and dehisced (e.g., see Figure 6(b)). The survival rate of control wild-type non-transgenic "Sedona" plants after 21 days of drought was 13.3% and for its HVA1 transgenic plants of the same variety was 33.3%. In case of "Matterhorn," the survival rate of control wild-type non-transgenic plants was 20%, and its transgenic counterpart of the same variety was 53.3%. Withdrawal of irrigation for more than 21 days resulted in the death of both control wild-type non-transgenic and HVA1 transgenic plants of "Sedona" and "Matterhorn" varieties.

The percent leaf wilting of transgenic "Sedona" plants was 78% and for its wild type was 91%. In the case of "Matterhorn," percent leaf wilting of transgenic plants of "Matterhorn" variety was 72% as compared to the wild-type control non-transgenic plants which were 88%.

Over all, the root growth of HVA1 transgenic plants with least percent wilting was more robust than wild-type plants under stress, but less developed than wild type plants under a normal moisture regime (e.g., see Figure 6(c)).

In a preliminary experiment, the average root length measurement after 21 days of water withdraw for "Sedona" HVA1 transgenic plants was 15 cm and for wild-type plants was 11 cm. For "Matterhorn" variety was 72% as compared to the wild-type control non-transgenic plants which was 88%. In contrast, for control wild type plants under normal irrigation without water withhold, the average root length was 28 cm.

The researchers of this paper exposed transgenic plants transcribing the HVA1 gene to drought prior to testing of plants for drought tolerance. The promoter deriving the HVA1 in this work is rice actin 1 promoter which is known to be a constitutive promoter. Transgenic plants might have shown more drought tolerance should the promoter used was an inducible one, such as *Arabidopsis* rd29 promoter [32].

4. Conclusions

GUS assay was essential to identify the relative location of the subepidermal area of explants as the target for Biolistic bombardment.

All plants transformed with the *bar* Liberty herbicide resistance gene showed stable expression of this gene because of continuous *in vitro* culture selections of explants, shootlets, and plantlets in media containing the active ingredient of this herbicide.

Our studies of transgenic P. *vulgaris* that expresses barley *HVA1* transgene agree with an earlier report [33] in which "Matterhorn" possesses a genotypic advantage over "Sedona" in terms of naturally tolerating drought.

The expression of barley *HVA1* gene in P. *vulgaris* resulting in drought tolerance agrees with results obtained from transfer of this gene into other crops and their tolerance to drought and/or salt [20–27].

Further studies are needed to locate the precise location of the subepidermal cell layer, possibly via the use of GUS monoclonal antibody followed by laser microscopy because GUS color easily diffuses from cell to cell.

Further studies are also needed to test *HVA1* transgenic P. *vulgaris* at the field level. The research presented here and the genes transferred into common bean varieties might improve the yield and economy of this important crop.

Acknowledgments

The authors wish to thank Prof. James Kelly of Michigan State University for availability of P. *vulgaris* seeds. The authors are appreciative of the generosity of past Prof. Ray Wu of Cornell University for the availability of pBY520 and pACT1F. Kingdom Kwapata was a Fulbright Scholar at Michigan State University.

References

[1] P. Delgado-Sánchez, M. Saucedo-Ruiz, S. H. Guzmán-Maldonado et al., "An organogenic plant regeneration system for common bean (*Phaseolus vulgaris* L.)," *Plant Science*, vol. 170, no. 4, pp. 822–827, 2006.

[2] F. J. L. Aragão, S. G. Ribeiro, L. M. G. Barros et al., "Transgenic beans (*Phaseolus vulgaris* L.) engineered to express viral antisense RNAs show delayed and attenuated symptoms to bean golden mosaic geminivirus," *Molecular Breeding*, vol. 4, no. 6, pp. 491–499, 1998.

[3] F. J. L. Aragão, G. R. Vianna, M. M. C. Albino, and E. L. Rech, "Transgenic dry bean tolerant to the herbicide glufosinate ammonium," *Crop Science*, vol. 42, no. 4, pp. 1298–1302, 2002.

[4] M. B. Sticklen and H. F. Oraby, "Invited review: shoot apical meristem: a sustainable explant for genetic transformation of cereal crops," *In Vitro Cellular and Developmental Biology*, vol. 41, no. 3, pp. 187–200, 2005.

[5] M. Veltcheva, D. Svetleva, S. Petkova, and A. Perl, "In vitro regeneration and genetic transformation of common bean (*Phaseolus vulgaris* L.)-problems and progress," *Scientia Horticulturae*, vol. 107, no. 1, pp. 2–10, 2005.

[6] Z. C. Liu, B. J. Park, A. Kanno, and T. Kameya, "The novel use of a combination of sonication and vacuum infiltration in Agrobacterium-mediated transformation of kidney bean (*Phaseolus vulgaris* L.) with lea gene," *Molecular Breeding*, vol. 16, no. 3, pp. 189–197, 2005.

[7] N. O. Amugune, B. Anyango, and T. K. Mukiama, "Agrobacterium-mediated transformation of common bean," *African Crop Science Journal*, vol. 19, no. 3, pp. 137–147, 2011.

[8] K. Bonfim, J. C. Faria, E. O. P. L. Nogueira, É. A. Mendes, and F. J. L. Aragão, "RNAi-mediated resistance to Bean golden mosaic virus in genetically engineered common bean (*Phaseolus vulgaris*)," *Molecular Plant-Microbe Interactions*, vol. 20, no. 6, pp. 717–726, 2007.

[9] F. J. L. Aragão and J. C. Faria, "First transgenic geminivirus-resistant plant in the field," *Nature Biotechnology*, vol. 27, no. 12, pp. 1086–1088, 2009.

[10] G. R. Vianna, M. M. C. Albino, B. B. A. Dias, L. D. M. Silva, E. L. Rech, and F. J. L. Aragão, "Fragment DNA as vector for genetic transformation of bean (*Phaseolus vulgaris* L.)," *Scientia Horticulturae*, vol. 99, no. 3-4, pp. 371–378, 2004.

[11] E. L. Rech, G. R. Vianna, and F. J. L. Aragão, "High-efficiency transformation by biolistics of soybean, common bean and cotton transgenic plants," *Nature Protocols*, vol. 3, no. 3, pp. 410–418, 2008.

[12] N. Colpaert, S. Tilleman, M. Van Montagu, G. Gheysen, and N. Terryn, "Composite *Phaseolus vulgaris* plants with transgenic roots as research tool," *African Journal of Biotechnology*, vol. 7, no. 4, pp. 404–408, 2008.

[13] J. Arellano, S. I. Fuentes, P. Castillo-España, and G. Hernández, "Regeneration of different cultivars of common bean (*Phaseolus vulgaris* L.) via indirect organogenesis," *Plant Cell, Tissue and Organ Culture*, vol. 96, no. 1, pp. 11–18, 2009.

[14] K. Kwapata, R. Sabzikar, M. B. Sticklen, and J. D. Kelly, "In vitro regeneration and morphogenesis studies in common bean," *Plant Cell, Tissue and Organ Culture*, vol. 100, no. 1, pp. 97–105, 2009.

[15] J. Tollefson, "Brazil cooks up transgenic bean," *Nature*, vol. 478, p. 168, 2011.

[16] M. Naderpour and I. E. Johansen, "Visualization of resistance responses in *Phaseolus vulgaris* using reporter tagged clones of Bean common mosaic virus," *Virus Research*, vol. 159, no. 1, pp. 1–8, 2011.

[17] J. K. Zhu, "Salt and drought stress signal transduction in plants," *Annual Review of Plant Biology*, vol. 53, pp. 247–273, 2002.

[18] B. Hong, R. Barg, and T. H. D. Ho, "Developmental and organ-specific expression of an ABA- and stress-induced protein in barley," *Plant Molecular Biology*, vol. 18, no. 4, pp. 663–674, 1992.

[19] G. Qian, Z. Han, T. Zhao, G. Deng, Z. Pan, and M. Yu, "Genotypic variability in sequence and expression of HVA1 gene in Tibetan hulless barley, Hordeum vulgare ssp. vulgare, associated with resistance to water deficit," *Australian Journal of Agricultural Research*, vol. 58, no. 5, pp. 425–431, 2007.

[20] X. Deping, X. Duan, B. Wang, B. Hong, T.-H. Ho, and R. Wu, "Expression of a late embryogenesis abundant protein gene, HVA1, from barley confers tolerance to water deficit and salt stress in transgenic rice," *Plant Physiology*, vol. 110, no. 1, pp. 249–257, 1996.

[21] E. Sivamani, A. Bahieldin, J. M. Wraith et al., "Improved biomass productivity and water use efficiency under water deficit conditions in transgenic wheat constitutively expressing the barley HVA1 gene," *Plant Science*, vol. 155, no. 1, pp. 1–9, 2000.

[22] A. Bahieldin, H. T. Mahfouz, H. F. Eissa et al., "Field evaluation of transgenic wheat plants stably expressing the HVA1 gene for drought tolerance," *Physiologia Plantarum*, vol. 123, no. 4, pp. 421–427, 2005.

[23] L. Zhang, A. Ohta, M. Takagi, and R. Imai, "Expression of plant group 2 and group 3 lea genes in *Saccharomyces cerevisiae* revealed functional divergence among LEA proteins," *Journal of Biochemistry*, vol. 127, no. 4, pp. 611–616, 2000.

[24] D. Fu, B. Huang, Y. Xiao, S. Muthukrishnan, and G. H. Liang, "Overexpression of barley hva1 gene in creeping bentgrass for improving drought tolerance," *Plant Cell Reports*, vol. 26, no. 4, pp. 467–477, 2007.

[25] S. Lal, V. Gulyani, and P. Khurana, "Overexpression of HVA1 gene from barley generates tolerance to salinity and water stress in transgenic mulberry (Morus indica)," *Transgenic Research*, vol. 17, no. 4, pp. 651–663, 2008.

[26] S. B. Maqbool, H. Zhong, Y. El-Maghraby et al., "Competence of oat (*Avena sativa* L.) shoot apical meristems for integrative transformation, inherited expression, and osmotic tolerance of transgenic lines containing hva1," *Theoretical and Applied Genetics*, vol. 105, no. 2-3, pp. 201–208, 2002.

[27] S. B. Maqbool, H. Zhong, H. F. Oraby, and M. B. Sticklen, "Transformation of oats and its application to improving osmotic stress tolerance," *Methods in Molecular Biology*, vol. 478, pp. 149–168, 2009.

[28] T. Murashige and F. Skoog, "A revised medium for rapid growth and bioassays with tobacco cultures," *Physiolgia Plantarum*, vol. 15, pp. 473–497, 1962.

[29] R. A. Jefferson, T. A. Kavanagh, and M. W. Bevan, "GUS fusions: beta-glucuronidase as a sensitive and versatile gene fusion marker in higher plants," *The EMBO Journal*, vol. 6, no. 13, pp. 3901–3907, 1987.

[30] M. A. Saghai-Maroof, K. M. Soliman, R. A. Jorgensen, and R. W. Allard, "Ribosomal DNA spacer-length polymorphisms in barley: mendelian inheritance, chromosomal location, and population dynamics," *Proceedings of the National Academy of Sciences of the United States of America*, vol. 81, no. 24, pp. 8014–8018, 1984.

[31] J. Sambrook, F. Fritsch, and T. Maniatis, *Molecular Cloning: A Laboratory Manual*, Cold Spring Harbor Laboratory, Cold Spring Harbor Laboratory Press, New York, NY, USA, 2nd edition, 1989.

[32] M. Kasuga, S. Miura, K. Shinozaki, and K. Yamaguchi-Shinozaki, "A combination of the Arabidopsis DREB1A gene and stress-inducible rd29A promoter improved drought- and low-temperature stress tolerance in tobacco by gene transfer," *Plant and Cell Physiology*, vol. 45, no. 3, pp. 346–350, 2004.

[33] S. P. Singh, "Drought resistance in the race Durango dry bean landraces and cultivars," *Agronomy Journal*, vol. 99, no. 5, pp. 1219–1225, 2007.

9

Line × Tester Mating Design Analysis for Grain Yield and Yield Related Traits in Bread Wheat (*Triticum aestivum* L.)

Zine El Abidine Fellahi,[1] **Abderrahmane Hannachi,**[1]
Hamenna Bouzerzour,[2] **and Ammar Boutekrabt**[3]

[1] National Institute of Agricultural Research, Setif Agricultural Research Unit, Setif 19000, Algeria
[2] Faculty of Life and Natural Sciences, Ecology and Plant Biology Department, University of Ferhat Abbas, Setif 1 19000, Algeria
[3] Faculty of Agro-Veterinary and Biological Sciences, Agronomy Department, University of Saad Dahlab, Blida 09000, Algeria

Correspondence should be addressed to Zine El Abidine Fellahi; zinou.agro@gmail.com

Academic Editor: David Clay

Nine bread wheat (*Triticum aestivum* L.) genotypes were crossed in a line × tester mating design. The 20 F_1's and their parents were evaluated in a randomized complete block design with three replications at the Field Crop Institute-Agricultural Experimental Station of Setif (Algeria) during the 2011/2012 cropping season. The results indicated that sufficient genetic variability was observed for all characters studied. $A_{899} \times$ Rmada, $A_{899} \times$ Wifak, and $A_{1135} \times$ Wifak hybrids had greater grain yield mean than the parents. A_{901} line and the tester Wifak were good combiners for the number of grains per spike. MD is a good combiner for 1000-kernel weight and number of fertile tillers. HD_{1220} is a good general combiner to reduce plant height; Rmada is a good general combiner to shorten the duration of the vegetative growth period. $A_{901} \times$ Wifak is a best specific combiner to reduce plant height, to increase 1000-kernel weight and number of grains per spike. AA × MD is a best specific combiner to reduce duration of the vegetative period, plant height and to increase the number of kernels per spike. $A_{899} \times$ Wifak showed the highest heterosis for grain yield, accompanied with positive heterosis for the number of fertile tillers and spike length, and negative heterosis for 1000-kernel weight and the number of days to heading. $\sigma_{gca}^2/\sigma_{sca}^2$, $(\sigma_D^2/\sigma_A^2)^{1/2}$ low ratios and low to intermediate estimates of h_{ns}^2 supported the involvement of both additive and nonadditive gene effects. The preponderance of non-additive type of gene actions clearly indicated that selection of superior plants should be postponed to later generation.

1. Introduction

Bread wheat (*Triticum aestivum* L.) is an important staple food in Algeria. This crop ranks third after durum wheat (*Triticum durum* Desf.) and barley (*Hordeum vulgare* L.), with a yearly cropped area of 0.8 million hectares, representing 24.2% of the 3.3 million hectares devoted to small grain cereals. Algeria imported 3.0 million tons of bread wheat in 2010/2011, to remedy the decline in the domestic production and to build stocks to meet the needs.

Increasing wheat production can be achieved by application of improved agronomic technics, developing and adopting high yielding varieties. Major emphasis, in breeding program, is put on the development of improved varieties with superior qualitative and quantitative traits and resilience to abiotic stresses. In fact, genetic improvement in bread wheat, having better tolerance against terminal heat and water stress, has a good promise to improve grain yield average and total wheat production.

However to breed high yielding varieties, breeders often face the problem of selecting parents and crosses. In this context various breeding approaches have been suggested. The line × tester analysis method introduced by Kempthorne [1] is one of the powerful tools available to estimate the combining ability effects and aids in selecting desirable parents and crosses for exploitation in pedigree breeding [2–4]. Performances *per se* do not necessarily reveal which parents are good or poor combiners. To surmount this difficulty, it is necessary to gather information on the nature of gene actions. General combining ability is attributed to additive type of gene effects, while specific combining ability is attributed to nonadditive type of gene actions. Nonadditive gene type of actions is not reliably fixable whereas additive type of gene actions or complementary type epistatic gene interactions are reliably fixable [5–7].

Heterosis estimates, for different morphological and yield related characters, are attributed to both additive and non-additive gene actions. Heritability gives information about genetic variation; it is useful for predicting the response to selection in the succeeding generations. Heritability is dependent upon the nature of gene action [8–10]. Better understanding of the underlying genetic control of important traits in bread wheat is useful in breeding for higher grain yield. Kamaluddin et al. [11] reported high contribution of general combining ability for genetic control of bread wheat characters. Kumar and Maloo [12] identified the best specific and general combiners that were efficient for breeding days to flowering and grain yield in bread wheat. Involvement of both additive and dominance gene actions was also reported for genetic control of heading time in wheat [7], grain yield [13], number of grains per spike, 1000-kernel weight [14], and fertile tillers per plant [15]. Ahmad et al. [16] reported that additive gene effect was important for days to heading. Khan and Habib [17] observed that grain weight was controlled by over dominance type of gene action.

The objectives of this study are to assess the combining ability, to determine the nature and magnitude of gene actions and to estimate heterosis and heritability for yield and yield-related traits in a line × tester mating design in bread wheat.

2. Materials and Methods

The present investigation was carried out at the Field Crop Institute-Agricultural Experimental Station of Setif (ITGC-AES, 36°12′N and 05°24′E, Algeria) during the 2010/2011 and 2011/2012 cropping seasons. The soil of the experimental site is silty clay, with $CaCO_3$ and organic matter contents of 35% and 1.35%, respectively. The experimental material comprises nine bread wheat (*Triticum aestivum* L.) genotypes. Five genotypes, $Acsad_{901}$, $Acsad_{899}$, $Acsad_{1135}$, $Acsad_{1069}$, and Ain Abid, were used as females, hereafter designated as lines; and four genotypes: Mahon Demias, Rmada, HD_{1220}, Wifak, designated as testers, were used as males. Mahon Demias is a genealogical selection from a land race introduced from Balearic Islands in the mid-forties of the past century. This cultivar is widely adapted to the arid and semiarid high plateaus of Algeria. HD_{1220} is a selection from CIMMYT segregating material; it was released as cultivar in the nineties. Drought tolerant and early maturing, this variety gained large acceptance from farmers due to its high yielding potential [18].

The nine parents were crossed to produce 20 F_1 hybrids according to the line × tester mating design developed by Kempthorne [1]. F_1 seeds were sown in the field, along with their parents, in a randomized complete block design with three replications. Each plot comprised one row of 2.5 m length with space of 30 cm between rows and seeds were placed 15 cm apart. Recommended cultural practices were followed to raise a good crop. Monoammonium phosphate

TABLE 1: ANOVA for line × tester analysis.

Source of variation	Degree of freedom (df)	Mean square
Replication (r)	$(r-1)$	
Genotypes (g)	$(g-1)$	MS_2
Parents (p)	$(p-1)$	
Parents versus crosses	1	
Crosses (c)	$(c-1)$	
Lines (l)	$(l-1)$	M_l
Testers (t)	$(t-1)$	M_t
Lines × testers	$(l-1)(t-1)$	$M_{l\times t}$
Error	$(r-1)(t-1)$	MS_1

Where MS_2, M_l, M_t, $M_{l\times t}$, and MS_1 were genotypic mean square, line mean square, tester mean square, line × tester mean square, and error mean square, respectively.

(52% P_2O_5 + 12% N) with 80 kg ha^{-1} was applied just before sowing and 75 kg ha^{-1} of Sulfate (26% N + 35% SO_3) was spread at tillering stage. Weeds were controlled by application of 12 g ha^{-1} of Granstar [*Methyl Triberunon*] herbicide mixed with water.

Five competitive plants (excluding border plants) were tagged before heading and data were recorded for the number of days to heading, plant height, spike length, number of fertile tillers per plant, number of grains per spike, 1000-kernel weight, and grain yield per plant. Data recorded were subjected to analysis of variance according to Steel and Torrie [19] to determine significant differences among genotypes. Combining ability effects are very effective genetic parameters in deciding the next phase of breeding programs. They were computed according to the line × tester method [20]. Line × tester analysis was performed as outlined in the format of ANOVA table given in Table 1.

The variances for general and specific combining ability were tested against their respective error variances, derived from the analysis of variance of the different traits as follows:

Covariance of half-sib of line

$$= \text{Cov.H.S. (line)} \tag{1}$$

$$= \frac{M_l - M_{l\times t}}{rt},$$

Covariance of half-sib of tester

$$= \text{Cov.H.S. (tester)} \tag{2}$$

$$= \frac{M_t - M_{l\times t}}{rl},$$

Covariance of full sib

$$= \text{Cov.F.S.}$$

$$= \frac{(M_l - M_e) + (M_t - M_e) + (M_{l\times t} - M_e)}{3r} \tag{3}$$

$$+ \frac{6r\text{Cov.H.S.} - r(l+t)\text{Cov.H.S.}}{3r}.$$

TABLE 2: Analysis of variance for combining ability effects of different bread wheat characters.

Source	df	DHE	PHT	SL	FT	TKW	NG	GY
Rep	2	5.7	54.1	0.3	13	12	52.3	62
Gen	28	15.6*	226.8*	2.9*	23.4*	33.2**	152.8**	39.7*
Par (P)	8	26.4*	348.0*	2.6*	29.9*	63.1**	279.1**	42.7*
Crosses (C)	19	10.7*	171.8*	2.5*	19.6*	22.3**	107.6**	33.8*
P versus C	1	21.4*	301.4*	11.9*	44.5*	0.9^{ns}	2.3^{ns}	129.0*
Lines (L)	4	21.2*	89.0*	3.3*	28.1*	26.8**	120.5*	65.5*
Testers (L)	3	39.5*	751.2*	2.5*	37.2*	91.1**	357.7**	25.8*
L versus T	1	7.6^{ns}	174.7*	0.3^{ns}	15.1*	45.3*	267.6*	2.2^{ns}
L × T	12	4.2^{ns}	46.8*	1.9*	10.0*	3.6^{ns}	40.8*	25.6*
Error	56	2.5	24.1	0.7	3.4	6	11.1	6.5

DHE: number of days to heading, PHT: plant height (cm), SL: spike length (cm), FT: number of fertile tillers, TKW: thousand-kernel weight (g), NG: number of grains per spike, GY: grain yield (g), ns, * and **: non-significant and significant effect at 0.05 and 0.01 probability, P: parents, C: crosses, L: lines, and T: testers.

While Cov. H.S. (average) was calculated by the formula

Cov.H.S. (average)

$$= \frac{1}{r(2lt - l - t)} \left[\frac{(l-1)(M_l) + (t-1)(M_t)}{l + t - 2} - M_{l \times t} \right]. \tag{4}$$

Assuming no epistasis, variance due to GCA (σ_{gca}^2) and variance due to SCA (σ_{sca}^2) were calculated as follows:

$$\sigma_{gca}^2 = \text{Cov.H.S.} = \left(\frac{1+F}{4} \right) \sigma_A^2$$

$$\sigma_{sca}^2 = \left(\frac{1+F}{2} \right)^2 \sigma_D^2. \tag{5}$$

Additive and dominance genetic variances (σ_A^2 and σ_D^2) were calculated by taking inbreeding coefficient (F) equal to one; that is, $F = 1$ because both lines and testers were inbred.

Significance test for general combining ability and specific combining ability effects were performed using t-test. Mid-parent heterosis (H_{PM}) is defined as the increased vigor of the F_1 over the mean of the parents. It was estimated from mean values and its significance was performed using t-test [21]. Narrow sense heritability was estimated, after derivation of the variance components [20]. ($\sigma_{gca}^2/\sigma_{sca}^2$), and ($\sigma_D^2/\sigma_A^2$)$^{1/2}$ ratios were used to rate the relative weight of additive versus nonadditive type of gene actions [22].

3. Results and Discussion

3.1. Genetic Variability among Parents and Hybrids. The analysis of variance revealed significant genotype effect for all the characters under study. This provides evidence of the presence of sufficient genetic variability among lines, testers, and hybrids and allows further assessment of general combining ability analysis (Table 2). Differences between the extreme mean values for the measured traits were 8.7 days, 36.6 cm, 3.4 cm, 10.4 tillers, 15.3 g, 33.7 grains, and 16.1 g for

days to heading, plant height, spike length, fertile tillers, 1000-kernel weight, grains per spike, and grain yield, respectively. These differences were 3 to 5 times higher than the LSD0.05 values. Parents and crosses showed significant effects for all traits. Mean square of the contrast "parents versus crosses" was significant for days to heading, plant height, spike length, fertile tillers, and grain yield and nonsignificant for 1000-kernel weight and number of grains per spike (Table 2). The differences between overall mean of parents and that of hybrids indicated that hybrids were 1.1 days earlier, 4.0 cm taller and had more effective tillers and a grain yield advantage of 2.7 g. Parents and hybrids showed similar averages for 1000-kernel weight and number of grains per spike (Table 3). Line and tester effects were significant for all traits (Table 2).

Among lines, the differences, between the extreme mean values for the measured traits, were 5.4 days, 14.7 cm, 2.5 cm, 7.2 tillers, 8.9 g, 15.6 grains, and 11.2 g for days to heading, plant height, spike length, fertile tillers, 1000-kernel weight, grains per spike, and grain yield, respectively (Table 3). The best grain yielding line is A_{1135} which is the tallest and had also the highest average for the number of fertile tillers and 1000-kernel weight (Table 3). A_{1069} was the earliest with an average of 132.3 days, while Ain Abid (AA) had the longest and most fertile spike (Table 3). Among testers, the differences, between the extreme mean values for the characters under study, were 7.3 days, 36.6 cm, 2.2 cm, 8.4 tillers, 11.9 g, 28.7 grains, and 6.9 g for days to heading, plant height, spike length, fertile tillers, 1000-kernel weight, grains per spike, and grain yield, respectively (Table 3). The best grain yielding tester was Rmada (20.4 g) which exhibited also the longest spike (13.6 cm). Wifak was the earliest with an average of 132.7 days, while Mahon Demias (MD) was the tallest with 99.8 cm and presented the highest mean for the number of fertile tillers and 1000-kernel weight. HD_{1220} expressed the best average of the number of grains per spike (Table 3). Compared to testers, lines showed shorter plant height (−5.1 cm), lighter 1000-kernel weight (−2.6 g), and higher number of grains per spike (+6.3 grains). The interaction lines × testers were significant for plant height, spike length, fertile tillers, number of grains per spike, and

TABLE 3: Means of the measured characters for 9 bread wheat parents and their 20 F_1 hybrids.

	DHE	PHT	SL	FT	TKW	NG	GY
Lines							
A_{901}	133.0^{efg}	66.2^{jk}	11.8^{hij}	7.8^{h}	23.9^{i}	55.1^{ab}	10.5^{i}
A_{899}	137.3^{b}	69.6^{hijk}	11.5^{ij}	13.7^{abcdef}	$32.7^{abcdefgh}$	44.9^{cdef}	$20.0^{abcdefgh}$
A_{1135}	132.7^{efg}	80.9^{bcdef}	13.5^{abcde}	15.0^{abcd}	$32.8^{abcdefg}$	44.4^{def}	$21.7^{abcdefg}$
A_{1069}	132.3^{efg}	73.1^{efghij}	12.7^{efghij}	$11.8^{bcdefgh}$	$32.4^{abcdefgh}$	49.9^{bcdef}	$18.8^{abcdefgh}$
AA	137.7^{ab}	71.6^{fghijk}	14.0^{abcde}	8.8^{efgh}	26.1^{ghi}	60.0^{a}	13.7^{ghi}
Testers							
MD	140.0^{a}	99.8^{a}	11.4^{j}	16.8^{ab}	39.2^{a}	26.3^{h}	$17.3^{bcdefghi}$
Rmada	132.7^{efg}	76.0^{efghi}	13.6^{abcde}	14.1^{abcd}	$29.5^{abcdefgh}$	47.6^{bcdef}	$20.4^{abcdefgh}$
HD_{1220}	137.3^{b}	63.2^{k}	12.2^{fghij}	$12.4^{bcdefgh}$	$27.3^{abcdefgh}$	55.0^{ab}	$18.8^{abcdefgh}$
Wifak	132.7^{efg}	70.6^{ghijk}	12.8^{defghi}	8.4^{fgh}	32.8^{hi}	49.3^{bcdef}	13.5^{hi}
Lines × testers							
A_{901} × MD	134.3^{def}	87.6^{bc}	13.2^{bcdefg}	14.7^{abcd}	32.1^{cdef}	42.6^{efg}	$20.2^{abcdefgh}$
A_{901} × Rmada	132.7^{efg}	75.2^{fghij}	13.5^{abcdef}	14.9^{abcd}	27.0^{hij}	51.8^{abcd}	$20.6^{abcdefgh}$
A_{901} × HD_{1220}	134.7^{cde}	69.8^{hijk}	11.9^{ghij}	10.9^{cdefgh}	25.0^{j}	50.6^{bcdef}	14.6^{fghi}
A_{901} × Wifak	133.7^{defg}	67.1^{ijk}	11.9^{ghij}	8.3^{gh}	29.3^{fghij}	60.0^{a}	15.4^{efghi}
A_{899} × MD	137.0^{bc}	89.6^{b}	13.1^{cdefg}	17.7^{a}	33.1^{bcde}	34.9^{gh}	$20.4^{abcdefgh}$
A_{899} × Rmada	132.0^{fg}	75.8^{efghi}	14.1^{abcd}	16.5^{ab}	29.7^{defghi}	49.3^{bcdef}	24.2^{abc}
A_{899} × HD_{1220}	135.7^{bcd}	70.6^{ghijk}	11.8^{ghij}	13.9^{abcde}	29.4^{efghi}	50.2^{bcdef}	$20.2^{abcdefgh}$
A_{899} × Wifak	132.3^{efg}	78.0^{cdefgh}	14.0^{abcde}	18.2^{a}	29.5^{efghi}	50.5^{bcdef}	26.6^{a}
A_{1135} × MD	138.0^{ab}	82.2^{bcde}	11.6^{ij}	14.0^{abcde}	35.2^{bc}	34.7^{gh}	$17.3^{bcdefghi}$
A_{1135} × Rmada	132.0^{fg}	80.3^{bcdef}	14.1^{abc}	$13.3^{abcdefg}$	30.7^{defgh}	47.5^{bcdef}	$19.4^{abcdefgh}$
A_{1135} × HD_{1220}	133.7^{defg}	74.5^{efghij}	14.0^{abcde}	15.1^{abcd}	32.0^{cdef}	47.8^{bcdef}	22.7^{abcde}
A_{1135} × Wifak	131.3^{g}	80.0^{cdefg}	14.5^{ab}	16.1^{abc}	31.4^{cdefg}	48.0^{bcdef}	24.3^{ab}
A_{1069} × MD	134.3^{def}	99.2^{a}	13.3^{bcdef}	15.1^{abcd}	36.8^{ab}	42.4^{fg}	23.6^{abcd}
A_{1069} × Rmada	132.0^{fg}	77.7^{defgh}	13.7^{abcde}	13.9^{abcde}	29.9^{defghi}	53.7^{abc}	22.2^{abcdef}
A_{1069} × HD_{1220}	133.7^{defg}	74.0^{efghij}	13.4^{bcdef}	$12.1^{bcdefgh}$	29.0^{fghi}	50.2^{bcdef}	$17.4^{bcdefghi}$
A_{1069} × Wifak	132.0^{fg}	78.4^{cdefgh}	14.0^{abcde}	13.7^{abcde}	30.3^{defgh}	48.1^{bcdef}	$19.7^{abcdefgh}$
AA × MD	135.7^{bcd}	86.1^{bcd}	13.8^{abcde}	$12.1^{bcdefgh}$	33.6^{bcd}	51.5^{abcdef}	$20.0^{abcdefgh}$
AA × Rmada	134.0^{def}	77.5^{defgh}	14.8^{a}	10.5^{defgh}	29.7^{defghi}	52.2^{abcd}	$16.2^{cdefghi}$
AA × HD_{1220}	137.0^{bc}	71.7^{fghijk}	13.9^{abcde}	11.1^{cdefgh}	27.8^{ghij}	52.1^{abcd}	15.7^{defghi}
AA × Wifak	134.0^{def}	76.5^{efghi}	13.6^{abcde}	10.8^{cdefgh}	30.1^{defghi}	51.6^{abcde}	$16.6^{bcdefghi}$
Over all mean	134.3	77.3	13.2	13.2	30.6	48.4	19.0
LSD0.05	1.9	6.9	1.0	2.2	2.9	6.6	3.0

DHE: number of days to heading, PHT: plant height (cm), SL: spike length (cm), FT: number of fertile tillers, TKW: thousand-kernel weight (g), NG: number of grains per spike, and GY: grain yield (g).
Means within each column followed by the same letter are not significantly different from each other based on the 0.05 probability level of LSD.

grain yield, suggesting that hybrids perform better than the parents for these traits (Table 2).

Mean values of the hybrids were within the limits of the means of the parents for the number of days to heading, plant height, spike length, and number of grain per spike. For grain yield, A_{899} × Rmada, A_{899} × Wifak and A_{1135} × Wifak cross-combinations presented higher mean values than the parents (Table 3). Genetic variability and mean performance of parents and hybrids are important criteria for genotypic evaluation; however, the parents with high mean value may not transmit this characteristic to their hybrids. These parental and hybrid abilities are estimated in terms of general combining ability (GCA) and specific combining ability (SCA) effects.

3.2. General and Specific Combining Ability Effects. The significance of mean squares, due to lines and testers, for the number of days to heading and 1000-kernel weight suggested the prevalence of additive genetic effects for these traits. While the simultaneous significance of mean squares due to lines, testers and lines × testers for plant height, spike length, fertile tillers, number of grains per spike and grain yield indicated that both additive and nonadditive type of gene action were involved in the genetic control of these characters. A_{901}, A_{899}, and AA lines and the tester HD_{1220} exhibited significant but negative GCA for grain yield. A_{901} line and the tester Wifak presented significant and positive GCA effects for the number of grains per spike (Table 4). MD is the best combiner for 1000-kernel weight, number

TABLE 4: General combining ability (gi) effects for characters in bread wheat parents.

Parents	DHE	PHT	SL	FT	TKW	NG	GY
Lines							
A_{901}	−0.25	−5.53**	−1.20**	−2.17*	−3.34**	4.17**	−3.26*
A_{899}	0.38	−0.12	−0.23	4.41**	−0.23	−3.41*	−4.47**
A_{1135}	−0.38	1	0.2	1.47	2.64**	−5.98**	1.61
A_{1069}	−1.50**	5.64**	0.28	0.09	1.34*	0.16	1.28
AA	1.75**	−0.99	0.95**	−3.80**	−0.4	5.05**	−4.11**
se (g_l)	0.45	1.64	0.23	0.9	0.61	0.45	1.35
Testers							
MD	2.80**	15.52**	−0.61**	1.64*	5.38**	−10.9**	0.63
Rmada	−2.20**	−1.92	0.96**	0.27	−1.76**	3.6**	0.97
HD_{1220}	1.40**	−9.74**	−0.63**	−1.58*	−2.92**	2.55	−2.58*
Wifak	−2.00**	−3.86**	0.28	−0.33	−0.71	4.7**	0.98
se (g_t)	0.39	1.42	0.2	0.78	0.53	1.29	1.17

DHE: number of days to heading, PHT: plant height (cm), SL: spike length (cm), FT: number of fertile tillers, TKW: thousand-kernel weight (g), NG: number of grains per spike, GY: grain yield (g), se (g_l): standard error for GCA effects for line, se (g_t): standard error for GCA effects for tester, ns, and * and **: nonsignificant and significant effect at 0.05 and 0.01 probability.

of fertile tillers, and the number of days to heading, with GCA effect of 5.38, 1.64, and 2.80, respectively. HD_{1220} is a good general combiner to reduce plant height (Table 4). Significant negative GCA effect for plant height is useful for the development of dwarf plant material. The tester Rmada had significant and negative GCA for the number of days to heading. This tester is a good general combiner to shorten the duration of the vegetative growth period.

Even though SCA effects do not contribute tangibly in the improvement of self-pollinated crops, except in situations where exploitation of heterosis is feasible, best hybrids are expected to generate transgressive segregants which could be selected as potent homozygous lines. $A_{901} \times MD$ hybrid exhibited significant SCA effects, simultaneously, for the number days to heading and spike length. This cross-combination is best suited to select among its offspring's the earliest ones, bearing long spike (Table 5). According to Kenga et al. [23], cross-combinations with high means, favorable SCA estimates and involving at least one of the parents with high GCA would likely enhance the concentration of favorable alleles to improve target traits. In the present study, it is worth noting that A_{901} presented a significant and negative GCA effect for spike length while MD presented significant GCA, negative for spike length and positive for days to heading (Table 4). The positive SCA effect for spike length of this cross-combination resulted from parents having both significant and negative GCA. One parent of this hybrid presented a positive and significant GCA for days to heading while the hybrid exhibited a significant and negative SCA for this trait (Tables 4 and 5). $A_{901} \times Rmada$ presented a significant and positive SCA effect for the number of fertile tillers per plant. One parent of this cross-combination, the line A_{901}, had a significant and negative GCA for the number of fertile tillers. $A_{901} \times Wifak$ exhibited significant SCA effects, negative for plant height, spike length, fertile tillers, and positive for 1000-kernel weight and number of grains per spike (Table 5). Both parents, A_{901} and Wifak, of this

cross presented significant GCA for the number of grains per spike and plant height; and at least one parent had significant GCA effect for days to heading (Wifak), spike length, and the number of fertile tillers (A_{901}) (Table 4). This hybrid is a best specific combiner to reduce plant height and to increase 1000-kernel weight and number of grains per spike. However this cross-combination presents the disadvantage to reduce the number of fertile tillers, which is a strong determinant of grain yield, under semiarid growth conditions [24].

$A_{1069} \times MD$ cross-combination has significant and positive SCA for plant height and 1000-kernel weight. This cross is a best specific combiner to increase plant height and 1000-kernel weight. AA × MD presented significant and negative SCA effect for days to heading and plant height and significant and positive SCA effect for the number of kernels per spike. This cross is a best specific combiner to reduce duration of the vegetative period, plant height and to increase the number of kernels per spike (Table 5). None of the hybrids exhibited significant SCA effect for grain yield, suggesting even though the parents varied widely for grain yield, they generated hybrids with grain yield averages within the grain yield limits of the parents.

3.3. Gene Action, Degree of Dominance, Heterosis, Heritability, and Contribution to the Total Variance.

The variance due to general combining ability (σ^2_{gca}) was lower than specific combining ability variance (σ^2_{sca}) for all traits studied, suggesting the preponderance of nonadditive gene action controlling these characters (Table 6). Dominance genetic variance was larger than additive genetic variance for all traits. These results are supported by ratio of variance of general to specific combining ability ($\sigma^2_{gca}/\sigma^2_{sca}$) which was smaller than unity and by the degree of dominance $(\sigma^2_D/\sigma^2_A)^{1/2}$ which takes values greater than unity (Table 6). Therefore, it appeared that the inheritance of all the studied characters was controlled by a preponderance of nonadditive gene effects. Such type

TABLE 5: Specific combining ability (sij) effects for characters in bread wheat hybrids.

Crosses	DHE	PHT	SL	FT	TKW	NG	GY
$A_{901} \times$ MD	-2.05^*	3.5	1.44^{**}	2.17	0.32	-2.05	3.1
$A_{901} \times$ Rmada	0.45	2.32	0.36	3.74^*	-0.33	-2.81	3.39
$A_{901} \times HD_{1220}$	-0.15	2.07	-0.49	-0.41	-2.14^*	-3.51	-2
$A_{901} \times$ Wifak	1.75	-7.89^{**}	-1.31^{**}	-5.50^{**}	2.14^*	8.36^{**}	-4.5
$A_{899} \times$ MD	1.32	1.15	0.41	0.09	-1.33	-6.11^*	-4.31
$A_{899} \times$ Rmada	-1.17	-2.14	0.31	-0.34	0.74	0.94	1
$A_{899} \times HD_{1220}$	0.72	-2.17	-1.50^{**}	-2.5	1.32	3.5	-1.35
$A_{899} \times$ Wifak	-0.87	3.17	0.77	2.76	-0.74	1.68	4.67
$A_{1135} \times$ MD	3.57^{**}	-11.14^{**}	-2.30^{**}	-2.57	-1.11	-3.84	3.7
$A_{1135} \times$ Rmada	-0.42	3.56	-0.06	-2.2	-0.64	0.9	1.22
$A_{1135} \times HD_{1220}$	-1.52	2.58	-1.26	2.26	2.44	2.39	-2.43
$A_{1135} \times$ Wifak	-1.62	5	-1.11	2.5	-0.69	0.55	-2.49
$A_{1069} \times$ MD	-0.8	9.79^{**}	0.23	0.52	2.56^*	1.59	6.1
$A_{1069} \times$ Rmada	0.7	-4.97	-0.82^*	-0.02	-0.65	4.04	1.37
$A_{1069} \times HD_{1220}$	-0.4	-2.81	0.29	-0.87	-0.8	-0.16	-3.83
$A_{1069} \times$ Wifak	0.5	-2.01	0.29	0.38	-1.11	-5.47^*	-3.64
AA × MD	-2.05^*	-3.30^{**}	0.22	-0.2	-0.45	10.41^{**}	3.62
AA × Rmada	0.45	1.23	0.2	-1.18	0.88	-3.08	-2.3
AA × HD_{1220}	1.35	0.34	0.44	1.52	-0.82	-2.21	0.49
AA × Wifak	0.25	1.73	-0.86^*	-0.13	0.39	-5.12	-1.8
Se (sij)	0.77	2.84	0.39	2.84	1.06	2.58	2.34

DHE: number of days to heading, PHT: plant height (cm), SL: spike length (cm), FT: number of fertile tillers, TKW: thousand-kernel weight (g), NG: number of grains per spike, GY: grain yield (g), se (sij): standard error for SCA effects for crosses, ns, and ∗ and ∗∗: non-significant and significant effect at 0.05 and 0.01 probability.

TABLE 6: Estimates of genetic components for the measured characters in bread wheat.

	DHE	PHT	SL	FT	TKW	NG	GY
σ^2_{gca}	1.93	27.65	0.07	1.68	4.10	14.69	1.49
σ^2_{sca}	12.17	173.48	0.84	12.27	23.80	98.03	15.28
σ^2_A	3.87	55.30	0.15	3.36	8.20	29.38	2.97
σ^2_D	12.17	173.48	0.84	12.27	23.80	98.03	15.28
$\sigma^2_{gca}/\sigma^2_{sca}$	0.16	0.16	0.09	0.14	0.17	0.15	0.10
$[\sigma^2_D/\sigma^2_A]^{1/2}$	1.77	1.77	2.39	1.91	1.70	1.83	2.27
h^2_{ns}	56.30	60.30	10.30	33.30	54.40	50.40	16.90

σ^2_A: additive genetic variance, σ^2_D: dominance genetic variance, h^2_{ns}: narrow sense heritability, σ^2_{gca}: estimate of GCA variance, σ^2_{sca}: estimate of SCA variance, and $\sigma^2_{gca}/\sigma^2_{sca}$: average degree of dominance. DHE: number of days to heading, PHT: plant height (cm), SL: spike length (cm), FT: number of fertile tillers, TKW: thousand-kernel weight (g), NG: number of grains per spike, and GY: grain yield (g).

of gene action clearly indicated that selection of superior plants, in terms of grain yield, plant height, fertile tillers, and duration of the vegetative growth period should be postponed to later generation, where these traits can be improved by making selections among the recombinants within the segregating populations.

Selection efficiency is related to the magnitude of heritability. In this study, low estimates of narrow-sense heritability were observed for grain yield and spike length, intermediate for the number of fertile tillers per plant, and high for the number of days to heading, plant height, 1000-kernel weight, and number of grains per spike (Table 6).

Heterosis is the process by which the performance of an F_1 is superior to that of the mean of the crossed parents. Nine hybrids among 20 exhibited a significant mid-parent heterosis for grain yield. Besides, heterosis for grain yield, $A_{901} \times$ MD, and $A_{899} \times$ Rmada expressed significant heterosis for the number of fertile tillers, spike length and the number of days to heading (Table 7). $A_{899} \times$ Wifak presented the highest heterosis for grain yield, accompanied with positive heterosis for the number of fertile tillers and spike length and negative heterosis for 1000-kernel weight and the number of days to heading (Table 7). Besides heterosis for grain yield, AA × MD hybrid exhibited significant and positive heterosis for

TABLE 7: Significant mid-parent heterosis (%) for seven traits in bread wheat genotypes.

Crosses	DHE	PHT	SL	FT	TKW	NG	GY
A_{901} × MD	−1.6		13.8	19.5			45.3
A_{901} × Rmada				36.1			33.3
A_{901} × HD_{1220}							
A_{901} × Wifak						14.9	28.3
A_{899} × MD			14.4	16.1			
A_{899} × Rmada	−2.2		12.4	18.7			19.8
A_{899} × HD_{1220}							
A_{899} × Wifak	−2.0	11.3	15.2	64.7	−9.9		58.8
A_{1135} × MD		−9.0					
A_{1135} × Rmada							
A_{1135} × HD_{1220}							
A_{1135} × Wifak			10.3	37.6			38.1
A_{1069} × MD		14.7	10.4			11.3	30.7
A_{1069} × Rmada						10.2	
A_{1069} × HD_{1220}							
A_{1069} × Wifak			9.8	35.6			22.0
AA × MD	−2.3		8.7			19.4	29.0
AA × Rmada							
AA × HD_{1220}						−9.4	
AA × Wifak							

DHE: number of days to heading, PHT: plant height (cm), SL: spike length (cm), FT: number of fertile tillers, TKW: thousand-kernel weight (g), NG: number of grains per spike, and GY: grain yield (g).

TABLE 8: Proportional (%) contribution of lines, testers, and lines × testers to total hybrids variation in bread wheat.

	DHE	PHT	SL	FT	TKW	NG	GY
Lines (L)	33.43	11.23	30.34	32.67	25.30	23.58	40.80
Testers (T)	46.71	71.06	17.24	32.44	64.50	52.49	12.05
L × T	19.87	17.71	52.41	34.88	10.20	23.95	47.84

DHE: number of days to heading, PHT: plant height (cm), SL: spike length (cm), FT: number of fertile tillers, TKW: thousand-kernel weight (g), NG: number of grains per spike, and GY: grain yield (g).

the number of grains per spike, spike length, and negative heterosis for the number of days to heading (Table 7). A_{1069} × MD hybrid exhibited significant and positive heterosis for the number of grains per spike, spike length, and plant height (Table 7).

Testers contributed more to the total sum square for number of days to heading, plant height, thousand kernel weight, and number of grains per spike. The contribution of lines was lower compared to the testers and lines × testers interaction for all traits under study. All three sources of variation contributed equally for the number of fertile tillers per plant. Contribution of line × tester was slightly greater than that of testers and lines for grain yield and spike length (Table 8). These results showed that testers and the interaction lines × testers brought much variation in the expression of the studied traits. The results of the present study revealed large variation between parents and hybrids for the seven traits under study. Compared to the parents, hybrids were earlier, taller, bearing more effective tillers, and had higher grain yielding. A_{1135} line and Rmada tester were high grain yielding. Compared to testers, lines were shorter and had low

1000-kernel weight and higher number of grains per spike. Mean values of the hybrids were within the limits of the parental means for the number of days to heading, plant height, spike length, and number of grain per spike. A_{899} × Rmada, A_{899} × Wifak, and A_{1135} × Wifak cross-combinations exhibited higher mean values for grain yield than the parents.

Concomitant significance of mean squares due to lines, testers, and lines × testers for plant height, spike length, fertile tillers, number of grains per spike, and grain yield suggested that both additive and nonadditive types of gene actions were involved in the genetic control of the characters. A_{901} line and the tester Wifak were good combiners for the number of grains per spike. MD is a good combiner for 1000-kernel weight and number of fertile tillers. HD_{1220} is a good general combiner to reduce plant height, while the tester Rmada is a good general combiner to shorten the duration of the vegetative growth period.

A_{901} × MD hybrid exhibited significant SCA effects, simultaneously, for the number of days to heading and spike length. This cross-combination is best suited and offers the opportunity to select, among the progenies, early plant with long spike. This cross-combination resulted from L × L parents for spike length and H × L for days to heading. Verma and Srivastava [22] mentioned that positive SCA effect was usually associated with crosses where at least one parent was a good general combiner. According to Singh et al. [25] the desirable performance of combination like H × L may be ascribed to the interaction between dominant alleles from good combiners and recessive alleles from poor combiners. A_{901} × Rmada presented a significant and positive SCA effect for the number of fertile tillers per plant. One parent of this cross-combination had significant and negative GCA for the number of fertile tillers. A_{901} × Wifak exhibited significant SCA effects, negative for plant height, spike length, fertile tillers, and positive for 1000-kernel weight and number of grains per spike. Both parents of this cross presented significant GCA for the number of grains per spike and plant height; and at least one parent had significant GCA effect for days to heading, spike length, and the number of fertile tillers. This hybrid is a best specific combiner to reduce plant height and to increase 1000-kernel weight and number of grains per spike. A_{1069} × MD cross-combination is a best specific combiner to increase plant height and 1000-kernel weight. AA × MD is a best specific combiner to reduce duration of the vegetative period, plant height and to increase the number of kernels per spike. None of the hybrids exhibited significant SCA effect for grain yield, suggesting even though the parents varied widely for grain yield, they generated hybrids with grain yield average within the grain yield limits of the parents. Tiwari et al. [26] mentioned that hybrid combinations, where at least one parent is a good general combiner, could be used to developing high yielding pure lines due to presence of additive gene action, even if these crosses showed nonsignificant SCA effects though. In this study A_{901}, A_{899}, AA, and HD_{1220} were poor general combiners for grain yield. σ^2_{gca} was lower than σ^2_{sca} and σ^2_D was larger than σ^2_A for all traits, suggesting the preponderance of nonadditive gene action. These results are supported by $\sigma^2_{gca}/\sigma^2_{sca}$ ratio which was smaller than unity and by $(\sigma^2_D/\sigma^2_A)^{1/2}$ ratio which takes values greater than unity. Premlatha et al. [27] reported the importance of nonadditive gene action for plant height and grain yield. Gnanasekaran et al. [28] reported nonadditive gene action for seed weight and plant height, while Sharma [29] reported an additive gene effect. Similar results of predominance of σ^2_{sca} variance over σ^2_{gca} have been reported by Verma et al. [30] for barley. The results of this study do not corroborated findings reported by Borghi et al. [31] and Borghi and Perenzin [32] who observed that σ^2_{gca} was of greater importance than σ^2_{sca} for majority of characters. Lucken [33] noted that nonadditive σ^2_{sca} is best expressed in space planting. The difference in the results of various pieces of research may be attributed to differences of breeding material and to genotype × environments. Betrán et al. [34] observed significant interactions for combining abilities under low and high nitrogen in maize. The preponderance of nonadditive type of gene actions clearly indicated that selection of superior plants, in terms of grain yield, plant height, fertile tillers, and duration of the vegetative growth period, should be postponed to later generation.

Low estimates of h^2_{ns} were observed for grain yield and spike length, intermediate for the number of fertile tillers per plant, and high for the number of days to heading, plant height, 1000-kernel weight, and number of grains per spike. This supported the involvement of both additive and nonadditive gene effects. Medium to high narrow sense heritability estimates were reported by Yadav et al. [35], for different traits. Nine hybrids among 20 exhibited a significant mid-parent heterosis for grain yield. Besides, heterosis for grain yield, A_{901} × MD and A_{899} × Rmada, expressed significant heterosis for the number of fertile tillers, spike length, and the number of days to heading. A_{899} × Wifak showed the highest heterosis for grain yield, accompanied with positive heterosis for the number of fertile tillers and spike length and negative heterosis for 1000-kernel weight and the number of days to heading. AA × MD hybrid exhibited significant and positive heterosis for the number of grains per spike, spike length, and negative heterosis for the number of days to heading. The results indicated that testers and the interaction lines × testers contributed more to the variation in the expression of the studied traits.

4. Conclusion

A_{899} × Rmada, A_{899} × Wifak, and A_{1135} × Wifak hybrids had greater grain yield mean than the parents. $\sigma^2_{gca}/\sigma^2_{sca}$, $(\sigma^2_D/\sigma^2_A)^{1/2}$ low ratios and low to intermediate estimates of h^2_{ns} supported the involvement of both additive and nonadditive gene effects with preponderance of nonadditive type of gene actions. The testers and the interaction lines × testers contributed more to the variation of the expression of the different traits. A_{901} line and the tester Wifak were good combiners for the number of grains per spike. MD is a good combiner for 1000-kernel weight and number of fertile tillers. HD_{1220} is a good general combiner to reduce plant height; Rmada is a good general combiner to shorten the duration of the vegetative growth period. A_{901} × Wifak is a best specific combiner to reduce plant height, to increase 1000-kernel weight and number of grains per spike. AA × MD is a best specific combiner to reduce duration of the vegetative period, plant height and to increase the number of kernels per spike. A_{899} × Wifak showed the highest heterosis for grain yield, accompanied with positive heterosis for the number of fertile tillers and spike length and negative heterosis for 1000-kernel weight and the number of days to heading. The preponderance of nonadditive type of gene actions clearly indicated that selection of superior plants should be postponed to later generations.

References

[1] O. Kempthorne, *An Introduction to Genetic Statistics*, John Wiley & Sons, New York, NY, USA, 1957.

[2] M. Rashid, A. A. Cheema, and M. Ashraf, "Line x tester analysis in basmati rice," *Pakistan Journal of Botany*, vol. 39, no. 6, pp. 2035–2042, 2007.

[3] S. Basbag, R. Ekinci, and O. Gencer, "Combining ability and heterosis for earliness characters in line x tester population of

Gossypium hirsutum L," *Hereditas*, vol. 144, no. 5, pp. 185–190, 2007.

[4] S. K. Jain and E. V. D. Sastry, "Heterosis and combining ability for grain yield and its contributing traits in bread wheat (*Triticum aestivum* L.)," *RRJAAS*, vol. 1, pp. 17–22, 2012.

[5] B. Xiang and B. Li, "A new mixed analytical method for genetic analysis of diallel data," *The Canadian Journal of Forest Research*, vol. 31, no. 12, pp. 2252–2259, 2001.

[6] W. Yan and L. A. Hunt, "Biplot analysis of diallel data," *Crop Science*, vol. 42, no. 1, pp. 21–30, 2002.

[7] M. Iqbal, A. Navabi, D. F. Salmon et al., "Genetic analysis of flowering and maturity time in high latitude spring wheat: genetic analysis of earliness in spring wheat," *Euphytica*, vol. 154, no. 1-2, pp. 207–218, 2007.

[8] P. G. Swati and B. R. Ramesh, "The nature and divergence in relation to yield traits in rice germplasm," *Annals of Agricultural Research*, vol. 25, pp. 598–5602, 2004.

[9] Z. Hasnain, G. Abbas, A. Saeed, A. Shakeel, A. Muhammad, and M. A. Rahim, "Combining ability for plant height and yield related traits in wheat (*Triticum aestivum* L.)," *Journal of Agricultural Research*, vol. 44, pp. 167–1175, 2006.

[10] M. A. Chowdhary, M. Sajad, and M. I. Ashraf, "Analysis on combining ability of metric traits in bread wheat (*Triticum aestivum* L.)," *Journal of Agricultural Research*, vol. 45, pp. 11–118, 2007.

[11] K. Kamaluddin, R. M. Singh, L. C. Prasad, M. Z. Abdin, and A. K. Joshi, "Combining ability analysis for grain filling duration and yield traits in spring wheat (*Triticum aestivum* L. Em. Thell)," *Genetics and Molecular Biology*, vol. 30, no. 2, pp. 411–416, 2007.

[12] V. Kumar and S. R. Maloo, "Heterosis and combining ability studies for yield components and grain protein content in bread wheat (*Triticum aestivum* L.)," *Indian Journal of Genetics and Plant Breeding*, vol. 71, no. 4, pp. 363–366, 2011.

[13] N. Mahmood and M. A. Chowdhry, "Inheritance of flag leaf in bread wheat genotypes," *Wheat Information Service*, vol. 90, pp. 7–12, 2000.

[14] J. Ahmadi, A. A. Zali, B. Y. Samadi, A. Talaie, M. R. Ghannadha, and A. Saeidi, "A study of combining ability and gene effect in bread wheat under stress conditions by diallel method," *Iranian Journal of Agricultural Sciences*, vol. 34, pp. 1–18, 2003.

[15] M. A. Chowdhry, M. T. Mahmood, and I. Khaliq, "Genetic analysis of some drought and yield related characters in Pakistani spring wheat varieties," *Wheat Information Service*, vol. 82, pp. 11–118, 1996.

[16] F. Ahmad, S. Khan, A. Latif, H. Khan, A. Khan, and A. Nawaz, "Genetics of yield and related traits in bread wheat over different planting dates using diallel analysis," *African Journal of Agricultural Research*, vol. 6, no. 6, pp. 1564–1571, 2011.

[17] A. S. Khan and I. Habib, "Gene action in a five parent diallel cross of spring wheat (*Triticum aestivum* L.)," *Pakistan Journal of Biological Sciences*, vol. 6, pp. 1945–11948, 2003.

[18] L. Benderradji, F. Brini, S. B. Amar et al., "Sodium transport in the seedlings of two bread wheat (*Triticum aestivum* L.) Genotypes showing contrasting salt stress tolerance," *Australian Journal of Crop Science*, vol. 5, no. 3, pp. 233–241, 2011.

[19] R. G. D. Steel and J. H. Torrie, *Principles and Procedures of Statistics: A Biometrical Approach*, McGraw Hill, New York, NY, USA, 1980.

[20] R. K. Singh and B. D. Chaudhary, *Biometrical Methods in Quantitative Genetic Analysis*, Kalyani, New Delhi, India, 1985.

[21] G. Oettler, S. H. Tams, H. F. Utz, E. Bauer, and A. E. Melchinger, "Prospects for hybrid breeding in winter triticale: I. Heterosis and combining ability for agronomic traits in European elite germplasm," *Crop Science*, vol. 45, no. 4, pp. 1476–1482, 2005.

[22] O. P. Verma and H. K. Srivastava, "Genetic component and combining ability analyses in relation to heterosis for yield and associated traits using three diverse rice-growing ecosystems," *Field Crops Research*, vol. 88, no. 2-3, pp. 91–102, 2004.

[23] R. Kenga, S. O. Alabi, and S. C. Gupta, "Combining ability studies in tropical sorghum (*Sorghum bicolor* (L.) Moench)," *Field Crops Research*, vol. 88, no. 2-3, pp. 251–260, 2004.

[24] A. Adjabi, H. Bouzerzour, C. Lelarge, A. Benmahammed, A. Mekhlouf, and A. Hanachi, "Relationships between grain yield performance, temporal stability and carbon isotope discrimination in durum wheat (*Triticum durum* Desf.) under Mediterranean conditions," *Journal of Agronomy*, vol. 6, no. 2, pp. 294–301, 2007.

[25] N. B. Singh, V. P. Singh, and N. Singh, "Variation in physiological traits in promising wheat varieties under late sown condition," *Indian Journal of Plant Physiology*, vol. 19, pp. 171–175, 2005.

[26] D. K. Tiwari, P. Pandey, S. P. Giri, and J. L. Dwivedi, "Prediction of gene action, heterosis and combining ability to identify superior rice hybrids," *International Journal of Botany*, vol. 7, no. 2, pp. 126–144, 2011.

[27] M. Premlatha, A. Kalamani, and A. Nirmalakumari, "Heterosis and combining ability for grain yield and quality in maize (*Zea mays* L.)," *Advances in Environmental Biology*, vol. 5, no. 6, pp. 1264–1266, 2011.

[28] M. Gnanasekaran, P. Vivekanandan, and S. Muthuramu, "Combining ability and heterosis for yield and grain quality in two line rice (*Oryza sativa* L.) hybrids," *Indian Journal of Human Genetics*, vol. 66, pp. 6–69, 2006.

[29] R. K. Sharma, "Studies on gene action and combining ability for yield and its component traits in rice (*Oryza sativa* L.)," *Indian Journal of Human Genetics*, vol. 66, pp. 227–2228, 2006.

[30] A. K. Verma, S. R. Vishwakarma, and P. K. Singh, "Line x tester analysis in barley (*Hordeum vulgare* L.) across environments," *Barley Genetics Newsletter*, vol. 37, pp. 29–233, 2007.

[31] B. Borghi, M. Perenzin, and R. J. Nash, "Combining ability estimates in bread wheat and performances of 100 F1 hybrids produced using a chemical hybridizing agent," *Journal of Genetics and Breeding*, vol. 43, pp. 11–116, 1989.

[32] B. Borghi and M. Perenzin, "Diallel analysis to predict heterosis and combining ability for grain yield, yield components and bread-making quality in bread wheat (*Triticum aestivum* L.)," *Theoretical and Applied Genetics*, vol. 89, no. 7-8, pp. 975–981, 1994.

[33] K. A. Lucken, "The breeding and production of hybrid wheat, in USA genetic improvement in yield of wheat," in *CSSA Spec. Pub. Crop Science Society of America and American Society of Agronomy*, vol. 13, pp. 87–107, Madison, Wis, USA, 1986.

[34] F. J. Betrán, D. Beck, M. Bänziger, and G. O. Edmeades, "Genetic analysis of inbred and hybrid grain yield under stress and nonstress environments in tropical maize," *Crop Science*, vol. 43, no. 3, pp. 807–817, 2003.

[35] A. K. Yadav, R. K. Maan, S. Kumar, and P. Kumar, "Variability, heritability and genetic advance for quantitative characters in hexaploid wheat (*Triticum aestivum* L.)," *Electronic Journal of Plant Breeding*, vol. 2, pp. 405–4408, 2011.

10

Impact of Poultry Litter Cake, Cleanout, and Bedding following Chemical Amendments on Soil C and N Mineralization

Dexter B. Watts,[1] Katy E. Smith,[2] and H. A. Torbert[1]

[1] National Soil Dynamics Laboratory, USDA-ARS, 411 S. Donahue Drive, Auburn, AL 36832, USA
[2] Depatement of Math, Science, and Technology, University of Minnesota-Crookston, 2900 University Avenue, Crookston, MN 56716, USA

Correspondence should be addressed to Dexter B. Watts, dexter.watts@ars.usda.gov

Academic Editor: Mark Reiter

Poultry litter is a great alternative N source for crop production. However, recent poultry litter management changes, and increased chemical amendment use may impact its N availability. Thus, research was initiated to evaluate the effect that broiler cake and total cleanout litter amended with chemical additives have on C and N mineralization. A 35-day incubation study was carried out on a Hartsells fine sandy loam (fine-loamy, siliceous, subactive, thermic Typic Hapludults) soil common to the USA Appalachian Plateau region. Three poultry litter components (broiler cake, total cleanout, and bedding material) from a broiler house were evaluated and compared to a soil control. Chemical amendments lime ($CaCO_3$), gypsum ($CaSO_4$), aluminum sulfate ($AlSO_4$), and ferrous sulfate ($FeSO_4$) were added to the poultry litter components to determine their impact on C and N mineralization. Litter component additions increased soil C mineralization in the order of broiler cake > total cleanout > bedding > soil control. Although a greater concentration of organic C was observed in the bedding, broiler cake mineralized the most C, which can be attributed to differences in the C : N ratio between treatments. Chemical amendment in addition to the manured soil also impacted C mineralization, with $AlSO_4$ generally decreasing mineralization. Nitrogen mineralization was also significantly affected by poultry litter component applications. Broiler cake addition increased N availability followed by total cleanout compared to soil control, while the bedding resulted in net N immobilization. Chemical amendments impacted N mineralization primarily in the broiler cake amended soil where all chemical amendments decreased mineralization compared to the no chemical amendment treatment. This short-term study (35-day incubation) indicates that N availability to crops may be different depending on the poultry litter component used for fertilization and chemical amendment use which could decrease N mineralization.

1. Introduction

Poultry litter is increasingly being demanded as an alternative nutrient source to commercial fertilizer in the southeastern US region. Poultry litter is regarded as one of the most valuable nutrient sources compared to other manures due to its relatively high N and phosphorus (P) content. However, poultry litter's N and P may be susceptible to environmental loss. This has resulted in the increased use of chemical amendments to specifically reduce surface water P loss and NH_3 volatilization. Concurrently, poultry litter management practices have also recently changed, with most poultry producers cleaning the litter out of their houses less often to reduce labor costs. The impact that increased chemical amendment use and the changes in poultry litter management practices will have on N availability to plants is largely unknown.

Previous studies have shown that the abatement of P transport from land applied manure to water bodies can be achieved with the use of chemical amendments containing aluminum (Al), iron (Fe), or calcium (Ca) [1–7]. These compounds (Al, Fe, or, Ca) work by binding to P in solution thereby forming insoluble compounds. For instance, Ca rich compounds (gypsum and lime) used as soil amendments tend to form insoluble P compounds, largely as a result of forming a Ca phosphate complex [8]. Aluminum and ferrous compounds (aluminum sulfate and ferrous sulfate) usually result in decreased pH. Consequently as pH decreases, Al and Fe salts result in the precipitation of aluminum hydroxy phosphate and ferrous hydroxy phosphate [9].

Similarly, Al, Fe, and Ca compounds are believed to reduce ammonia volatilization from poultry litter. Ammonia volatilization occurs through enzymatic conversion and decomposition of organic nitrogenous compounds contained in poultry litter. Reece et al. [10] reported that NH_3 volatilization can be reduced when the litter pH falls below 7, while it is greatly increased when pH is above 8. Thus, chemical additives that manipulate poultry litter's pH are believed to be the most effective. Kithome et al. [11] evaluated effects of the chemical amendments gypsum ($CaSO_4$) and aluminum sulfate (Al_2SO_4) as reducing agents, influencing ammonia volatilization. A mixture of 20% Al_2SO_4 effectively reduced ammonia emissions by 74% compared to control, while gypsum was somewhat ineffective. Moore Jr. et al. [12] also reported significant reductions in ammonia volatilization from poultry manure amended with Al_2SO_4 and $FeSO_4$. Although these chemical additives show promise for reducing environmental degradation from litter, the question of what effect these compounds have on manure decomposition rate has not been explored. This leads to the following question: what effect do these amendments have on agricultural production? Understanding how soil amendments containing Al, Fe, and Ca affect the N mineralization capacity of manure will aid in developing better management practices for manure application.

Periodic removal of litter from poultry houses is important to promote bird health and limit manure buildup [13]. Traditionally, poultry producers cleaned their houses to the ground (total cleanout) each year. While this management is still practiced by some, most producers no longer follow this procedure. Decaking (removal of the cake or top layer of harden manure) is becoming the most popular practice utilized by producers to save money and labor. This procedure involves removing the top portion of the litter, leaving behind the old bedding material. As a result, producers do not replace bedding materials as often. Thus, total cleanout of the litter is typically reduced to once every three to five years. Given that the cake contains a more concentrated manure component and less bedding than litter removed during total cleanout, the difference between the two poultry litter sources may influence N mineralization rates following land application for fertilization.

Continual changes in management practices can impact the sustainability of agricultural production. In order to keep producers abreast of the benefits and drawbacks from these changing practices, research must be done to evaluate their impact on agricultural productivity. A better understanding of how some recent manure management practices affect the decomposition and the availability of N is required. Therefore, the objective of this study was to determine impacts of chemical amendments added to poultry litter components, on C and N mineralization in a laboratory incubation study.

2. Material and Methods

2.1. Site Description for Incubated Soil. Soil for this study was collected from a bermudagrass (Cynodon dactylon) pasture located at the Sand Mountain Agriculture Research and Extension Center in the Appalachian Plateau region of northeast Alabama, USA. Soil was from plots that had not received manure within the last 10 years. The soil was a Hartsells fine sandy loam (fine-loamy, siliceous, subactive, thermic Typic Hapludults). The regional climate is subtropical with no dry season; mean annual rainfall is 1325 mm, and mean annual temperature is 16°C [14]. Bulk soil samples were collected from the 0–20 cm depth and sieved through a 2 mm mesh screen to remove rocks and roots. Samples were stored in a cold room at 4°C until laboratory incubations were performed.

2.2. Experimental Treatments. A 35-day incubation study was conducted to measure C and N mineralization of soil amended with poultry litter containing different chemical amendments that had been previously reported to reduce P loss. Poultry litter components consisted of broiler cake, total cleanout, and bedding. Theses component were collected from a local broiler production facility in the Sand Mountain Region of north Alabama, USA. Broiler cake was collected following the decaking process during broiler house cleaning. Total cleanout litter was collected following the cleaning of a broiler house to the ground. Bedding material consisted of pine wood shavings. In order to apply small-uniform quantities of poultry litter for incubation, samples were air dried at 40°C and ground to pass through a 2 mm sieve. Characteristics of the poultry litter components are presented in Table 1. Laboratory grade chemicals consisting of gypsum ($CaSO_4$), lime ($CaCO_3$), ferrous sulfate ($FeSO_4$), and aluminum sulfate ($AlSO_4$) were used as the chemical amendments for this study.

2.3. Laboratory Analysis. Chemical analysis of soil and poultry litter component sources was performed by the Auburn University Soil Testing Laboratory as described by Hue and Evans [15]. Specifically, soil pH was determined on 1 : 1 soil/water suspensions with a glass electrode pH meter. Total C and N for the soil and poultry litter components were determined by dry combustion using a LECO TruSpec CN analyzer (LECO Corp., St. Joseph, MI). Concentrations of P, K, Ca, Mg, and so forth were determined using a Mehlich 1 (double acid) extracting solution for soil [16] and with the dry ash procedure for poultry litter components [17]; both were measured by inductively coupled Argon plasma emission spectrometry [18] using an ICAP 9000 (Thermo Jarrell Ash, Franklin, MA). Soil textural analysis (percentage sand, silt, and clay) for the incubated soil was determined using the hydrometer method [19].

2.4. Incubation Study. Methods described by Torbert et al. [20] were utilized for quadruple determinations of potential C and N mineralization. Twenty-five grams of soil (oven-dried weight basis), passed through a 2 mm sieve, were

TABLE 1: Nutrient properties of soil and poultry litter components bedding, total cleanout, and broiler cake used for incubations, dry weight basis.

Variables	pH	CEC	C	N	P P_2O_5	K K_2O	Ca	Mg	Al	B	Cu	Fe	Mn	Zn
					g kg^{-1}						mg kg^{-1}			
Poultry litter components														
Broiler cake	7.6	—	375	42.1	41.3	37.4	25.0	60.0	997	39.0	461	903	751	572
Total cleanout	8.5	—	262	31.3	48.4	32.3	37.5	64.0	2813	39.0	332	2115	509	439
Bedding	4.9	—	456	0.50	1.0	1.4	1.3	0.20	203	<0.1	13.36	4116	62	15
Soil														
Sandy loam	6.1	4.9	7.81	0.83	0.003	0.08	0.49	0.05	103	<0.1	1	9	11	1

placed in 118 mL (4-oz) plastic containers. The poultry litter component source (broiler cake, total cleanout, or bedding material) was added to soils at a rate of 350 kg total N ha^{-1} (based on application to a 15 cm soil furrow slice) and mixed homogeneously to ensure uniformity within and between samples. Following poultry litter component addition, chemical amendments ($CaSO_4$, $CaCO_3$, $FeSO_4$, and $AlSO_4$) were added to the soil at a rate that provided either Al, Fe, or Ca at an amount equivalent to the molar P content of the poultry litter component source (i.e., 651 kg ha^{-1} gypsum, 480 kg ha^{-1} lime, 930 kg ha^{-1} ferrous sulfate, and 700 kg ha^{-1} aluminum sulfate) with the greatest concentration. To facilitate uniform addition, chemical amendments were dissolved with deionized water and added to samples using a micropipette. Deionized water was added to bring the soil matric potential to approximately -20 kPa at a bulk density of 1.2 Mg m^{-3}. After sample preparation, the containers were placed in 1.06 L wide-mouth incubation jars, and 10 mL of water was added to the bottom of each jar (not sample) for humidity control. Jars were incubated in the dark at 25°C and removed after 15 and 35 days for analysis of C and N mineralization.

Soil C mineralization was determined using a 10 mL CO_2 trap (vial containing 10 mL 1 N NaOH). The CO_2 trap was placed in each jar and hermetically sealed. After removal, 1 mL of a saturated $BaCl_2$ solution ($\sim$1N) was added to each sample to stop CO_2 adsorption, and the NaOH was then backtitrated with 1 N HCl using phenolphthalein as an indicator to determine the amount of CO_2 released from soil samples. Soil C mineralization was determined as the difference between CO_2-C captured in sample traps and from blanks (sealed jar without soil). The concentrations of CO_2 determined on day 15 and 35 days after incubation were added together to determine the total amount of C mineralized for the 35 d incubation period, as described by Anderson (1982).

Soil N mineralization was determined by evaluating inorganic N concentrations during incubation. Concentrations of ammonium (NH4) and nitrite (NO2)+ nitrate (NO$_3$) were determined by extraction using 2 M KCl as described by Keeney and Nelson [21] and measured colorimetrically using

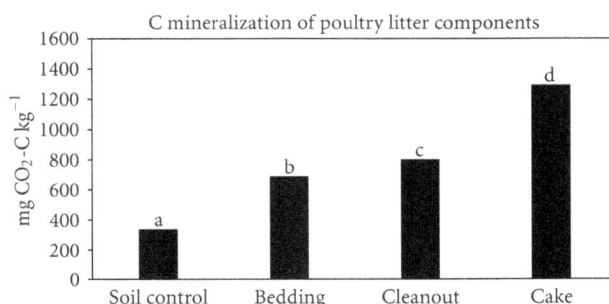

FIGURE 1: Effect of poultry litter components bedding, total cleanout, and broiler cake compared to a soil control on C mineralization following a 35-day incubation.

automated laboratory equipment (Bran-Luebbe, Norderstedt, Germany). Soil N mineralization was determined as the difference between final and initial inorganic N contents of the incubated sample.

2.5. Statistics. The incubation study was analyzed as a completely randomized design with four replications. There were 3 poultry litter sources X 5 chemical amendments X 4 replications compared to a control (4 replicates) for a total of 64 experimental units. Statistical analyses were performed using the GLM procedure of SAS [22], and means were separated using least significant difference (LSD). A significance level of $\alpha < 0.05$ was established *a priori.*

3. Results and Discussion

Poultry litter is generally applied as a readily available N source to pastures in the southeastern USA. Thus, a common upland, well-drained, low microbial activity soil managed under bermudagrass pasture from the southeastern USA region was used to evaluate C and N mineralization for this study. The soil used for this incubation historically received minimal fertilization, no grazing, or harvesting for hay. Background nutrient concentrations for the soil

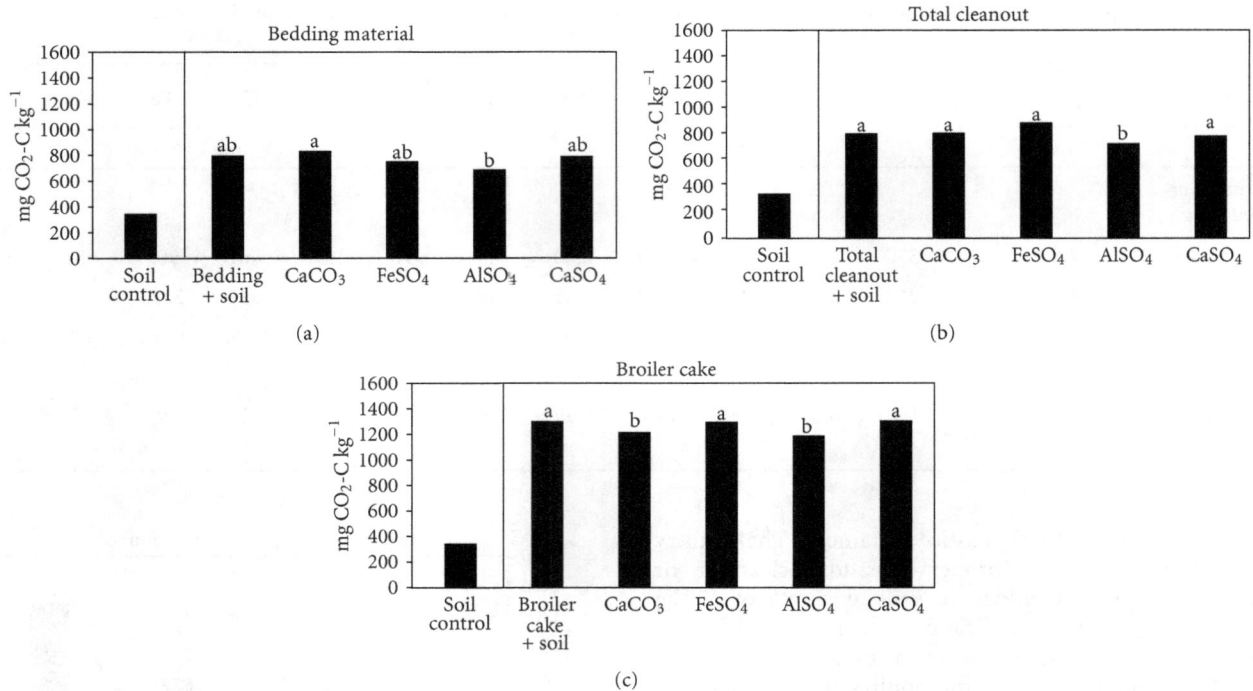

(a)

(b)

(c)

FIGURE 2: Effect of chemical amendments $CaCO_3$, $FeSO_4$, $AlSO_4$, and $CaSO_4$ on C mineralization of soil amended with poultry litter components bedding, total cleanout, and broiler cake following a 35-day incubation. Soil control is presented for comparison purposes of background levels.

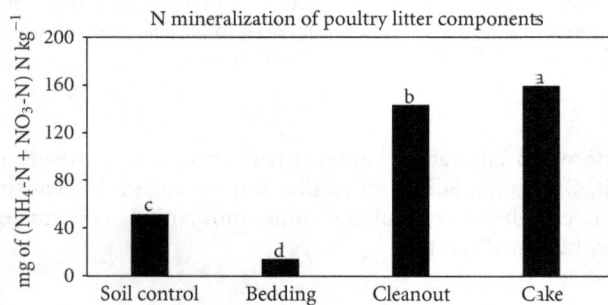

FIGURE 3: Effect of poultry litter components bedding, total cleanout, and broiler cake compared to a soil control on N mineralization following a 35-day incubation.

and poultry litter component sources used in this study are presented in Table 1. Nutrient concentrations varied among the poultry litter component sources. The bedding material had the greatest C concentrations compared to broiler cake and total cleanout. On the other hand, the bedding material N concentration was the lowest among the poultry litter component sources. Between the poultry litter component sources containing manure (broiler cake and total cleanout), broiler cake had greater C and N concentrations compared to the total cleanout, but C:N ratios were similar between the two sources. Total C, N, and C:N ratios were within the range reported by others in previous research [13, 23]. The focus of this incubation study

was to evaluate C and N cycling in soil following poultry litter addition amended with chemical amendments. Information from this study will aid in a better understanding of the N-supplying potential of broiler cake and total cleanout amended with chemical additives containing Al, Fe, and Ca. For discussion purposes of this manuscript, C and N mineralization comparisons were made between the poultry litter components alone (without chemical amendments) and between chemical amendments for each poultry litter component source (broiler cake, total cleanout, and bedding) separately.

3.1. Soil Carbon Mineralization.

Carbon mineralization is the conversion of organic C material into inorganic forms through soil microbial oxidation. This process occurs during decomposition. An evaluation of C mineralization can be used as an index to understand the decomposition rate of organic substances. Carbon evolution values observed on 15 d and 35 d were summed together and expressed as total C mineralized. Significant differences were observed in C mineralization rates resulting from the different poultry litter components (Figure 1). Cumulative soil C mineralization for the 35-day incubation period was 335 mg kg^{-1} for soil, 687 mg kg^{-1} for bedding, 801 mg kg^{-1} for total cleanout, and 1293 mg kg^{-1} for broiler cake. As expected, amending soil with the poultry litter components sources significantly increased C evolution, with broiler cake having the greatest C mineralization capacity compared to the other treatments. Carbon mineralization rates following the 35-day incubation was in the order of broiler cake > total cleanout > bedding > soil control. Although bedding contained the most C, the percentage of total C mineralized was less than the other poultry litter component sources. This suggests that the bedding is a more stable material than the litter sources containing manure (broiler cake and total cleanout). Also, the difference in C mineralization between the bedding and the other poultry litter component sources was most likely due to the amount of available N. Although the litter sources containing manure had a similar C : N ratio, broiler cake had the highest C mineralization. Thus, since N was not a limiting factor, the poultry litter component source with the greatest C concentration mineralized the most when both broiler cake and total cleanout were applied at 350 kg N ha^{-1}. Also, it is important to note that the total cleanout contained a greater percentage of bedding material compared to the broiler cake. Thus, the total cleanout most likely contained more stable C compared to the cake. Since broiler cake consisted of a lower recalcitrant form of organic C, it contributed to a greater mineralization capacity. Although C mineralization data reported for this study represent the total C mineralized during the two sampling periods (15 and 35 days after incubation), it is important to note that a significantly greater C mineralization rate was observed during the second portion of the incubation for the bedding material (data not shown). The other poultry litter components generally resulted in a slightly lower C mineralization rate.

Addition of chemical amendments to the incubated soil also impacted carbon mineralization. Differences in C mineralization rates for each chemical compound added to the broiler cake, total cleanout, and bedding amended soil are illustrated in Figure 2. Chemical amendment additions impacted C mineralization differently depending on the poultry litter component (total cleanout, broiler cake, and bedding) evaluated. For instance, significant differences were observed among chemical amendments in the bedding treatment (i.e., CaCO$_3$ > AlSO$_4$) with CaCO$_3$ producing the greatest C mineralization; however when each chemical amendment was compared to the no chemical amendment bedding treatment (bedding + soil only), no significant differences were observed. Cumulative C mineralization in the total cleanout treatment was significantly increased with FeSO$_4$, producing significantly greater C mineralization rates compared to the other treatments. No significant differences were observed between the CaSO$_4$, FeSO$_4$, AlSO$_4$, and the no chemical amendment treatment (total cleanout + soil only). In the broiler cake treatment, no significant differences were observed between CaSO$_4$, FeSO$_4$, and the no chemical amendment broiler cake treatment (broiler + soil only), while CaCO$_3$ and AlSO$_4$ produced statistically lower C mineralization.

3.2. Soil Nitrogen Mineralization. Nitrogen in manure is primarily in an organic form. Before plant uptake can occur, it must be converted to inorganic forms of N. This is achieved via microbes as a by-product of organic matter decomposition in a process called N mineralization. Understanding of how recent changes in poultry litter management practices affect N mineralization can help improve N management for crop demands.

Similar to C mineralization, addition of the different poultry litter components impacted N mineralization. Mean N mineralization observed during the 35-day incubation is presented in Figure 3. Cumulative soil N mineralizations for the 35-day incubation period were 51 mg kg^{-1} for soil, 13 mg kg^{-1} for bedding material, 143 mg kg^{-1} for total cleanout, and 159 mg kg^{-1} for broiler cake. Addition of bedding resulted in a lower mineralization compared to the soil control (no chemical amendment or poultry litter addition), while broiler cake and total cleanout amended soils resulted in greater N availability. This was to be expected since wood shavings (bedding) are highly stable carbonaceous material with a C : N ratio of 143 : 1. As rule of thumb, a C : N ratio of 20 : 1 is the breakeven point between immobilization and N release. When organic materials with C : N ratios greater than 30 : 1 are added to soil, there is N immobilization during the initial decomposition process. Thus, the highly carbonaceous bedding material supplied the microbes with a new energy source, but insufficient N to build protein. Thus the soil microbes scavenged N from the soil, thereby reducing N availability, or, in other words, caused net N immobilization. Broiler cake produced the highest N mineralization capacity compared to the other poultry litter components evaluated during this incubation study. Although the broiler cake and total cleanout had similar C : N ratios (8.9 broiler cake and 8.4 total cleanout), broiler cake most likely contains a more energy-rich supply of C compared to the total cleanout. For instance, the carbon source (although not quantified) in the cake portion of the litter in broiler houses generally contains a higher concentration of manure while the total cleanout contains both manure and bedding.

Use of chemical amendments with the different poultry litter components resulted in N availability differences following the 35-day incubation (Figure 5). Generally, all of the chemical amendments negatively impacted N availability when compared with poultry litter component sources. There was a significant interaction between the poultry litter components and the chemical amendments. For the bedding material, adding chemical amendments produced a statistically similar N rate compared to the no chemical amendment bedding treatment (bedding + soil only). The CaCO$_3$ amendment produced the greatest N mineralization rate compared to the other chemical amendments applied to the bedding material. Nitrogen mineralization for CaCO$_3$, FeSO$_4$, and CaSO$_4$ was statistically similar to the no chemical amendment bedding treatment, while the AlSO$_4$ produced a statistically lower N mineralization rate. Addition of AlSO$_4$ to the bedding material, which was slightly acidic, could have caused Al toxicity to the soil microbes that were responsible for decomposition. Chander and Brookes [24] reported that Al concentrations at low pH can directly affect microbial biomass and indirectly decrease inputs of soil plant-derived substrates.

Results of N mineralization for chemical amendment additions to total cleanout are reported in Figure 4. Addition of chemical amendments FeSO$_4$, AlSO$_4$, and CaSO$_4$ to soil amended with total cleanout resulted in N mineralization

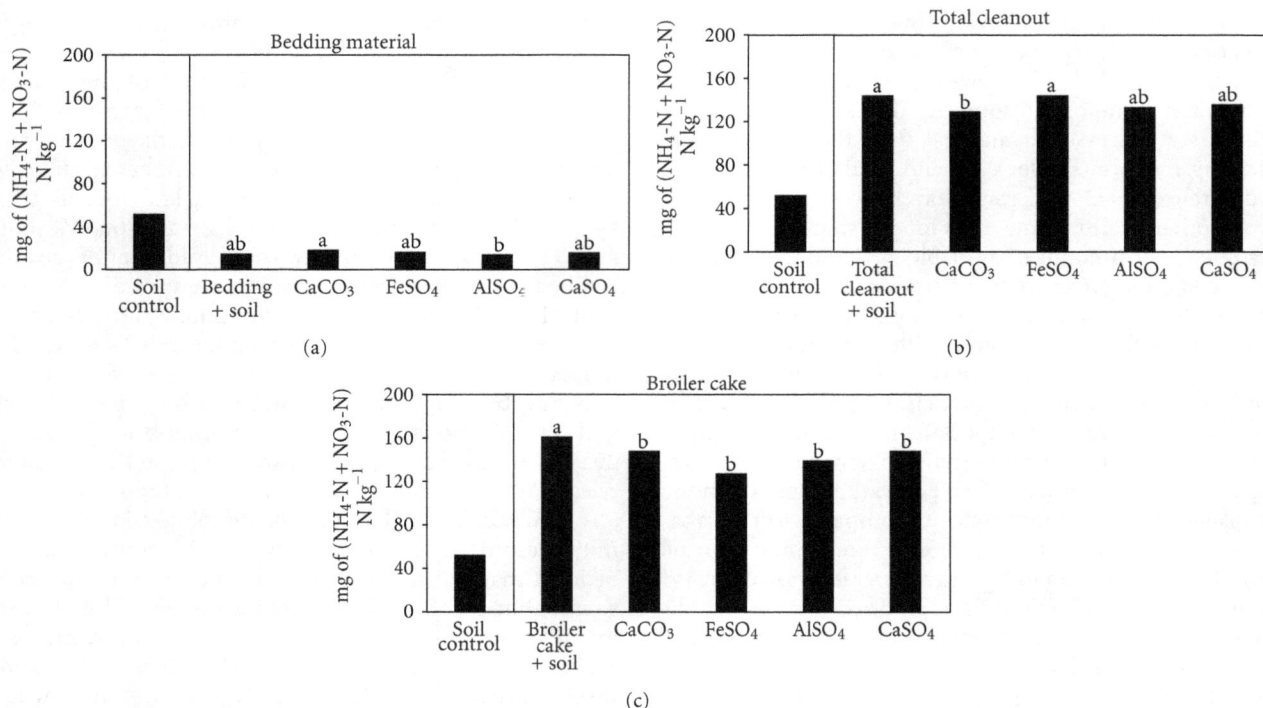

(a)

(b)

(c)

FIGURE 4: Effect of chemical amendments $CaCO_3$, $FeSO_4$, $AlSO_4$, and $CaSO_4$ on N mineralization of soil amended with poultry litter components bedding, total cleanout, and broiler cake following a 35-day incubation. Soil control is presented for comparison purposes of background levels.

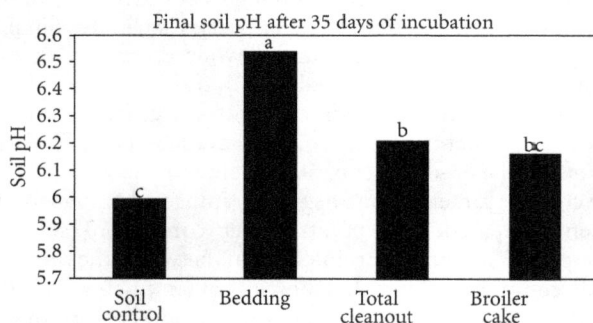

FIGURE 5: Effect of poultry litter components bedding, total cleanout, and broiler cake compared to a soil control on soil pH.

statistically similar to the no chemical amendment total cleanout treatment. However, addition of $CaCO_3$ to the total cleanout amended soil resulted in a statistically lower N mineralization rate compared to when no chemical amendments were applied. Originally, the pH of the total cleanout was 8.5; thus addition of the liming agent to an already basic material had a negative effect on the N mineralization.

When evaluating the effect of chemical additions to broiler cake amended soil, a slightly different effect was observed in the N mineralization. All of the chemical amendments evaluated resulted in statistically lower plant available N during the 35-day incubation compared to the

no chemical amendment broiler cake treatment. Although all the chemical amendments decreased N mineralization, the rate of decrease among treatments with chemical amendments was statistically similar. These results suggest that chemical amendments added to land-applied poultry litter to decrease P loss with surface water runoff or in the poultry production facilities to reduce NH_3 volatilization during growout could affect N availability to crops.

3.3. Soil pH Change. Addition of the poultry litter component sources significantly impacted pH of the incubated soil. Soil pH during the 35-day incubation containing the poultry litter component sources and soil control are shown in Figure 5. Initial soil pH prior to the incubation was 6.1. The soil control pH was essentially unchanged following the 35-day incubation (6.0). Poultry litter additions significantly increased soil pH compared to the soil control. The greatest pH change was observed in soil amended with bedding material followed by total cleanout, broiler cake, and soil control. Generally, poultry litter is not a liming material, but because of the alkaline pH and large Ca content, the broiler cake and total cleanout amendments increased the incubated soil's pH. Increases in soil pH following poultry litter addition have been reported by others [25–27]. Moore and Edwards [27] reported an increase in pH from 5.1 to 5.8 after poultry litter was applied for 7 years. On the other hand, an unexpected increase was also observed with the bedding material, but this phenomenon cannot be explained from previous reports.

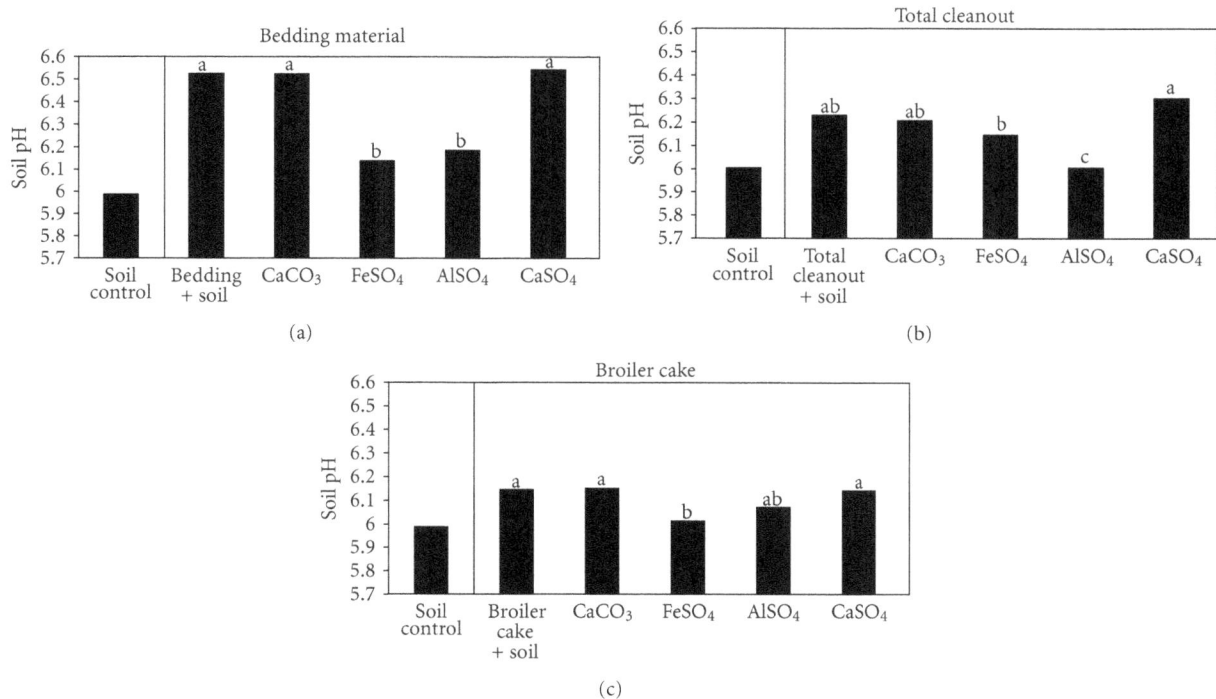

FIGURE 6: Effect of chemical amendments $CaCO_3$, $FeSO_4$, $AlSO_4$, and $CaSO_4$ on the pH of soil amended with poultry litter components bedding, total cleanout, and broiler cake following a 35-day incubation. Soil control is presented for comparison purposes of background levels.

Chemical amendment addition to soil with the poultry litter components also impacted pH levels (Figure 6). The general pH trend resulting from chemical amendment additions to the incubated soil containing the different poultry litter components was statistically similar between the no chemical amendment, $CaSO_4$, and $CaCO_3$ treatments; lower levels were observed for $AlSO_4$ and $FeSO_4$. Evaluation of the chemical amendment additions to the incubated soil with bedding was similar to the general trend with no significant difference being observed between the no chemical amendment, $CaSO_4$, and $CaCO_3$ treatments, while $AlSO_4$ and $FeSO_4$ resulted in significantly lower soil pH. Significant differences were also observed for the total cleanout. No significant difference was observed between the no chemical amendment, $CaSO4$, $FeSO_4$, and $CaCO_3$ treatments. The addition of $AlSO_4$ to the total cleanout treatment resulted in decreases to soil pH. An evaluation of broiler cake showed that $AlSO_4$ and $FeSO_4$ resulted in the lowest pH values similar to the bedding and total cleanout poultry litter component. However, $AlSO_4$ was the only chemical amendment that was statistically different from the no chemical amendment broiler cake treatment.

4. Conclusions

Nitrogen is the most limiting nutrient in a crop production system. When manure sources are used for N fertilization, N mineralization accounts for most of the crop N needs. An understanding of how manure management practices affect N availability is important to maintain crop yields. The information obtained from this incubation study may be useful when considering fertilization with poultry litter. For instance, should poultry litter cake or cleanout be used for N fertilization and which chemical amendment provides the best N availability? Carbon mineralization, which is a representation of microbial activity, was increased by the addition of the poultry litter components. Even though the bedding material had the greatest C source, broiler cake had the highest C mineralization rate. Carbon mineralization was in the order of broiler cake, total cleanout, and bedding. Evaluation of N mineralization showed that the bedding material resulted in severe N immobilization. On the other hand, broiler cake had the highest N mineralization rate followed by total cleanout. Addition of chemical amendments to the poultry litter components also impacted C mineralization. The greatest differences in C mineralization were observed with $AlSO_4$, generally decreasing mineralization. Chemical amendment additions also resulted in significant N mineralization differences. The greatest differences were observed in the broiler cake treatment. All chemical amendments applied to the broiler cake amended soil resulted in decreased N mineralization compared to the no chemical amendment treatment. For the total cleanout treatment, $CaCO_3$ was the only chemical amendment to decrease N mineralization compared to the no chemical amendment treatment. These results suggest that while the use of chemical amendments has been shown to reduce P loss and decrease NH_3 volatilization, a reduction in N mineralization may also occur. Also changes in poultry litter management practices are likely to impact N availability in soil.

References

[1] P. A. Moore Jr. and D. M. Miller, "Decreasing phosphorus solubility in poultry litter with aluminum, calcium, and iron amendments," *Journal of Environmental Quality*, vol. 23, no. 2, pp. 325–330, 1994.

[2] D. L. Anderson, O. H. Tuovinen, A. Faber, and I. Ostrokowski, "Use of soil amendments to reduce soluble phosphorus in dairy soils," *Ecological Engineering*, vol. 5, no. 2-3, pp. 229–246, 1995.

[3] T. H. Dao, "Coamendments to modify phosphorus extractability and nitrogen/phosphorus ratio in feedlot manure and composted manure," *Journal of Environmental Quality*, vol. 28, no. 4, pp. 1114–1121, 1999.

[4] Z. Dou, G. Y. Zhang, W. L. Stout, J. D. Toth, and J. D. Ferguson, "Efficacy of alum and coal combustion by-products in stabilizing manure phosphorus," *Journal of Environmental Quality*, vol. 32, no. 4, pp. 1490–1497, 2003.

[5] M. Kalbasi and K. G. Karthikeyan, "Phosphorus dynamics in soils receiving chemically treated dairy manure," *Journal of Environmental Quality*, vol. 33, no. 6, pp. 2296–2305, 2004.

[6] H. A. Torbert, K. W. King, and R. D. Harmel, "Impact of soil amendments on reducing phosphorus losses from runoff in sod," *Journal of Environmental Quality*, vol. 34, no. 4, pp. 1415–1421, 2005.

[7] D. B. Watts and H. A. Torbert, "Impact of gypsum applied to grass buffer strips on reducing soluble p in surface water run-off," *Journal of Environmental Quality*, vol. 38, no. 4, pp. 1511–1517, 2009.

[8] D. Brauer, G. E. Aiken, D. H. Pote et al., "Amendment effects on soil test phosphorus," *Journal of Environmental Quality*, vol. 34, no. 5, pp. 1682–1686, 2005.

[9] Y. Ann, K. R. Reddy, and J. J. Delfino, "Influence of chemical amendments on phosphorus immobilization in soils from a constructed wetland," *Ecological Engineering*, vol. 14, no. 1-2, pp. 157–167, 1999.

[10] F. N. Reece, B. J. Bates, and B. D. Lott, "Ammonia control in broiler houses," *Poultry Science*, vol. 58, pp. 754–755, 1979.

[11] M. Kithome, J. W. Paul, and A. A. Bomke, "Reducing nitrogen losses during simulated composting of poultry manure using adsorbents or chemical amendments," *Journal of Environmental Quality*, vol. 28, no. 1, pp. 194–201, 1999.

[12] P. A. Moore Jr., T. C. Daniel, D. R. Edwards, and D. M. Miller, "Effect of chemical amendments on ammonia volatilization from poultry litter," *Journal of Environmental Quality*, vol. 24, no. 2, pp. 293–300, 1995.

[13] K. R. Sistani, G. E. Brink, S. L. McGowen, D. E. Rowe, and J. L. Oldham, "Characterization of broiler cake and broiler litter, the by-products of two management practices," *Bioresource Technology*, vol. 90, no. 1, pp. 27–32, 2003.

[14] R. H. Shaw, "Climate of the United States," in *Handbook of Soils and Climate in Agriculture*, V. J. Kilmer, Ed., CRC Press, Boca Raton, Fla, USA, 1982.

[15] N. V. Hue and C. E. Evans, *Procedures used for soil and plant analysis by the Auburn University Soil Testing Laboratory*, Auburn University, Auburn, Ala, USA, 1986.

[16] S. R. Olsen and L. E. Sommers, "Phosphorus," in *Methods of Soil Analysis Part 2*, A. L. Page, Ed., Agronomy Monograph 9, pp. 403–430, ASA and SSSA, Madison, Wis, USA, 1932.

[17] S. J. Donahue, *Reference soil test methods for the Southern Region of the United State*, Southern Cooperative Service, Bulletin 289, Georgia Agricultural Experiment Station, Athens, Ga, USA, 1983.

[18] P. N. Soltanpour, J. B. Jones, and S. M. Workman, "Optical emission spectrometry," in *Methods of Soil Analysis. Part 2*, A. L. Page et al., Ed., Agronomy Monograph 9, pp. 29–65, ASA and SSSA, Madison, Wis, USA, 2nd edition, 1982.

[19] G. J. Bouyoucos, "Hydrometer method for making particle analysis of soil," *Agronomy Journal*, vol. 54, pp. 464–465, 1962.

[20] H. A. Torbert, S. A. Prior, and D. W. Reeves, "Land management effects on nitrogen and carbon cycling in an ultisol," *Communications in Soil Science and Plant Analysis*, vol. 30, no. 9-10, pp. 1345–1359, 1999.

[21] D. R. Keeney and D. W. Nelson, "Nitrogen: inorganic forms," in *Methods of Soil Analysis. Part 1*, A. L. Page et al., Ed., Agronomy Monograph 9, pp. 643–698, ASA and SSSA, Madison, Wis, USA, 2nd edition, 1982.

[22] SAS Institute, *SAS, version 9.1*, SAS Institute, Cary, NC, USA, 2002.

[23] P. L. Ward, J. E. Wohlt, P. K. Zajac, and K. R. Cooper, "Chemical and physical properties of processed newspaper compared to wheat straw and wood shavings as animal bedding," *Journal of Dairy Science*, vol. 83, no. 2, pp. 359–367, 2000.

[24] K. Chander and P. C. Brookes, "Synthesis of microbial biomass from added glucose in metal-contaminated and non-contaminated soils following repeated fumigation," *Soil Biology and Biochemistry*, vol. 24, no. 6, pp. 613–614, 1992.

[25] E. E. Codling, R. L. Chaney, and C. L. Mulchi, "Use of aluminum- and iron-rich residues to immobilize phosphorus in poultry litter and litter-amended soils," *Journal of Environmental Quality*, vol. 29, no. 6, pp. 1924–1931, 2000.

[26] G. Gupta and S. Charles, "Trace elements in soils fertilized with poultry litter," *Poultry Science*, vol. 78, no. 12, pp. 1695–1698, 1999.

[27] P. A. Moore and D. R. Edwards, "Long-term effects of poultry litter, alum-treated litter, and ammonium nitrate on aluminum availability in soils," *Journal of Environmental Quality*, vol. 34, no. 6, pp. 2104–2111, 2005.

11

Influence of *Verticillium dahliae* Infested Peanut Residue on Wilt Development in Subsequent Cotton

Shilpi Chawla,[1] Jason E. Woodward,[1, 2] and Terry A. Wheeler[3]

[1] *Department of Plant and Soil Science, Texas Tech University, Lubbock, TX 79409, USA*
[2] *Texas AgriLIFE Extension Service, Texas A&M System, Lubbock, TX 79403, USA*
[3] *Texas AgriLIFE Research, Texas A&M System, Lubbock, TX 79403, USA*

Correspondence should be addressed to Jason E. Woodward, jewoodward@ag.tamu.edu

Academic Editor: M. Tejada

Texas ranks first in cotton production in the United States and accounts for approximately 40% of the total production. Most of the cotton production is concentrated in the Texas High Plains where cotton and peanut are commonly grown in rotation. With peanut being a legume crop, farmers routinely leave residue on the soil surface to improve soil fertility; however, *V. dahliae* can survive in the crop residue contributing inoculum to the soil. A microplot study was conducted to investigate the impact of peanut residue infested with *V. dahliae* on subsequent microsclerotia density in soil and Verticillium wilt development in cotton. The effects of infested peanut residue rate on percent germination of cotton seeds and on wilt incidence were monitored in 2008 and 2009. In both years microplots were planted with a susceptible cotton cultivar, Stoneville (ST) 4554B2RF. Increasing infested peanut residue rate was positively correlated with wilt incidence in cotton and negatively correlated with germination of cotton seeds. Density of microsclerotia in the soil increased significantly with increasing rates of infested peanut residue over time. Results indicate infested peanut residue serve as a source of *V. dahliae* inoculum, and removing infested residue can reduce disease development in subsequent cotton crops.

1. Introduction

Verticillium wilt, caused by the soilborne fungus *Verticillium dahliae* Kleb., is an economically important disease of cotton (*Gossypium hirsutum* L.) and peanut (*Arachis hypogaea* L.). The pathogen has a broad host range of more than 400 plant species including field crops and most vegetables [1]. Several factors, including cultivar selection, pathogen aggressiveness, inoculum density, and environmental conditions, influence disease development [2, 3]. Cotton and peanut plants affected by Verticillium wilt show stunting and epinasty [4]. Their leaves exhibit interveinal chlorosis, necrosis, curling, and die back from the margins inward. Plants develop characteristic mosaic patterns on foliage, starting from the base of the plant and progressing towards the top [5]. Ramification of the fungus in the xylem vessels leads to a tan-to-brown discoloration, a decrease in hydraulic conductance, wilting, and eventually death [6]. In infested cotton, plants can appear stunted, defoliate prematurely, and have fewer fruiting positions; bolls may abscise or not open [7], whereas in peanut, pegs are formed in less numbers and have fewer seeds [8].

The fungus is capable of infecting plant roots directly or through wounds throughout the growing season between temperatures 21 and 27°C whereas temperatures between 24 and 27°C are best suited for *V. dahliae* growth and survival in the plant [6, 9]. According to Huisman [10], *V. dahliae* primarily colonizes the rhizoplane of the host plant, penetrates roots early in the growing season, and then infects the vascular system and grows systemically throughout the plant. Microsclerotia, the survival structures of *V. dahliae*, are produced once the plant dies. Microsclerotia are composed of masses of melanized hyphae and are considered the principal source of inoculum for Verticillium wilt development. Microsclerotia can survive for more than 20 years in the soil [11], and root exudates stimulate germination initiating infection [6, 12]. Microsclerotia are formed depending on temperature and moisture availability with the decay of plant tissues [13]. Microsclerotia are dispersed in the soil, and only a single cycle of inoculum is produced during a growing season. Inoculum density in soil at planting has been shown to play a critical role in disease development [14]. Production of *V. dahliae* microsclerotia on infected plant parts of many field crops has been found [15, 16]. Formation of microsclerotia on dying host debris in the soil can cause an increase in inoculum density in the following year, especially if the increase is greater than the reduction in microsclerotia due to mortality [15, 16].

Cotton is economically the most important crop in Texas High Plains. Cotton is commonly grown in rotation with peanut in this region. Farmers in this region routinely utilize peanut residue on the soil surface to improve soil fertility. Typical rates of peanut residue left in the field after harvest are approximately 3500 kg/ha (Woodward, *unpublished data*). This practice can help reduce soil erosion, conserve energy, maintain soil moisture, improve organic matter content and soil fertility [17]. However, *V. dahliae* can survive in the crop residue, and disease problems may be more severe by protecting the residue from microbial degradation and lowering soil temperature.

Both peanut and cotton are suitable hosts for *V. dahliae* allowing for a continued increase in microsclerotia production. Currently, there is no quantitative data available regarding the influence of *V. dahliae* infested peanut residue on wilt development in cotton. The objective of this study was to determine the effect of *V. dahliae* infested peanut residue rate on release of microsclerotia in the soil and its implications for Verticillium wilt development in cotton over time. The hypothesis was that peanut residue infested with *V. dahliae* will increase microsclerotia density in soil and Verticillium wilt development in cotton.

2. Materials and Methods

2.1. Microplot Experiment. A microplot experiment was conducted in 2008 and 2009 to examine the effect of increasing rate of peanut residue infested with *V. dahliae* on microsclerotia production and disease development in cotton over time. Microplots were constructed out of cylindrical galvanized aluminum rings (90 cm diameter and 60 cm height) and buried at the depth of 50 cm. Treatments (0, 370, 925, 1850, 2775, 3700, 18,495, and 37,000 kg/ha infested peanut residue) were arranged in a randomized complete block design with nine replications. Peanut residue was collected from a field that was infested with *V. dahliae* and had experienced severe Verticillium wilt for several growing seasons. In 2008, two months prior to planting infested residues were incorporated in microplots at the rate of assigned treatments by mixing the residues with top soil by hand tilling to mimic the peanut residue naturally left on the field. Microplots were planted with a susceptible cotton cultivar, Stoneville (ST) 4554B2RF at the rate of 25 seeds per microplot in a circular pattern. Irrigation, fertilizer, and weeding practices were conducted as needed, according to local extension recommendations for both seasons. In 2008 and 2009, cotton plants were hand harvested leaving about

six inch stems above ground, and remaining cotton residues were removed.

2.2. Soil Sampling and Data Collection. All microplots were sampled in February 2008, prior to the assigning of residue treatments, to determine baseline populations of *V. dahliae* within the soil. Subsequent soil samples were taken in April and November 2009 and April 2010. A 2.5 cm diameter auger to a depth of 20 cm was used in taking soil samples from each microplot. Each sample consisted of four cores and had a total soil weight of approximately 250 g. The samples were air-dried at room temperature for 14 days. A soil dilution plating technique [18], utilizing Sorensen's NP-10 semiselective medium [19] amended with 0.025 N NaOH as suggested by Kabir et al. [20], was used for enumeration of microsclerotia in soil. Air-dried soil was ground with a roller pin; a 20 cm^3 soil sample was combined with 80 ml of deionized water and stirred using a magnetic stir plate. A 1 ml aliquot of the soil solution was distributed on each Petri dish (10 replications) containing Sorensen's NP-10 semiselective medium and was spread with a glass rod. After 14 days of incubation at room temperature in the dark, the soil was rinsed from the Petri dishes by gentle rubbing and then air-dried for 2 hours prior to counting. The numbers of colonies of *V. dahliae* were counted under a stereo dissecting microscope and expressed as the number of microsclerotia (ms)/cm^3 of dry soil. Percent germination of cotton seed was recorded in June, and disease incidence was assessed in September as percent symptomatic plants in each microplot for both seasons.

2.3. Statistical Analysis. Data for percent germination of cotton seeds, percent disease incidence and inoculum density of *V. dahliae* in soil (ms/cm^3) were analyzed using Proc MIXED (SAS Institute Inc., 2008, Ver. 9.2, Cary, NC, USA). Infested peanut residue rate (treatments) were log transformed. Data were analyzed as a split-plot in time where the sub-plots were four sampling dates as described in Steel and Torrie [21]. The method used to adjust the degrees of freedom (df) to match adjustments in the sums of square was the Satterthwaite option in the LSMEANS statement in Proc MIXED. Standard error were determined from the PDIFF (probability of difference of two means) option. Regression analysis was carried out for percent germination, percent disease incidence and inoculum density of *V. dahliae* in soil (ms/cm^3) with log_{10} transformed infested peanut residue rates using Proc MIXED. Linear regression ($f = y^0 + ax + bx^2$) with quadratic terms and the slopes were analyzed.

3. Results

Soil samples collected in February 2008 were void of *V. dahliae* inoculum in all the microplots prior to artificial incorporation of infested peanut residue (Figure 1). There was a positive correlation between increasing infested peanut residue rates and microsclerotia densities in soil (Figure 1). Linear regression with a quadratic term and slope represented the overall effect as we go from low to high values

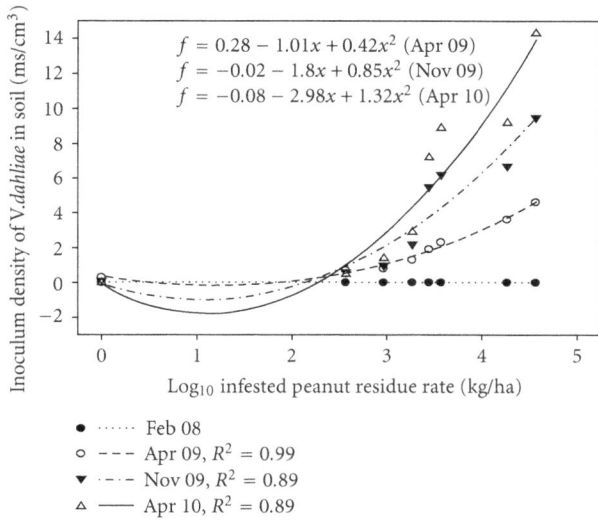

$f = 0.28 - 1.01x + 0.42x^2$ (Apr 09)
$f = -0.02 - 1.8x + 0.85x^2$ (Nov 09)
$f = -0.08 - 2.98x + 1.32x^2$ (Apr 10)

● ······ Feb 08
○ --- Apr 09, $R^2 = 0.99$
▼ -·- Nov 09, $R^2 = 0.89$
△ —— Apr 10, $R^2 = 0.89$

FIGURE 1: Effect of infested peanut residue rates on inoculum density of *Verticillium dahliae* in soil (microsclerotia/cm³). Infested peanut residue rates 0, 370, 925, 1850, 2775, 3700, 18495, and 37000 kg/ha were log$_{10}$ transformed and were expressed as 0, 2.6, 3.0, 3.3, 3.4, 3.6, 4.3, and 4.6 on *X*-axis. Soil samples collected in February 2008 were void of *V. dahliae* inoculum prior to artificial incorporation of infested peanut residue. Linear quadratic equation $f = y^0 + ax + bx^2$ was used to calculate intercept (y^0), slope (a) and curvature (b).

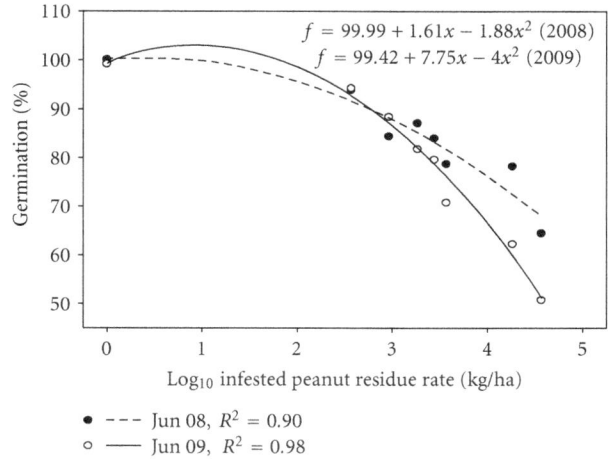

$f = 99.99 + 1.61x - 1.88x^2$ (2008)
$f = 99.42 + 7.75x - 4x^2$ (2009)

● --- Jun 08, $R^2 = 0.90$
○ —— Jun 09, $R^2 = 0.98$

FIGURE 2: Effect of infested peanut residue rates on germination of cotton seeds (%). Infested peanut residue rates 0, 370, 925, 1850, 2775, 3700, 18495, and 37000 kg/ha were log$_{10}$ transformed and were expressed as 0, 2.6, 3.0, 3.3, 3.4, 3.6, 4.3, and 4.6 on *X*-axis. Linear quadratic equation $f = y^0 + ax + bx^2$ was used to calculate intercept (y^0), slope (a) and curvature (b).

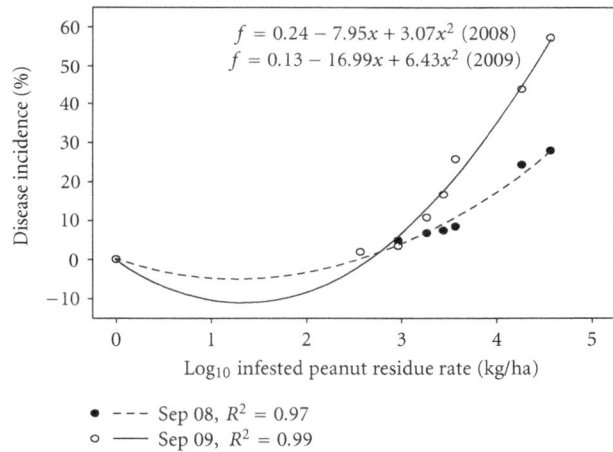

$f = 0.24 - 7.95x + 3.07x^2$ (2008)
$f = 0.13 - 16.99x + 6.43x^2$ (2009)

● --- Sep 08, $R^2 = 0.97$
○ —— Sep 09, $R^2 = 0.99$

FIGURE 3: Effect of infested peanut residue rates on Verticillium wilt incidence (%). Infested peanut residue rates 0, 370, 925, 1850, 2775, 3700, 18495, and 37000 kg/ha were log$_{10}$ transformed and were expressed as 0, 2.6, 3.0, 3.3, 3.4, 3.6, 4.3, and 4.6 on *X*-axis. Linear quadratic equation $f = y^0 + ax + bx^2$ was used to calculate intercept (y^0), slope (a) and curvature (b).

of the inoculum density. The slope of the quadratic term increased from 0.42 ($R^2 = 0.99$) in April 2009 to 0.85 ($R^2 = 0.89$) in November 2009 to 1.32 ($R^2 = 0.89$) in April 2010. Overall, microsclerotia densities were found to increase with higher rates of *V. dahliae* infested peanut residue and with time (Figure 1).

Increasing peanut residue rates had a negative effect on percent germination of cotton seeds (Figure 2). Cotton germination in microplots amended with the lowest rate of residue averaged 93.8 and 94.2% for 2008 and 2009, respectively, whereas microplots amended with the highest rate of infested peanut residue resulted in 64.4 and 50.7% over the two years (Figure 2). Germination of cotton seeds in the nonamended controls averaged 99.6%, whereas germination for all other treatments was intermediate (Figure 2). Linear regression with a quadratic term and slope represented the overall effect as we go from high to low values of the germination of cotton seeds. The reduction in germination per unit peanut residue increased from 2008 to 2009 (quadratic curvature of −1.88 ($R^2 = 0.90$) in June 2008 and −4.00 ($R^2 = 0.98$) in June 2009).

A positive correlation occurred between peanut residue rates and Verticillium wilt incidence on cotton (Figure 3). Linear regression with a quadratic term and slope represented the overall effect as we go from low to high values of the disease incidence. The increase in wilt incidence per unit of peanut residue rate was higher in 2009 than in 2008 (slope of 3.07 ($R^2 = 0.97$) in September 2008 and 6.43 ($R^2 = 0.99$) in September 2009). Disease incidence increased

from 1.9% to 27.8% when comparing the lowest and highest residue rates, respectively, in 2008 (Figure 3). A similar trend was observed for the 2009 growing season. Disease incidence increased substantially between years in microplots amended with 1850 kg/ha residue rate or greater (Figure 3).

4. Discussion

Microsclerotia of *V. dahliae* are produced in large numbers on senescing parts of host plants and may remain viable in the soil for many years [12]. Davis et al. [22] found a strong correlation between microsclerotia population in the soil and the number of microsclerotia in stem tissue, which can further contribute to soil inoculum for the next season. Peanut is commonly used as a rotation crop with cotton in parts of the United States, and both crops are suitable hosts for *V. dahliae*. This study demonstrates that microsclerotia can survive in infested peanut residue and then can infect susceptible cotton in the next season. Root exudates stimulate germination of microsclerotia leading to a decline in the population of viable propagules in soil until the incorporation of new inoculum from infected tissues at the end of the crop [16, 23]. Inoculum potential is related to the density and distribution of infested residues as well as to the susceptibility of the host crop [2]. Being a monocyclic disease, disease incidence in Verticillium wilt is positively related to the concentration of primary inoculum; thus understanding the relationship between microsclerotia density in soil at planting and wilt development is essential for developing a disease risk assessment based on preplant soil assays and also for disease management [14].

The beneficial effects of leaving crop residue as a nutrient source can be offset by the negative effects of providing shelter for survival, growth, and reproduction of plant pathogens and raise concerns over the role of crop residue in epidemics caused by soilborne pathogens. Crop residue at the soil surface is a principal source of inoculum. Infested peanut residue increased soil inoculum density and disease incidence on subsequent cotton in this study and similar results were found by Adee et al. [24] for other pathosystems. Tjamos [25] and Zilberstein et al. [26] found that the degree of pathogenicity of *V. dahliae* was related to the plant species from which the isolate was obtained and was also dependent on the previous cropping history. Continual uses of susceptible host cultivars and cultural practices that leave abundant infested residue on the soil surface have been observed to increase the damage caused by other soilborne pathogens in subsequent crops [27].

Microsclerotia mainly occur in the aerial parts of the crop [12], and, therefore, removal of the aerial crop debris from the field is potentially an effective measure to prevent the accumulation of microsclerotia in the soil. In the present study, lower rates of infested peanut residue had lower ms/cm^3 of soil, which agrees with results of Hoekstra [28], who found that many microsclerotia can be produced in plant residue of field bean, and removing the debris of field bean resulted in a lower microsclerotia population in the next spring. The apparent increase in microsclerotia population, a year after incorporation of infested peanut residue, may be explained by the disintegration of plant debris containing microsclerotia or reproduction on the susceptible cotton cultivar in 2008. Higher numbers of *V. dahliae* microsclerotia were found on cauliflower roots eight weeks after harvest [29]. Huisman and Ashworth [15] and Joaquim et al. [30] found a sharp increase in microsclerotia population in the second year after growth of a susceptible crop, despite host susceptibility.

Peanut residue is composed of vines and leaves remaining after harvest. These materials provide a good source of

nutrition [31] and are often used as a feed supplement. According to Darrell [32], the cost of feed represents the largest single cost item in most beef operations. This is an additional way to add economic benefits to removing peanut residue from the field. Balkcom et al. [17] found that peanut residue does not contribute significant amounts of nitrogen to the subsequent cotton crop; however, these interactions need to be investigated further under the arid conditions. While removing residue may remove nitrogen from the soil, the potential benefits are twofold reducing soil populations of *V. dahliae* and sale of hay for feedstock.

Results from the present study suggest the importance of removing the peanut residue infested with *V. dahliae*, for managing the Verticillium wilt in the subsequent cotton crop. Due to the lack of effective fungicides and truly resistant cultivars, disease management will likely rely on an integrated management program using a number of management options. Reducing inoculum of *V. dahliae* in host debris and other reservoirs may be a key to Verticillium wilt management. Cropping systems have a changing dynamics of disease and soilborne pathogens that are influenced by cultural practices such as residue management. The importance of primary inoculum in Verticillium wilt development, justifies the use of cultural management practices, including the use of partially resistant cotton cultivars and destruction of infested peanut residues for effective disease management.

Acknowledgments

This research was funded in part by the Texas Peanut Producers Board and the National Peanut Board. The authors would like to thank B. Mullinix for statistical analysis and C. Rowland for providing infested peanut residue. They also thank M. Batla, M. Ratliff, and I. Yates for technical support.

References

[1] A. H. McCain, R. D. Raabe, and S. Wilhelm, "Plants resistant or susceptible to Verticillium wilt," in *Cooperative Extension*, pp. 1–12, U.S. Department of Agriculture, University of California, Berkeley, Calif, USA, 1981.

[2] R. G. Bhat and K. V. Subbarao, "Host range specificity in *Verticillium dahliae*," *Phytopathology*, vol. 89, no. 12, pp. 1218–1225, 1999.

[3] E. A. Markakis, S. E. Tjamos, P. P. Antoniou, P. A. Roussos, E. J. Paplomatas, and E. C. Tjamos, "Phenolic responses of resistant and susceptible olive cultivars induced by defoliating and nondefoliating *Verticillium dahliae* pathotypes," *Plant Disease*, vol. 94, no. 9, pp. 1156–1162, 2010.

[4] T. E. Smith, "Occurrence of Verticillium wilt on peanuts," *Plant Disease Report*, vol. 44, article 435.

[5] I. J. Misaghi, J. E. DeVay, and J. M. Duniway, "Relationship between occlusion of xylem elements and disease symptoms in leaves of cotton plants infected with *Verticillium dahliae*," *Canadian Journal of Botany*, vol. 56, pp. 339–342, 1978.

[6] W. C. Schnathorst, "Life cycle and epidemiology of Verticillium," in *Fungal Wilt Diseases of Plants*, M. E. Mace, A. A. Bell, and C. H. Beckman, Eds., pp. 81–111, Academic Press, New York, NY, USA, 1981.

[7] G. S. Pullman and J. E. DeVay, "Epidemiology of Verticillium wilt of cotton: effects of disease development on plant phenology and lint yield," *Phytopathology*, vol. 72, pp. 554–559, 1982.

[8] H. A. Melouk, D. F. Wadsworth, and J. L. Sherwood, "Effect of Verticillium wilt on root and top weight of peanut cultivar Tamnut 74," *Plant Disease*, vol. 67, pp. 1349–1350, 1983.

[9] R. C. Rowe, J. R. Davis, M. L. Powelson, and D. I. Rouse, "Potato early dying: causal agents and management strategies," *Plant Disease*, vol. 71, pp. 482–489, 1987.

[10] O. C. Huisman, "Colonization of field-grown cotton roots by pathogenic and saprophytic soilborne fungi," *Phytopathology*, vol. 78, pp. 716–722, 1988.

[11] S. Wilhelm, "Longevity of the Verticillium wilt fungus in the laboratory and field," *Phytopathology*, vol. 45, pp. 180–181, 1955.

[12] L. Mol and K. Scholte, "Formation of microsclerotia of *Verticillium dahliae* Kleb. on various plant parts of two potato cultivars," *Potato Research*, vol. 38, no. 2, pp. 143–150, 1995.

[13] H. Schneider, "Susceptibility of guayule to Verticillium wilt and influence of soil temperature and moisture on development of infection," *Journal of Agricultural Research*, vol. 76, pp. 129–143, 1948.

[14] E. J. Paplomatas, D. M. Bassett, J. C. Broome, and J. E. DeVay, "Incidence of Verticillium wilt and yield losses of cotton cultivars (*Gossypium hirsutum*) based on soil inoculum density of *Verticillium dahliae*," *Phytopathology*, vol. 82, pp. 1417–1420, 1992.

[15] O. C. Huisman and L. J. Ashworth Jr., "Influence of crop rotation on survival of *Verticillium albo-atrum* in soils," *Phytopathology*, vol. 66, pp. 978–981, 1976.

[16] L. Mol, K. Scholte, and J. Vos, "Effects of crop rotation and removal of crop debris on the soil population of two isolates of *Verticillium dahliae*," *Plant Pathology*, vol. 44, no. 6, pp. 1070–1074, 1995.

[17] K. S. Balkcom, C. W. Wood, J. F. Adams, and B. H. Wood, "Composition and decomposition of peanut residue in Georgia," *Peanut Science*, vol. 31, pp. 6–11, 2004.

[18] I. Isaac, P. Fletcher, and J. A. C. Harrison, "Quantitative isolation of *Verticillium* spp. from soil and moribund potato haulm," *Annals of Applied Biology*, vol. 67, pp. 177–83, 1971.

[19] L. H. Sorensen, A. T. Scheider, and J. R. Davis, "Influence of sodium polygalacturonate sources and improved recovery of *Verticillium* spp. from soil (abstract)," *Phytopathology*, vol. 81, p. 1347, 1991.

[20] Z. Kabir, R. G. Bhat, and K. V. Subbarao, "Comparison of media for recovery of *Verticillium dahliae* from soil," *Plant Disease*, vol. 88, no. 1, pp. 49–55, 2004.

[21] R. G. D. Steel and J. H. Torrie, *Principles and Procedures of Statistics*, McGraw-Hill Book Company, 1960.

[22] J. R. Davis, J. J. Pavek, and D. L. Corsim, "A sensitive method for quantifying *Verticillium dahliae* colonization in plant tissue and evaluating resistance among potato genotypes," *Phytopathology*, vol. 73, pp. 1009–1014, 1983.

[23] L. Mol, "Effect of plant roots on the germination of microsclerotia of *Verticillium dahliae*. II. Quantitative analysis of the luring effect of crops," *European Journal of Plant Pathology*, vol. 101, no. 6, pp. 679–685, 1995.

[24] E. A. Adee, C. R. Grau, and E. S. Oplinger, "Population dynamics of *Phialophora gregata* in soybean residue," *Plant Disease*, vol. 81, no. 2, pp. 199–203, 1997.

[25] E. C. Tjamos, "Virulence of *Verticillium dahliae* and *V. albo-atrum* isolates in tomato seedlings in relation to their host of origin and the applied cropping system," *Phytopathology*, vol. 71, pp. 98–100, 1981.

[26] Y. Zilberstein, I. Chet, and Y. Henis, "Influence of microsclerotia source of *Verticillium dahliae* on inoculum quality," *Transactions of the British Mycology Society*, vol. 81, pp. 613–617, 1983.

[27] S. A. Pereyra and R. Dill-Macky, "Colonization of the residues of diverse plant species by *Gibberella zeae* and their contribution to fusarium head blight inoculum," *Plant Disease*, vol. 92, no. 5, pp. 800–807, 2008.

[28] O. Hoekstra, "Effects of leguminous crops on potato production and on incidence of *Verticillium dahliae* in various crop rotations with potatoes," in *Effects of Crop Rotation on Potato Production in the Temperate Zones*, J. Vos, C. D. Van Loon, and G. J. Bollen, Eds., pp. 223–235, Kluwer Academic Publishers, Dordrecht, The Netherlands, 1989.

[29] C. L. Xiao, K. V. Subbarao, K. F. Schulbach, and S. T. Koike, "Effects of crop rotation and irrigation on *Verticillium dahliae* microsclerotia in soil and wilt in cauliflower," *Phytopathology*, vol. 88, no. 10, pp. 1046–1055, 1998.

[30] T. R. Joaquim, V. L. Smith, and R. C. Rowe, "Seasonal variation and effects of wheat rotation on populations of *Verticillium dahliae* Kleb. in Ohio potato field soils," *American Potato Journal*, vol. 65, no. 8, pp. 439–447, 1988.

[31] J. Parish, "Alternative feedstuffs for beef cattle operations—part II," TheBeefSite.com, 2006, http://www.thebeefsite.com/articles/663/alternative-feedstuffs-for-beef-cattle-operations-part-ii.

[32] L. R. Darrell Jr., "By-product feeds for Alabama beef cattle," Alabama Cooperative Extension System, 2004, http://www.aces.edu/pubs/docs/A/ANR-1237.pdf.

12

Grain Sorghum Response to Row Spacing and Plant Populations in the Texas Coastal Bend Region

Carlos J. Fernandez,[1] Dan D. Fromme,[2] and W. James Grichar[1]

[1] Texas AgriLife Research, Corpus Christi, TX 78406, USA
[2] Texas AgriLife Extension Service, Corpus Christi, TX 78406, USA

Correspondence should be addressed to Carlos J. Fernandez, cjfernandez.work@me.com

Academic Editor: David Clay

Two grain sorghum (Sorghum bicolor L. Moerch) studies were conducted in the Coastal Bend Region of Texas over a two-year period. In one study, sorghum growth and yield were compared when planted in a single row on beds or planted in twin rows on beds with different plant populations under dryland or irrigation. Above average rainfall occurred in May 2000 which resulted in twin rows at any plant population producing higher yields than the single row at lower plant population. In 2001, single-row plantings with either plant population (124,000–160,000 or 161,000–198,000 plants/ha) produced higher yield than twin rows planted at 161,000–198,000 plants/ha. Under irrigation, twin rows planted at 161,000–198,000 plants/ha produced higher yields than single row at the same population; however, no other yield differences were noted when row systems or plant populations were compared. In another study, 38 cm row spacings were compared with 76 cm row spacings under two plant populations. In 2000, when rains fell at an opportune time, no yield differences were noted; however, in 2001 with below average rainfall, the 76 cm plantings at 170,000–200,000 and 210,000–240,000 plants/ha produced higher yield than the 38 cm plantings at those same plant populations.

1. Introduction

Row spacing and plant populations are variables that can have a significant impact on the net returns of sorghum producers. Grain sorghum along the Texas Gulf Coast is commonly cultivated in rows 76–102 cm apart [1]. Recent technological developments in farming equipment and improved herbicides open new doors for using rows narrower than 76 cm or twin rows on a single bed in grain sorghum production. The use of narrow rows in grain sorghum production is not new. A number of previous studies published in the late 1950s and 1960s showed yield increases when grain sorghum was planted in narrow rows [2, 3].

Though optimal plant densities for grain sorghum differ from region to another, previous research has indicated that grain yields generally increase as plant populations increase [4–6]. At lower than suggested plant densities, grain sorghum head number per plant or seed number per head increased when compared to the recommended plant density [7–10].

In other crops, Grichar [11] reported variable results of the effects of different seeding rates in soybean (Glycine max L.). He reported that the effect of seeding rate on soybean yields varied from year to another depending on variety and rainfall received during the growing season. Brown et al. [12] showed a 34% yield increase in corn (Zea mays L.) grown on 51 cm rows compared with 102 cm rows. Fulton [13] reported that under conditions of adequate soil moisture, higher corn plant densities (54,360 plants/ha) produced greater yields than lower densities (39,540 plants/ha), and rows spaced at 50 cm produced higher yields than rows spaced 100 cm apart.

The row spacing in a crop can also impact crop yield potential [6, 14–16]. Weed-grain sorghum competition is intensified by open canopy structures [17], while narrow-row planting gives grain sorghum a competitive advantage over weeds [18]. Staggenborg [6] reported that crop row spacings of less than 76 cm would increase grain yield in areas with high yield potential with little risk of reduced yield in areas with lower yield potential. Grichar [19] reported that soybeans in a twin-row configuration yielded more than the single-row system 50% of the time. Reducing the distance between rows can also improve weed control by increasing crop competitiveness and reducing light transmittance to the soil [20–22]. Johnson et al. [23] reported that total weed densities were less when peanut (*Arachis hypogaea* L.) rows were spaced 30 cm apart compared with rows spaced 91 cm apart. Teasdale [24] showed that reduced row spacing and increased corn populations decreased weed growth in the absence of herbicides and shortened the time of canopy closure by one week. Significant yield increases were reported when grain sorghum was planted in double rows 38 cm apart on 97 cm beds [25]. This planting pattern increased grain yield 24% (1174 kg/ha) compared to single-row planting across two deficit-irrigation levels (76 and 152 mm of in-season irrigation) and two planting densities (148,000 and 222,000 plants/ha). Although the increase in yield was 26% with the higher level of irrigation and 22% with the lower level of irrigation, there were no significant differences between the two planting densities.

Seedlings in close proximity to each other express phytochrome-mediated responses by developing narrow leaves, long stems, and less massive roots [26]. Planting a crop in a pattern that reduces the spacing of plants within and between rows can increase plant biomass and leaf area index [27]. Work by Bullock et al. [27] showed that reduced row spacing increased the total interception of photosynthetic active radiation by the corn canopy and redistributed the radiation toward the top of the canopy.

Until recently, profitability of grain sorghum production, as well as that of other crops, has been in decline as production costs continue to increase and crop prices remain low [28]. Although grain sorghum has a comparative advantage over other summer crops with regard to its adaptability to make use of limited soil moisture conditions, its dryland production is characterized by variable yield as crop performance depends highly on soil moisture availability [29]. Higher and more stable yields are obtained under irrigation, but at higher production costs [30]. An increased and more stable profitability of grain sorghum production is essential to improve and secure the sustainability of the farming industry in the region.

More research is needed in the Coastal Bend region of Texas to further understand the effects of narrow-row planting (row spacing as well as twin rows on a bed) on grain sorghum production and its interactions with planting density. Many producers question whether there is a different optimum plant density for grain sorghum grown in narrow rows. For these reasons, the objectives of this study were to determine (a) plant density and (b) row spacing on grain sorghum growth and yield.

2. Materials and Methods

Two different studies were conducted in the Coastal Bend region of Texas. The first study was conducted under irrigated conditions in 2000 and dryland conditions in 2000 and 2001 to determine grain sorghum response to twin rows

planted on a bed compared to a single row planted on a bed using three different plant populations. This will be called Study 1. Another study was conducted under dryland conditions in 2000 and 2001 to determine grain sorghum response to two different row spacings (38 and 76 cm) and two different plant populations. This will be referred to as Study 2.

2.1. Research Sites. Study 1 was conducted at Texas AgriLife and Extension Center located approximately 3.2 km north of Corpus Christi (27.77°N and 97.51°W). This study was done under dryland conditions in both years but only under irrigation in 2000. Study 2 was conducted at Perry Foundation Farm located near Robstown (27.79°N and 97.51°W), which is approximately 13 km from Study 1 location. The soils at both locations were a Victoria clay (fine, montmorillonitic, and thermic Udic Pellusterts) with less than 1.0% organic matter and pH 7.5. Each year, the studies were moved to different locations within the same field.

2.2. Planting Dates and Plot Layout. In Study 1, in 2000 for both the dryland and irrigated study, Asgrow 459 was planted on March 25 and in 2001, Pioneer 84G62 was planted on March 13 at the dryland location. Due to issues with the irrigation system, the irrigated portion of this study was not planted in 2001. In Study 2, Pioneer 8313CG was planted on March 3, 2000, and Pioneer 84G62 was planted on March 7, 2001.

In both studies, a randomized complete block design with a 2 × 2 factorial treatment arrangement with 4 replicates was used. In Study 1, the two factors included two row configurations, single rows on a bed spaced 96.5 cm apart or twin rows spaced 30 cm apart on a bed. Beds were approximately 36–41 cm wide. In the dryland study, three plant populations (124,000–160,000, 161,000–198,000, 199,000–235,000 plants/ha) were planted in 2000 or two plant populations (124,000–160,000 and 161,000–198,000 plants/ha) in 2001. Under irrigation, the plant populations were the same as those for the 2000 dryland study. Row spacing and plant populations plots were 4 beds (9.8 m) wide and 47 m in length. In the irrigation study, water was applied with aboveground drip lines at panicle formation (122 mm) and at flowering (43 mm).

For Study 2, the two factors included two row spacings, 38 and 76 cm apart, and two plant populations of 170,000–200,000 and 210,000–240,000 plants/ha. Row spacing and plant populations plots were 16 rows for the wider spacing or 32 rows for the narrower spacing (12.7 m wide) and 152 to 170 m long.

For all studies, fertilizer was applied 45–60 days prior to planting based on the soil test recommendations provided by the Texas AgriLife Extension Service Soil and Plant Testing Laboratory. All treatments were planted on a flat seedbed. Grain sorghum was planted with a Monosem vacuum planter (Monosem ATI, Inc., 17135 West 116th Street, Leneka, KS 66219, USA) equipped with precision seed meters calibrated to deliver the desired seeding rate. Plant counts taking 4–6 wk after planting assured that each plot

TABLE 1: Monthly rainfall in the Corpus Christi area in 2000 and 2001.

Month	Monthly rainfall (mm)		
	2000	2001	62 yr average
February	15.5	18.5	45.7
March	93.5	51.6	35.6
April	26.0	0.8	47.2
May	121.9	34.8	71.8
June	66.3	77.2	81.8
July	0.3	26.4	59.7
August	24.4	142.8	76.2
Total	347.9	352.1	418.0

was within the desired plant populations. Atrazine (Aatrex (Drexel Chemical Company, Memphis, TN 38113, USA)) at 2.9 L/ha, S-metolachlor (Dual Magnum (Syngenta Crop Protection, Wilmington, DE 19810, USA)) at 1.22 L/ha, and glyphosate (Roundup Ultra (Monsanto Chemical Company, St. Louis, MO 63167, USA))at 1.22 L/ha were applied before emergence within two days after planting for weed control.

2.3. Data Collection and Analysis. In the single- and twin-row studies (Study 1), grain heads were hand harvested in early to mid-August, counted from 9.4 m of a center row, and threshed for grain yield determination. For the row spacing study (Study 2), grain was harvested using a commercial combine (John Deere 9600 (John Deere, Moline, IL 61265, USA)). Prior to harvest, the number of heads was counted in two 4.8 m section of center rows. Also, a sample was taken from each plot for measurement of moisture and 1000-grain weight. Crop weights were adjusted to 14% moisture.

Data were analyzed using PROC GLM with SAS (SAS Institute, Inc., Cary, NC, USA) and a model statement appropriate for a factorial design. Treatments means were separated by Fisher's protected least significant difference test at $P = 0.05$. Data for the two years were analyzed separately due to year-by-treatment interactions for all variables.

3. Results

3.1. Rainfall. Rainfall amounts in the Corpus Christi area were variable for the two years (Table 1). Rainfall in 2000 can be characterized as below average for February, April, June, and July and above average rainfall for March and May. In 2001, below average rainfall was received for February, April, May, and July. Total rainfall for the February through July growing season (plots were harvested early to mid-August) was below average for both years (Table 1).

3.2. Grain Sorghum Response to Single and Twin Rows and Plant Populations—Study 1

3.2.1. Heads/Plant. Under dryland conditions in 2000, the twin rows planted at 124,000–160,000 plants/ha produced fewer heads than the single row planted at 124,000–160,000 or 199,000–235,000 plants/ha (Table 2). The single row planted at 124,000–160,000 plants/ha produced a greater

number of heads/plant than the twin rows planted at any populations.

In 2001, the twin rows planted at 161,000–198,000 plants/ha produced less heads/plant than either population planted on a single row or the twin rows planted at 124,000–160,000 plants/ha (Table 2). Mascagni and Bell [31] reported that under dryland conditions, the yield component that contributed the most to a twin-row yield response was heads/ha. They concluded that there may have been more tillering for twin rows since the intrarow spacing was greater than the single row for a given seeding rate.

Under irrigation, no differences in heads/plants were noted (Table 3). Mascagni and Bell [31] reported that under irrigation, the number of heads/ha had no effect on sorghum yield.

3.2.2. Number of Seed/Head. The number of seed/head account for 70% of the grain sorghum yield and therefore play an important role in yield determination [32]. In 2000, under dryland conditions, both single- and twin-row plantings at 199,000–235,000 plants/ha produced the lowest number of seed/head (Table 2). The twin rows planted at 124,000–160,000 plants/ha produced the greatest number of seed/head which was greater than for all row spacing and plant populations with the exception of the twin rows planted at 161,000–198,000 plants/ha. In 2001, the single row planted at 124,000–160,000 plants/ha produced the greatest number of seed/head (Table 2). No differences were noted with any other row spacings or plant populations.

Under irrigation, the twin rows planted at 124,000–160,000 plants/ha produced the greatest number of seed/head, and this was greater than any other plant populations (Table 3). The single row planted at 199,000–235,000 plants/ha produced the lowest number of seed/head.

3.2.3. Grain Weight/Head. Under dryland conditions in 2000, the twin rows planted at 124,000–160,000 plants/ha produced the greatest grain weight/head (Table 2). This was greater than the single-row planting at 161,000 or greater plants/ha or the twin rows planted at 199,000 or greater plants/ha. In 2001, the single row planted at 124,000–160,000 plants/ha produced the greatest grain weight/head, and both single-row populations produced higher grain weight/head than either twin rows (Table 2). Karchi and Rudich [32] stated that yields were directly associated with number of heads per unit area and inversely associated with head weights.

3.2.4. Weight of 1000 Seed. Under dryland conditions, in neither year, there were not any differences in the weight of the 1000 seed with either row spacing or plant populations (Table 2). However, under irrigated conditions, the single-row planting at 124,000–160,000 seed/ha produced the highest weight, and this was greater than the weight of twin rows planted at any population (Table 3).

3.2.5. Yield. Under dryland conditions in 2000, the single-row planting at 124,000–160,000 plants/ha produced a lower

TABLE 2: Grain sorghum response to single and twin rows and plant populations under dryland conditions.

Rows	Plant populations	Heads/plant		Number of seed/head		Grain weight/head (g)		Weight of 1,000 seed (g)		Yield (kg/ha)	
	(1000s/ha)	2000	2001	2000	2001	2000	2001	2000	2001	2000	2001
Single row	124–160	0.98	0.71	1610	776	40.3	23.8	25.0	18.9	4400	2140
	161–198	0.88	0.63	1436	592	33.1	14.0	23.1	19.8	5110	1910
	199–235	0.90	—	1272	—	34.0	—	26.9	—	5400	—
Twin rows	124–160	0.77	0.68	1937	531	48.1	10.4	24.8	19.6	5700	1810
	161–198	0.81	0.42	1650	542	40.4	10.1	24.5	18.7	5640	1430
	199–235	0.84	—	1449	—	34.4	—	23.8	—	5780	—
LSD (0.05)		0.12	0.13	316	150	11.2	2.3	NS	NS	1010	570

TABLE 3: Grain sorghum response to single and twin rows and plant populations under irrigated conditions in 2000.

Rows	Plant population (1000s/ha)	Heads/plant	Number of seed/head	Weight of 1,000 seed (g)	Yield (kg/ha)
Single row	124–160	0.86	1386	26.7	5980
	161–198	0.84	1206	25.7	5900
	199–235	0.89	1005	26.4	6040
Twin rows	124–160	0.82	1639	23.1	6390
	161–198	0.86	1393	24.0	6840
	199–235	0.88	1161	23.8	6730
LSD (0.05)		NS	222	2.0	880

yield than twin rows planted at any plant population (Table 2). This is atypical of what usually occurs. Rainfall was above normal in May (Table 1), and this contributed to the greater yield with the twin-row system. This rainfall occurred during panicle formation and early flowering. This growth stage is the period when the plants are especially sensitive to any type of stress as water deficits, and this stage is considered the most critical period for grain production [33]. In 2001, the single row planting at 124,000–160,000 plants/ha produced yields that were greater than the twin rows planted at 161,000–198,000 plants/ha. No other differences in yield were noted in 2001.

Under irrigated conditions, the twin rows planted at 161,000–198,000 plants/ha produced greater yield than the single row planted at the same plant population. No other differences in yield was noted with any row spacing or plant population (Table 3). Since irrigation was applied during panicle formation, the plant was not allowed to stress, which would have reduced the number of seeds/plant [33].

3.3. Grain Sorghum Response to Row Spacing and Plant Populations—Study 2

3.3.1. Heads/Plant.
In 2000, no differences were noted in heads/plants when row spacing and plant populations were compared (Table 4). In 2001, the 76 cm row spacing planted at 210,000–240,000 plants/ha produced the greatest number of heads/plant, and this was greater than the 38 cm row spacing at either plant population or the 76 cm row spacing planted at the lower plant population.

3.3.2. Number of Seed/Head.
No differences were noted in either year with any row spacing or plant population (Table 4).

3.3.3. Weight of 1000 Seed.
In 2000, the 76 cm row spacing planted at 170,000–200,000 plants/ha produced the greatest weight, and this was greater than the 38 cm row spacing at either plant population (Table 4). In 2001, the 38 cm row spacing planted at 210,000–240,000 plants/ha produced the lowest weight, and trends were similar to those seen in 2000.

3.3.4. Yield.
In 2000, no differences in yield were noted with either row spacing or plant population (Table 4). In 2001, both 76 cm row spacings at either plant population produced higher yields than the 38 cm planting at either plant population. It has been reported that the yield response to narrow rows in corn and grain sorghum is affected by many environmental, spatial, and temporal field interactions [4–6, 13, 21]. It has also been suggested that a positive yield response to narrow rows is more likely to occur in the presence of environmental yield-limiting factors. Andrade et al. [22] reported that the narrow-row yield response was inversely proportional to the radiation interception achieved with wider rows. Under very favorable growing conditions, when radiation interception for wide rows was optimized, the yield response to narrowing the rows was minimized.

4. Summary

In the year when rainfall fell at the most opportune time (panicle formation), the twin rows produced higher yields

TABLE 4: Grain sorghum response to row spacing and plant populations under dryland conditions.

Row spacing (cm)	Plant population (1000s/ha)	Heads/plant		Number of seed/head		Weight of 1,000 seed (g)		Yield (kg/ha)	
		2000	2001	2000	2001	2000	2001	2000	2001
38	170–200	0.93	0.79	744	1234	25.3	14.0	4610	1665
	210–240	0.91	0.81	702	969	24.8	13.3	4630	1726
76	170–200	0.95	0.80	846	1220	28.5	15.3	4666	2104
	210–240	0.96	1.03	772	733	27.4	15.7	4576	2104
LSD (0.05)		NS	0.18	NS	NS	2.1	2.4	NS	339

than single-row plantings. Under less rainfall, neither system (twin or single) produced yield differences within their respective plant population. Under irrigation, no differences in yield were noted between twin- and single-row plantings at 124,000–160,000 and 199,000–235,000 plants/ha; however, at 161,000–198,000 plants/ha, the twin-row plantings produced higher yields than the single-row planting. Comparing narrow-row plantings (38 cm) with a more conventional row spacing (76 cm), when rainfall occurred during panicle formation or beginning of flowering and the early boot stage, no yield difference between row spacing was noted; however, when rainfall was limiting, the narrow-row plantings produced less yield than conventional row spacing. Results for the differences in plant populations were mixed. The lower plant populations did not always produce the greater yields especially in dryer years.

Responses to narrow-row spacing in grain sorghum have been varied and inconsistent. Conley et al. [5] reported that grain yield response to row spacing was variable and dependent on environment. Welch et al. [34] reported that in the presence of adequate nitrogen, production of grain and residue increased with increasing populations. They concluded that under dryland conditions, optimum populations for both grain and residue production were between 100,000 and 150,000 plants/ha and that at populations of 100,000 plant/ha, grain and residue yields in 40 cm rows equaled or exceeded those in 102 cm rows. In contrast, Katchi and Rudich [32] in Israel reported, under dryland conditions, that greater yields resulted from narrow rows combined with lower plant populations. They found that the greater grain yields were primarily due to increased number of heads per unit area rather than to changes in head weight. They also stated that heads per unit area and the number of kernels per head were largely free of environmental effects.

Also, grain sorghum grown under dryland conditions and deficit irrigation is commonly exposed to water stress. The effect of water stress on yield, however, depends on its timing and intensity [35]. Narrow-row planting affects the canopy structure and, therefore, would affect the rate of development of soil water deficits and the timing of the onset of plant water stress. An early report regarding the interaction of soil moisture and row width indicated that the optimum row width increases as soil moisture becomes more limiting [29]. However, Steiner [36] concluded that narrow-row planting appeared to increase the transpiration component of evapotranspiration, thus increasing the production efficiency of dryland grain sorghum. Supporting Steiner's

conclusions, Sanabria et al. [37] concluded that narrow-row, north-south planting patterns resulted in water conservation through enhanced stomatal control of transpiration under conditions of high evaporative demand.

References

[1] D. D. Fromme, C. J. Fernandez, W. J. Grichar, and R. L. Jahn, "Grain sorghum response to hybrid, row spacing, and plant populations along the upper Texas Gulf Coast," *International Journal of Agronomy*, vol. 2012, Article ID 930630, 5 pages, 2012.

[2] K. B. Porter, M. E. Johnson, and W. H. Stetten, "The effect of row spacing, fertilizer, and planting rates on yield and water use of irrigated grain sorghum," *Agronomy Journal*, vol. 52, pp. 431–433, 1960.

[3] F. C. Stickler and S. Wearden, "Yield and yield components of grain sorghum as affected by row width and stand density," *Agronomy Journal*, vol. 57, pp. 564–567, 1965.

[4] O. R. Jones and G. L. Johnson, "Row width and plant density effects on Texas High Plains sorghum," *Journal Production Agriculture*, vol. 4, pp. 613–621, 1991.

[5] S. P. Conley, W. G. Stevens, and D. D. Dunn, "Grain sorghum response to row spacing, plant density, and planter skips," *Crop Management*, 2005.

[6] S. A. Staggenborg, "Grain sorghum response to row spacings and seeding rates in Kansas," *Journal of Production Agriculture*, vol. 12, no. 3, pp. 390–395, 1999.

[7] T. J. Gerik and C. L. Neely, "Plant density effects on main culm and tiller development of grain sorghum," *Crop Science*, vol. 27, pp. 1225–1230, 1987.

[8] T. A. Lafarge and G. L. Hammer, "Predicting plant leaf area production: shoot assimilate accumulation and partitioning, and leaf area ratio, are stable for a wide range of sorghum population densities," *Field Crops Research*, vol. 77, no. 2-3, pp. 137–151, 2002.

[9] T. A. Lafarge and G. L. Hammer, "Tillering in grain sorghum over a wide range of population densities: modelling dynamics of tiller fertility," *Annals of Botany*, vol. 90, no. 1, pp. 99–110, 2002.

[10] Y. O. M'Khaitir and R. L. Vanderlip, "Grain sorghum and pearl millet response to date and rate of planting," *Agronomy Journal*, vol. 84, pp. 579–582, 1992.

[11] W. J. Grichar, "Planting date, cultivar, and seeding rate effects on soybean production along the Texas Gulf Coast," *Crop Management*, 2007.

[12] R. H. Brown, E. R. Beaty, W. J. Ethredge, and D. D. Hays. ", "Influence of row width and plant population on yield of two varieties of corn (*Zea mays* L.)," *Agronomy Journal*, vol. 62, pp. 767–770, 1970.

[13] J. M. Fulton, "Relationships among soil moisture stress, plant populations, row spacing, and yield of corn," *Canadian Journal Plant Science*, vol. 50, pp. 31–38, 1970.

[14] B. A. Besler, W. J. Grichar, S. A. Senseman, R. G. Lemon, and T. A. Baughman, "Effects of row pattern configurations and reduced (1/2x) and full rates (1x) of imazapic and diclosulam for control of yellow nutsedge (*Cyperus esculentus*) in peanut," *Weed Technology*, vol. 22, no. 3, pp. 558–562, 2008.

[15] A. Limon-Ortega, S. C. Mason, and A. R. Martin, "Production practices improve grain sorghum and pearl millet competitiveness with weeds," *Agronomy Journal*, vol. 90, no. 2, pp. 227–232, 1998.

[16] H. H. Bryant, J. T. Touchton, and D. P. Moore, "Narrow rows and early planting produce top grain sorghum yields," *Highlights Agriculture Research Alabama Agricultural Experiment Station*, vol. 33, article 5, 1986.

[17] A. P. Everaarts, "Effects of competition with weeds on the growth, development and yield of sorghum," *Journal of Agricultural Science*, vol. 120, no. 2, pp. 187–196, 1993.

[18] R. H. Walker and G. A. Buchanan, "Crop manipulation in integrated weed management systems," *Weed Science*, vol. 30, pp. 17–24, 1982.

[19] W. J. Grichar, "Row spacing, plant populations, and cultivar effects on soybean production along the Texas Gulf Coast," *Crop Management*, 2007.

[20] B. E. Tharp and J. T. Kells, "Effect of glufosinate-resistant corn (*Zea mays*) population and row spacing on light interception, corn yield, and common lambsquarters (*Chenopodium album*) growth," *Weed Technology*, vol. 15, pp. 413–418, 2001.

[21] K. D. Thelen, "Interaction between row spacing and yield: why it works," *Crop Management*, 2006.

[22] F. H. Andrade, P. Calviño, A. Cirilo, and P. Barbieri, "Yield responses to narrow rows depend on increased radiation interception," *Agronomy Journal*, vol. 94, no. 5, pp. 975–980, 2002.

[23] W. C. Johnson, E. P. Prostko, and B. G. Mullinix, "Improving the management of dicot weeds in peanut with narrow row spacings and residual herbicides," *Agronomy Journal*, vol. 97, no. 1, pp. 85–88, 2005.

[24] J. R. Teasdale, "Influence of narrow row/high population corn (*Zea mays*) on weed control and light transmittance," *Weed Technology*, vol. 9, no. 1, pp. 113–118, 1995.

[25] C. J. Fernandez, T. Foutz, and R. Schawe, "Increasing irrigated grain sorghum yield through double-row planting," in *4th Australian Sorghum Conference*, A. K. Borrell and R. G. Hensell, Eds., Department of Primary Industries-Queensland Government, Grains and Research Development Corporation, Queensland, Australia, 2001.

[26] M. J. Kasperbauer and D. L. Karlen, "Plant spacing and reflected far-red light effects on phytochrome-regulated photosynthate allocation in corn seedlings," *Crop Science*, vol. 34, no. 6, pp. 1564–1569, 1994.

[27] D. G. Bullock, R. L. Nielsen, and W. E. Nyquist, "A growth analysis comparison of corn growth in conventional and equidistant plant spacing," *Crop Science*, vol. 28, pp. 254–258, 1988.

[28] H. C. Dethloff and G. L. Nall, "AGRICULTURE," Handbook of Texas Online, Texas State Historical Association, 2012, http://www.tshaonline.org/handbook/online/articles/ama01.

[29] P. L. Brown and W. D. Shrader, "Grain yields, evapotranspiration and water use efficiency of grain sorghum under different cultural practices," *Agronomy Journal*, vol. 51, pp. 339–343, 1959.

[30] C. Stichler, M. McFarland, and C. Coffman, "Irrigated and dryland grain sorghum production," 2012, http://publications.tamu.edu/CORN_SORGHUM/PUB_Irrigated%20and%20Dryland%20Grain%20Sorghum%20Production.pdf.

[31] H. J. Mascagni and B. Bell, "Plant patterns for different grain sorghum hybrids," *Louisiana Agriculture Magazine*, 2005, http://www.Isuagcenter.com/en/communications/publications/agmag/Archive/2005/Winter/Plant+Patterns+for+Different+Grain+Sorghum+Hybrids.htm.

[32] Z. Karchi and Y. Rudich, "Effects of row width and seedling spacing on yield and its components in grain sorghum grown under dryland conditions," *Agronomy Journal*, vol. 58, pp. 602–604, 1966.

[33] J. Kelley, "Chapter 1: growth and development," Grain Sorghum Handbook, MP 297, http://www.uaex.edu/Other_areas/publications/PDF/MP297/.

[34] N. H. Welch, E. Burnett, and H. V. Eck, "Effect of row spacing, plant population, and nitrogen fertilization on dryland grain sorghum production," *Agronomy Journal*, vol. 58, pp. 160–163, 1966.

[35] M. M. Jones and H. M. Rawson, "Influence of rate of development of leaf water deficits upon photosynthesis, leaf conductance, water use efficiency, and osmotic potential in sorghum," *Physiologia Plantarum*, vol. 45, no. 1, pp. 103–111, 1979.

[36] J. L. Steiner, "Dryland grain sorghum water use, light interception, and growth responses to planting geometry," *Agronomy Journal*, vol. 78, pp. 720–726, 1986.

[37] J. R. Sanabria, J. F. Stone, and D. L. Weeks, "Stomatal response to high evaporative demand in irrigated grain sorghum in narrow and wide row spacing," *Agronomy Journal*, vol. 87, no. 5, pp. 1010–1017, 1995.

13

Nectarine Fruit Ripening and Quality Assessed Using the Index of Absorbance Difference (I_{AD})

E. Bonora,[1] D. Stefanelli,[2] and G. Costa[1]

[1] *Department of Agricultural Sciences, 46 Via Fanin, 40127 Bologna, Italy*
[2] *Department of Primary Industries, Knoxfield Centre, Private Bag 15, 621 Burwood Highway, Ferntree Gully 3156, Australia*

Correspondence should be addressed to E. Bonora; elisa.bonora4@unibo.it

Academic Editor: Anish Malladi

Consistency of fruit quality is extremely important in horticulture. Fruit growth and quality in nectarine are affected by fruit position in the canopy, related to the tree shape. The "open shaped" training systems, such as Tatura Trellis, improve fruit growth and quality. The Index of Absorbance Difference (I_{AD}) is a new marker that characterizes climacteric fruit during ripening. A study on fruit ripening was performed by using the I_{AD} on nectarine to monitor fruit maturity stages of two cultivars trained as Tatura Trellis in Victoria, Australia. Fruit of cv "Summer Flare 34" ("SF34") grown in different positions on the tree showed high ripening homogeneity. Fruit harvested at a similar ripening stage showed fruit firmness and soluble solid content homogeneity. Fruits from hand-thinned variety "Summer Flare 26" ("SF26") were larger in size, had advanced ripening, and showed greater homogeneity. For "SF26", a weak correlation between I_{AD} and SSC was observed. The experiment showed that the Tatura Trellis training system is characterized by high homogeneity of nectarine fruit when coupled with a proper management of fruit density. It also confirmed that the I_{AD} could be used as new nondestructive maturity index for nectarine fruit quality assessment in the field.

1. Introduction

A tree training system is defined as a method of manipulating the tree structure and canopy geometry to improve the interception and distribution of light, for the purpose of optimizing fruit quality and yield [1]. In 1970, a group of Australian researchers developed the Tatura Trellis [2], suitable for the complete mechanization of harvest in intensive peach orchards. Despite of the higher light available and photosynthetic rate that this tree shape allows, it was judged too expensive because of the intensive work needed to maintain the complex scaffold. Keeping the same open canopy design, simplified and cheaper tree shapes were developed during the following decades, such as the "KAC V" [3] and "Y" [1]. Several aspects of the Tatura Trellis training system on apple and cherry trees were studied [4], but only a few experiments on tree productivity were available regarding peach fruit [5]. Numerous studies on different tree architectures pointed out that fruit position in the canopy represents one of the most critical factors for peach fruit quality development and homogeneity of fruit characteristics [6–8] related to the light availability [9]. The open center training systems increase the light available in the inner canopy, giving rise to a gradient of quality traits. Fruit that develops in the periphery and center of the canopy obtains higher light levels and is characterized by better quality attributes, while fruits located halfway between the tree center and periphery are more shaded and developed lower quality [8, 9]. Final fruit size and quality may also depend on shoot length, fruit distribution on the shoot, and number of fruit per centimeter of shoot length [10]. The correct management of the fruit density in relation to the position and light exposure is required to get optimal fruit size [6, 11]. Several studies in the past attempted to evaluate crop load and fruit quality distribution in different training systems [12, 13]. For tree shapes that allow a uniform light distribution, fruit thinning has to be consistently performed in every part of the tree [14]. Farina et al. [6] showed that a balanced peach fruit number on an open shape produced a greater number of large-sized fruit.

As well as the final commercial diameter, the quality traits commonly used as indicators of peach and nectarine maturity stage into the orchard are the changes in fruit firmness and background color turning from green to yellow [15]. The changes observed in the appearance and quality traits are related to the time course of ethylene production in ripening, since peaches and nectarines are climacteric fruits. Peach fruit characteristics such as soluble solids content, red color, and background color show a clear gradient related to fruit position in the canopy [6, 7, 9]. Farina et al. [6] reported different gradients of peach fruit firmness in different training systems, while other authors found that light intensity did not affect fruit firmness [8]. Changes in the background color and fruit firmness in peaches are generally linked, but light interception or canopy position may alter the relationship between these two parameters [7]. In fact, while recent studies on peach fruit observed that as firmness declines, background color became more yellow and less green; it was also pointed out that fruit with similar background color harvested from different positions into the canopy may not have the same fruit firmness [7, 16]. Iglesias and Echeverría [17] reported that peach fruit firmness alone is not a satisfactory minimum maturity index because it varies between nectarine cultivars, and for a given cultivar firmness varies in relation to fruit size, climatic conditions, and agronomical practices. Instead, background color is an informative harvest index as it reflects the chlorophyll content of the fruit [18]. Cascales et al. [19] found that changes in peach fruit background color due to chlorophyll degradation are proportional to those perceived by a panel of assessors. Recently, based on the vis-NIR spectroscopy a new measurement, the "index of absorbance difference" (I_{AD}) that strongly correlates with the chlorophyll-a content and the ethylene production of peach and nectarine fruit was introduced [20]. The I_{AD} could be used for individual cultivars to define the ideal time to harvest in accordance with consumer preferences, as shown by its higher correlation with consumer acceptance than with the traditional quality parameters found by Gottardi et al. [21]. However, few results are available regarding the use of the I_{AD} as a ripening index for peach and nectarine fruit.

The objectives of this study were (a) to evaluate the possible application of the I_{AD} as a nondestructive maturity index to follow fruit ripening in the field and objectively define the ideal harvesting time, (b) to characterize the performance of the Tatura Trellis training system for nectarine in affecting fruit quality, maturity, and homogeneity.

2. Materials and Methods

Trials were conducted in 2010 on two six-year-old yellow flesh nectarines (*Prunus persica* [L.] Batsch) cultivar "Summer Flare 34" and "Summer Flare 26" grafted onto "GF677" (*P. persica x P. dulcis*). The orchards were located in Ardmona, Victoria, Australia (−36.38 N, 145.32 S), and trees were trained to a North-South oriented Tatura Trellis system with spacing of 4.5 m between rows and 3.0 m within row (4.5 m × 3.0 m) and a planting density of 740 trees/ha. Industry standard management techniques were applied throughout the

TABLE 1: Fruit maturity stage, I_{AD} value and corresponding ethylene production (nl L^{-1} h^{-1} g^{-1} FW) of the two nectarine varieties SF34 and SF26.

Variety	Maturity stage	I_{AD} value	Ethylene emission (nl L^{-1} h^{-1} g^{-1} FW)	
SF34	PM[z]	0.3–0.6	2.13	a[y]
	CM	0.6–1.3	0.26	b
	I	1.3–1.6	0.06	b
SF26	PM	0.3–0.6	2.14	a
	CM	0.6–1.0	0.40	b
	I	1.0–1.3	0.01	b

[z] Fruit maturity stages: physiological maturity (PM), commercial maturity (CM), and immature (I).
[y] Numbers with different letters would be statistically significant at $P < 0.005$; LSD = 0.05.

season in terms of pruning, irrigation, fertilization, and pest control. No summer pruning was applied, neither reflective mulches were used in the orchard under study. Full bloom dates recorded for the two cultivars were the 14th and 16th of October 2010 for "SF34" and "SF26", respectively. Three similar trees for "SF34" and six similar trees for "SF26" were randomly selected within each orchard. To understand intracanopy variability, the canopies were divided in three parallel horizontal areas of equal size representing the top (T), middle (M), and bottom (B) canopy layers as described by He et al. [8].

Fruit maturity was assessed by measuring the Index of Absorbance Difference (I_{AD}) with the DA-Meter (TR, Forlì, Italy), a portable vis/NIRs that correlates with chlorophyll-a content and ethylene production as described by Ziosi et al. [20]. Fruits of the two cultivars "SF34" and "SF26" were catalogued into three ripening classes at harvest representing physiological maturity (PM), that corresponded to the ethylene production; commercial maturity (CM); at the onset of climacteric and immature (I), the time before the climacteric (Table 1).

To detect the maturity stage at which the ethylene climacteric occurred in cultivars "SF34" and "SF26", ethylene production was assessed on a sample of 10 to 15 fruits per I_{AD} class (forty fruits in total), randomly picked one week before the main harvest. Fruits were individually placed in sealed 1 L jars, and a 1.0 mL gas sample removed and injected into a Shimadzu GC-14B packed-gas chromatograph (column = packed alumina SS 80/100 180 cm; 140°C; Inj/Det = 180°C, Shimadzu, Kyoto, Japan). Fruits were left to incubate for at least one hour at 20°C prior to a second gas sample being removed and injected into the GC. The ethylene production was calculated as the difference between the result of the second and the first injection.

For both cultivars at harvest, twenty to fifty fruits per I_{AD} class were assessed with the standard quality trait measured: fruit firmness (FF), soluble solids content (SSC), percent blush, and a* and b* on both blush and background color of the peel. Fruit firmness was measured on the two opposite cheeks using a FT011 hand-operated Effegi penetrometer (Effegi, Ravenna, Italy) equipped with an 8 mm diameter

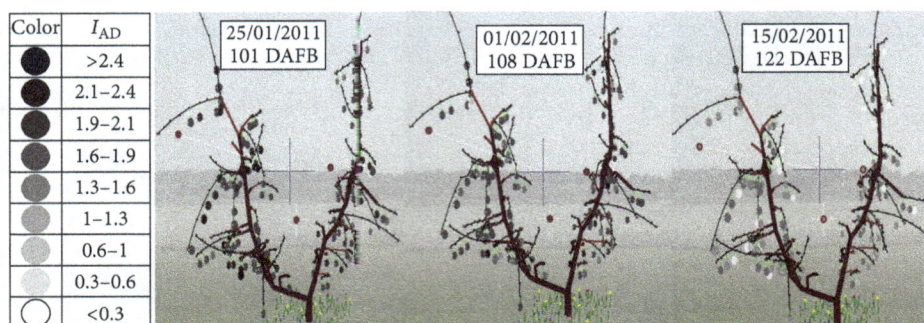

FIGURE 1: PlantToon image of the fruit ripening distribution (I_{AD}) at 101, 108, and 122 DAFB (cv SF34). The white circles ($I_{AD} < 0.3$) as well as the circles colored with the lighter shade of grey ($0.3 < I_{AD} < 0.6$) represent fruits at their physiological maturity stage (PM). The higher the I_{AD} value and the more unripe the fruit, the darker the shade of grey. The most unripe fruits are represented by black circles ($I_{AD} > 1.4$).

Magness-Taylor probe and mounted on a hand-operated drill press. Central portion of each fruit cheek was squeezed, and SSC was determined with a digital hand-held refractometer (PAL-1, Atago, Tokyo, Japan). The percent of blush was visually evaluated and expressed as percentage of the fruit surface covered with a uniform red color (0% corresponded to a fruit that did not develop any blush; 100% corresponded to a fruit completely red). The a^* and b^* color-component dimensions, based on nonlinearly compressed coordinates, were measured with a CR400 Minolta digital colorimeter (Konica-Minolta, Tokyo, Japan).

"SF34" was used to evaluate the influence of the Tatura Trellis system on fruit maturity and quality. Five fruits per each canopy layer (bottom, middle, and top) from the east and west sides of the canopy of every tree were tagged (ninety fruits in total) and followed during the growing season. To evaluate the influence that fruit position within the canopy had on nectarine development, fruit growth (diameter) and ripening (I_{AD}) were weekly monitored on tagged fruits from 83 to 130 days after full bloom (DAFB). The first harvest was performed at 122 DAFB and the main harvest at 130 DAFB. Fruit ripening distribution was measured with the DA-meter on a total population of 100 randomly selected fruits picked from the trees under study at the main harvest. The previously described standard laboratory quality assessments were performed on a sample of twenty to fifty fruits per I_{AD} class.

On one of the trees in trial, the spatial position of each fruit, as well as the complete canopy structure, was identified using a "woody stick-compass system" (Costa et al. [22]; personal communication) to obtain length, direction (°N), and horizontal projection of each element following. By inputting the collected data in the 3D graphic software PlantToon [23], the architecture of the tree was recreated and modeled in order to link the relative position of each fruit with the information collected from the field (I_{AD}).

"SF26" was used to evaluate the effect of fruit density on fruit maturity of Tatura Trellis grown trees. Six branches with similar length (around 40 to 50 cm), one branch per each canopy layer (bottom, middle, and top), and orientation (East, West) were selected and tagged on each of the six trees in trial (thirty-six branches in total), based on the assumption

that peach tree branches behaved as functionally autonomous units, as demonstrated by Volpe et al. [24]. All tagged branches from three trees were hand thinned to 4 fruits per branch (1 fruit every 10–12 cm of shoot length) 15 to 20 DAFB (as suggested [10]), while all the tagged branches from the remaining three trees were left unthinned with roughly 8 fruits per branch (1 fruit every 5-6 cm of shoot length).

Fruit growth (diameter) and ripening (I_{AD}) were monitored weekly from 68 to 89 DAFB on the fruit from all tagged branches. As previously described, to assess the correlation between the fruit ripening stage (I_{AD}) and SSC, during fruit growth a sample of fifteen fruits was collected weekly. Harvest was performed in two picks one week apart (main harvest was at 89 DAFB). Because of an overwhelming infection of *Monilia laxa* near harvest, the standard laboratory quality assessments (FF and SSC) at harvest of "SF26" were not performed.

The study was organized as completely randomized design. All the collected data were statistically evaluated using the Duncan's multiple range t-test at $P < 0.05$. The interactions between factors were assessed with a multiple factor ANOVA test. Both the statistical evaluations were performed with the software STATISTICA 7 (StatSoft. Inc., Tulsa, OK, USA).

3. Results

Table 1 shows the I_{AD} values at which the fruit maturity stage ranges start affecting the physiology of the fruit for both "SF34" and "SF26" nectarines. I_{AD} values were different for the two cultivars even when inside the same maturity stage. Fruits at an immature stage (>1.3 I_{AD} value for "SF34" and >1.0 for "SF26") were preclimacteric with negligible ethylene production. Fruits at commercial maturity show the onset of the climacteric with the starting of ethylene production (I_{AD} values of 0.6 to 1.3 for SF34 and 0.6 to 1.0 for SF26). Below a 0.6 I_{AD} value in both cultivars, fruits were at the physiological maturity stage with high ethylene production.

As shown in Figure 1, at 101 and 108 DAFB fruits of the "SF34" were immature, with I_{AD} values greater than 1.3–1.6. At 122 DAFB, fruits in the outer canopy appeared riper than

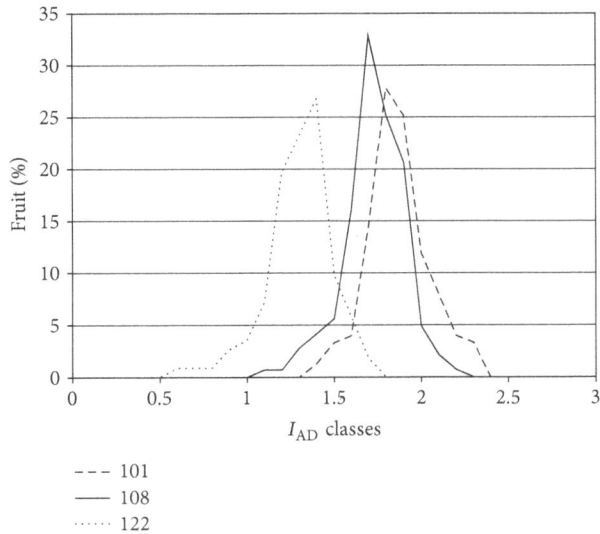

FIGURE 2: Fruit ripening distribution curves between I_{AD} classes at 101, 108, and 122 DAFB (SF34).

fruits in the inner and bottom canopies as shown by the light gray and white circles representing the riper fruit.

At every sampling, fruit ripening distribution between I_{AD} classes was concentrated in a narrow range of values (Figure 2), showing a high fruit ripening homogeneity. The three curves seemed to maintain the same shape over time and only sliding toward lower I_{AD} values when the fruit became riper (122 DAFB).

As shown in Table 2, significant differences were observed monitoring fruit ripening of the "SF34" every week from 93 to 130 DAFB during which the I_{AD} values progressively decreased from around 1.8 to roughly 0.8 in a month.

Fruit reached the onset of climacteric at 122 DAFB and harvested at 122 and 130 DAFB for "SF34" cultivar. The maturity stage of the fruit was not different between the three horizontal canopy layers, bottom, middle, and top, over time (data not shown), while fruit growth during the season was significantly affected by fruit positioning in the canopy (Table 3). At the first sampling (73 DAFB), fruits from the T canopy layer were larger in diameter than fruits in the M and B canopy layers (44.5 mm, 42.8 mm, and 41.2 mm, resp.). For the rest of the season and up to the first harvest (122 DAFB), fruits in the B canopy layer had on average 2 to 4 mm smaller diameters than the M and T canopy layers (Table 3). At the main harvest (130 DAFB) of "SF34", no additional differences were observed between fruit diameters from the three canopy layers (average of 71 to 73 mm).

Fruits of "SF34" showed a high ripening homogeneity at the main harvest (130 DAFB) at which more than 80% of the fruits were included in the CM class, whereas only 3% of the fruits were in the I class (I_{AD} value greater than 1.3), and the remaining (17%) were at the physiological maturity stage (I_{AD} 0.3–0.6).

Riper fruit (PM) did not show any significant differences between bottom, middle, and top canopy layers in term of percent of blush at harvest (55 to 60%). Fruits from top

TABLE 2: Average of the fruit maturity stage (I_{AD}) on the tree at 93, 101, 108, 122, and 130 DAFB (cv SF34).

DAFB	I_{AD}	
93	1.87	a[z]
101	1.81	b
108	1.65	c
122	1.26	d
130	0.78	e

[z]Numbers with different letters would be statistically significant at $P < 0.005$; LSD = 0.05.

TABLE 3: Average of the fruit diameter in the bottom (B), middle (M), and top (T) canopy layers of the tree at 73, 80, 93, 101, 108, 122, and 130 DAFB (cv SF34).

DAFB	Layer	Diameter (mm)	
73	B	41.2	b[z]
	M	42.8	b
	T	44.5	a
80	B	42.7	b
	M	44.6	a
	T	46.0	a
93	B	50.4	b
	M	53.3	a
	T	55.3	a
101	B	54.5	b
	M	57.1	ab
	T	57.7	a
108	B	58.7	b
	M	61.2	a
	T	62.3	a
122	B	65.9	b
	M	69.4	a
	T	69.2	a
130	B	71.1	a
	M	71.3	a
	T	73.2	a

[z]Numbers with different letters would be statistically significant at $P < 0.005$.

canopy layer of the CM class were more colored (60% blush) than bottom fruit (40% blush) of the same class, while no differences were shown between fruits of the I class, that developed only 10% blush, independent of the canopy layer. Immature fruits that were east exposed had less blush than west-oriented fruits (data not shown). No significant differences between ripening classes and canopy layers in term of a* and b* components of both blush and background color were observed (data not shown), while traditional destructive quality parameters were differently affected by the fruit ripening stage and the position in the canopy, as shown in Table 4.

No differences were observed for fruit firmness between fruits within the same ripening class, coming from the three canopy layers. If we consider the canopy layers, only the

TABLE 4: Average of firmness (FF kg/cm^2) and soluble solids content (SSC °Brix) of fruits at the I_{AD} classes of physiological maturity (PM), commercial maturity (CM), and immature (I) in the three canopy layers; bottom (B), middle (M), and top (T) of "SF34".

I_{AD} class	FF (kg/cm^2)														SSC (°Brix)												
	B			M			T				B			M			T			
PM	6.6	a^z	A^y	6.5	a	A	5.9	a	B	12.0	b	A	12.3	ab	A	13.4	a	A
CM	6.7	a	A	6.9	a	A	6.8	a	AB	12.4	b	A	12.6	b	AB	13.6	a	A
I	6.5	a	A	6.5	a	A	7.1	a	A	11.9	a	A	11.6	a	B	12.3	a	B

^{z}Small letters represent significant differences between canopy layers within the same I_{AD} class at $P < 0.05$.
^{y}Capital letters represent significant differences between I_{AD} values within the same canopy layer at $P < 0.05$.

TABLE 5: Average fruits diameters in the hand-thinned and unthinned fruit densities in the bottom (B), middle (M), and top (T) canopy layers at 68, 75, 82, and 89 DAFB ("SF26").

DAFB	Layer	Hand-Thinned (diameter-mm)			Unthinned (diameter-mm)		
	B	46.9	a^z	A^y	35.8	b	B
68	M	47.3	a	A	37.3	b	B
	T	45.2	a	A	40.5	b	A
	B	53.3	a	A	40.4	b	B
75	M	54.2	a	A	41.4	b	B
	T	51.7	a	A	46.0	b	A
	B	59.1	a	A	43.1	b	B
82	M	57.3	a	A	45.6	b	AB
	T	56.3	a	A	49.4	b	A
	B	62.9	a	A	45.9	b	A
89	M	61.9	a	A	46.3	b	A
	T	56.8	a	A	48.2	b	A

^{z}Small letters represent significant differences between fruit densities within the same canopy layer (B, M, and T) at $P < 0.05$.
^{y}Capital letters represent significant differences between canopy layers within the same fruit density (hand thinned or unthinned) at $P < 0.05$.

TABLE 6: Average of fruit ripening stage (I_{AD}) in the hand-thinned and unthinned treatments, in the three canopy layers (B, M, and T), at 68, 75, 82, and 89 DAFB ("SF26").

DAFB	Layer	Hand-Thinned (ripening-I_{AD})			Unthinned (ripening-I_{AD})		
	B	1.84	b^z	A^y	1.99	a	A
68	M	1.82	b	A	1.96	a	A
	T	1.89	a	A	1.91	a	A
	B	1.64	b	A	1.76	a	A
75	M	1.64	b	A	1.71	a	A
	T	1.66	a	A	1.69	a	A
	B	1.33	b	A	1.58	a	A
82	M	1.28	b	A	1.48	a	A
	T	1.18	b	A	1.45	a	A
	B	0.92	b	A	1.29	a	A
89	M	0.72	b	A	1.28	a	A
	T	0.88	b	A	1.21	a	A

^{z}Small letters represent significant differences between fruit densities within the same canopy layer (B, M, and T) at $P < 0.05$.
^{y}Capital letters represent significant differences between canopy layers within the same fruit density (hand thinned or unthinned) at $P < 0.05$.

top showed variation between ripening classes, with riper fruit (PM) measuring the lowest fruit firmness and immature fruit I the highest (Table 4). Fruits of both the PM and CM classes developed the highest SSC at the top of the trees, while no differences were noticed between tree canopy layers within the immature I_{AD} class. When comparing fruits within the same canopy layer, fruits at the PM and CM ripening stages showed higher SSC values than (I) fruits, while no differences were noticed between ripening classes in the bottom canopy layer (Table 4). Both fruits' firmness and SSC were not affected by fruit orientation (East West) in the canopy (data not shown).

As shown in Table 5 for "SF26", hand-thinned fruits were bigger than the unthinned at every sampling date and in all tree canopy layers. When considering fruit density (hand thinned and unthinned) inside each canopy layer, the diameter of the hand-thinned fruit did not differ between canopy layers at any sampling date. The unthinned fruits were bigger in the top than in the other canopy layers at most sampling dates. Only at 89 DAFB, all fruits from the three

canopy layers reached the same diameter in the unthinned trees, and no statistical differences were observed.

Table 6 shows the I_{AD} values decreased during the season for both the hand-thinned and thinned treatments. Within fruit densities for every sampling time, no differences were observed between the three canopy layers. Fruit density had an interactive effect with canopy layer on fruit I_{AD} values. Higher fruit densities at 68 and 75 DAFB resulted in delayed ripening values in fruit from the middle and bottom canopy layers but not from the top canopy layer. In all subsequent sampling dates, unthinned fruits showed delayed maturity (lower I_{AD} values) when compared with the hand-thinned fruits reaching the point at 89 DAFB (harvest) in which unthinned fruits were still at a preclimacteric stage while hand-thinned fruits were already at the onset of climacteric (Tables 1 and 6). The east or west orientation did not affect the fruit growth or ripening (data not shown).

Figure 3 describes the correlation, with a coefficient of determination of $R^2 = 0.60$ ($P < 0.01$), between the I_{AD} values of fruit of "SF26" and the respective SSC during fruit

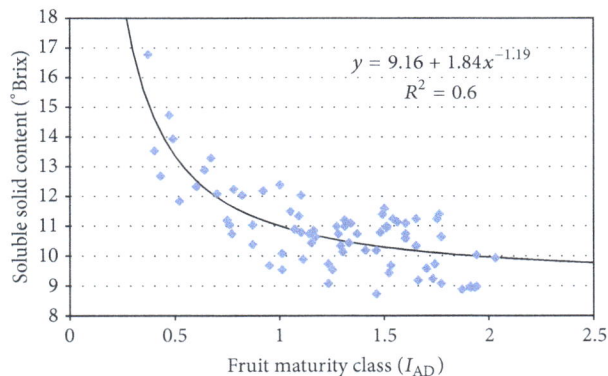

FIGURE 3: Correlation between fruit ripening stage (I_{AD}) and soluble solids content (°Brix) "SF26".

growth. Unripe fruit with an I_{AD} between 1.0 and 2.0 had a lower SSC than the riper fruit. Fruits that reached the PM showed the highest SSC (12–15°Brix).

4. Discussion

Recently, Reig et al. [25] used firmness instead of ethylene production to establish a correlation between the I_{AD} and fruit maturity stage, but their findings were not satisfactory as different I_{AD} values were obtained at different firmness values, and the relationship was cultivar dependent. In fact, as demonstrated by Ziosi et al. [20] on "Stark Red Gold" nectarine, I_{AD} had a higher correlation with ethylene production than with fruit firmness. Both "SF34" and "SF26" showed a clear and different trend in ethylene production at the respective I, CM, and PM fruit maturity stages (Table 1), which agreed with Ziosi et al. [20] who defined the relationship between ethylene production and ripening stage (I_{AD}) as cultivar specific. The I_{AD} can be regarded as a marker for peach fruit ripening that is more sensitive and confident than the physicochemical parameters commonly used to describe physiological condition including firmness, which was the most reliable measurement until now [26].

The I_{AD} value measured on fruit of "SF34" decreased following ripening from four-six weeks after full bloom (Table 2), even if at the onset of climacteric (CM) the ethylene production still remained very low (Table 1). Prior to the CM ripening stage, at the immature stage, the I_{AD} probably better correlates with chlorophyll content than with the ethylene production, though still remaining cultivar specific [27, 28]. Ripening assessment on fruit of "Stark Red Gold" [20] as well as on eleven different nectarine cultivars by Reig et al. [25] confirmed the same behavior.

The nondestructive DA-meter, coupled with the 3D representation of the tree, permitted objective observations of fruit ripening in their exact location within the canopy (Figure 1), without removing them from the tree (Costa et al. [29], personal communication). Our experiment showed that fruit ripening (I_{AD}) of "SF34" trained on a Tatura Trellis was not affected by fruit position inside the canopy (bottom, middle, and top canopy layers) during the season as well

as at harvest (Figures 1 and 2). This is probably due to the open shape of the training system that allows better exposure of fruits in the inner and bottom parts of the canopy to direct sunlight, especially during the latter stages of fruit development [30]. A similar behavior was observed on peach and apple fruits grown on a Y-trellis [1], characterized by a wider angle between branches (27.5° instead of the 17.5° from vertical for the Tatura Trellis) and "perpendicular V" also called "Kearney-V" or "KAC-V" [3], a hybrid between the traditional open-vase system and the Tatura Trellis. All these training systems showed greater levels of intercepted radiation than the delayed vase and free palmette for the life of the orchard [31].

Fruit ripening seemed not to be affected by fruit position, while growth and fruit final size for "SF34" (Table 3) appeared to be strongly affected by the fruit position in the canopy (Table 3). Several studies on peach trees have demonstrated that the fruit position in the canopy was an important factor affecting fruit growth and size [32, 33]. At every sampling, fruits of SF34 located at the top of the canopy were consistently bigger than the fruits in the bottom (Table 3). Also Lewallen and Marini [9] observed that the fruit size was the largest in fruits located in the outside of the canopy, and a similar pattern was reported in peach trees trained to a perpendicular-Y and "delayed vase" [6]. Basile et al. [34] showed that at harvest fruit size increased moving from the top to the bottom of the canopy, while at the beginning of the growing season fruits showed an opposite trend. Likewise, only at the first sampling (73 DAFB) on "SF34", no differences were observed between the diameters of fruits from the middle and bottom canopy layers, while afterwards the diameters of fruits from the middle and top canopy layers were similar until harvest (Table 3). This behaviour could be due to a change in fruit diameter gradient in the canopy described by Basile et al. [34]. A possible explanation of the opposite trend early in the season of fruit growth could be related to the time of blooming that starts from the tree bottom to the top of the tree [35]. Alternatively, part of the variability in fruit growth appeared to be related to carbon (C) source limitation due to the insufficient area of leaves per fruit early in the season [10, 14]. In peach, which carries vegetative and reproductive buds at most nodes, the competition may be stronger for young fruit, and this may cause stronger early fruit-to-fruit competition in the top compared to the bottom of the canopy and a slow growth in the upper part of the trees [10]. Subsequently, when fruits become a stronger competitor for the photosynthates, they start to use the leaves in the vicinity as C-sources. Thus, fruits in the tree top are at an advantage because they are more exposed to light [34]. Fruit competition and use of leaves as C-sources could explain our findings that after the main harvest (corresponding to fruits reaching the commercial maturity stage) of cv "SF34", fruit diameters were similar in the three considered canopy layers, probably because fruit removal caused a redistribution of the photosynthates between the remaining fruits, which continued their growth throughout the last stages of maturation [36]. An additional explanation could be that the removal of larger fruits, often harvested in the first pick and mainly located in the top or outside of the canopy, allows the remaining fruits to reach similar diameters.

Our results showed that fruit of "SF34", trained on Tatura Trellis, with the same ripening stage at harvest, were very homogeneous also in terms of firmness (Table 4) establishing a loose correlation between fruit firmness and ethylene production. This observation is in accord with other authors, who reported a rapid decline of fruit firmness after ethylene production inside the fruit has begun [37, 38]. Conversely, Lewallen and Marini [9] observed that fruit with similar background color, as an indication of fruit ripening, harvested from different positions within the canopy did not have the same fruit firmness, with firmer fruit in the inside positions of which the nearby leaves would be the least exposed to light. Our findings were somewhere in the middle since fruit from the bottom and middle canopy layer were found having similar firmness independently of their ripening stages (Table 4) while fruit from the top of the canopy showed that less ripened fruit were more firm, probably also due to a combined effect of light and position as suggested by Marini and Trout [39].

Ziosi et al. [20] described SSC as ethylene independent and did not observe strong differences in soluble solids between ripening stages (I_{AD}) for the "Stark Red Gold." Our results on "SF26" seemed to validate these findings. In fact, a relatively low correlation ($R^2 = 0.60$) between I_{AD} and SSC was observed (Figure 3). These results were also in agreement with Hale et al. [40] for "August Fire" but in contrast to a recent publication by Infante et al. [18] on two cultivars of Japanese plums, "Angeleno" and 'Autumn Beauty". Infante [41] described the I_{AD} as an index having high correlation with the most common parameters used for monitoring ripening, such as fruit firmness and SSC with $R^2 > 0.89$ and >0.70, respectively. Our results for "SF34" (Table 4), however, showed that only fruit from the bottom canopy layer appeared to have the same SSC, independent from the ripening stage. Overall in our experiments, it seems that there was a low interaction between canopy position and fruit ripening stage in regards to SSC (Table 4), and most of the effects were probably due to the higher exposure to light for fruit in the upper parts of the canopy than to their specific ripening stages, since only the immature or less exposed fruit of the bottom canopy layers had lower SSC. This hypothesis is supported by other research that found a strong influence of light on peach fruit quality [1] and, consequently, of tree growth trends, reproductive habits, training systems, and pruning techniques for light distribution [42, 43]. Despite the variation in SSC fruit content found in our experiment, trees trained in the Tatura Trellis system seemed to have a good uniformity in SSC distribution since 97% of the fruit at harvest were at the CM and PM maturity stage. Only 30% came from the bottom canopy layer, with over 80% of the total fruit harvested having similar soluble solids content. There could also be a variety component influencing the overall correlation between SSC and I_{AD} and more research is necessary to validate this.

The highly uniform tree structure created by the Tatura Trellis system seemed to be the reason for the relatively high fruit uniformity found in our experiment, in terms of fruit maturity level, SSC, and firmness. In fact, as suggested by DeJong et al. [44], the uniform tree structure of Tatura Trellis also allows for an easy regulation of fruit density which can be summed by just leaving about four fruit per fruiting shoot during stage I of fruit growth [10]. From our experiment on "SF26", it was observed that maintaining fruit numbers at the suggested density resulted in uniform fruit within the canopy, both in terms of diameter and ripening stage (Tables 5 and 6). These results are in contrast with previous studies that showed gradients of fruit sizes within peach tree canopies both in commercially and heavily-thinned peach trees [6, 33, 45]. The higher light availability to the fruit, coupled with a balanced crop load, probably allowed the Tatura Trellis to reduce the fruit-to-fruit competition with a greater distribution of the photosynthates between vegetative and reproductive structures [46–48], which would explain the high fruit variability in terms of size and maturity stage that we found when fruit density was doubled (Tables 5 and 6). Our results were confirmed by other authors that observed that leaving too many fruit on a tree reduces SSC as well as fruit sizes at harvest [10, 49].

5. Conclusions

The present study showed that nectarine trees trained to Tatura Trellis produced fruit with high homogeneity in terms of growth, maturation, and SSC content, when fruit density is balanced. Our results also confirmed that the I_{AD} can be regarded as a sensitive, confident, and nondestructive marker of nectarine fruit maturity stage that allows for an early assessment of fruit ripening still on the tree. Further investigations are required to better define the relationship between I_{AD} and the traditional quality traits, fruit firmness and SSC.

Acknowledgments

This paper is a publication from a University of Bologna, College of Agriculture, Italy PhD dissertation thesis performed in Australia during the required international study period abroad. The authors thank the DPI Knoxfield Centre, for hosting and cofunding the experiment under Premium Fruit, a Victorian Department of Primary Industries project. Also thanks go to Rick Varapodio for making his orchards available, John Lopresti for helping in the field measurements, and the rest of the staff of the DPI Knoxfield Centre postharvest group for help and support in the laboratory.

References

[1] T. Caruso, C. Di Vaio, F. Guarino, A. Motisi, and V. Nuzzo, "Peach varieties for intensive plantations in Southern Italy," in *4th Congresso Nazionale Sulla Peschicoltura Meridionale*, F. P. Marra and F. Sottile, Eds., pp. 44–51, Panuzzo Prontostampa, Caltanissetta, Italy, 2003.

[2] D. J. Chalmers and I. B. Wilson, "Productivity of peach trees: tree growth and water stress in relation to fruit growth and

assimilate demand," *Annals of Botany*, vol. 42, no. 2, pp. 285–294, 1978.

[3] T. M. DeJong, "Developmental and environmental control of dry-matter partitioning in peach," *HortScience*, vol. 34, no. 6, pp. 1037–1040, 1999.

[4] E. D. Cittadini, N. de Ridder, P. L. Peri, and H. van Keulen, "Relationship between fruit weight and the fruit-to-leaf area ratio, at the spur and whole-tree level, for three sweet cherry varieties," *Acta Horticulturae*, vol. 795, pp. 669–672, 2008.

[5] B. van den Ende, "The tatura trellis," *Compact Fruit Tree*, vol. 27, article 97, 1994.

[6] V. Farina, R. Lo Bianco, and P. Inglese, "Vertical distribution of crop load and fruit quality within vase- and Y-shaped canopies of "Elegant Lady" peach," *HortScience*, vol. 40, no. 3, pp. 587–591, 2005.

[7] M. Dani, "Connection between the light availability and the peach fruit quality," *Cereal Research Communications*, vol. 35, no. 2, pp. 337–340, 2007.

[8] F. L. He, F. Wang, Q. P. Wei, X. W. Wang, and Q. Zhang, "Relationships between the distribution of relative canopy light intensity and the peach yield and quality," *Agricultural Sciences in China*, vol. 7, no. 3, pp. 297–302, 2008.

[9] K. S. Lewallen and R. P. Marini, "Relationship between flesh firmness and ground color in peach as influenced by light and canopy position," *Journal of the American Society for Horticultural Science*, vol. 128, no. 2, pp. 163–170, 2003.

[10] L. Corelli Grappadelli and D. C. Coston, "Thinning pattern and light environment in peach tree canopies influence fruit quality," *HortScience*, vol. 26, no. 12, pp. 1464–1466, 1991.

[11] T. Caruso, A. de Michele, F. Sottile, and F. P. Marra, "La peschicoltura siciliana nel contesto italiano: ambiente, cultivar e tecniche colturali," in *2nd Atti Convegno sulla Peschicoltura Meridionale*, pp. 83–88, Paestum, Italy, July 1998.

[12] S. Sansavini, L. Corelli, and L. Giunchi, "Peach yield efficiency as related to tree shape," *Acta Horticulturae*, vol. 173, pp. 139–158, 1985.

[13] L. C. Grappadelli and S. Sansavini, "Light interception and photosynthesis related to planting density and canopy management in apple," *Acta Horticulturae*, vol. 243, pp. 159–174, 1989.

[14] G. Costa, M. Noferini, G. Fiori, and A. Orlandi, "Nondestructive technique to assess internal fruit quality," *Acta Horticulturae*, vol. 603, pp. 571–575, 2003.

[15] M. Delwiche and R. A. Baumgardner, "Ground color measurements of peach," *Journal of the American Society for Horticultural Science*, vol. 108, pp. 1012–1016, 1983.

[16] R. P. Marini, D. Sowers, and M. C. Marini, "Peach fruit quality is affected by shade during final swell of fruit growth," *Journal of the American Society for Horticultural Science*, vol. 116, no. 3, pp. 383–389, 1991.

[17] I. Iglesias and G. Echeverría, "Differential effect of cultivar and harvest date on nectarine colour, quality and consumer acceptance," *Scientia Horticulturae*, vol. 120, no. 1, pp. 41–50, 2009.

[18] R. Infante, C. Pía, M. Noferini, and G. Costa, "Determination of harvest maturity of D'Agen plums using the chlorophyll absorbance index," *Ciencia e Investigación Agraria*, vol. 38, no. 2, pp. 199–203, 2011.

[19] A. I. Cascales, E. Costell, and F. Romojaro, "Effects of the degree of maturity on the chemical composition, physical characteristics and sensory attributes of peach (*Prunus persicas*) cv. caterin," *Food Science and Technology International*, vol. 11, no. 5, pp. 345–352, 2005.

[20] V. Ziosi, M. Noferini, G. Fiori et al., "A new index based on vis spectroscopy to characterize the progression of ripening in peach fruit," *Postharvest Biology and Technology*, vol. 49, no. 3, pp. 319–329, 2008.

[21] F. Gottardi, M. Noferini, G. Fiori, M. Barbanera, C. Mazzini, and G. Costa, "The index of absorbance difference (I_{AD}) as a tool for segregating peaches and nectarines into homogeneous classes with different shelf-life and consumer acceptance," in *Proceedings of the 8th Pangborn Sensory Science Symposium*, Firenze, Italy, 2009.

[22] Costa et al., Proceedings of the 7th International Symposium on Kiwifruit, University of Bologna, 2010.

[23] E. Magnanini, E. Bonora, and G. Vitali, "PlantToon—drawing and pruning fruit trees," in *Proceedings of the 6th International Workshop on Functional-Structural Plant Models*, p. 255, Davis, Calif, USA, September 2010.

[24] G. Volpe, R. Lo Bianco, and M. Rieger, "Carbon autonomy of peach shoots determined by 13C-photoassimilate transport," *Tree Physiology*, vol. 28, no. 12, pp. 1805–1812, 2008.

[25] G. Reig, S. Alegre, I. Iglesias, G. Echeverría, and F. Gatius, "Fruit quality, colour development and index of absorbance difference (I_{AD}) of different nectarine cultivars at different harvest dates," in *Proceedings of the 28th IHC International Symposium on Postharvest Technology*, vol. 934, pp. 1117–1126, Acta Horticulturae, 2012.

[26] C. Valero, C. H. Crisosto, and D. Slaughter, "Relationship between nondestructive firmness measurements and commercially important ripening fruit stages for peaches, nectarines and plums," *Postharvest Biology and Technology*, vol. 44, no. 3, pp. 248–253, 2007.

[27] R. Cubeddu, C. D'Andrea, A. Pifferi et al., "Nondestructive quantification of chemical and physical properties of fruits by time-resolved reflectance spectroscopy in the wavelength range 650–1000 nm," *Applied Optics*, vol. 40, no. 4, pp. 538–543, 2001.

[28] R. Cubeddu, C. D'Andrea, A. Pifferi et al., "Time-resolved reflectance spectroscopy applied to the nondestructive monitoring of the internal optical properties in apples," *Applied Spectroscopy*, vol. 55, no. 10, pp. 1368–1374, 2001.

[29] Costa et al., Proceedings of the 10 International Symposium on Integrating Canopy, Rootstock and Environmental Physiology in Orchard Systems, University of Bologna, 2012.

[30] V. Nuzzo, B. Dichio, A. M. Palese, and C. Xiloyannis, "Sviluppo della chioma ed intercettazione radiativa in piante di pesco allevate ad Y trasversale ed a vaso ritardato nei primi tre anni dall'impianto," in *Proceedings of 5th Giornate Scientifiche SOI*, O. Failla and I. Piagnani, Eds., pp. 319–320, Edizioni Tecnos, Milan, Italy, 2000.

[31] L. C. Grappadelli and R. P. Marini, "Orchard planting systems," in *The Peach: Botany, Production and Uses*, D. R. Layne and D. Bassi, Eds., pp. 264–288, 2008.

[32] M. Génard and C. Bruchou, "A functional and exploratory approach to studying growth: the example of the peach fruit," *Journal of the American Society for Horticultural Science*, vol. 118, pp. 317–323, 1993.

[33] A. Weibel, *Effect of size-controlling rootstocks on vegetative and reproductive growth of peach [Prunus persica (L.) Batsch] [M.S. thesis]*, University of California, Davis, Calif, USA, 1999.

[34] B. Basile, L. I. Solari, and T. M. Dejong, "Intra-canopy variability of fruit growth rate in peach trees grafted on rootstocks with different vigour-control capacity," *Journal of Horticultural Science and Biotechnology*, vol. 82, no. 2, pp. 243–256, 2007.

[35] I. R. Dann and P. H. Jerie, "Gradients in maturity and sugar levels of fruit within peach trees," *Journal of the American Society for Horticultural Science*, vol. 113, pp. 27–31, 1988.

[36] R. P. Marini and D. L. Sowers, "Peach fruit weight is influenced by crop density and fruiting shoot length but not position on the shoot," *Journal of the American Society for Horticultural Science*, vol. 119, no. 2, pp. 180–184, 1994.

[37] P. Tonutti, C. Bonghi, and A. Ramina, "Fruit firmness and ethylene biosynthesis in three cultivars of peach (*Prunus persica* L. Batsch)," *Journal of Horticultural Science and Biotechnology*, vol. 71, no. 1, pp. 141–147, 1996.

[38] D. A. Brummell, V. D. Cin, S. Lurie, C. H. Crisosto, and J. M. Labavitch, "Cell wall metabolism during the development of chilling injury in cold-stored peach fruit: association of mealiness with arrested disassembly of cell wall pectins," *Journal of Experimental Botany*, vol. 55, no. 405, pp. 2041–2052, 2004.

[39] R. P. Marini and J. R. Trout, "Sampling procedures for minimizing variation in peach fruit quality," *Journal of the American Society for Horticultural Science*, vol. 109, no. 3, pp. 361–364, 1984.

[40] G. Hale, J. Lopresti, E. Bonora, D. Stefanelli, and R. Jones, "Using non-destructive methods to correlate chilling injury with fruit maturity," in *Proceedings of the 7th International Post-Harvest Symposium, Malaysia, 2012*, Acta Horticolturae, March 2013.

[41] R. Infante, "Harvest maturity indicators in the stone fruit industry," *Stewart Postharvest Review*, pp. 1–6, 2012.

[42] T. M. DeJong and J. F. Doyle, "Leaf gas exchange and growth response of mature "Fantasia" nectarine trees to paclobutrazol," *Journal of the American Society for Horticultural Science*, vol. 109, pp. 878–882, 1984.

[43] R. Scorza, L. Zailong, G. W. Lightner, and L. E. Gilreath, "Dry matter distribution and responses to pruning within a population of standard, semidwarf, compact, and dwarf peach seedlings," *Journal of the American Society for Horticultural Science*, vol. 111, pp. 541–545, 1986.

[44] T. M. DeJong, K. R. Day, and J. F. Doyle, "The Kearney agricultural center perpendicular "V" (KAC-V) orchard system for peaches and nectarines," *HortTechnology*, vol. 4, no. 4, pp. 362–367, 1994.

[45] M. Forlani, B. Basile, C. Cirillo, and C. Iannini, "Effects of harvest date and fruit position along the tree canopy on peach fruit quality," *Acta Horticulturae*, vol. 2, pp. 459–466, 2002.

[46] G. Costa and G. Vizzotto, "Fruit thinning of peach trees," *Plant Growth Regulation*, vol. 31, no. 1-2, pp. 113–119, 2000.

[47] M. Faust, *Physiology of Temperate Zone Fruit Trees*, John Wiley & Sons, New York, NY, USA, 1989.

[48] E. W. Pavel and T. M. DeJong, "Source- and sink-limited growth periods of developing peach fruits indicated by relative growth rate analysis," *Journal of the American Society for Horticultural Science*, vol. 118, no. 6, pp. 820–824, 1993.

[49] C. H. Crisosto, T. DeJong, K. R. Day et al., "Studies on stone fruit internal breakdown," in *1994 Research Reports for California Peaches and Nectarines*, California Tree Fruit Agreement, Sacramento, Calif, USA, 1995.

14

Effect of the Soil pH on the Alkaloid Content of
Lupinus angustifolius

Gisela Jansen, Hans-Ulrich Jürgens, Edgar Schliephake, and Frank Ordon

Institute for Resistance Research and Stress Tolerance, Julius Kühn-Institut (JKI), Federal Research Centre for Cultivated Plants, 18190 Sanitz, Germany

Correspondence should be addressed to Gisela Jansen, gisela.jansen@jki.bund.de

Academic Editor: O. Mario Aguilar

Field studies were conducted in growing seasons 2004, 2005, and 2010 to investigate the effect of different soil pH values on the alkaloid content in seeds of *Lupinus angustifolius*. Two-year experiments with eleven cultivars were carried out in acid soils with an average of pH = 5.8 (Mecklenburg-Western Pomerania) and on calcareous soils with an average pH of 7.1 (Bavaria), respectively. In addition, in 2010, eight cultivars were grown in field experiments in soils with pH values varying between pH = 5.3 and pH = 6.7. In all experiments conducted on soils with a higher pH (pH = 6.7 and pH = 7.1), a significantly lower alkaloid content was detected in all *Lupinus angustifolius* cultivars than on soils with a lower pH (pH = 5.3 and pH = 5.8). Results clearly show that the alkaloid content is significantly influenced by the soil pH but genotypic differences regarding the reaction to different pH values in the soil were observed.

1. Introduction

Lupins as protein crops can be used in many ways. They are grown for green manure, for animal feed, or for human nutrition. Unlimited feeding of bitter seeds led to the disease "lupinose" in former times. It was caused by the alkaloids contained in bitter lupins [1]. Only breeding of the so-called sweet lupins [2] facilitates the use of lupins to a greater extent in animal feed and for human nutrition. But besides genetics the alkaloid content in sweet lupins, for example, *Lupinus angustifolius,* is influenced by different environmental factors. Apart from drought, heat, and nutrient deficiencies, plants are largely affected by the soil pH [3]. Lupin species are differing concerning their demands for optimal growth, but in general commercial cultivars of lupins grow poorly on alkaline or neutral soils [4]. In general blue lupins (*Lupinus angustifolius*) and especially yellow lupins (*Lupinus luteus*) are less sensitive to calcareous soils than white lupins (*Lupinus albus*) [5, 6]. The soil pH considerably influences yield and protein content of *Lupinus angustifolius* [7, 8], that is, the higher the soil pH the lower the kernel yield and protein content. Besides this, the pH value has an impact on the production of secondary metabolites [9–11], for example, out of many soil parameters analysed,

the highest correlation of the production of the glycoside salidroside in *Rhodiola sachalinensis* was observed to the soil pH [9]. Similar results were obtained for the alkaloid production, for example, when the pH value in cell culture media of *Lupinus polyphyllus* decreased from pH = 5.5 to pH = 3.5, the alkaloid production increased [12].

The objective of our field studies was to assess the influence of the soil pH on the alkaloid content of narrow-leafed lupin cultivars (*Lupinus angustifolius*).

2. Materials and Methods

2.1. Field Trials. Seeds of *Lupinus angustifolius* cultivars Borlana, Borweta, Bordako, Boruta, Borlu, Bora, Boregine, Boltensia, Bolivio, Vitabor, and Haags Blaue were supplied by the seed company Saatzucht Steinach (Bornhof, Germany). The variety Sonet was provided by Kruse Saatzucht (Münster, Germany).

In 2004 and in 2005, field experiments with 11 cultivars of *Lupinus angustifolius* (except Haags Blaue) were carried out under organic farming conditions in Bogen (Bavaria, northern latitude: 48.912925, eastern longitude: 12.692792) and in Gross Luesewitz (Mecklenburg-Western Pomerania,

TABLE 1: Characteristic of locations.

	Mecklenburg-Western Pomerania		Bavaria	
	Gross Luesewitz	Bornhof	Bogen	Gründl
Soil type	Loamy sand	Sand	Sandy clay loam	Sandy loam
pH value	5.8	5.3	7.2	6.7
Mean annual rainfall (mm)	620	558	803	822
Mean annual temperature (°C)	8.2	8.2	7.7	8.6

Source: [8, 13].

TABLE 2: Content of alkaloids in different cultivars of *Lupinus angustifolius* at different locations (2004 and 2005).

Genotype	Year	13-Hydroxylupanine		Angustifoline		Isolupanine		Lupanine	
		Bo	GL	Bo	GL	Bo	GL	Bo	GL
Bolivio	2004	45.5	215.5	20.8	102.8	7.0	23.7	16.8	179.2
	2005	52.2	157.8	41.8	87.3	10.2	26.6	6.6	204.7
	Mean	**48.9**	**186.7**	**31.3**	**95.1**	**8.6**	**25.2**	**11.7**	**192.0**
Boltensia	2004	2.4	73	2.3	52.1	2.3	11.3	7.5	136.5
	2005	1.8	58.1	1.5	50.5	0.0	16.8	1.2	216.8
	Mean	**2.1**	**65.6**	**1.9**	**51.3**	**1.2**	**14.1**	**4.4**	**176.7**
Bora	2004	15.5	119.9	13.3	78.3	6.8	24.2	27.7	199.1
	2005	2.1	94.3	0.6	72.1	1.0	34.4	4.5	293.8
	Mean	**8.8**	**107.1**	**7.0**	**75.2**	**3.9**	**29.3**	**16.1**	**246.5**
Bordako	2004	13.8	135.2	9.3	98.7	3.6	17.5	10.5	231.9
	2005	4.9	80.1	4.6	80.7	1.4	26.8	2.6	340.8
	Mean	**9.4**	**107.7**	**7.0**	**89.7**	**2.5**	**22.2**	**6.6**	**286.4**
Boregine	2004	12.7	67.6	7.4	46.3	4.6	18.8	10.5	113.6
	2005	1.8	75.5	0.8	66.7	0.5	33.6	0.3	245.7
	Mean	**7.3**	**71.6**	**4.1**	**56.5**	**2.6**	**26.2**	**5.4**	**179.7**
Borlana	2004	5.0	67.4	2.8	49.6	1.9	11.1	4.1	103.9
	2005	5.1	43.7	5.5	41.1	1.1	14.1	0.9	171.9
	Mean	**5.1**	**55.6**	**4.2**	**45.4**	**1.5**	**12.6**	**2.5**	**137.9**
Borlu	2004	12.6	118.6	5.7	79.5	3.3	18.8	9.6	151.2
	2005	3.8	36.4	5.3	31.6	1.5	11.2	5.5	106.3
	Mean	**8.2**	**77.5**	**5.5**	**55.6**	**2.4**	**15.0**	**7.6**	**128.8**
Boruta	2004	18.7	53.3	8.3	31.0	3.3	6.9	14.8	82.0
	2005	2.7	34.7	0.5	32.0	0.0	7.2	2.4	115.0
	Mean	**10.7**	**44.0**	**4.4**	**31.5**	**1.7**	**7.1**	**8.6**	**98.5**
Borweta	2004	11.2	119.2	6.4	87.0	3.0	12.4	17.3	167.4
	2005	5.0	19.5	2.9	23.9	0.0	8.2	4.2	136.7
	Mean	**8.1**	**69.4**	**4.7**	**55.5**	**1.5**	**10.3**	**10.8**	**152.1**
Sonet	2004	32.7	135.6	22.7	96.1	4.9	14.8	32.2	215.3
	2005	35.7	89.8	36.8	98.1	3.7	15.1	55.4	364.0
	Mean	**34.2**	**112.7**	**29.8**	**97.1**	**4.3**	**15.0**	**43.8**	**289.7**
Vitabor	2004	4.5	19.9	2.0	11.3	1.7	3.4	3.0	25.7
	2005	2.2	14.7	0.7	10.1	0.0	3.1	0.6	36.9
	Mean	**3.4**	**17.3**	**1.4**	**10.7**	**0.9**	**3.3**	**1.8**	**31.3**

BO: Bogen, GL: Gross Luesewitz.

TABLE 3: Mean content (±SD) of alkaloids in different cultivars of *Lupinus angustifolius* at different locations (2004 and 2005).

Alkaloid	Bogen		Gross Luesewitz	
13-Hydroxylupanine	13.3 ± 14.9	a	83.2 ± 50.5	b
Angustifoline	9.2 ± 11.5	a	60.3 ± 29.5	b
Isolupanine	2.8 ± 2.6	a	16.4 ± 8.9	b
Lupanine	10.8 ± 13.1	a	174.5 ± 87.6	b

Means for the same alkaloid between the locations with different letters are significantly different from each other (alpha = 0.05).

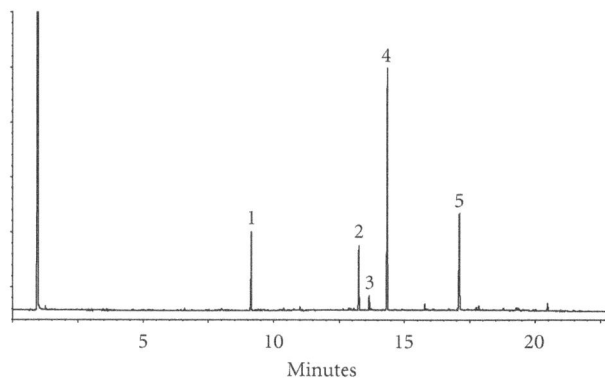

FIGURE 1: Gas chromatogram for different alkaloids (1 ISTD-Caffeine, 2 Angustifoline, 3 Iso-Lupanine, 4 Lupanine, 5 13-Hydroxy-Lupanine) present in cultivar Haags Blaue grown at Bornhof.

northern latitude: 54.071955, eastern longitude: 12.321031). Field experiments were conducted in a randomized block design with four replications with a plot size of 9.6 m². The alkaloid content in these trials was analysed on a mixed sample of each cultivar. In 2010 the variety Boruta and the variety Haags Blaue as well as six newly developed breeding lines (51–56) of the Saatzucht Steinach were grown in field experiments in fourfold replications in 4.2 m² plots at different soils at Gründl (Bavaria, northern latitude: 48.519305, eastern longitude: 11.816175) and Bornhof (Mecklenburg-Western Pomerania, northern latitude: 53.477371, eastern longitude: 12.911754) under conventional growing conditions. In this experiment the alkaloid content was measured separately for each plot. In Table 1 the characteristics of the locations are given.

2.2. Determination of the Alkaloid Content. Grain samples of about 250 g were randomly taken and grounded as described by [14]. All wholemeal samples revealed a dry matter content of about 90% and were stored at 20°C until analysis. The subsequent alkaloid analysis in lupin whole meal flour was carried out according to [14–16]. Main alkaloids were calculated as the sum of alkaloids shown in Figure 1. The main alkaloids in narrow-leafed lupins are angustifoline, isolupanine, lupanine, and 13-hydroxylupanine.

The determination of the alkaloid content was performed twice per sample with a coefficient of variation lower than 4%.

2.3. Statistical Analysis. To assess the effects of the location on the alkaloid content, a generalized linear model for the analysis of variance (ANOVA) was applied, using the GLM procedure of the software package SAS (version 9.3) followed by a Tukey test ($\alpha = 0.05$) for comparing the means. The two datasets were analysed separately.

3. Results and Discussion

The alkaloid content in sweet narrow-leaf lupin cultivars is in general very low. Already Sengbusch [17] suggested to call lupins alkaloid-poor (0.05% alkaloids in seeds) or alkaloid-free (0.025% alkaloids in seeds). Nevertheless, the seed alkaloid content of sweet lupins is influenced by different environmental factors such as fertilizers [18–22], ambient temperature during initiation of flowering up to pod ripening [14], and drought stress [23]. In 2004

and 2005, the mean daily temperature at the beginning of flowering until harvest in August was very similar [14]. The experiments were also carried out under uniform agronomic management (fertilizer, herbicides, etc.), so that the abiotic stress factors temperature, drought, and nutrient deficiency as well as mechanical damage can be neglected. In 2004 at Bogen the main alkaloid content of the cultivars tested ranged between 11.2 µg/g for the cultivar Vitabor and 92.5 µg/g for the cultivar Sonet and 90.1 µg/g for Bolivio. At Gross Luesewitz the alkaloid content ranged between 60.3 µg/g for cultivar Vitabor and 521.2 µg/g for cultivar Bolivio. The alkaloid content of all varieties tested is shown in Table 2. In 2005, as expected, the lowest alkaloid content was also found in the cultivar Vitabor (Bogen 3.5 µg/g and Gross Luesewitz 64.8 µg/g) and the highest content in Sonet (Bogen 131.6 µg/g and Gross Luesewitz 567.0 µg/g).

Out of all alkaloids analyzed the lupanine content shows the largest increase with decreasing soil pH. Christiansen et al. [23] reported that drought stress during the vegetative phase reduces mostly the concentration of lupanine, 13-hydroxylupanine, and angustifoline, whereas isolupanine is affected to a much smaller extent.

The differences in the alkaloid content between the locations in Mecklenburg-Western Pomerania and Bavaria are significant (Table 3).

Figure 2 clearly demonstrates that the alkaloid production is significantly higher when lupins are cultivated at a lower pH (mean value of two years is shown). However, also clear differences are observed between cultivars opening the opportunity to breed cultivars with low alkaloid content under low pH conditions, for example, Vitabor.

In 2010 also two cultivars and six newly developed breeding lines of *Lupinus angustifolius* were analyzed concerning the alkaloid content at locations with different soil pH. At Gründl (pH = 6.7) the total alkaloid content ranged between 0.0166% and 0.1293% while at Bornhof (pH = 5.3) the alkaloid content was in general higher and ranged between 0.029% and 0.1810% (Figure 3). The variety Boruta and

TABLE 4: Mean alkaloid content estimated on soil with pH = 5.8 and pH = 7.2 observed in cultivars of *Lupinus angustifolius* at different locations (2004 and 2005, $n = 22$).

Location	Main alkaloid content [%]	
	2004	2005
Mecklenburg-Western Pomerania (pH = 5.8)	0.0330	0.0339
Bavaria (pH = 7.2)	0.0043	0.0029
Least significant difference (Tukey, $\alpha = 0.05$)	0.0084	0.01

TABLE 5: Mean alkaloid content estimated on soil with pH = 5.3 and pH = 6.7 observed in actual varieties and breeding lines of *Lupinus angustifolius L.* at different locations (2010, $n = 32$).

Location	Alkaloid content [%]
	2010
Mecklenburg-Western Pomerania (pH = 5.3)	0.0687
Bavaria (pH = 6.7)	0.0355
Least significant difference (Tukey, $\alpha = 0.05$)	0.0081

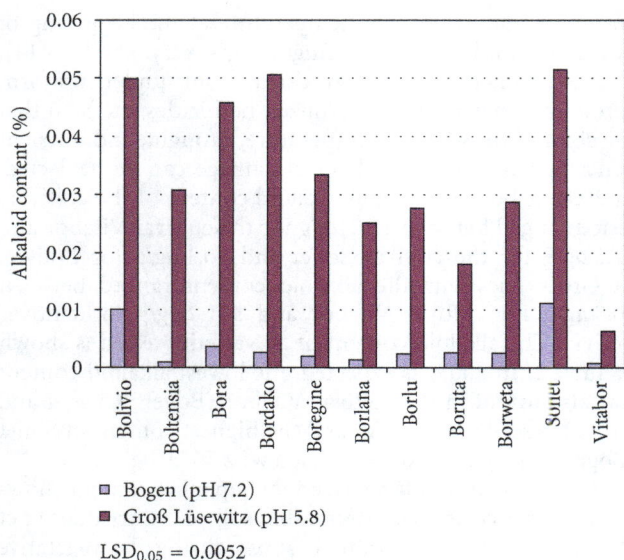

FIGURE 2: Influence of the soil pH on the main alkaloid content of *Lupinus angustifolius* (mean of 2004 and 2005).

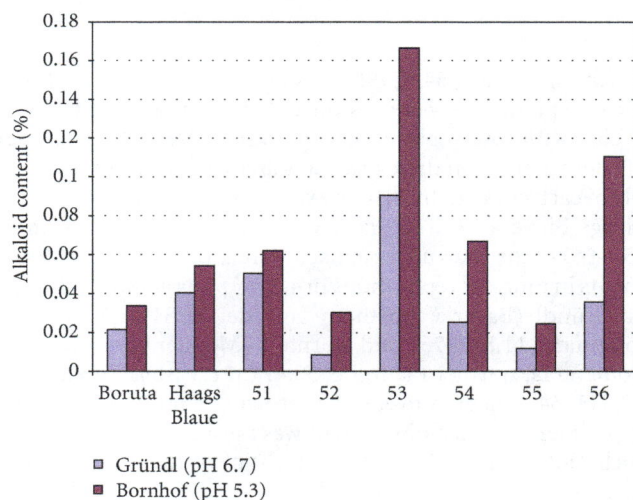

FIGURE 3: Influence of the soil pH on the main alkaloid content of *Lupinus angustifolius* cultivars and breeding lines (2010).

two new breeding lines revealed the lowest alkaloid content (Figure 3).

Concerning the average main alkaloid content in cultivars and breeding lines significant differences between locations were observed in 2010 (Table 5) as it was previously observed in 2004 and 2005 (Table 4). In Germany, there is no law concerning the upper threshold of alkaloids in lupins for animal feed and human nutrition, but in general the upper threshold for use in animal and human nutrition is 0.05% and 0.02%, respectively. In 2010 this threshold has been exceeded in general at the soil with the lower pH (Mecklenburg-Western Pomerania) and also by some cultivars in 2004/2005.

For lupin cultivation a rather low soil pH (pH = 5.0–6.8) is recommended [5]. Jansen et al. [16] reported that the grain yield at pH = 7.2 is lower than at pH = 5.8. On the other hand the alkaloid content of *Lupinus angustifolius* decreases significantly at a higher pH (Figures 2 and 3). Therefore, for a high yield combined with low alkaloid content, the soil pH is of prime importance, although it has to be taken into account that also additional environmental factors, for example, drought stress, fertilization, and temperature, can have an adverse effect on the alkaloid content in lupin seeds. Yaber Grass and Leicach [24] noted that a significant increase in the total alkaloid content was observed from samples of *Senecio grisebachii* growing in highly deteriorated soil compared to those from samples grown in less deteriorated ones.

The results presented show that additional breeding efforts are needed to achieve low alkaloid content also under low soil pH. As shown in Figures 2 and 3 genetic variation concerning this trait is present in cultivars and new breeding lines which could be exploited in the future.

Acknowledgments

The authors thank Ch. Peters and M. Jennerjahn for technical assistance. This work was supported by the BMBF with the Project no. 03WKBV01B and BLE with the Project no. 03OE355.

References

[1] W. Hondelmann, "Die Lupine-Geschichte und Evolution einer Kulturpflanze," *Landbauforschung Völkenrode*, vol. 162, pp. 10–18, 1996.

[2] R. von Sengbusch, "Bitterstoffarme Lupinen III," *Der Züchter Zeitschrift für Theoretische und Angewandte Genetik*, vol. 10, no. 4, pp. 91–95, 1938.

[3] 2011, http://www.biosicherheit.de/lexikon/1277.abiotische-stressfaktoren.html.

[4] C. Tang, B. J. Buirchell, N. E. Longnecker, and A. D. Robson, "Variation in the growth of lupin species and genotypes on alkaline soil," *Plant and Soil*, vol. 155-156, no. 1, pp. 513–516, 1993.

[5] U. Schmiechen, Anbauratgeber Blaue Süßlupine, 2011, http://www.ufop.de/files/9113/4080/9714/PI_Blaue_Suesslupine_240611.pdf.

[6] J. S. Gladstones, "Lupins as crop plants," *Field Crop Abstracts*, vol. 23, pp. 123–148, 1970.

[7] G. Jansen, H. U. Jürgens, and W. Flamme, *Züchterische Bearbeitung von Süßlupinen für den ökologischen Landbau—Qualitätsuntersuchungen im Hinblick auf Futtereignung*, Bericht orgprints, 2006, http://orgprints.org/11087/1/11087-03 OE355-jki-jansen-2006-suesslupinen.pdf.

[8] G. Jansen, F. Eickmeyer, and V. Michel, "Einfluss von Kalkung und pH-Wert im Boden auf Kornertrag und Eiweißgehalt von *Lupinus angustifolius* L," *Journal für Kulturpflanzen*, vol. 62, no. 10, pp. 367–375, 2010.

[9] X. Yan, S. Wu, Y. Wang, X. Shang, and S. Dai, "Soil nutrient factors related to salidroside production of Rhodiola sachalinensis distributed in Chang Bai Mountain," *Environmental and Experimental Botany*, vol. 52, no. 3, pp. 267–276, 2004.

[10] A. G. Medentsev, A. Y. Arinbasarova, and V. K. Akimenko, "Biosynthesis of naphthoquinone pigments by fungi of the genus Fusarium," *Applied Biochemistry and Microbiology*, vol. 41, no. 5, pp. 503–507, 2005.

[11] P. Babula, R. Mikelova, V. Adam, R. Kizek, L. Havel, and Z. Sladky, "Using of liquid chromatography coupled with diode array detector for determination of naphthoquinones in plants and for investigation of influence of pH of cultivation medium on content of plumbagin in Dionaea muscipula," *Journal of Chromatography B*, vol. 842, no. 1, pp. 28–35, 2006.

[12] R. Endres, "Influence of culture conditions on secondary metabolite accumulation. Plant cell biotechnology," in *Plant Cell Biotechnology*, pp. 187–242, Springer, New York, NY, USA, 1994.

[13] A. K. Klamroth and R. Dieterich, "Genotyp-Umwelt-Wechselwirkung bei der Blauen Süßlupine am Beispiel ausgewählter Ertragsparameter," in *Tagungsband 61. Tagung der Vereinigung der Pflanzenzüchter und Saatgutkaufleute Österreichs*, pp. 101–104, Gumpenstein, Austria, 2010.

[14] M. Wink, C. Meissner, and L. Witte, "Patterns of quinolizidine alkaloids in 56 species of the genus Lupinus," *Phytochemistry*, vol. 38, no. 1, pp. 139–153, 1995.

[15] K. Bermúdez Torres, N. Robledo Quintos, L. L. Barrera Necha, and M. Wink, "Alkaloid profile of leaves and seeds of Lupinus hintonii C. P. Smith," *Zeitschrift für Naturforschung C*, vol. 57, no. 3-4, pp. 243–247, 2002.

[16] G. Jansen, H. U. Jürgens, and F. Ordon, "Effects of temperature on the alkaloid content of seeds of *Lupinus angustifolius* cultivars," *Journal of Agronomy and Crop Science*, vol. 195, no. 3, pp. 172–177, 2009.

[17] R. von Sengbusch, "Süßlupinen und Öllupinen. Teil III. Die Entstehungsgeschichte einiger neuer Kulturpflanzen," *Landwirtschaftliches Jahrbuch*, vol. 91, pp. 793–880, 1942.

[18] P. Barlóg, "Effect of magnesium and nitrogenous fertilisers on the growth and alkaloid content in *Lupinus angustifolius* L," *Australian Journal of Agricultural Research*, vol. 53, no. 6, pp. 671–676, 2002.

[19] D. Ciesiołka, M. Muzquiz, C. Burbano et al., "An effect of various nitrogen forms used as fertilizer on Lupinus albus L. yield and protein, alkaloid and α-galactosides content," *Journal of Agronomy and Crop Science*, vol. 191, no. 6, pp. 458–463, 2005.

[20] P. Gremigni, M. T. F. Wong, N. K. Edwards, D. Harris, and J. Hamblin, "Potassium nutrition effects on seed alkaloid concentrations, yield and mineral content of lupins (*Lupinus angustifolius*)," *Plant and Soil*, vol. 234, no. 1, pp. 131–142, 2001.

[21] P. Gremigni, J. Hamblin, and D. Harris, "Genotype x Environment Interactions and Lupin Alkaloids," in *Proceedings of the 9th International Lupin Conference*, 2002.

[22] P. Gremigni, J. Hamblin, D. Harris, and W. A. Cowling, "The interaction of phosphorus and potassium with seed alkaloid concentrations, yield and mineral content in narrow-leafed lupin (*Lupinus angustifolius* L.)," *Plant and Soil*, vol. 253, no. 2, pp. 413–427, 2003.

[23] J. L. Christiansen, B. Joørnsgård, S. Buskov, and C. E. Olsen, "Effect of drought stress on content and composition of seed alkaloids in narrow-leafed lupin, *Lupinus angustifolius* L," *European Journal of Agronomy*, vol. 7, no. 4, pp. 307–314, 1997.

[24] M. A. Yaber Grass and S. R. Leicach, "Changes in Senecio grisebachii pyrrolizidine alkaloids abundances and profiles as response to soil quality," *Journal of Plant Interactions*, vol. 7, no. 2, pp. 175–182, 2011.

15

Influence of Simulated Imazapic and Imazethapyr Herbicide Carryover on Cotton (*Gossypium hirsutum* L.)

W. James Grichar,[1] Peter A. Dotray,[2] and Todd A. Baughman[3]

[1] *Texas AgriLife Research, 3507 Hwy 59E, Beeville, TX 78102, USA*
[2] *Texas Tech University, Texas AgriLife Research, and Texas AgriLife Extension Service, Lubbock, TX 79403, USA*
[3] *Texas AgriLife Extension Service, Vernon, TX 76384, USA*

Correspondence should be addressed to W. James Grichar, w-grichar@tamu.edu

Academic Editor: David Clay

Field studies were conducted during the 2001 and 2002 growing seasons in the Texas peanut growing regions to simulate residual concentrations of imazapic and imazethapyr in the soil and subsequent effects on cotton (*Gossypium hirsutum* L.). Simulated imazapic or imazethapyr rates included 0, 1/64X (1.09 g ai/ha), 1/32X (2.19 g ai/ha), 1/16X (4.38 g ai/ha), 1/8X (8.75 g ai/ha), 1/4X (17.5 g ai/ha), and 1/2X (35 g ai/ha) of the full labeled rate for peanut (*Arachis hypogaea* L.) and incorporated prior to cotton planting. Cotton stunting with imazapic or imazethapyr was more severe at Denver City than other locations. All rates of imazapic and imazethapyr resulted in cotton stunting at Denver City while at Munday and Yoakum the 1/8X, 1/4X, and 1/2X rates of imazapic resulted in reduced cotton growth when compared with the untreated check. At all locations imazapic caused more stunted cotton than imazethapyr. Cotton lint yield was reduced by imazapic or imazethapyr at 1/4 X and 1/2 X rates at all locations when compared with the untreated check.

1. Introduction

Imazethapyr and imazapic are imidazolinone herbicides registered for use in peanut (*Arachis hypogaea* L.) and are used extensively in the various peanut growing regions of Texas. Imazethapyr may be applied preplant incorporated (PPI), preemergence (PRE), ground cracking (GC), or postemergence (POST) for effective weed control [1]. Imazethapyr applied PPI or PRE controls many troublesome weeds such as coffee senna (*Cassia occidentalis* L.), common lambsquarter (*Chenopodium album* L.), morningglory species (*Ipomoea* spp.), pigweed species (*Amaranthus* spp.) including Palmer amaranth (*Amaranthus palmeri* S. Wats), prickly sida (*Sida spinosa* L.), purple and yellow nutsedge (*Cyperus rotundus* L. and *C. esculentus* L., resp.), spurred anoda [*Anoda cristata* (L.) Schlecht.], and wild poinsettia (*Euphorbia heterophylla* L.) [2–6].

Imazethapyr applied POST provides broad spectrum and most consistent control when applied within 10 d of weed emergence [3, 7–9]. Imazethapyr and imazapic are the only POST herbicides to effectively control both yellow and purple nutsedge [5, 10]. Control is most effective when imazethapyr is applied to the soil or to yellow nutsedge that is no more than 13 cm tall [1, 10, 11].

Imazapic is similar to imazethapyr and controls all the weeds controlled by imazethapyr [1, 9, 12–14]. In addition, imazapic provides control or suppression of Florida beggarweed [*Desmodium tortuosum* (S.W.) D.C.] and sicklepod [*Senna obtusifolia* (L.) Irwin and Barneby], which are not adequately controlled by imazethapyr [15]. Imazethapyr provides consistent control of many broadleaf and sedge species if applied within 10 d after emergence, but imazapic has a longer effectiveness period when applied POST [1, 10, 14, 16]. Imazapic also is effective for control of rhizome and seedling johnsongrass [*Sorghum halepense* (L.) Pers.], Texas millet [*Urochloa texana* (Buckl.) R. Webster], large crabgrass [*Digitaria sanguinalis* (L.) Scop.], southern crabgrass [*Digitaria ciliaris* (Retz.) Koel.], and broadleaf signalgrass [*Brachiaria platyphylla* (Griseb.) Nash] [14].

In crop rotations, the imidazolinone herbicides must be used with caution. Monks and Banks [17] observed slight corn (*Zea mays* L.) injury and severe cotton injury from imazaquin (another imidazolinone herbicide) applied to soybean [*Glycine max* (L.) Merr.] the previous year. Renner et al. [18] observed significant corn injury from imazaquin applied the previous year in one of two years. In Arkansas, cotton yield was reduced 7 to 42% as the soil concentration of imazaquin increased from 7.5 to 26 g ai/ha [19]. Imazethapyr has been observed to moderately injure corn [20]. Johnson et al. [21] reported slight but significant injury to rice (*Oryza sativa* L.) from imazethapyr applied to soybean the previous year. Rotational crops such as sugarbeet (*Beta vulgaris* L.), canola (*Brassica napus* L.), cauliflower (*Brassica oleracea* L.), broccoli (*Brassica oleracea* L.), and lettuce (*Lactuca sativa* L.) may also be damaged when planted following imazethapyr [22, 23].

Previous research on imazapic carryover has shown varying results. In North Carolina, imazapic applied PPI at 35 g ai/ha reduced cotton yield 43% the following year while imazapic at the same rate applied at emergence caused 20% cotton injury but no yield reduction the following season [24]. In Georgia, imazapic at 35 g ai/ha reduced cotton yield an average of 34% the year following application regardless of application method [24].

A Mississippi study indicated no reduction in shoot weight when corn, grain sorghum [*Sorghum bicolor* (L.) Moench], cotton, rice, wheat (*Triticum aestivum* L.), soybean, and Italian ryegrass (*Lolium multiflorum* L.) were planted directly into soil treated and incorporated with imazapic at rates up to 35 g ai/ha [25]. In that study, all crops were more sensitive in the greenhouse with rates of 11.6 g ai/ha reducing corn and grain sorghum shoot weights. However, cotton, rice, and wheat tolerated rates of 19 to 38 g ai/ha. Grymes et al. [26] reported that imazapic at 69 g ai/ha or imazapic plus imazethapyr each at 35 g ai/ha reduced rice yield the year following application. Grymes et al. [26] felt that imazapic injury to rice grown in rotation with soybean may be reduced by implementing a later rice planting date. They hypothesized that the later date allowed time for more herbicide degradation in the soil. Also, herbicide metabolism by the rice plant may be greater at the later planting date due to warmer temperatures [26].

The persistence of the imidazolinones in soil is influenced by the degree of adsorption to soil, soil moisture content, temperature, and amount of exposure to sunlight [27–29]. The degree of soil adsorption increases as organic matter content increases and pH decreases [30, 31].

The primary mode of herbicidal decomposition is by microbial degradation, and degradation is most rapid in soils with temperatures and moisture contents that favor microbial activity [32, 33]. Photodecomposition accounts for a small amount of imidazolinone degradation when the herbicide is on the soil surface but rainfall or incorporation removes the herbicide from exposure to light [32, 34].

Above soil pH 4.0, the carboxyl groups on imazethapyr dissociate, and soil adsorption of the resulting herbicide anion is negligible [29]. However, in the presence of clay at pH 5.0, fluorescence emission spectra indicate that imazethapyr is adsorbed in the neutral form [30]. At pH 8.0, only the ionized form was observed even in the presence of clay. Increased adsorption and persistence were observed

as the pH decreased from 6.5 to 4.5 [33]. Injury to crops seeded following imidazolinone herbicide application also increased as soil pH decreased from 7.7 to 6.0 [35]. This indicated that increased adsorption did not protect crops from imidazolinone herbicide residue at pH 6.0.

Most peanut soils in south and central Texas have a pH of 6.5 to 7.5 and organic matter content ≤1%. Therefore, in these soils, imidazolinone herbicides are readily available for microbial degradation. However, in the Texas High Plains, the pH may range from 7.0 to 8.5 resulting in reduced microbial degradation. With soils low in organic matter and near neutral pH, little imidazolinone herbicide should be adsorbed onto soil particles. Crops with low tolerance to the imidazolinone herbicides such as cotton are grown in rotation with peanut in many areas of Texas where imazethapyr or imazapic may be used. Evaluating imazethapyr or imazapic in the different regions will provide a more relevant understanding of the persistence issue. Therefore, the objective of this research was to evaluate cotton tolerance to imazethapyr and imazapic concentrations when planted at several locations in the peanut growing areas of Texas.

2. Materials and Methods

Field studies were conducted in Knox County (Munday), Lavaca County (Yoakum), and Yoakum County (Denver City), Texas during the 2001 and 2002 growing seasons to evaluate cotton response to sublabeled imazapic and imazethapyr rates to simulate carryover. Soil characteristics are presented in Table 1. The soils selected are representative of the soils found in different areas of Texas where a peanut-cotton rotation may be found.

The experimental design was a randomized complete block with a factorial arrangement of two herbicide treatments and seven rates with four replications. One factor was herbicide which included imazapic (Cadre, BASF Corporation, P.O. Box 13528, Research Triangle Park, NC 27709) and imazethapyr (Pursuit, BASF Corporation). The other factor was herbicide rate applied at 0, 1.09 g ai/ha (1/64X), 2.19 g ai/ha (1/32X), 4.38 g ai/ha (1/16X), 8.75 g ai/ha (1/8X), 17.5 g ai/ha (1/4X), and 35 g ai/ha (1/2X). These rates were chosen as a representation of the dissipation of imazapic and imazethapyr over time with respect to estimated dissipation time. The normal use rate of imazapic and imazethapyr in peanut is 69 g ai/ha with a half-life of 120 d [36].

Herbicides at the Yoakum location were applied with a CO_2 pressurized backpack sprayer equipped with Teejet 11002 DG flat fan spray tips (Spraying Systems Company, P.O. Box 7900, North Avenue, Wheaton, IL 60188) which delivered a spray volume of 190 L/ha at 180 kPa. At the Denver City location, herbicides were applied with a CO_2 pressurized backpack sprayer using Teejet 110015 TT flat fan spray tips calibrated to deliver a spray volume of 94 L/ha at 207 kPa. At the Munday location, herbicides were applied with a CO_2 pressurized backpack sprayer equipped with Teejet 110015 AI flat fan spray tips which delivered 94 L/ha at 180 kPa. After application, herbicides at Yoakum were incorporated approximately 5 to 6 cm deep with a tractor-driven power tiller while at the Denver City location,

TABLE 1: Cotton varieties, planting dates, and soil characteristics of each site.

Variables	2001			2002	
	Denver City	Munday	Yoakum	Denver City	Yoakum
Herbicides applied	April 26	May 8	April 27	April 18	April 23
Planting date	May 18	May 8	April 27	June 3	April 23
Soil texture	LFS	FSL	SL	LFS	SL
Soil name	Brownfield	Miles	Hallettsville	Brownfield	Hallettsville
pH	7.6	8.1	6.8	7.6	7.2
OM (%)	<1.0	0.1	1.2	<1.0	1.0
Sand (%)	80	75	65	80	65
Silt (%)	3	16	18	3	17
Clay (%)	17	9	17	17	18
Cotton variety	PM1218RR	PM1218RR	ST 4793RR	PM 2280RR	SG 215RR

[a] Abbreviations: FSL: fine sandy loam; LFS: loamy fine sand; SL: sandy loam; OM: organic matter.

herbicides were incorporated into the soil using a tandem disk set to incorporate 10 to 15 cm deep. At the Munday location, herbicides were applied and incorporated twice 2.5 to 5 cm deep with a rolling cultivator. Cotton was planted at Yoakum and Munday within 24 h of herbicide incorporation while at Denver City herbicides were applied approximately 6 wk prior to cotton planting. At Yoakum, each plot contained two rows, 91 cm apart and 7.9 m long while at Denver City and Munday each plot contained four rows spaced 102 cm apart and 9.5 m long. All plots were maintained weed-free using standard herbicides recommended by the Texas AgriLife Extension Service.

Visual estimates of crop stunting were determined 7 to 9 wk after cotton planting using a scale of 0 to 100, where 0 equals no crop stunting and 100 equals complete crop death. Cotton was either hand-picked or mechanically harvested using commercial harvesting equipment modified for plot harvest. Data were analyzed using the general linear models and means separated using Fisher's protected LSD at $P <$ 0.05.

3. Results and Discussion

3.1. Cotton Emergence. No stand reduction was noted with any rate of imazethapyr or imazapic at any location (data not shown). In previous work in Texas, Matocha et al. [37] reported that cotton stand was not affected by imazapic applied at rates up to 144 g ai/ha the previous season. Wixson and Shaw [25] reported that imazapic did not reduce the emergence of cotton with rates up to 35 g ai/ha on a silty clay soil with pH of 7.2 and 3.2% organic matter while Walsh et al. [38] found that imazethapyr at 48 to 96 g ai/ha did not cause a loss of cotton stand. Wiatrak et al. [39] noted that imazapic at the 1X rate (70 g ai/ha) reduced cotton stand in one of two years in Florida while Grey et al. [40] reported no stand reduction at Tifton, GA on a loamy sand with pH of 6.0 and 1.3% organic matter. However, at Plains, GA on a sandy loam with pH of 5.8 to 6.0 and 1.0% organic matter, cotton plants emerged at all imazethapyr and imazapic rates, but by 14 days after treatment (DAT), cotton began to die with sporadic plants exhibiting distended growth. Similar effects

were seen in cotton with imazaquin carryover at 70 g ai/ha in Arkansas [41]. None of these effects were observed at any of the Texas locations.

3.2. Cotton Injury and Stunting. Cotton injury observed at all locations included malformation and chlorosis of leaf tissue and plant stunting, typical of imidazolinone herbicides [19, 42, 43]. There was a herbicide, rate, and location interaction; therefore, data are presented individually by herbicide rate and location.

2001. Stunting with imazapic and/or imazethapyr was more severe at Denver City than the other locations (Table 2). At Denver City, imazapic at 1/16 to 1/2X resulted in 81 to 100% cotton stunting while imazethapyr at 1/16 to 1/2X resulted in 60 to 100% cotton stunting. At all rates, with the exception of the 1/64 and 1/2X rates, imazethapyr was less injurious to cotton than imazapic.

At Munday, the high rate of imazapic and imazethapyr caused 48 and 16% cotton stunting, respectively. No significant stunting was observed at Munday with imazapic rates 1/16X or lower or imazethapyr rates 1/4X or lower. The 1/4X rate of imazapic resulted in over 20% cotton stunting.

At Yoakum, cotton stunting was at least 45% when the rate of either imazapic or imazethapyr was 1/4X or greater (Table 2). Although imazapic at 1/8X rate caused 15% stunting, only 4% cotton stunting was noted with the same rate of imazethapyr. None of the other rates of imazapic or imazethapyr resulted in cotton stunting that was different from the untreated check.

2002. At Denver City, all rates of imazapic or imazethapyr resulted in at least 32% cotton stunting (Table 2). No difference in cotton stunting was noted between imazapic and imazethapyr. The highest applied rate of 35 g ai/ha (1/2X) of imazapic or imazethapyr resulted in at least 85% cotton stunting.

At Yoakum, imazapic caused no greater than 18% cotton stunting while imazethapyr resulted in 8% or less cotton stunting at all rates (Table 2). The 1/4 and 1/2X rates of imazapic were the only treatments that resulted in cotton stunting that was different from the untreated check. Rainfalls for May, June, July, August, and September were 33,

TABLE 2: Cotton stunting as affected by simulated rates of imazapic and imazethapyr.[a]

Herbicide	Rate[b]	2001			2002	
		Denver City	Munday	Yoakum	Denver City	Yoakum
		%				
Untreated	—	0	0	0	0	0
Imazapic	1/64X	14	0	3	37	6
	1/32X	48	3	0	43	3
	1/16X	81	4	8	54	10
	1/8X	84	8	15	54	13
	1/4X	98	23	62	73	18
	1/2X	100	48	75	85	15
Imazethapyr	1/64X	8	0	3	32	8
	1/32X	17	3	0	33	8
	1/16X	60	1	0	53	5
	1/8X	70	1	4	64	4
	1/4X	89	6	45	77	3
	1/2X	100	16	53	89	4
LSD (0.05)		9	7	14	15	13

[a] Stunting ratings taken 7 weeks after herbicide application at Yoakum in 2001; 8 weeks after herbicide application at Denver City in 2001, Munday, and Yoakum in 2002; 9 weeks after herbicide application at Denver City in 2002.
[b] Herbicide rate: 1.09 g ai/ha (1/64X), 2.19 g ai/ha (1/32X), 4.38 g ai/ha (1/16X), 8.75 g ai/ha (1/8X), 17.5 g ai/ha (1/4X), and 35 g ai/ha (1/2X).

TABLE 3: Cotton lint yield as affected by simulated rates of imazapic and imazethapyr.

Herbicide	Rate[a]	2001			2002	
		Denver City	Munday	Yoakum	Denver City	Yoakum
		Kg/ha				
Untreated	—	1080	1800	925	830	1455
Imazapic	1/64X	1205	1905	995	860	1565
	1/32X	1050	1630	1210	910	1690
	1/16X	800	1920	700	975	1425
	1/8X	785	1845	845	1020	1350
	1/4X	195	1615	275	805	1040
	1/2X	0	1645	100	400	1385
Imazethapyr	1/64X	1195	1890	935	955	1635
	1/32X	1160	1660	1180	885	1310
	1/16X	1073	1720	1220	1100	1690
	1/8X	830	1685	895	975	1675
	1/4X	485	1600	435	780	1645
	1/2X	100	1685	375	700	1740
LSD (0.05)		280	NS	370	310	535

[a] Herbicide rate: 1.09 g ai/ha (1/64X), 2.19 g ai/ha (1/32X), 4.38 g ai/ha (1/16X), 8.75 g ai/ha (1/8X), 17.5 g ai/ha (1/4X), and 35 g ai/ha (1/2X).

114, 136, 106, and 114 mm, respectively. Normal rainfalls for these months are 112, 109, 66, 79, and 102 mm, respectively. The above normal rainfall for July and August may have accounted for the lack of cotton response to imazapic and imazethapyr. Microbial degradation is the primary degradation mechanism of imidazolinones and is accentuated by warm, moist soil conditions [44]. In contrast, dry conditions can prolong carryover effects of these herbicides [44]. Wixson and Shaw [25] reported that in soils with a pH 7.2 and 3.2% organic matter, corn and cotton tolerated imazapic up to 35 g ai/ha. Crop injury was observed with imazethapyr in both crops at rates from 5.5 to 17 g ai/ha. The authors indicated that the injury noted with low rates of imazethapyr could be related to the increase of adsorption of the imidazoline herbicides with increasing organic matter content. Wiatrek et al. [39] reported that cotton height was reduced with the high rates of imazapic in one year but not another. Grey et al. [40] reported a negative exponential trend where cotton height decreased as imazapic rate increased. Matocha et al. [37] reported a reduction in cotton height with imazapic applied at 140 and 210 g ai/ha the previous year.

3.3. Cotton lint Yield. There was a herbicide (imazapic and imazethapyr), rate, and location interaction; therefore, data are presented separately by herbicide, rate, and location.

2001. Lint yields at Denver City were reduced following the 1/8X, 1/4X, and 1/2X rates of imazethapyr or imazapic (Table 3). No cotton was produced from plots treated with imazapic at the 1/2X rate while imazethapyr at the 1/2X rate produced cotton yield that was 8% of the untreated check.

At Munday, none of the herbicide treatments reduced cotton yield when compared with the untreated check. An explanation for the lack of yield differences may be due to soil characteristics. The other locations all had a clay content of at least 17% while this site had a clay content of less than 10%. The pH at the Munday site was 8.1 which was greater than the pH values of 7.6 or less at the other locations (Table 1). Imazapic is weakly adsorbed in high pH soils and adsorption increases as the pH decreases and with increasing clay and organic matter content [19, 37, 40, 42, 45]. At Yoakum, cotton lint yields were reduced by the 1/4X and 1/2X rates of imazapic and imazethapyr (Table 3). Imazapic at the 1/16X rate resulted in lower yields than imazethapyr at the 1/16 or imazapic and imazethapyr at the 1/32X rates.

2002. At Denver City, only the 1/2X rate of imazapic resulted in lower cotton yields than the untreated check (Table 3). No negative response was noted with any of the imazethapyr treatments.

At Yoakum, no reduction in cotton yield was noted when the untreated check was compared with any imazapic or imazethapyr treatments (Table 3). However, plots which received imazethapyr, with the exception of the 1/32X rate, produced higher yields than those that received 1/4X rate of imazapic. The above average rainfall amounts for July and August may help explain a lack of yield reduction observed with the higher rates of imazapic and imazethapyr. The imidazolinones are soluble in water and are not degraded hydrolytically in aqueous solution [34]. However, in water, these herbicides are rapidly photodegraded by sunlight with a half-life of one to two days [29, 34].

Previous research on imazapic carryover has shown varying results. In North Carolina, imazapic applied PPI to peanut at 36 g ai/ha reduced cotton yield 43% the following year while the same rate of imazapic applied at peanut ground cracking resulted in 20% injury but no yield reduction [24]. In Georgia, imazapic at 36 g ai/ha reduced cotton yield an average of 34% the following year regardless of application timing [24]. Grey et al. [40] also reported that there were no detectable differences in cotton variety response to the imidazolinone herbicides.

4. Conclusion

Although different cotton cultivars were used in this study over locations and years, no previous work could be found that reported differential response of cotton cultivars to any of the imidazolinone herbicides. Cotton stunting did not

always result in reduced yield, and this may be the result of soil characteristics. However, when stunting was greater than 50% there was almost always a decrease in cotton yield when compared with the untreated check. This study reveals that several factors are involved in the persistence of imazethapyr and imazapic in the soil and helps to explain the various results observed under varying conditions. By possibly knowing the level of imazapic or imazethapyr residual in the soil, producers could have some flexibility in crop rotations if sensitive crops such as cotton are to be planted following imidazolinone use on peanut.

References

[1] J. W. Wilcut, A. C. York, W. J. Grichar, and G. R. Wehtje, "The biology and management of weeds in peanut (*Arachis hypogaea*)," in *Advances in Peanut Science*, H. E. Pattee and H. T. Stalker, Eds., pp. 207–244, American Peanut Research Education Society, Stillwater, OK, USA, 1995.

[2] T. A. Cole, G. R. Wehtje, J. W. Wilcut, and T. V. Hicks, "Behavior of imazethapyr in soybeans (*Glycine max*), peanuts (*Arachis hypogaea*), and selected weeds," *Weed Science*, vol. 37, pp. 639–644, 1989.

[3] J. W. Wilcut, F. R. Walls Jr., and D. N. Norton, "Imazethapyr for broadleaf weed control in peanuts (*Arachis hypogaea*)," *Peanut Science*, vol. 18, pp. 26–30, 1991.

[4] J. W. Wilcut, F. R. Walls Jr., and D. N. Norton, "Weed control, yield, and net returns using imazethapyr in peanuts (*Arachis hypogaea*)," *Weed Science*, vol. 39, pp. 238–242, 1991.

[5] W. J. Grichar and P. R. Nester, "Nutsedge (*Cyperus* spp.) control in peanut (*Arachis hypogaea*) with imazethapyr," *Weed Technology*, vol. 6, pp. 396–400, 1992.

[6] A. C. York, J. W. Wilcut, C. W. Swann, D. L. Jordan, and F. R. Walls, "Efficacy of imazethapyr in peanut (*Arachis hypogaea*) as affected by time of application," *Weed Science*, vol. 43, pp. 107–116, 1995.

[7] T. L. Grey, G. R. Wehtje, R. H. Walker, and K. P. Paudel, "Comparison of imazethapyr and paraquat-based weed control systems in peanut," *Weed Technology*, vol. 9, pp. 813–818, 1995.

[8] J. W. Wilcut, J. S. Richburg, E. F. Eastin, G. R. Wiley, F. R. Walls Jr., and S. Newell, "Imazethapyr and paraquat systems for weed management in peanut (*Arachis hypogaea*)," *Weed Science*, vol. 42, pp. 601–607, 1994.

[9] J. W. Wilcut, J. S. Richburg III., G. Wiley et al., "Imidazolinone herbicide systems for peanut (*Arachis hypogaea*)," *Peanut Science*, vol. 21, pp. 23–28, 1994.

[10] J. S. Richburg III., J. W. Wilcut, and G. R. Wehtje, "Toxicity of foliar and/or soil applied AC 263,222 to purple (*Cyperus rotundus*) and yellow nutsedge (*C. esculentus*)," *Weed Science*, vol. 42, pp. 398–402, 1993.

[11] J. W. Wilcut, J. W. A. C. York, and G. R. Wehtje, "The control and interaction of weeds in peanut (*Arachis hypogaea*)," *Review Weed Science*, vol. 6, pp. 177–205, 1994.

[12] P. R. Nester and W. J. Grichar, "Cadre combinations for broadleaf weeds control in peanut," in *Proceedings of the Southern Weed Science Society*, vol. 46, p. 317, 1993.

[13] W. J. Grichar, A. E. Colburn, and P. R. Nester, "Weed control in Texas peanut with Cadre," in *Proceedings of the American Peanut Research and Education Society*, vol. 26, p. 70, 1994.

[14] J. W. Wilcut, E. F. Eastin, J. S. Richburg III. et al., "Imidazolinone systems for southern weed management in resistant corn," *Weed Science Society America*, vol. 33, p. 5, 19933

[15] T. L. Grey, D. C. Bridges, E. F. Eastin et al., "Residual weed control for peanut (Arachis hypogaea) with imazapic, diclosulam, flumioxazin, and sulfentrazone in Alabama, Georgia, and Florida: a multi-state and year summary," in Proceedings of the American Peanut Research and Education Society, vol. 33, p. 19, 2001.

[16] J. S. Richburg, J. W. Wilcut, D. L. Colvin, and G. R. Wiley, "Weed management in southeastern peanut (Arachis hypogaea) with AC 263,222," Weed Technology, vol. 10, pp. 145–152, 1996.

[17] C. D. Monks and P. A. Banks, "Rotational crop response to chlorimuron, clomazone, and imazaquin applied the previous year," Weed Science, vol. 39, pp. 629–633, 1991.

[18] K. A. Renner, W. F. Meggitt, and D. Penner, "Response of corn (Zea mays) cultivars to imazaquin," Weed Science, vol. 36, pp. 625–628, 1988.

[19] C. J. Barnes, A. J. Goetz, and T. L. Lavy, "Effects of imazaquin residues on cotton (Gossypium hirsutum)," Weed Science, vol. 37, pp. 820–824, 1989.

[20] J. A. Mills and W. W. Witt, "Efficacy, phytotoxicity, and persistence of imazaquin, imazethapyr, and clomazone in no-till double-crop soybeans (Glycine max)," Weed Science, vol. 37, pp. 353–359, 1989.

[21] D. H. Johnson, R. E. Talbert, J. D. Beaty, C. B. Guy, and R. J. Smith, "Rice response following imazaquin, imazethapyr, chlorimurion, and clomazone use," in Proceedings of the Southern Weed Science Society, vol. 45, p. 371, 1992.

[22] S. D. Miller and H. P. Alley, "Weed control and rotational crop response with AC 222,293," Weed Technology, vol. 1, pp. 29–33, 1987.

[23] B. R. Tickes and K. Umeda, "The effect of imazethapyr upon crops grown in. rotation with alfalfa," in Proceedings of the Society Western Weed Science, vol. 44, p. 97, 1991.

[24] A. C. York and J. W. Wilcut, "Potential for Cadre and Pursuit applied to peanuts to carryover to cotton in North Carolina and Georgia," in Proceedings of the Beltwide Cotton Conference, vol. 1, p. 602, 1995.

[25] M. B. Wixson and D. R. Shaw, "Effects of soil-applied AC 263,222 on crops rotated with soybean (Glycine max)," Weed Technology, vol. 6, pp. 276–279, 1992.

[26] C. F. Grymes, J. M. Chandler, and P. R. Nester, "Response of soybean (Glycine max) and rice (Oryza sativa) in rotation to AC 263,222," Weed Technology, vol. 9, pp. 504–511, 1995.

[27] R. Allen and J. C. Casely, "The persistence and mobility of AC 222,293 in cropped and fallow land," in Proceedings of the British Crop Protection Conference, pp. 569–576, 1987.

[28] N. D. Malik, E. Cole, A. L. Darwent, and J. R. Moyer, "Imazethapyr (Pursuit)—a promising new herbicide for forage legumes," Forage Notes, vol. 32, pp. 42–45, 1988.

[29] G. Mangels, "Behavior of the imidazolinone herbicide in soil—a review of the Literature," in The Imidazolinone Herbicides, D. L. Shaner and S. L. O'Connor, Eds., pp. 191–209, CRC Press, Boca Raton, FL, USA, 1991.

[30] M. Che, M. M. Loux, S. J. Traina, and T. J. Logan, "Effect of pH on sorption and desorption of imazaquin and imazethapyr on clays and humic acid," Journal of Environmental Quality, vol. 21, pp. 698–703, 1992.

[31] M. M. Loux, R. A. Liebl, and F. W. Slife, "Adsorption of imazaquin and imazethapyr on soils, sediments, and selected adsorbents," Weed Science, vol. 37, pp. 712–718, 1989.

[32] A. J. Goetz, T. L. Lavy, and E. E. Gbur, "Degradation and field persistence of imazethapyr," Weed Science, vol. 38, pp. 421–428, 1990.

[33] M. M. Loux and K. D. Reese, "Effect of soil type and pH on persistence and carryover of imidazolinone herbicides," Weed Technology, vol. 7, pp. 452–458, 1993.

[34] W. S. Curran, M. M. Loux, R. A. Liebl, and F. W. Simmons, "Photolysis of imidazolinone herbicides in aqueous solutions and on soil," Weed Science, vol. 40, pp. 143–148, 1992.

[35] G. M. Fellows, P. K. Kay, G. R. Carlson, and V. R. Stewart, "Effect of AC 222,293 soil residues on rotational crops," Weed Technology, vol. 4, pp. 48–51, 1990.

[36] S. A. Senseman, Herbicide Handbook, Weed Science Society of America, Lawrence, KA, 9th edition, 2007.

[37] M. A. Matocha, W. J. Grichar, S. A. Senseman, C. A. Gerngross, B. J. Brecke, and W. K. Vencill, "The persistence of imazapic in peanut (Arachis hypogaea) crop rotations," Weed Technology, vol. 17, pp. 325–329, 2003.

[38] J. D. Walsh, M. S. Defelice, and B. D. Sims, "Soybean (Glycine max) herbicide carryover to grain and fiber crops," Weed Technology, vol. 7, pp. 625–632, 1993.

[39] P. J. Wiatrak, D. L. Wright, and J. J. Marois, "Influence of imazapic herbicide simulated carryover on cotton growth, yields, and lint quality," Crop Management. In press.

[40] T. L. Grey, E. P. Prostko, C. W. Bednarz, and J. W. Davis, "Cotton (Gossypium hirsutum) response to simulated imazapic residues," Weed Technology, vol. 19, pp. 1045–1049, 2005.

[41] G. W. Basham, T. L. Lavy, L. R. Oliver, and H. D. Scott, "Imazaquin persistence and mobility in three Arkansas soils," Weed Science, vol. 35, pp. 576–582, 1987.

[42] D. H. Johnson, R. E. Talbert, and D. R. Horton, "Carryover potential of imazaquin to cotton, grain sorghum, wheat, rice, and corn," Weed Science, vol. 43, pp. 454–460, 1995.

[43] A. C. York, D. L. Jordan, R. B. Batts, and A. S. Culpepper, "Cotton response to imazapic and imazethapyr applied to a preceding peanut crop," Journal of Cotton Science, vol. 4, pp. 210–216, 2000.

[44] D. L. Shaner and S. L. O'Connor, "Behavior of the imidazolinone herbicides in the soil-a review of the literature," in The Imidazolinone Herbicides, pp. 191–209, CRC Press, Boca Raton, FL, USA, 1991.

[45] R. N. Stougaard, P. J. Shea, and A. R. Martin, "Effect of soil type and pH on absorption, mobility, and efficacy of imazaquin and imazethapyr," Weed Science, vol. 38, pp. 67–73, 1990.

16

The Effect of Microwave Radiation on Prickly Paddy Melon (*Cucumis myriocarpus*)

Graham Brodie, Carmel Ryan, and Carmel Lancaster

Melbourne School of Land and Environment, Dookie Campus, University of Melbourne, Nalinga Road, Dookie, VIC 3647, Australia

Correspondence should be addressed to Graham Brodie, grahamb@unimelb.edu.au

Academic Editor: David Clay

The growing list of herbicide-resistant biotypes and environmental concerns about chemical use has prompted interest in alternative methods of managing weeds. This study explored the effect of microwave energy on paddy melon (*Cucumis myriocarpus*) plants, fruits, and seeds. Microwave treatment killed paddy melon plants and seeds. Stem rupture due to internal steam explosions often occurred after the first few seconds of microwave treatment when a small aperture antenna was used to apply the microwave energy. The half lethal microwave energy dose for plants was 145 J/cm^2; however, a dose of at least 422 J/cm^2 was needed to kill seeds. This study demonstrated that a strategic burst of intense microwave energy, focused onto the stem of the plant is as effective as applying microwave energy to the whole plant, but uses much less energy.

1. Introduction

Interest in the effects of high frequency electromagnetic waves on biological materials dates back to the late 19th century [1], while interest in the effect of high frequency waves on plant material began in the 1920s [1]. More recent studies have been motivated by a growing list of herbicide-resistant biotypes [2] and environmental concerns over herbicide use [3].

Many of the earlier experiments on plant material focused on the effect of radio frequencies (RF) on seeds [1]. In many cases, short exposure resulted in increased germination and vigour of the emerging seedlings [4]; however, long exposure usually resulted in seed death [1]. For example, Headlee [5] observed that exposure to electromagnetic fields in the frequency range between 750 kHz and 3 MHz caused no injury to wheat seeds even after 80 minutes; however, exposure to 5 MHz electromagnetic fields reduced germination to 54% compared with the control samples which maintained an 88.6% germination rate. McKinley [6] exposed seeds of Golden Bantam corn to high frequency fields. Seeds were killed after 5 minutes. A one-minute exposure did not kill the seeds, but slightly retarded their germination.

Exposures of 30 to 40 seconds resulted in accelerated growth of the seedlings in their early germination period.

Davis et al. [7, 8] were among the first to study the lethal effect of microwave heating on seeds. They treated seeds, with and without any soil, in a microwave oven and showed that seed damage was mostly influenced by a combination of seed moisture content and the energy absorbed per seed.

Other findings suggested that both the specific mass and specific volume of the seeds were strongly related to a seed's susceptibility to damage by microwave fields [8]. The association between the seed's volume and its susceptibility to microwave treatment may be linked the "*radar cross-section*" [9] presented by seeds to propagating microwaves. Large radar cross sections allow the seeds to intercept, and, therefore, absorb, more microwave energy.

Several patents dealing with microwave treatment of weeds and their seeds have been registered [10–12]; however, none of these systems appear to have been commercially developed. This may be due to concerns about the energy requirements to manage weed seeds in the soil using microwave energy. In a theoretical argument based on the dielectric and density properties of seeds and soils, Nelson [13] demonstrated that using microwaves to selectively heat seeds in the soil "*cannot be expected*". He concluded that seed susceptibility to damage from microwave treatment is a purely thermal effect, resulting from soil heating and thermal conduction into the seeds. This has been confirmed experimentally by Brodie et al. [14].

Microwave treatment of soil may be useful for niche applications where small areas of soil need to be intensively treated, and soil fumigation is not an option; however, the use of microwaves for killing soil-borne seeds in a broad acre situation may be prohibitively expensive [13]; however, seedlings are many times more susceptible to microwave damage than seeds in the soil [7, 15].

Bigu-Del-Blanco et al. [16] exposed 48-hour-old seedlings of *Zea mays* (var. Golden Bantam) to 9 GHz radiation for 22 to 24 hours. The power density levels were between 10 and 30 mW cm^{-2} at the point of exposure. Temperature increases of only 4°C, when compared with control seedlings, were measured in the treated specimens. The authors concluded that the long exposure to microwave radiation, even at very low power densities, was sufficient to dehydrate the seedlings and inhibit their development. Rapid dehydration of the plant tissue appears to be the cause of death and growth inhibition. This is because microwave heating results have been coupled with rapid diffusion of moisture [17] through porous materials, which in plants may accelerate moisture loss in spite of the very small rise in plant temperature observed during these trials.

Horn antennas (Figure 1) are popular for microwave communication systems [18]. Horn antennas are also very useful for projecting microwave energy onto objects that cannot be placed inside an enclosed microwave cavity, such as a standard microwave oven.

The vertical plane of the horn antenna is usually referred to as the *E*-plane, because of the orientation of the electrical field (or *E*-field) in the antenna's aperture. The horizontal plane is referred to as the *H*-plane, because of the orientation of the magnetic field (or *H*-field) of the microwave energy. Typical temperature distributions, created by a pyramidal horn antennas, can be estimated using mathematical equations [19, 20]. The unique feature of microwave heating is that the maximum temperature is located well below the surface of the heated material (Figure 2).

There is a concern about the effect of microwave treatment on soil biota. Ferriss [21] reported that microwave treatment reduced populations of soil microorganisms with increasing treatment time; however, these effects decreased with increasing amounts of soil and decreased with increasing soil water content between 16 and 37% (wt. water/dry wt. soil). No pronounced effect of soil type was noted in this experiment. Treatment of 1 kg of soil, at 7% to 37% water content, for 150 s in a 653 W microwave oven, operating at 2.45 GHz, eliminated populations of *Pythium*, *Fusarium*, and all nematodes except *Heterodera glycines*. Marginal survival of *Rhizoctonia*, cysts of *H. glycines*, and *mycorrhizal* fungi was observed in some treated soils. Treatment of 4 kg of soil for 425 s gave comparable results [21].

In a previously reported experiment using a horn antenna, Cooper and Brodie [22] discovered that microwave

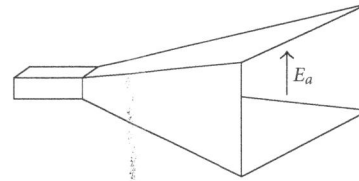

FIGURE 1: A typical horn antenna showing the orientation of the electrical field component of the microwave energy in the antenna's aperture.

treatment of clay loam soil reduced bacterial populations by 78% in the top 2 cm of soil after 16 minutes of microwave treatment; however, populations at 10 cm depth were not significantly affected by treatment. This study also investigated the effect of microwave heating on key soil nutrients and pH. The findings from this earlier work [22] suggest that microwave treatment of up to 16 minutes will have little effect on soil nutrients and pH and will not sterilise the soil of bacteria [22]; therefore, treatment durations of a few seconds or minutes are unlikely to change any soil properties.

Two important features of microwave heating in moist materials are (1) simultaneous heat and moisture diffusion through the irradiated material [17] and (2) an uncontrolled temperature increase called "*thermal runaway*" [23–25].

Plant materials with large moisture content have greater dielectric constants (Figure 3). As plant materials dry during microwave heating, their dielectric properties change significantly. These changes are too slow to affect individual cycles of the microwave fields; however, they affect microwave heating on the much longer thermal time scale.

The effect of changing dielectric properties on microwave heating can be studied using mathematical equations for microwave heating [20] and equations for the dielectric properties of the plant materials [26]. The heating rate for 10 mm and 12 mm diameter cylinders of plant material is relatively constant with time (Figure 4); however, there is a sudden 80°C jump in the 15 mm diameter case at 5 seconds of microwave heating when there is no change in the applied microwave power (Figure 4). This sudden jump in temperature is the result of "*thermal runaway*".

Vriezinga has studied this phenomenon and concluded that thermal runaway is caused by (1) the specific characteristic of the dielectric loss factor of water, which decreases with increasing temperature [27] and (2) resonance of the electromagnetic waves within the irradiated medium due to changes in the electromagnetic wavelength inside the irradiated medium as the dielectric properties change during heating [27, 28]. Resonance will only occur when the object's dimensions are similar to the wave length of the microwave fields inside the object. That is why thermal runaway only becomes evident in the 15 mm diameter stem (Figure 4), while the smaller stems are too narrow to allow field resonance.

In most cases, thermal runaway is a problem during microwave heating. It usually leads to undesirable destruction of the microwave-heated material [29]; however, it has been very effectively used in applications such as the microwave

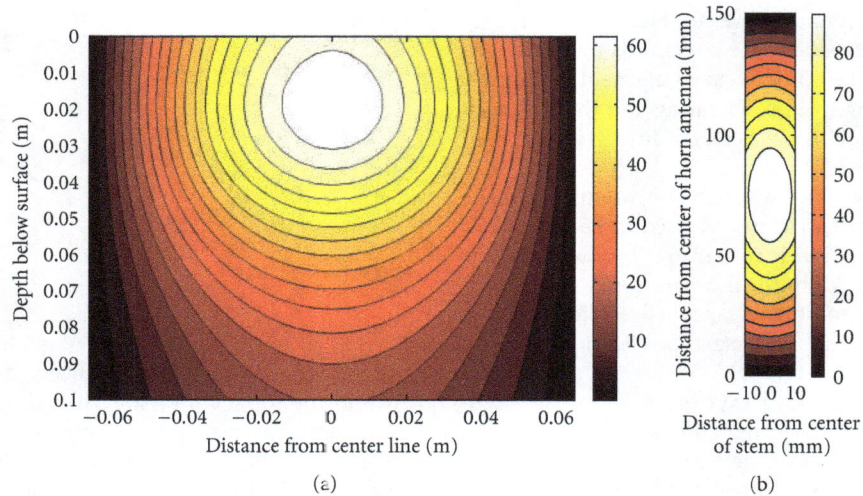

(a)   (b)

FIGURE 2: Typical temperature distribution created by a pyramidal horn with aperture dimensions of 130 mm by 43 mm and a length of 100 mm in soil profile (a) and the cross section of a 10 mm thick plant stem (b) calculated using equations presented in [20]. (Note: the temperature scales are in °C).

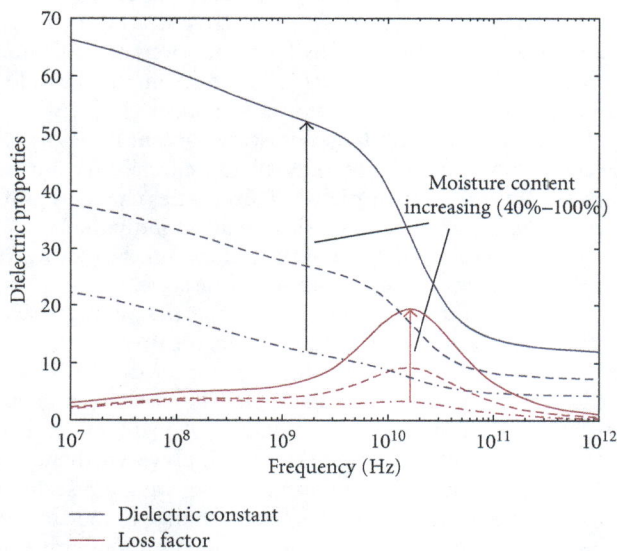

FIGURE 3: Dielectric properties of plant materials as a function of frequency and moisture content.

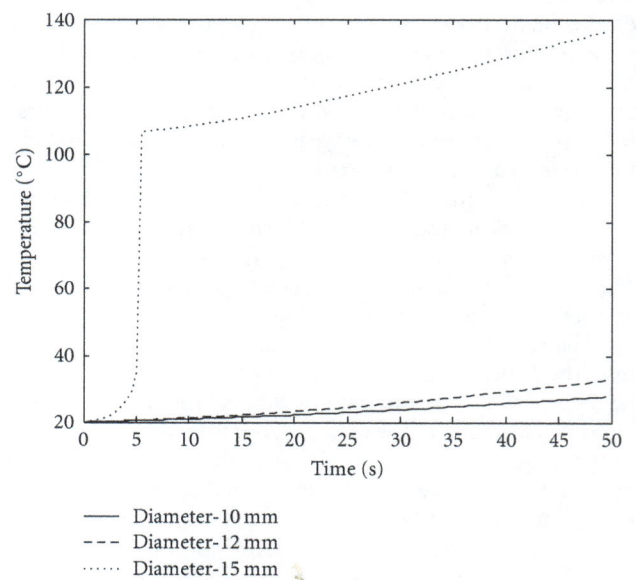

FIGURE 4: Temperature response, at a constant microwave power density, in the centre of a cylinder of plant-based material, as a function of plant stem diameter, assuming a constant moisture loss from a moisture content of 0.87 to 0.10 during microwave heating.

drill [30, 31], which can drill holes through ceramics by superheating a very small section of the material. In the case of weeds, sudden destruction of the plant tissue due to microwave-induced thermal runaway will be very desirable.

Cucumis myriocarpus is an annual prostrate vine belonging to the family *Cucurbitaceae* that is native to southern Africa [32]. *Cucumis myriocarpus* has multiple stems that can grow up to 1.6 m long and produces fruit in the form of small spiky yellow-green or green-striped melons that are approximately 20–25 mm in diameter. These melons contain many seeds that are 3.5–4 mm long [33]. The fruit and foliage are toxic due to the presence of cucurbitacin, which is a cytotoxic steroid produced by the plant as a defence against herbivores [32]. The plant has been known to kill livestock

[32]. *Cucumis myriocarpus* is a weed in Australia and in California, where it may also be known as prickly paddy melon, bitter apple, and gooseberry gourd [32]. The plant also occurs in Spain, where it is known by the common names of "*habanera*" or "*sandía habanera*" [32].

Cucumis myriocarpus germinates in spring or early summer depending on rainfall. They have the ability to germinate over an extended period particularly if soil disturbance occurs [19]. Growth can be rapid if moisture is adequate and plants can quickly establish a substantial tap root [29]. Established plants produce new growth from the large crown in spring to early summer with trailing branches or vines, growing rapidly in the summer months with a single mature plant being able to occupy an area of up to $7\,m^2$ [33]. Plants normally flower and fruit in mid to late summer and growth can progress well into autumn if the weather is warm. Chemicals are commonly used for summer weed control, but Leys et al. [30] have found that *Cucumis myriocarpus* is very tolerant of glyphosate [30]. Resistance to other herbicides may develop as it has in other species [31]; therefore, it is important to explore alternative methods of weed management before conventional options fail. The objective of this study was to explore the effect of microwave energy on prickly paddy melon (*Cucumis myriocarpus*) plants and seeds.

2. Method

2.1. Study Strategy. This study was subdivided into a series of separate but linked experiments that investigated the effect of microwave treatment on paddy melon plants, fruits, and seeds. Some experiments were completed in the field using a prototype microwave system (Figure 5), energised from the magnetron of a microwave oven, operating at 2.45 GHz. It had an 86 mm by 43 mm rectangular wave-guide channeling the microwaves from the oven's magnetron to a horn antenna outside of the oven.

The horn antenna allowed microwave energy to be focused onto sample plants in pots and *in situ*, while the oven's timing and power circuitry was used to control the activity of the magnetron. Other experiments in this study were completed using a standard microwave oven, also operating at 2.45 GHz.

Therefore, two pyramidal horn applicators (Figure 6), with varying aperture dimensions (130 mm by 43 mm and 86 mm by 20 mm), were developed and tested during these experiments.

2.2. Calibration Tests. The deliverable power from a microwave system depends on many parameters including the impedance match between all the components of the wave-guide system. In this case, no attempt to match the impedance along the wave-guide was made; therefore, the delivered microwave power was determined experimentally so that energy analyses could be conducted.

The deliverable microwave power can be determined using two samples of water. One acts as a control to determine the energy balance associated with the ambient conditions, while the other is heated by the microwave system. The power absorbed by the treated water sample (P_a) can be calculated from the combination of sensible and latent heat observed in the two samples using the following equation:

$$
\begin{aligned}
&P_a \\
&= \frac{\left\{ \left(4.18 \Delta T_m + \dfrac{2260 \Delta m_m}{m_m} \right) - \left(4.18 \Delta T_c + \dfrac{2260 \Delta m_c}{m_c} \right) \right\} m_m}{\tau \times t_h}.
\end{aligned}
\tag{1}
$$

FIGURE 5: A laboratory prototype system based on a modified microwave oven.

FIGURE 6: Horn antennas used in these microwave experiments.

The water in the treated sample will reflect some microwave energy from its surface. The portion of the microwave energy that is transmitted through the water surface (τ) is determined by the dielectric properties of the air and water at their interface:

$$
\tau = \frac{2\sqrt{\varepsilon_a}}{\sqrt{\varepsilon_a} + \sqrt{\varepsilon_w}}.
\tag{2}
$$

The microwave oven used in the prototype has a nominal rating of 750 W; however, the delivered power may be much less than this due to internal reflections in the magnetron coupling antenna that feeds the energy into the wave guide. The system was calibrated using water at 19°C. Three calibration runs were used to determine the output power of the system.

Water at 19°C has a complex dielectric constant of $78.9 - j11.0$, and air has a dielectric constant of $1 + j0$. Therefore, the transmission coefficient at the water surface was 0.2. The prototype oven was producing an average output power of 541.3 W.

This calibration procedure was also applied to a microwave oven that was used in Experiments 3 and 4 described later in this section. The calibration procedure revealed that the microwave oven was producing an average output power of 644.4 W.

Experiment 1 (Microwave Treatment of Potted Paddy Melon). Paddy melon plants, with 10–20 fully opened leaves, were collected from a field site near the township of St James, Victoria, Australia (36° 17′ S Latitude and 145° 53′ East Longitude) and transplanted into 13 cm pots with two plants in each pot. The pots were filled with commercial potting mix. Plants were watered every 2 or 3 days for 7 days to allow recovery from transplanting.

Each plant was individually exposed to microwave energy. The microwave energy was directed onto the plants using one of the horn antennas, depending on the particular treatment being applied. The experiment was set up as a 2 by 5 factor randomised design, where Factor A was the two different horn antenna designs and Factor B was the exposure time (0, 5, 15, 30, and 60 seconds). Each treatment combination was applied to 10 plants, with each plant acting as an individual replicate. The second plant in the treated pot was shielded from microwave energy during treatment using aluminium foil.

Radiation measurements, using a hand held microwave leakage detector, revealed that the magnetron of the microwave oven took approximately 3 seconds from initial power up to begin generating microwave energy. Therefore, an additional 3 seconds was added to the treatment times. For example, to achieve 5 seconds of microwave treatment, the microwave oven's timer was set to 8 seconds to allow for this starting delay.

All plants were placed in a sunlit area and watered every 2 or 3 days to maintain the potting mix at field moisture capacity. Eleven days after treatment, the plants were evaluated to estimate the portion of the plant's foliage and stem tissue that was either unaffected or had recovered from the microwave treatment, with the data being expressed as a fraction of the whole plant (i.e., a value of 0.25 indicated that 25% of the plant's foliage and stem tissue appeared to be healthy 11 days after treatment). The resulting data was analysed using a 2 by 5 factor Friedman analysis of variance. The Friedman analysis of variance is a nonparametric variation of the analysis of variance that can analyse data that is continuous but not necessarily normally distributed [34].

Experiment 2 (Microwave Treatment of Paddy Melon in the Field). Paddy melon is a multistemmed prostrate plant where individual plants can have an effective surface area of several square metres. It could be far too energy expensive to treat the entire plant [13]. The motivating interest in Experiment 2 was to determine whether focusing microwave energy onto a small section of the plant's stem could effectively kill the rest of the stem beyond the point of treatment.

Two sites were selected for this experiment; one site was near the township of St James, where the potted samples for Experiment 1 were sourced, and the other site was at the Dookie Campus of The University of Melbourne (36° 23′ S Latitude and 145° 42′ East Longitude).

Twenty paddy melon plants, with 30–50 fully opened leaves, were randomly selected at each site and treated using the small aperture horn antenna attached to the prototype microwave system. The applied treatments were 5, 10, 15, 30,

and 60 seconds of microwave exposure. Four control plants were also labelled at each site. The treatment that each plant received was identified using metal tags on pegs driven into the soil adjacent to each plant. All plants at both sites were assessed in the same fashion as the potted plants 16 days after treatment. The resulting data was analysed using a single factor Kruskal-Wallis analysis of variance, with the treatment factor being microwave exposure time. The Kruskal-Wallis analysis of variance is another nonparametric variation of the analysis of variance that can analyse data that is not normally distributed [34].

Experiment 3 (Microscopic Analysis). The purpose of this experiment was to determine if there was any obvious cellular damage created in the treated plant that could be attributed to microwave irradiation. Paddy melon plants, with 10–20 fully opened leaves, were collected from the field at the Dookie Campus site and transplanted into pots using the same arrangements as in Experiment 1. After acclimation in the pots for 5 days, two of the plants were selected at random and treated for 5 seconds using the small aperture horn antenna. After another 5 days, two of the remaining untreated plants were randomly selected and carefully removed from the pots to ensure that the root material was removed with the plant. The roots were washed to remove any potting mix.

One of the plants from the second selection was randomly selected to be the control plant, and the other plant from the second selection was treated in a standard microwave oven, operating at 2.45 GHz, for 5 seconds. A leaf stem on each plant was selected for microscopic comparisons. The stems were of similar sizes and the selected leaf stem on each plant was marked with a black marker for easy identification after treatment.

Three cross-sections of the selected leaf stem, a few cell lengths thick, were shaved from the stem using a sharp razor blade. These sections were mounted on a microscope slide. For comparison, three cross-sections from the control plant's leaf stem were also cut and mounted on the same microscope slide next to the freshly treated cross-sections.

After selecting a different leaf stem on each plant, the freshly treated plant was placed in the microwave oven again and treated for a further five seconds, before another three cross-sections were cut and mounted onto the slide. This process was repeated for a third time, so that the slide contained three different treatments (5, 5+5 and 5+5+5 seconds) with cross-sections from the control plant near each treated sample for easier comparison; however, the discontinuous treatments imposed on these samples may not have the same effect as continuous treatments of the same total time duration.

A dye, made from a mixture of dilute Fuchin 1% Phenol 5% and Methylene Blue 1%, was used to highlight the cell structures in the stems.

Sections of plant stem that were treated using the small aperture horn antenna 5 days earlier were also examined. Sections were taken from part of the stem that was clearly affected by the microwave treatment and other sections were taken from the same stem in areas that were not directly affected by microwave treatment. Microscope slides were prepared using the same process as described above.

Microscopic examination of the cross-sections was performed using a dissecting microscope with a video camera attached to the eye piece. Unfortunately, the video equipment did not allow image capture, so photographs of the cross-sections were captured from the video display using a 2 MPixel digital camera.

Experiment 4 (Seed Viability Testing). Paddy melon seeds, collected from the Dookie Campus field site, were treated in the same microwave oven that was used in Experiment 3. There were six microwave treatment groups plus two untreated groups of seeds. Three of the treatments involved extracting the seeds from ripe melons and treating them on a microwave-resistant dish. These seeds were exposed to 644.4 W of microwave power for 10, 20, or 30 seconds.

Three other treatments involved treating whole melons for 10, 20, or 30 seconds. Melons were placed in a beaker and exposed to microwave energy for the prescribed amount of time. The seeds were extracted from the melons after they had cooled. One set of melons was left untreated. There were approximately sixty seeds in each of the eight treatment groups.

Using a scalpel, under a dissecting microscope, a small sliver of the seed coat was cut from each seed to reveal the seed kernel. The seeds were then soaked in 1% 2, 3, 5 Triphenyl Tetrazolium Chloride and placed in an incubator at 35°C.

After 18 hours, the seeds were removed from the incubator and, using a scalpel under a dissecting microscope, the seeds were cut in half. Seeds were assessed as being viable; nonviable; or having uncertain viability, depending on the completeness of seed staining caused by the Tetrazolium [35].

Seeds that were fully stained were assessed as being viable and allocated a viability score of 1.0. Seeds that had no staining were assessed as being nonviable and allocated a viability score of 0.0. Some seeds were only partially stained, so these were classed as having uncertain viability and allocated a viability score of 0.5.

The resulting seed viability data was analysed using a single factor analysis of variance. The resulting data was analysed using a 2 by 3 factor Friedman two-way analysis of variance, where Factor A was whether the seeds were in the melons or free from the melons during treatment and Factor B was the microwave exposure time (0 s, 10 s, 20 s, or 30 s).

Additional 20 fruits were randomly selected from the field site at Dookie Campus. These were weighed and had their three major diameters measured using callipers. This data was used to calculate energy density requirements to effectively treat the seeds in the fruit. Energy density can be estimated by assuming that all of the microwave energy in the chamber is absorbed by the fruit. Therefore, the energy per fruit is given by

$$E_f = \frac{P \times t_h}{N}. \tag{3}$$

Similarly, the energy density as a function of fruit cross-sectional area can be calculated using

$$E_A = \frac{P \times t_h}{A}. \tag{4}$$

Microwave energy density was calculated for each of the treatments used in Experiments 1 and 2. The data from all treatments was pooled in terms of this energy density to determine a dose response curve for paddy melon plants. The probability of survival can be described by

$$P_{\text{survival}} = \frac{1.0}{1 + (\psi/D_{50})^{20}}. \tag{5}$$

3. Results

Experiment 1. Microwave heating affected the melon plants quickly. Audible clicking sounds emanated from the plant during the treatment process. It was also possible to see the plant wilt during the first 10 to 15 seconds of treatment when the antenna with the larger aperture was used (Figure 7). The only visible difference in the plants induced by microwave treatment was drooping of the leaves. The small aperture antenna caused much louder acoustic emissions and visible rupture of the plant stems (Figure 8) within a few seconds of initial microwave treatment. In the case of the larger aperture antenna, increasing treatment time significantly reduces survival of the plants (Table 1). Using the antenna with the smaller aperture killed all the paddy melon plants for all treatment times.

Experiment 2. The control plants at both field sites had grown during the 16-day interval between applying the microwave treatments at the site and the evaluation of results. All treated plants at both sites, irrespective of the treatment duration, died. Therefore, these results were the same as those listed for the small aperture antenna in Table 1. This confirms that microwave treatment is effective in the field as well as in the more controlled situation where potted plants were used.

Experiment 3. The microscopic images of the freshly treated plant material (Figure 9) showed no obvious signs of cellular rupture due to microwave treatment, and no other obvious differences in cell structure could be seen between the control plant samples and the microwave-treated plant (Figure 9); however, at the macroscopic level the microwave treated plant had wilted during treatment.

For plants that were treated with microwave energy from the small aperture horn antenna five days prior to examination, no obvious cell structures were visible in the section of stem that was affected by microwave treatment (Figure 10(b)); however, some cell structures were visible in the unaffected section of the same leaf stem (Figure 10(a)). This suggests that the cells in the affected section of the stem have completely collapsed due to microwave treatment.

(a)   (b)

FIGURE 7: Paddy melon plants (a) immediately after microwave treatment for 15 seconds and (b) 11 days after treatment.

TABLE 1: Mean percentage of unaffected or recovered paddy melon plants according to treatment combination.

Antenna design	Treatment time (s)					
	0	5	10	15	30	60
Large aperture	100%[a]	100%[a]	100%[a]	60%[b]	0%[c]	0%[c]
Small aperture	100%[a]	0%[c]	0%[c]	0%[c]	0%[c]	0%[c]
			LSD ($P = 0.05$): 13%			

Note: Means with different superscripts are significantly different from one another.

FIGURE 8: Section of plant stem ruptured during microwave treatment using the small aperture antenna.

Experiment 4. There was no significant difference in seed survival between the various microwave treatment combinations, compared with each other; however, all combinations of microwave treatment had a significant effect on seed survival compared with the controls. Based on data from Experiments 1 and 2, the half lethal microwave energy dose, across all treatments, was 145 J cm^{-2} (Figure 11).

4. Discussion

Microwave treatment kills paddy melon plants and seeds (Tables 1 and 2). Although there is no immediate evidence of plant cell damage (Figure 9) during microwave treatment that is not intense enough to cause localised steam explosions in the stem (Figure 8), there must be a mechanism that causes the stems to collapse (Figure 10) leading to permanent wilting and death of the plant stem within a few days (Figure 7).

Unlike conventional heating, microwave heating produces its highest temperature in the core of the plant stem (Figure 3). Water in the xylem tissue is usually under tensile stresses [36]. The high core temperatures associated with microwave heating may lead to cavitation in the water-filled xylem tissue [36]. This may account for the acoustic emissions from the plant during microwave treatment. Extensive cavitation in the xylem tissue may lead to irrecoverable embolisms that block sap flow and ultimately lead to permanent wilting and death of the stem (Figure 7). In extreme cases, there may also be cell rupture due to localised steam explosions inside the stem (Figure 8).

The temperature inside the plant's stem (5) depends on the strength of the microwave's electric field (E) and the dielectric properties of the plant material. The small aperture antenna often caused stem rupture within the first few seconds of microwave treatment because the temperature inside the core of the stem may have been sufficient to create steam inside the plant cells. Thus a strategically applied short burst of intense microwave energy onto the plant's stem was sufficient to kill the plants (Table 1). This is because microwave heating is proportional to the square of the electric field strength. Very intense microwave fields have also been linked to thermal runaway.

The dose response curve (Figure 11) indicates that 100% control of paddy melon should be achieved using 200 J cm^{-2} (or 20 GJ ha^{-1}) of microwave energy. In a study of various cropping systems by Mari and Chengying [37] estimated

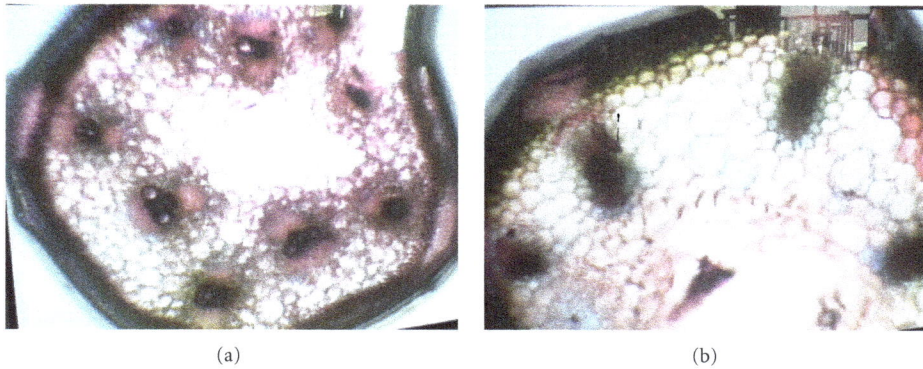

FIGURE 9: Comparison of stem cross-sections for (a) fresh untreated plant and (b) a plant after 15 seconds of treatment.

FIGURE 10: Comparison of stem cross-sections for (a) unaffected part of microwave-treated plant and (b) a part of the same stem affected by 10 seconds of microwave treatment using the small aperture horn antenna (microscopic observations made 5 days after microwave treatment).

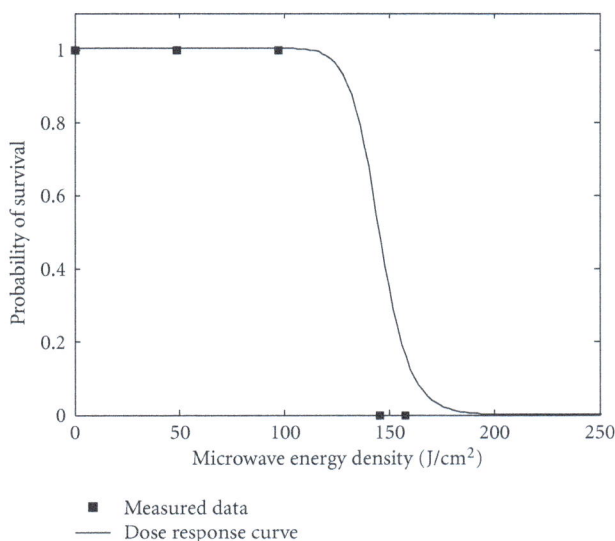

FIGURE 11: Microwave energy density dose response curve for paddy melon plants.

that the embodied energy of herbicide usage, which included manufacture, transport, and application, was $2.2\,GJ\,ha^{-1}$; therefore, uniformly applied microwave treatment is an order of magnitude higher than the embodied energy associated with chemical plant protection [37–39].

A strategic burst of intense energy, focused onto the stem of the plant, is as effective as, but much quicker at treating the plant, than an application of microwave energy to the whole plant. This may be of great interest for controlling chemically resistant Paddy Melon plants [30]. For example, if the microwave field from a travelling microwave weed control system cuts across any of the prostrate stems of a Paddy Melon plant, this research has demonstrated that the stem tissue beyond the point of treatment should die within a few days of treatment.

If a mature plant has an average of 5 long stems and is assumed to occupy an area of $7\,m^2$, as mentioned earlier, then a 5-second burst of microwave energy from the small aperture horn antenna, applied to each of the 5 stems near the crown of the plant, equates to an energy density of $0.2\,J\,cm^{-2}$ (or $0.02\,GJ\,ha^{-1}$) over the whole area of the plant.

TABLE 2: Mean fraction of paddy melon seeds that survived microwave treatment.

Seed condition	Microwave treatment			
	Control	10 sec	20 sec	30 sec
Seeds in melons	0.79[a]	0.20[b]	0.18[b]	0.07[b]
Free seeds	0.81[a]	0.18[b]	0.15[b]	0.19[b]
	LSD ($P = 0.05$): 0.27			

Note: Means with different superscripts are significantly different from one another.

The practicalities of applying microwave energy to the stems of plants have yet to be determined; however, various remote sensing systems that strategically apply chemical treatments to individual plants already exist [40]. Eventually, these systems may be adapted to apply microwave energy onto the stems of individual plants.

"Scale up" calculations in microwave heating systems are difficult to discuss because they are notoriously nonlinear. This is due to the nonlinear temperature/microwave field strength relationships, redistribution of energy associated with the coupling of heat and moisture movement [17], and phenomena such as thermal runaway, which all occur when the applied microwave power is increased. Other impediments to using microwave energy for controlling weeds in the field include nonuniform energy distributions in the microwave fields; protection of operators from microwave field leakage; loss of microwave energy due to transmission through the soil, beyond the treated plants.

Microwave treatment also reduced paddy melon seed viability by between 60% and 70% (Table 2); however, analysis of the microwave energy used during the 10 second treatment inside the microwave oven revealed that the melons were exposed to approximately 1610 J/fruit. In terms of energy density over the minimal projected area of the fruit, this equates to 422 J/cm^2, which is twice the energy density needed to kill the parent plants. This is consistent with other studies [7, 15], which show that plants are more susceptible to microwave damage than seeds [15].

If thermal runaway can be induced in plant tissues, fruits, and seeds, treatment time and the associated treatment energy may be drastically reduced for a small increase in applied microwave field density. This would make microwave treatment energy comparable with the embodied energy in chemical systems; however, this can only be explored by further research using a more powerful prototype system.

5. Conclusion

Microwave treatment kills paddy melon plants and significantly reduces the number of viable seeds.

Nomenclature

ψ: Microwave energy density ($J\,cm^{-2}$)
A: Cross-sectional area of melon fruit
$\quad A = \pi \times r_1 \times r_2$ (m^2)
D_{50}: Half lethal dose
m_c: Initial masses of the control samples (kg)

m_m: Initial masses of the microwave-treated samples (kg)
N: Number of treated fruit
P: Applied microwave power (W)
t_h: Microwave heating time for experiment (s)
Δm_c: Changes in mass of the control samples (kg)
Δm_m: Changes in mass of the microwave-treated samples (kg)
ΔT_c: Change in temperature of the control samples (K)
ΔT_m: Change in temperature of the microwave-treated samples (K)
ε_a: Complex dielectric constant of air
ε_w: Complex dielectric constant of water
τ: Transmission coefficient for the transfer of microwave energy into the material.

Acknowledgments

The authors thank the Grains Research and Development Corporation (GRDC) and the Rural Industries Research and Development Corporation (RIRDC) for their generous financial support during this study.

References

[1] P. A. Ark and W. Parry, "Application of high-frequency electrostatic fields in agriculture," *The Quarterly Review of Biology*, vol. 15, no. 2, pp. 172–191, 1940.

[2] I. M. Heap, "The occurrence of herbicide-resistant weeds worldwide," *Pesticide Science*, vol. 51, no. 3, pp. 235–243, 1997.

[3] I. Sartorato, G. Zanin, C. Baldoin, and C. De Zanche, "Observations on the potential of microwaves for weed control," *Weed Research*, vol. 46, no. 1, pp. 1–9, 2006.

[4] V. N. Tran, "Effects of microwave energy on the strophiole, seed coat and germination of acacia seeds," *Australian Journal of Plant Physiology*, vol. 6, no. 3, pp. 277–287, 1979.

[5] T. J. Headlee, "The difference between the effect of radio waves on insects and on plants," *Journal of Economic Entomology*, vol. 24, no. 2, pp. 427–437, 1931.

[6] G. M. McKinley, "Some biological effects of high-frequency electrostatic fields," *Proceedings of the Pennsylvania Academy of Science*, vol. 4, pp. 43–46, 1930.

[7] F. S. Davis, J. R. Wayland, and M. G. Merkle, "Ultrahigh-frequency electromagnetic fields for weed control: phytotoxicity and selectivity," *Science*, vol. 173, no. 3996, pp. 535–537, 1971.

[8] F. S. Davis, J. R. Wayland, and M. G. Merkle, "Phytotoxicity of a UHF electromagnetic field," *Nature*, vol. 241, no. 5387, pp. 291–292, 1973.

[9] W. W. Wolf, C. R. Vaughn, R. Harris, and G. M. Loper, "Insect radar cross-sections for aerial density measurements and target classification," *Transactions of the American Society of Agricultural Engineers*, vol. 36, no. 3, pp. 949–954, 1993.

[10] H. E. Haller, "Microwave Energy Applicator," patent, 2002/0090268A1, 20020090268A1, United States, 2002.

[11] W. J. Clark and C. W. Kissell, "System and Method for In Situ Soil Sterilization, Insect Extermination and Weed Killing," patent, 2003/0215354A1, 20030215354A1, United States, 2003.

[12] G. R. Grigorov, "Method and System for Exterminating Pests, Weeds and Pathogens," patent, 20030037482A1, 20030037482A1, United States, 2003.

[13] S. O. Nelson, "A review and assessment of microwave energy for soil treatment to control pests," *Transactions of the American Society of Agricultural Engineers*, vol. 39, no. 1, pp. 281–289, 1996.

[14] G. Brodie, C. Botta, and J. Woodworth, "Preliminary investigation into microwave soil pasteurization using wheat as a test species," *Plant Protection Quarterly*, vol. 22, no. 2, pp. 72–75, 2007.

[15] G. Brodie, S. Hamilton, and J. Woodworth, "An assessment of microwave soil pasteurization for killing seeds and weeds," *Plant Protection Quarterly*, vol. 22, no. 4, pp. 143–149, 2007.

[16] J. Bigu-Del-Blanco, J. M. Bristow, and C. Romero-Sierra, "Effects of low level microwave radiation on germination and growth rate in corn seeds," *Proceedings of the IEEE*, vol. 65, no. 7, pp. 1086–1088, 1977.

[17] G. Brodie, "Simultaneous heat and moisture diffusion during microwave heating of moist wood," *Applied Engineering in Agriculture*, vol. 23, no. 2, pp. 179–187, 2007.

[18] F. R. Connor, *Antennas*, Edward Arnold, London, UK, 1972.

[19] G. I. Brodie, "The influence of load geometry on temperature distribution during microwave heating," *Transactions of the American Society of Agricultural and Biological Engineers*, vol. 51, no. 4, pp. 1401–1413, 2008.

[20] G. Brodie, "Microwave heating in moist materials," in *Advances in Induction and Microwave Heating of Mineral and Organic Materials*, pp. 553–584, InTech, Vienna, Austria, 2011.

[21] R. S. Ferriss, "Effects of microwave oven treatment on microorganisms in soil," *Phytopathology*, vol. 74, no. 1, pp. 121–126, 1984.

[22] A. Cooper and G. Brodie, "The effect of microwave radiation on key soil parameters," *Plant Protection Quarterly*, vol. 24, no. 2, pp. 67–70, 2009.

[23] C. A. Vriezinga, "Thermal runaway and bistability in microwave heated isothermal slabs," *Journal of Applied Physics*, vol. 79, no. 3, pp. 1779–1783, 1996.

[24] P. E. Parris and V. M. Kenkre, "Thermal runaway in ceramics arising from the temperature dependence of the thermal conductivity," *Physica Status Solidi B*, vol. 200, no. 1, pp. 39–47, 1997.

[25] M. J. Ward, "Thermal runaway and microwave heating in thin cylindrical domains," *IMA Journal of Applied Mathematics*, vol. 67, no. 2, pp. 177–200, 2002.

[26] F. T. Ulaby and M. A. El-Rayes, "Microwave dielectric spectrum of vegetation: part II: dual-dispersion model," *IEEE Transactions on Geoscience and Remote Sensing*, vol. GE-25, no. 5, pp. 550–557, 1987.

[27] C. A. Vriezinga, "Thermal profiles and thermal runaway in microwave heated slabs," *Journal of Applied Physics*, vol. 85, no. 7, pp. 3774–3779, 1999.

[28] C. A. Vriezinga, S. Sánchez-Pedreno, and J. Grasman, "Thermal runaway in microwave heating: a mathematical analysis," *Applied Mathematical Modelling*, vol. 26, no. 11, pp. 1029–1038, 2002.

[29] J. Borger and R. Madin, *Paddy and Afghan Melons*, Department of Agriculture Western Australia, Perth, Australia, 2005.

[30] A. R. Leys, R. L. Amor, A. G. Barnett, and B. Plater, "Evaluation of herbicides for control of summer-growing weeds on fallows in south-eastern Australia," *Australian Journal of Experimental Agriculture*, vol. 30, no. 2, pp. 271–279, 1990.

[31] Q. Yu, A. Cairns, and S. Powles, "Glyphosate, paraquat and ACCase multiple herbicide resistance evolved in a Lolium rigidum biotype," *Planta*, vol. 225, no. 2, pp. 499–513, 2007.

[32] R. A. McKenzie, R. D. Newman, A. C. Rayner, and P. J. Dunster, "Prickly paddy melon (Cucumis myriocarpus) poisoning of cattle," *Australian Veterinary Journal*, vol. 65, no. 6, pp. 167–170, 1988.

[33] G. M. Cunningham, W. E. Mulham, P. L. Milthorpe, and J. H. Leigh, *Plants of Western New South Wales*, Inkata Press, Melbourne, Australia, 1992.

[34] J. D. Gibbons, *Nonparametric Statistical Inference*, Marcel Dekker, New York, NY, USA, 1985.

[35] M. K. Pasha and R. K. Das, "Quick viability test of soybean seeds by using tetrazolium chloride," *Seed Science and Technology*, vol. 10, no. 3, pp. 651–655, 1982.

[36] F. B. Salisbury and C. W. Ross, *Plant Physiology*, Wadsworth Publishing Company, Belmont, Calif, USA, 4th edition, 1992.

[37] G. R. Mari and J. Chengying, "Energy analysis of various tillage and fertilizer treatments on corn production," *American-Eurasian Journal of Agricultural and Environmental Science*, vol. 2, no. 5, pp. 486–497, 2007.

[38] Z. R. Helsel, "Energy and alternatives for fertilizer and pesticide use," in *Energy in Farm Production*, vol. 6, pp. 177–201, Elsevier, New York, NY, USA, 1992.

[39] K. J. Hülsbergen, B. Feil, S. Biermann, G. W. Rathke, W. D. Kalk, and W. Diepenbrock, "A method of energy balancing in crop production and its application in a long-term fertilizer trial," *Agriculture, Ecosystems and Environment*, vol. 86, no. 3, pp. 303–321, 2001.

[40] H. R. Langner, H. Böttger, and H. Schmidt, "A special vegetation index for the weed detection in sensor based precision agriculture," *Environmental Monitoring and Assessment*, vol. 117, no. 1, pp. 505–518, 2006.

17

Will the *Amaranthus tuberculatus* Resistance Mechanism to PPO-Inhibiting Herbicides Evolve in Other *Amaranthus* Species?

Chance W. Riggins and Patrick J. Tranel

Department of Crop Sciences, University of Illinois, 1201 West Gregory Drive, Urbana, IL 61801, USA

Correspondence should be addressed to Chance W. Riggins, criggins@life.illinois.edu

Academic Editor: Robert J. Kremer

Resistance to herbicides that inhibit protoporphyrinogen oxidase (PPO) has been slow to evolve and, to date, is confirmed for only four weed species. Two of these species are members of the genus *Amaranthus* L. Previous research has demonstrated that PPO-inhibitor resistance in *A. tuberculatus* (Moq.) Sauer, the first weed to have evolved this type of resistance, involves a unique codon deletion in the *PPX2* gene. Our hypothesis is that *A. tuberculatus* may have been predisposed to evolving this resistance mechanism due to the presence of a repetitive motif at the mutation site and that lack of this motif in other amaranth species is why PPO-inhibitor resistance has not become more common despite strong herbicide selection pressure. Here we investigate inter- and intraspecific variability of the *PPX2* gene—specifically exon 9, which includes the mutation site—in ten amaranth species via sequencing and a PCR-RFLP assay. Few polymorphisms were observed in this region of the gene, and intraspecific variation was observed only in *A. quitensis*. However, sequencing revealed two distinct repeat patterns encompassing the mutation site. Most notably, *A. palmeri* S. Watson possesses the same repetitive motif found in *A. tuberculatus*. We thus predict that *A. palmeri* will evolve resistance to PPO inhibitors via the same *PPX2* codon deletion that evolved in *A. tuberculatus*.

1. Introduction

Herbicides that inhibit protoporphyrinogen oxidase (PPO) have been used for many years for control of broadleaf weeds in large-scale crop production systems in the United States. Their use began to slowly decline during the late 1990s due to the introduction and subsequent widespread adoption of glyphosate-resistant crop varieties, such as Roundup Ready soybean, corn, and cotton. Glyphosate-resistant crops currently dominate throughout much of the United States and elsewhere in the world, and these systems rely almost exclusively on glyphosate as the sole means for weed management [1]. Unfortunately, the continuous broad-scale use of glyphosate over time has triggered the evolution of glyphosate-resistant biotypes among an increasing diversity of weed species [2]. In many cases, these glyphosate-resistant biotypes have been selected from weed populations that already had resistance to one or more other herbicide families, such as ALS inhibitors, triazines, and, less frequently, PPO inhibitors. As glyphosate resistance continues to increase in

frequency, distribution, and the number of species [2], growers are once again relying on PPO-inhibiting herbicides as an alternative approach to control weeds.

The enzyme protoporphyrinogen oxidase (EC 1.3.3.4) is one of the most important targets for herbicide development [3, 4]. PPO is the last common enzyme in the tetrapyrrole biosynthesis pathway and is responsible for converting protoporphyrinogen IX (Protogen) to protoporphyrin IX (Proto). In plants, two isoforms of the PPO enzyme are encoded by two different nuclear genes, *PPX1* and *PPX2*. These two enzymes share little sequence identity and are functionally compartmentalized, with *PPX1* being targeted to plastids and *PPX2* targeted to the mitochondria. Inhibition of PPO disrupts the synthesis of chlorophylls and hemes, which results in the damaging photodynamic effect characteristic of PPO inhibiting herbicides [3].

Natural resistance to PPO-inhibitors has been slow to evolve [3, 5], yet it has been confirmed in four weed species [2]. The first weed to evolve resistance to PPO herbicides was *Amaranthus tuberculatus* (Moq.) Sauer (waterhemp) in 2000

[6]. PPO resistance has subsequently been confirmed in *Euphorbia heterophylla* L. (wild poinsettia) and *Amaranthus quitensis* Kunth, both from South America, and *Ambrosia artemisiifolia* L. (common ragweed). The mechanism of PPO-inhibitor resistance, a unique target-site amino acid deletion, was first elucidated in a biotype of *A. tuberculatus* from Illinois [7]. This mechanism involves the loss of a glycine at position 210 in the mitochondrial isoform of the PPO enzyme. Loss of this amino acid is considered to have occurred via a slippage-like mechanism within a trinucleotide repeat of the *PPX2* gene [7, 8]. Specifically, the sequence motif spanning position 210 (i.e., ...TGTGGTGGA...) contains both a GTG and a TGG bi-repeat. Loss of either one of these repeat elements results in a loss of a glycine codon (GGT) without affecting the reading frame. This glycine deletion alters the binding domain of the enzyme without negatively affecting substrate affinity, and thus overall sensitivity to PPO-inhibiting herbicides is reduced by at least 100-fold [3]. The presence of a short repetitive motif at the mutation site, together with the favorable biochemical consequences of the deletion, seems to have predisposed *A. tuberculatus* to evolving this unique resistance mechanism, which thus far is the only identified mechanism of evolved PPO-inhibitor resistance. Despite being an unusual mutation, it is found commonly in *A. tuberculatus* populations across the Midwestern United States [9–11]. Assuming a slippage mechanism is responsible for evolved PPO resistance in *A. tuberculatus*, the question arises whether other weedy amaranth species possess the same sequence repeat and thus are also predisposed to acquiring this mutation.

It is unknown how conserved the *PPX2* gene is among weed species or whether other weedy members of the genus *Amaranthus*, such as *A. hybridus* L., *A. retroflexus* L., *A. powellii* S. Watson, and most notably Palmer amaranth (*A. palmeri* S. Watson), share the same repeat motif as in *A. tuberculatus*. These species are aggressive and pernicious weeds in their own right and have evolved resistances to multiple herbicides [2], though as of yet not to PPO inhibitors. However, there is growing concern that over-reliance on the PPO herbicides for controlling glyphosate-resistant *A. palmeri* in Roundup Ready crops (soybean, cotton, and corn) in the Southeastern United States will promote PPO resistance in this species [12]. In addition to glyphosate, populations of *A. palmeri* have evolved resistance to ALS inhibitors, dinitroanilines, and triazines [2, 13, 14], so the threat of a multiple-resistant individual or population is real and will be much more difficult to control once PPO resistance occurs in this species.

In this paper, we investigate inter- and intraspecific variability of the *PPX2* gene among weedy amaranths with conventional PCR and sequencing methods in conjunction with a newly developed PCR-RFLP (restriction fragment length polymorphism) assay. Our primary objective was to determine if the repeat motif encompassing Gly210 of *PPX2* from *A. tuberculatus* is shared among other weedy amaranth species. A second objective was to obtain baseline sequence information for future genetic and molecular studies of evolved PPO-inhibitor resistance in weedy amaranths.

2. Materials and Methods

Plant species and populations sampled for this study are listed in Table 1. Genomic DNA of some accessions utilized in this study was previously isolated for analysis of genetic similarity [15] and herbicide target-site genes [16, 17]. Source material for additional *Amaranthus* accessions was obtained either from herbarium specimens (for *A. acanthochiton* Sauer and *A. caudatus* L.) or from germplasm collections of the authors. In the case of *A. quitensis*, seed was obtained from the North Central Regional Plant Introduction Station (NCRPIS) in Ames, Iowa, and represented semi domesticated types from the Pacific side of South America and field weeds collected from Brazil. Seeds were initially sown in containers with $3:1:1:1$ mixture of commercial potting mix to soil to peat to sand. When seedlings exhibited true leaves, the plants were thinned and transplanted into new containers of the same size to ensure ample material for DNA extractions. Plants were fertilized as necessary with slow-release fertilizer and grown under mercury halide and sodium vapor lamps along with incident sunlight. Lamps were programmed for a 16-hour photoperiod and the temperature maintained at 22 C at night and 28 C daytime. Leaf material for DNA extraction was harvested from one individual mature plant per accession and flash frozen in liquid nitrogen prior to DNA extraction using the modified CTAB method [18].

PCR amplifications were performed using MJ Research thermal cyclers in 25 μL volumes with 1x buffer (GoTaq Flexi Buffer, Promega Corp., Madison, WI, USA), 2.5 mM MgCl$_2$, 200 μM dNTPs, 0.4 μM each primer, 1.0 μL total genomic DNA (10–50 ng), and 1.25 units of GoTaq polymerase. Parameters for PCR featured an initial denaturation of 95 C for 2 min, followed by 37 cycles (95 C for 45 sec, 55 C for 30 sec, and 72 C for 90 sec) and a final extension of 72 C for 5 min. Three primer sets were used for amplifications of the exon 9 region of *PPX2*. Since genomic *PPX2* information with intron sizes and sequence identity was limited, we initially used the primer set ARMS7-F (5′-TCTGAT-GAGCATGTTCAGGAAAGGCAAG) and *PPX2*ex10-R (5′-CTGGAAATGTATGGTGCATC) to amplify a large fragment (~1200 bp) between exons 7 and 10 for all species. Following sequence comparisons to identify conserved motifs in the introns adjacent to exon 9, the primer sets *PPX2*int8-F1 (5′-CAACTTGCCATGCTCTATTCC) and *PPX2*int9-R1 (5′-ATGGCGAAATGAGTTAAGGTTC), or *PPX2*int8-F2 (5′-ATTGCCATGCTCTATTCATTCC) and *PPX2*int9-R2 (5′-CGCCTATTCAAATCAAATGTCC), were used to amplify smaller fragments of ~500 bp and ~400 bp, respectively. PCR products were visualized on a 1% agarose gel containing 0.5 μg mL^{-1} ethidium bromide and cleaned using an E.Z.N.A. Cycle-Pure Kit (Omega Bio-Tek, Inc., Norcross, GA, USA) following the manufacturer's instructions. Purified PCR products were directly sequenced using an ABI Prism BigDye Terminator Kit v3.1 (Applied Biosystems, Foster City, CA, USA), and run on an ABI 3730XL capillary sequencer at the W. M. Keck Center for Comparative and Functional Genomics at the University of Illinois. Sequencing reactions were prepared in 13–16 μL volumes and

TABLE 1: *Amaranthus* accessions tested for *PPX2* sequence variation.

Species	Accession no.	Location information	PCR-RFLP result
A. acanthochiton Sauer	398443	New Mexico, USA	Uncleaved
A. albus L.	MH36	Washington, USA	Uncleaved
(Tumble pigweed)	MH38	Mississippi, USA	Uncleaved
	PT70	Illinois, USA	Uncleaved
A. caudatus L.	CAU	Bolivia	Cleaved
A. hybridus L.	MH154	North Carolina, USA	Cleaved
(Smooth pigweed)	PT12	Illinois, USA	Cleaved
	MH165	Ohio, USA	Cleaved
A. palmeri S. Watson	Alex 1	Illinois, USA	Uncleaved
(Palmer amaranth)	Alex 2	Illinois, USA	Uncleaved
	S1 union	Illinois, USA	Uncleaved
	S1 pul	Illinois, USA	Uncleaved
	Mass 1	Illinois, USA	Uncleaved
	GA-5	Georgia, USA	Uncleaved
	GA-7	Georgia, USA	Uncleaved
	GA-8	Georgia, USA	Uncleaved
	GA-10	Georgia, USA	Uncleaved
	MH253	New Mexico, USA	Uncleaved
	MH247	California, USA	Uncleaved
	MH254	Texas, USA	Uncleaved
A. powellii S. Watson	MH234	Washington, USA	Cleaved
(Powell's amaranth)	MH237	New York, USA	Cleaved
	MH242	Ontario, Canada	Cleaved
A. quitensis Kunth	511736	Bolivia	Cleaved
	511738	Ecuador	Cleaved
	511745	Ecuador	Cleaved
	511751	Peru	Cleaved
	568154	Tarija, Bolivia	Cleaved
	652421	Goias, Brazil	Cleaved
	652423	Goias, Brazil	Cleaved
	652426	Federal District, Brazil	Cleaved
	652429	Federal District, Brazil	Cleaved
	652430	Federal District, Brazil	Cleaved
A. retroflexus L.	MH84	New Mexico, USA	Cleaved
(Redroot pigweed)	PT25	Illinois, USA	Cleaved
	PT67	Illinois, USA	Cleaved
A. spinosus L.	MH267	Puerto Rico	Uncleaved
(Spiny amaranth)	MH203	Texas, USA	Uncleaved
	MH205	Louisiana, USA	Uncleaved
A. tuberculatus (Moq.) Sauer	PT43	Illinois, USA	Uncleaved
(Waterhemp)	ACR	Adams Co., IL, USA	Uncleaved
	WCS	Wayne Co., IL, USA	Uncleaved
	MH320	Ohio, USA	Uncleaved
	MO-a	Missouri, USA	Uncleaved
	MO-b	Missouri, USA	Uncleaved
	MO-c	Missouri, USA	Uncleaved
	MO-1	Missouri, USA	Uncleaved

contained 1.8 μL ddH₂O, 5.2 μL 12.5% glycerol, 2.0 μL 5x sequencing buffer, 2.0 μL 10 μM primer, 1.0 μL BigDye Terminator v3.1, and 1.0–4.0 μL PCR product. Cycle sequencing conditions started at 96 C for 1 min, followed by 30 cycles of (96 C for 30 sec, 50 C for 15 sec, 60 C for 4 min) and final extension of 60 C for 4 min. Forward and reverse sequences were manually edited and assembled into contiguous sequences (contigs) using the Alignment Explorer in MEGA5 [19]. Alignment of coding regions and exon-intron boundaries were determined by comparison with published genomic and cDNA *PPX2* sequences of *A. tuberculatus* (DQ394875, DQ394876, DQ386113, DQ386114, DQ386116, DQ386117, DQ386118) and *A. hypochondriacus* L. (EU024569) from GenBank.

Comparative sequence alignments of exon 9 of the *PPX2* gene showed nucleotide variations at the Gly210 mutation site. To facilitate the screening of additional accessions of each species for variation at this site, a PCR-RFLP assay was developed using the restriction enzyme EciI (New England Biolabs Inc., USA). Amplified products from all primers sets were digested and analyzed for fragment patterns. Following PCR, 10 μL of each reaction was added to 10 μL of a digestion mixture and incubated at 37 C for 1.5 hrs. The digestion mixture contained EciI (1 unit μL⁻¹) and 1x concentrations of the supplied buffer (NEB2) and BSA. After digestion, the fragments were separated on a 1% agarose gel and visualized under UV with ethidium bromide staining. PCR amplifications and digestions for each sample were replicated at least twice to validate the assay.

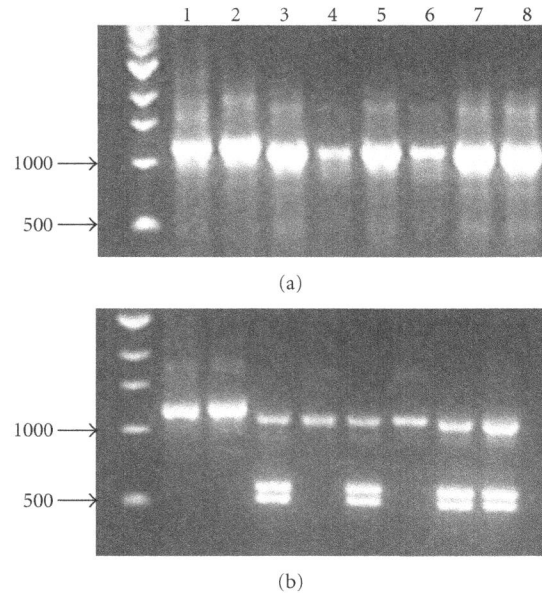

FIGURE 1: (a) Amplification of a segment of the *PPX2* gene with the primers ARMS7-F and *PPX2*ex10-R. (b) Digested PCR products with the enzyme EciI. Separation of fragments on a 1% w/v agarose gel with a 1-kb DNA ladder (NEB). Lanes 1–8 represent the following accessions: (1) *A. tuberculatus* PT43, (2) *A. tuberculatus* MH320, (3) *A. retroflexus* PT25, (4) *A. palmeri* MH253, (5) *A. hybridus* MH154, (6) *A. palmeri* MH247, (7) *A. powellii* MH242, and (8) *A. powellii* MH237.

3. Results and Discussion

Sequence comparisons between a PPO-inhibitor-sensitive biotype (i.e., wild-type) of *A. tuberculatus* (GenBank accession DQ394875) and *A. hypochondriacus* (GenBank accession EU024569) revealed the presence of a single-nucleotide polymorphism (SNP) in the third position of the Gly210 codon in exon 9 of the genomic *PPX2* gene. Although this SNP is a synonymous substitution, its position is significant in being part of a codon that, when deleted, confers resistance to PPO-inhibiting herbicides in *A. tuberculatus* [7]. In biotypes of *A. tuberculatus*, the thymine present at this site creates a bi-GTG or bi-TGG nucleotide repeat that spans the Gly210 codon. In the nucleotide sequence of *A. hypochondriacus*, a cytosine rather than a thymine is present, so no repeat motifs are formed. Repetitive sequences are thought to be more prone to slippage during replication, and it would not matter in this case if the deleted triplet was a TGG or a GTG since the reading frame is maintained in both instances [8]. Moreover, the resultant loss of a glycine at this position imparts herbicide resistance without adversely affecting the normal functions of the PPO enzyme [3].

The initial PCR experiments tested for cross-species amplification of the *PPX2* gene using primers based on sequence data from *A. tuberculatus*. The primers ARMS7-F and *PPX2*ex10-R produced a single product of approximately 1200 bp in the five amaranth species tested (Figure 1(a)). The amplified products for each species were fairly uniform in size with only minor differences in length being attributable

to short indel (insertion/deletion) mutations located in the introns. Digestion of these PCR products with the restriction enzyme EciI produced two smaller fragments of approximately 500 and 580 bp in *A. hybridus*, *A. retroflexus*, and *A. powellii*, but no fragments in the multiple accessions of *A. tuberculatus* and *A. palmeri* (Figure 1(b)). Sequencing these PCR products confirmed the digestion patterns by showing that both accessions of *A. palmeri* had the same repeat motif as *A. tuberculatus* and thus lacked the restriction site. Conversely, a single restriction site was present in the sequences of the three other species since they possessed a cytosine substitution similar to *A. hypochondriacus* rather than a thymine as in *A. tuberculatus* and *A. palmeri*. The uncut bands observed for *A. hybridus*, *A. retroflexus*, and *A. powellii* in Figure 1(b) were determined to be the result of incomplete digestion rather than heterozygosity as their intensity decreased, but was not completely eliminated, through additional digestion experiments with different enzyme and template concentrations and longer incubation times. Sequencing evidence also confirmed that none of these accessions were heterozygous for the C↔T polymorphism. Subsequent PCR experiments using the primers ARMS7-F and *PPX2*ex10-R produced single bands of ~1200 bp in *A. albus* L., *A. quitensis*, and *A. spinosus* L., which again suggested that this portion of the gene is relatively conserved across amaranth species. Digestion with EciI produced cleaved products only for the accessions of *A. quitensis* (data not shown).

(a)

| Exon 7 | Exon 8 | Exon 9 | | Exon 10 |
| 77 bp | 37 bp | 66 bp | | 43 bp |

488 bp   94 bp   531 bp

PPX2ex7-F   PPX2int8-F1   PPX2int9-R1   PPX2ex10-R
                 PPX2int8-F2   PPX2int9-R2

(b)

```
          F  V  D  Y  V  I  D  P  F  V  A  G  T  C  G  G  D  P  Q  S  L  S
                                                      (V)
Consensus             TTTGTTGATTATGTTATTGACCCTTTTGTTGCGGGTACATGYGGYGKAGAYCCTCAATCGCTWTCY
A.tuberculatus PPO-S  ...................................................T..T.G...T..........T..C
A.tuberculatus PPO-R  ...................................................T---.G...T..........T..C
A.hypochondriacus     ...................................................T..C.G...T..........A..T
A.palmeri Mass1       ...................................................T..T.G...T.......R..A..T
A.palmeri MH247       ...................................................T..T.G...T..........A..T
A.acanthochiton       ...................................................T..T.T...C..........A..T
A.quitensis 511751    ...................................................C..C.G...T..........A..T
A.quitensis 511738    ...................................................T..C.G...T..........A..T
A.quitensis 652421    ...................................................T..C.G...T..........A..T
A.quitensis 652426    ...................................................T..C.G...T..........A..T
A.retroflexus MH84    ...................................................T..C.G...T..........A..T
A.powellii MH237      ...................................................T..C.G...T..........A..T
A.powellii MH242      ...................................................T..C.G...T..........A..T
A.hybridus MH154      ...................................................T..C.G...T..........A..T
A.caudatus            ...................................................T..C.G...T..........A..T
```

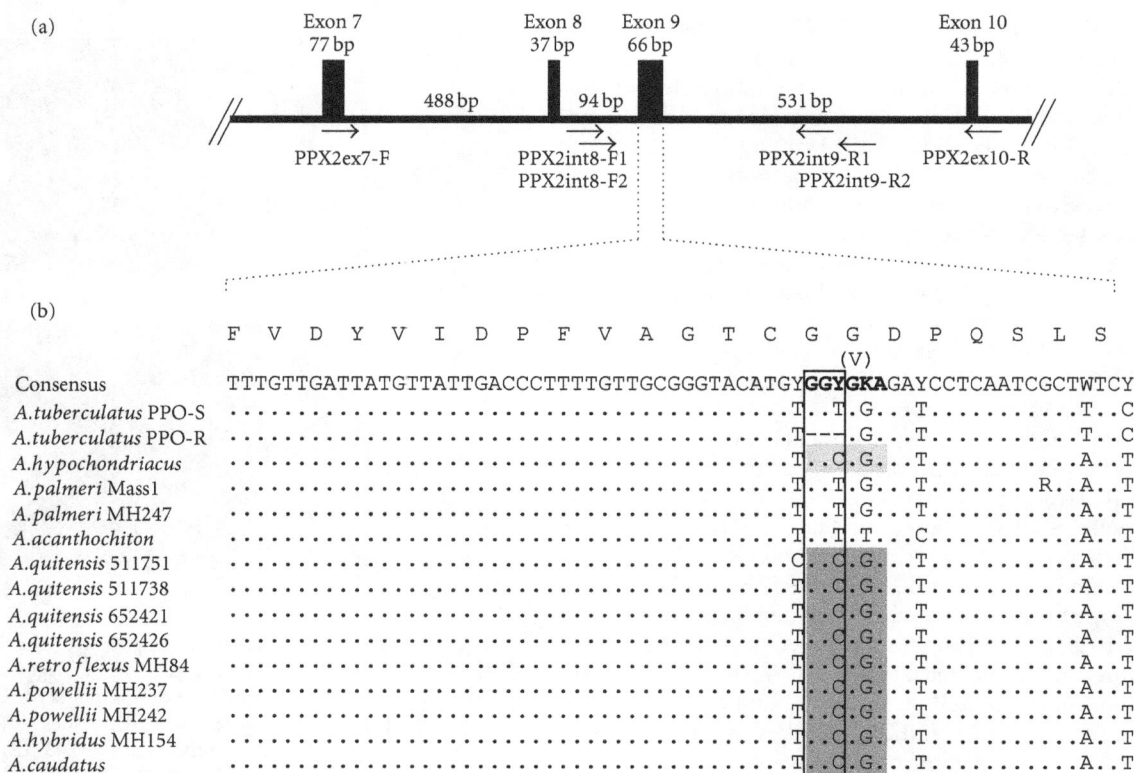

FIGURE 2: (a) Partial *PPX2* gene structure showing exon/intron positions and sizes. (b) Alignment of exon 9 from different *Amaranthus* species. The consensus nucleotide sequence and translated amino acid sequence are at the top. Conserved nucleotides are represented by dots. The boxed codon indicates the site of the Gly210 deletion mutation that is present in the PPO-resistant biotype of *A. tuberculatus* (GenBank DQ394876) and not in the wild-type (GenBank DQ394875) or any of the other species. Sequences that are cleaved in the PCR-RFLP assay are indicated by the shaded box which depicts the EciI recognition site. An inferred amino acid change of Gly211 to Val211 is also shown that results from a G to T nucleotide polymorphism in the Gly211 codon in the sequence of *A. acanthochiton*. The *PPX2* sequence of *A. hypochondriacus* is from GenBank (EU024569).

Based on sequence data from the first series of experiments, two additional primer sets were designed that are positioned in relatively conserved regions of the introns immediately flanking exon 9 (Figure 2(a)). The first primer set, *PPX2*int8-F1 and *PPX2*int9-R1, worked well for amplifying and sequencing exon 9 in all species except some accessions of *A. tuberculatus*. In this case, the modified primers *PPX2*int8-F2 and *PPX2*int9-R2 were successfully used. PCR experiments with both primer sets resulted in single amplified products of the expected size (~500 bp) in all accessions tested (data not shown). Results of the digestion analyses conducted with the different primer sets were in complete agreement with one another and with the sequencing evidence for every accession tested (Table 1). No infraspecific variation was observed in the PCR-RFLP fragment patterns for any species, although infraspecific variation was detected in *A. quitensis* after sequencing four randomly chosen accessions. In accession 511751 of this species, sampled from a Peruvian population, a silent T → C nucleotide mutation was observed in the codon immediately preceding Gly210, which in effect produced a

bi-CGG sequence repeat spanning the resistance mutation site (Figure 2(b)). The position of this polymorphism has not been directly implicated in resistance to PPO herbicides, but it does open the possibility for an alternative repeat motif that could lead to the same Gly210 mutation.

Sequence comparisons among the remaining accessions, including wild-type and confirmed PPO-resistant *A. tuberculatus* (i.e., biotypes without and with the deletion mutation), revealed six additional SNPs in exon 9, but only one resulted in a nonsynonymous substitution (Figure 2(b)). This substitution occurred at position 47 in the sequence of *A. acanthochiton* (a dioecious species) and resulted in an inferred change of glycine to valine (amino acid position 211). Only one accession sequenced appeared heterozygous, and this too was from a sample of a dioecious species (i.e., *A. palmeri* Mass1). This polymorphism was in the third position of codon 215 (serine) and thus did not result in an amino acid change. These observations are not surprising as dioecious species are obligately outcrossing and thus expected to have higher levels of sequence polymorphisms compared to the primarily self-pollinated monoecious species [20, 21].

Nonetheless, additional sampling of these widespread and ecologically diverse species is needed to determine the extent of sequence variability present in natural populations.

Although the experimental results were not in conflict with one another, it is important to point out some limitations of the PCR-RFLP assay. This test cannot be used to screen plants for the same target-site deletion mutation responsible for PPO-inhibitor resistance in *A. tuberculatus*. For example, PCR-amplified products for both the wild-type and PPO-resistant biotypes of *A. tuberculatus* were uncut by EciI. However, mutated alleles can be detected using an allele-specific PCR method [9]. This allele-specific marker should theoretically work in other amaranth species (e.g., Palmer amaranth) that have the same repeat motif and, of course, the same deletion mechanism as operative in *A. tuberculatus*. On the other hand, supposing a similar codon deletion mutation occurred at the Gly210 site in an aberrant biotype of *Amaranthus* that normally possesses a cytosine SNP in the wild-type (*A. quitensis*, for example), then the PCR-RFLP test would be able to distinguish between the mutated and non-mutated alleles. There is a risk, however, of generating false positives or negatives due to other mutations within the EciI recognition site. Furthermore, the likelihood of mutations occurring outside the recognition site and endowing resistance must also be considered [22]. Even with these caveats, the PCR-RFLP assay is one more molecular tool that can be used in conjunction with other molecular markers to monitor the evolution of PPO-inhibitor resistance in weedy amaranths.

4. Conclusions

Even though resistance to PPO-inhibiting herbicides has been slow to evolve, it may be expected to occur in weedy species with large populations that are under strong and continuous selection pressure [23]. Inherently high levels of genetic variation also help to increase the likelihood that mutated alleles conferring resistance will be selected under strong herbicide pressure. Although *PPX2* variability was relatively low among the ten species investigated in this study, the results do show that two groups can be recognized: species with and species without a repeat motif. It is noteworthy that the *PPX2* gene of Palmer amaranth shared the same repeat motif as waterhemp, which suggests that Palmer amaranth will evolve resistance to PPO inhibitors via the Gly210 deletion mutation. Of course, the possibility of Palmer amaranth evolving another resistance mechanism (e.g., a different target-site mutation or a non-target-site mechanism) cannot be ruled out. In fact, preliminary evidence for a different target-site mutation in the *PPX2* gene of common ragweed was recently reported [24]. The occurrence of a second repeat pattern in *A. quitensis* is also noteworthy, especially since this species is one of the four known with PPO-inhibitor resistance. Further studies with resistant and sensitive biotypes of this species will be useful in linking sequence patterns with the likelihood of evolving PPO-inhibitor resistance. Finally, the information provided in this study will facilitate the use and further development of molecular markers for screening amaranth populations suspected of PPO-inhibitor resistance for target-site-based mechanisms.

References

[1] R. G. Wilson, B. G. Young, J. L. Matthews et al., "Benchmark study on glyphosate-resistant cropping systems in the United States—part 4: weed management practices and effects on weed populations and soil seedbanks," *Pest Management Science*, vol. 67, no. 7, pp. 771–780, 2011.

[2] I. Heap, "International Survey of Herbicide Resistant Weeds," 2011, http://www.weedscience.org/.

[3] F. E. Dayan, P. R. Daga, S. O. Duke, R. M. Lee, P. J. Tranel, and R. J. Doerksen, "Biochemical and structural consequences of a glycine deletion in the α-8 helix of protoporphyrinogen oxidase," *Biochimica et Biophysica Acta*, vol. 1804, no. 7, pp. 1548–1556, 2010.

[4] K. Grossmann, G. Hutzler, J. Caspar, J. Kwiatkowski, and C. L. Brommer, "Saflufenacil (Kixor): biokinetic properties and mechanism of selectivity of a new protoporphyrinogen ix oxidase inhibiting herbicide," *Weed Science*, vol. 59, no. 3, pp. 290–298, 2011.

[5] S. B. Powles and Q. Yu, "Evolution in action: plants resistant to herbicides," *Annual Review of Plant Biology*, vol. 61, pp. 317–347, 2010.

[6] D. E. Shoup, K. Al-Khatib, and D. E. Peterson, "Common waterhemp (*Amaranthus rudis*) resistance to protoporphyrinogen oxidase-inhibiting herbicides," *Weed Science*, vol. 51, no. 2, pp. 145–150, 2003.

[7] W. L. Patzoldt, A. G. Hager, J. S. McCormick, and P. J. Tranel, "A codon deletion confers resistance to herbicides inhibiting protoporphyrinogen oxidase," *Proceedings of the National Academy of Sciences of the United States of America*, vol. 103, no. 12, pp. 329–334, 2006.

[8] J. Gressel and A. A. Levy, "Agriculture: the selector of improbable mutations," *Proceedings of the National Academy of Sciences of the United States of America*, vol. 103, no. 12, pp. 215–216, 2006.

[9] R. M. Lee, A. G. Hager, and P. J. Tranel, "Prevalence of a novel resistance mechanism to PPO-inhibiting herbicides in waterhemp (*Amaranthus tuberculatus*)," *Weed Science*, vol. 56, no. 3, pp. 371–375, 2008.

[10] K. A. Thinglum, C. W. Riggins, A. S. Davis, K. W. Bradley, K. Al-Khatib, and P. J. Tranel, "Wide distribution of the waterhemp (*Amaranthus tuberculatus*) ΔG210 PPX2 mutation, which confers resistance to PPO-inhibiting herbicides," *Weed Science*, vol. 59, no. 1, pp. 22–27, 2011.

[11] P. J. Tranel, C. W. Riggins, M. S. Bell, and A. G. Hager, "Herbicide resistances in *Amaranthus tuberculatus*: a call for new options," *Journal of Agricultural and Food Chemistry*, vol. 59, no. 11, pp. 5808–5812, 2011.

[12] L. M. Sosnoskie, J. M. Kichler, R. D. Wallace, and A. S. Culpepper, "Multiple resistance in palmer amaranth to glyphosate and pyrithiobac confirmed in Georgia," *Weed Science*, vol. 59, no. 3, pp. 321–325, 2011.

[13] T. A. Gaines, D. L. Shaner, S. M. Ward, J. E. Leach, C. Preston, and P. Westra, "Mechanism of resistance of evolved glyphosate-resistant palmer amaranth (*Amaranthus palmeri*)," *Journal of Agricultural and Food Chemistry*, vol. 59, no. 11, pp. 5886–5889, 2011.

[14] P. Neve, J. K. Norsworthy, K. L. Smith, and I. A. Zelaya, "Modelling evolution and management of glyphosate resistance in *Amaranthus palmeri*," *Weed Research*, vol. 51, no. 2, pp. 99–112, 2011.

[15] J. J. Wassom and P. J. Tranel, "Amplified fragment length polymorphism-based genetic relationships among weedy *Amaranthus* species," *Journal of Heredity*, vol. 96, no. 4, pp. 410–416, 2005.

[16] R. M. Lee, J. Thimmapuram, K. A. Thinglum et al., "Sampling the waterhemp (*Amaranthus tuberculatus*) genome using pyrosequencing technology," *Weed Science*, vol. 57, no. 5, pp. 463–469, 2009.

[17] C. W. Riggins, Y. Peng, C. N. Stewart, and P. J. Tranel, "Characterization of de novo transcriptome for waterhemp (*Amaranthus tuberculatus*) using GS-FLX 454 pyrosequencing and its application for studies of herbicide target-site genes," *Pest Management Science*, vol. 66, no. 10, pp. 1042–1052, 2010.

[18] J. J. Doyle and J. L. Doyle, "Isolation of plant DNA from fresh tissue," *Focus*, vol. 12, pp. 13–15, 1990.

[19] K. Tamura, D. Peterson, N. Peterson, G. Stecher, M. Nei, and S. Kumar, "MEGA5: molecular evolutionary genetics analysis using maximum likelihood, evolutionary distance, and maximum parsimony methods," *Molecular Biology and Evolution*, vol. 28, no. 10, pp. 2731–2739, 2011.

[20] M. J. Murray, "The genetics of sex determination in the family Amaranthaceae," *Genetics*, vol. 25, pp. 409–431, 1940.

[21] M. Costea, S. E. Weaver, and F. J. Tardif, "The biology of Canadian weeds. 130. *Amaranthus retroflexus* L., *A. powellii* S. Watson and *A. hybridus* L," *Canadian Journal of Plant Science*, vol. 84, no. 2, pp. 631–668, 2004.

[22] C. Délye, K. Boucansaud, F. Pernin, and V. Le Corre, "Variation in the gene encoding acetolactate-synthase in *Lolium* species and proactive detection of mutant, herbicide-resistant alleles," *Weed Research*, vol. 49, no. 3, pp. 326–336, 2009.

[23] M. Jasieniuk, A. L. Brule-Babel, and I. N. Morrison, "The evolution and genetics of herbicide resistance in weeds," *Weed Science*, vol. 44, pp. 176–193, 1996.

[24] S. L. Rousonelos, J. L. Luecke, J. M. Stachler, and P. J. Tranel, "Resistance to PPO-inhibiting herbicides in common ragweed: one mechanism or many?" *North Central Weed Science Society*, vol. 39, abstract 65, 2010.

18

Responses of Metabolites in Soybean Shoot Apices to Changing Atmospheric Carbon Dioxide Concentrations

Richard Sicher

Crop Systems & Global Change Laboratory, Agricultural Research Service-USDA, Room 342, Building 001, BARC-west, 10300 Baltimore Avenue, Beltsville, MD 20705, USA

Correspondence should be addressed to Richard Sicher, richard.sicher@ars.usda.gov

Academic Editor: Bernd Lennartz

Soybean seedlings were grown in controlled environment chambers with CO_2 partial pressures of 38 (ambient) and 72 (elevated) Pa. Five or six shoot apices were harvested from individual 21- to 24-day-old plants. Metabolites were analyzed by gas chromatography and, out of 21 compounds, only sucrose and fructose increased in response to CO_2 enrichment. One unidentified metabolite, Unk-21.03 decreased up to 80% in soybean apices in response to elevated CO_2. Levels of Unk-21.03 decreased progressively when atmospheric CO_2 partial pressures were increased from 26 to 100 Pa. Reciprocal transfer experiments showed that Unk-21.03, and sucrose in soybean apices were altered slowly over several days to changes in atmospheric CO_2 partial pressures. The mass spectrum of Unk-21.03 indicated that this compound likely contained both an amino and carboxyl group and was structurally related to serine and aspartate. Our findings suggested that CO_2 enrichment altered a small number of specific metabolites in soybean apices. This could be an important step in understanding how plant growth and development are affected by carbon dioxide enrichment.

1. Introduction

Atmospheric CO_2 partial pressures are increasing due to human activities that include industrialization, fossil fuel combustion, and deforestation [1]. Since CO_2 is an important substrate for photosynthesis, elevated atmospheric CO_2 has the potential to alter the productivity of terrestrial plants and that of natural or managed ecosystems [2]. Single leaf gas exchange rates of higher plants were affected by CO_2 enrichment, and this often resulted in larger plants with increased reproductive capacity [3–5]. Due to accelerated rates of net CO_2 assimilation, concentrations of various leaf components including starch, soluble carbohydrates, amines, organic acids, pigments, and important photosynthetic proteins were affected by plant growth in CO_2-enriched atmospheres [6–8]. Increased biomass accumulation in response to CO_2 enrichment impacted the demand for soil nutrients, and in some cases this resulted in nutritionally limited growth conditions [9]. Nutrient limitations under CO_2 enrichment also decreased leaf photosynthetic capacity and further altered leaf constituents [8].

In comparison to source leaves, much less attention has been given to the effects of elevated CO_2 on the growth and development of sinks. Sink organs are dependent upon source leaves for assimilates to provide the carbon, nitrogen, and energy needed for growth and development. In general, metabolite levels in sink tissues were altered in concert with changes in source leaves on the same plant. For example, Geiger et al. [8] reported that starch, sucrose, and reducing sugars were increased by CO_2 enrichment in unopened sink leaves of tobacco. Hexoses and sucrose also were increased by CO_2 enrichment in studies of roots from seedlings of tobacco and barley [8, 10]. Components of reproductive tissues and seeds also were affected by CO_2 enrichment. For example, doubling ambient CO_2 levels throughout plant development increased soybean seed oil content by 1 or 2% with a commensurate decrease in seed protein content [11, 12]. Ziska et al. [13] also reported that omega-3 fatty acids were increased in mungbean seeds by doubling the ambient CO_2 partial pressure, and Wang et al. [14] observed that antioxidant levels in strawberry fruit were increased by CO_2 enrichment.

The present study examined metabolite changes in soybean apices in response to varying CO_2 partial pressures during plant growth. Apical tissue contains the meristem, leaf primordia, and other rapidly differentiating tissues, that are critical determinants of shoot growth and development. Kinsman et al. [15] previously showed that CO_2 enrichment shortened cycling times of rapidly dividing cells in the shoot and root meristems of Dactylis glomerata. Since specific metabolites are capable of modifying gene expression, we hypothesized that changes of metabolites in shoot apices in response to CO_2 enrichment could be important in regulating shoot growth. The current study describes changes of three metabolites that varied in soybean shoot apices in response to CO_2 enrichment.

2. Materials and Methods

2.1. Plant Materials. Soybean [Glycine max (L.) Merr. cv. Williams] plants were grown in matching pairs of controlled environment chambers (model M-2, Environmental Growth Chamber Corp., Chagrin Falls, OH, USA) as described previously [16]. Individual seeds were planted in 1.8 dm^3 plastic pots filled with vermiculite and the pots were watered once daily with a complete mineral nutrient solution containing 12 mM nitrate and 2.5 mM ammonium. After 17 ds of growth, plants were watered twice daily to insure adequate nutrient supply. The air temperature was $27 \pm 1°C$, the PPFD was $850 \pm 40 \mu mol\,m^{-2}\,s^{-1}$, and the photoperiod used a 14 h day/10 h night cycle. Ambient and elevated chamber air CO_2 partial pressures were normally 38 ± 10 and 72 ± 10 Pa, respectively and were varied as indicated in the text. Reciprocal transfer experiments were performed 21 DAS by switching half of the ambient and elevated CO_2 grown plants to the opposite CO_2 treatment. Apices from switched plants were harvested daily for the next 3 ds. Apical tissue was normally harvested from the main shoot and from 4 or 5 lateral branches. A total of 4 or 5 apices from an individual plant were combined and transferred to 1.5 mL Eppendorf tubes. Collecting apices from an individual plant normally took about 1 min. Tubes containing the harvested apices were quickly sealed and immersed in liquid N_2 to quench metabolism. Apical samples could be stored at $-80°C$ for up to 1 month prior to analysis without altering results. Preliminary experiments showed that metabolite concentrations in apical tissue did not differ between the main shoot and lateral branches. At each time point, four individual plants were harvested from either CO_2 treatment. Metabolite measurements from four individual plants were combined, and experiments were replicated at least once. Results are presented as means from the combined experiments and significant differences were determined using Student's t-test.

2.2. Component Analysis. Metabolite concentrations were determined by gas chromatography according to Roessner et al. [17]. Isolated apices from a single plant (~25 mg FW) were extracted at 4°C with 1.4 mL methanol using a ground glass tissue homogenizer. Prior to extraction, 0.1 mg of

adonitol (ribitol) was added to each sample, and this served as an internal standard. The homogenates were incubated in a H_2O bath at 70°C for 15 min and allowed to cool before dilution with an equal volume of deionized H_2O. The diluted extracts were centrifuged at 6000 g, and 20 μL of supernatant was transferred to a 1 mL Reacti-Vial and dried overnight in a desiccator under vacuum. Dried samples and appropriate standards were dissolved in 100 μL of pyridine containing 2 mg of methoxyamine and were then incubated in a H_2O bath at 30°C for 90 min with continuous shaking. Subsequently, 50 mL of MSTFA [N-methyl-N-(trimethylsilyl)fluoroacetamide] was added to each vial, which was then incubated as above for 30 min at 37°C. Derivatized samples and standards were separated by gas chromatography (model 6890A, Hewlett Packard), and metabolites were detected with a mass selective detector (model 7125, Agilent Technologies, Wilmington, DE, USA) coupled to Agilent MSD Chemstation Software. Separations were performed with a 30 m $\times$ 0.25 mm Supelco SPB-50 column (Sigma-Aldrich, St. Louis, MO, USA) using high-purity helium as a carrier gas at 1.2 mL min^{-1}. The oven temperature was increased at 5°C min^{-1} from 70 to 310°C and a solvent delay of 8.5 min was used. The detector was operated in full-scan mode at 50 scans min^{-1} with a range of 0–550 m/z. Total ion chromatograms were quantified using peak identification and calibration parameters within the Chemstation program. Standard curves were prepared with a mixture of known concentrations of specific compounds, and sample quantitation was performed using slopes derived by linear regression.

3. Results

3.1. Isolated Apical Tissue. Figure 1 shows the magnified images of the soybean shoot tip and of an excised shoot apex. The shoot apex was readily separated from the uppermost node on the stem and, upon visual inspection, this tissue was composed mostly of nascent leaves that were covered by numerous trichomes. Individual soybean plants possessed 6 or 7 lateral branches when harvested between 21 and 24 DAS. Total mass of the isolated shoot apex averaged about 5 mg FW each, whereas apices from lateral branches were 3 to 4 mg FW each.

3.2. Effects of CO_2 Enrichment on Metabolite Levels in Soybean Apices. Concentrations of 21 individual compounds in shoot apices are shown in Figure 2. Sucrose and fructose levels in isolated apical tissue increased by 23% and 41%, respectively, in response to CO_2 enrichment ($P \leq 0.05$). In addition to the known compounds discussed above, an unknown compound with a retention time of 21.03 min (Unk-21.03) decreased 50 to 80% when the ambient CO_2 partial pressure was doubled (compare Figures 3(a) and 3(b)). Levels of Unk-21.03 in soybean apices decreased in proportion to the CO_2 partial pressure when chamber air CO_2 levels were increased from 26 to 100 Pa (Figure 4(a)). The mass spectrum of this unidentified compound contained major mass fragments at

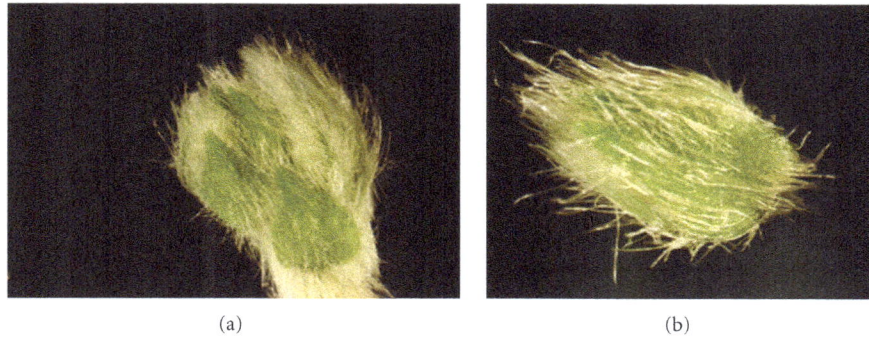

(a)   (b)

FIGURE 1: Manual isolation of soybean apices from the shoot tip. (a) Image of soybean shoot tip (10x); (b) image of isolated apical tissue from the main shoot (40x).

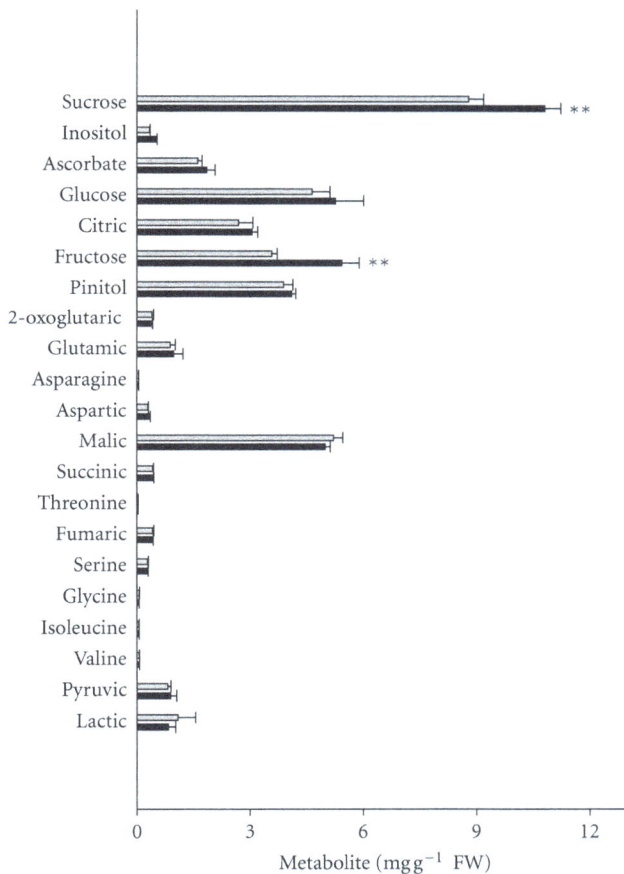

FIGURE 2: Responses of metabolites in soybean apices to CO_2 enrichment. Measurements were performed on apices harvested from plants grown with 38 (no fill) or 72 Pa CO_2 (black fill). (**) denotes significant differences at $P \leq 0.01$.

m/z 147, 205, 234, and 306 and the signature mass had an m/z of 423 (Figure 4(b)).

3.3. Effects of Switching CO_2 Treatments on Metabolite Levels in Soybean Apices.
The effects of reversing the ambient and elevated CO_2 treatments on sucrose and on Unk-21.03 levels in soybean apices are shown in Figure 5. Soybean apices were initially sampled 21 DAS and harvests continued daily for the next 3 ds. When averaged over all measurement dates, sucrose levels in soybean apices were enhanced 28% in the elevated compared to the ambient CO_2 treatment. Differences in sucrose due to CO_2 enrichment disappeared 3 ds after plants were switched from the elevated to the ambient CO_2 treatment. Conversely, transferring plants from the ambient to the elevated CO_2 treatment did not affect sucrose concentrations in soybean apices over the duration of the 3-day experiment. As shown in Figure 3, the peak area attributable to Unk-21.03 was 80% less in soybean apices from the elevated compared to the ambient CO_2 treatment (Figure 5(b)). Transferring plants grown at 72 Pa CO_2 to ambient CO_2 increased levels of Unk-21.03 in soybean apices by 19% in 3 ds and the reciprocal transfer decreased levels of Unk-21.03 by 50% over the same time period. Similar effects of CO_2 enrichment on Unk-21.03 concentrations were observed in soybean root tissue, but levels of Unk-21.03 were variable in leaf tissue and were not affected by the elevated CO_2 treatment (data not shown).

4. Discussion

Current atmospheric CO_2 levels do not saturate rates of photosynthesis by terrestrial plants having the C_3 photosynthetic pathway [2, 18]. Therefore, elevated CO_2 usually increases rates of photosynthesis and inhibits photorespiration by source leaves. These effects on gas exchange usually result in increased carbohydrate synthesis and enhanced C/N ratios of many higher plant species [8]. Since sucrose and other soluble sugars are exported from source leaves, CO_2 enrichment often elevates soluble carbohydrates and starch in sink tissues. Although very little is known about the biochemical components of soybean apices, the finding that this tissue contained increased concentrations of sucrose or fructose was not altogether surprising. Glucose and fructose are equimolar constituents of sucrose, and these two hexoses occur in plant tissues when sucrose is hydrolyzed enzymically [19]. The finding that glucose was unaffected by CO_2 enrichment suggested that it was preferentially metabolized by soybean apices in comparison to fructose. No other

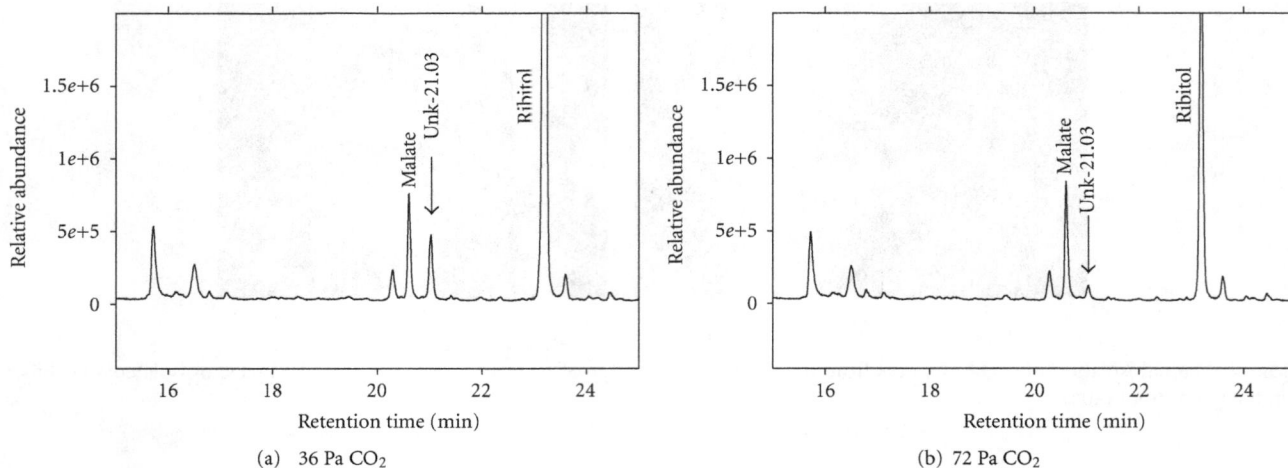

(a) 36 Pa CO_2

(b) 72 Pa CO_2

FIGURE 3: Total ion chromatogram showing changes of an unidentified compound (Unk-21.03) that decreased in soybean apices in response to CO_2 enrichment. Chromatograms were prepared with similar amounts of extracts of soybean apices from the ambient (a) and elevated (b) CO_2 treatments. The unidentified compound eluted from the column at 21.03 minutes and is marked with a descending arrow. Values are means $\pm$ SE ($n = 8$).

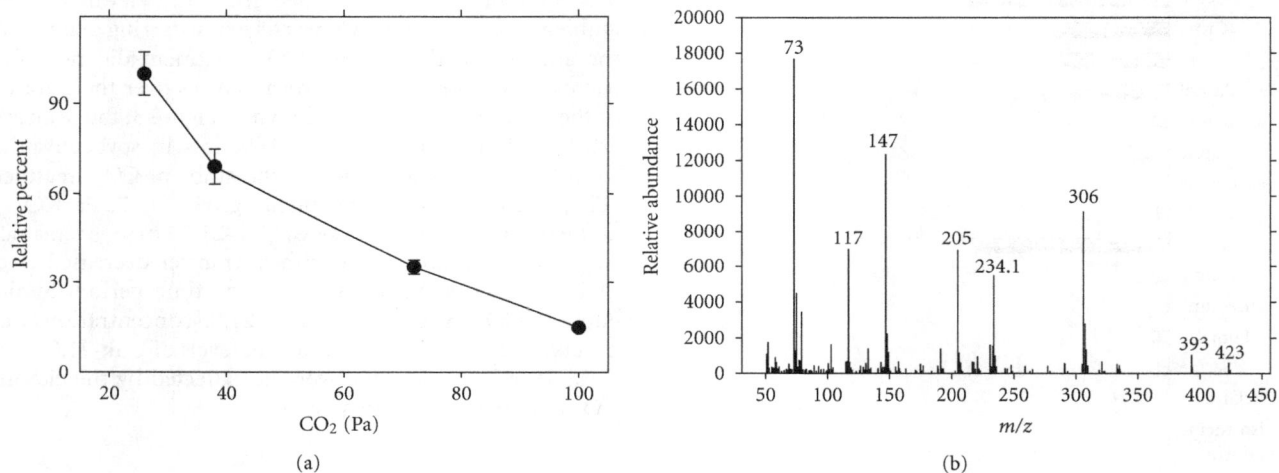

(a)

(b)

FIGURE 4: Inverse relationship between Unk-21.03 and increasing partial pressures of atmospheric CO_2. (a) Plants were grown for 21 ds at four different partial pressures of CO_2 prior to analyzing for Unk-21.03. Values on the y-axis are integrated peak areas normalized by the mass of apical tissue in each sample. (b) The mass spectrum of Unk-21.03.

compound that we measured in soybean apices increased in response to CO_2 enrichment.

Growth rates of soybean are usually accelerated by CO_2 enrichment, and this creates an increased demand for soil nutrients [4, 8, 10]. Nutrient limiting conditions can occur in experiments employing CO_2 enrichment, particularly, if soil fertility is not monitored carefully [8, 9]. Nitrogen-limiting conditions inhibit the synthesis of essential aminoacids and proteins and decreases of inorganic N in leaves and other plant parts. Results of Figure 1 showed that important soluble aminoacids in soybean apices were unaffected by CO_2 enrichment. This finding suggested that nutrient limiting conditions did not influence the results of this study. In a prior report [16], asparagine was the most abundant amino

acid in soybean trifoliolates. Therefore, it was interesting that asparagine was a minor constituent of soybean apices.

Only one metabolite in soybean apices, Unk-21.03, decreased in response to CO_2 enrichment, and Unk-21.03 was progressively decreased by CO_2 partial pressures that ranged from 26 to 100 Pa. Overall, Unk-21.03 the most CO_2 responsive metabolite observed in this investigation. Reciprocal transfer experiments showed that changes of Unk-21.03 in soybean apices occurred over a period of several days in response to an abrupt change in CO_2 partial pressure. A more rapid response to a change in CO_2 partial pressure would have occurred, if changes of Unk-21.03 in the shoot apex were the direct result of an internal CO_2 fixation reaction or were due to CO_2 effects on photorespiration. If

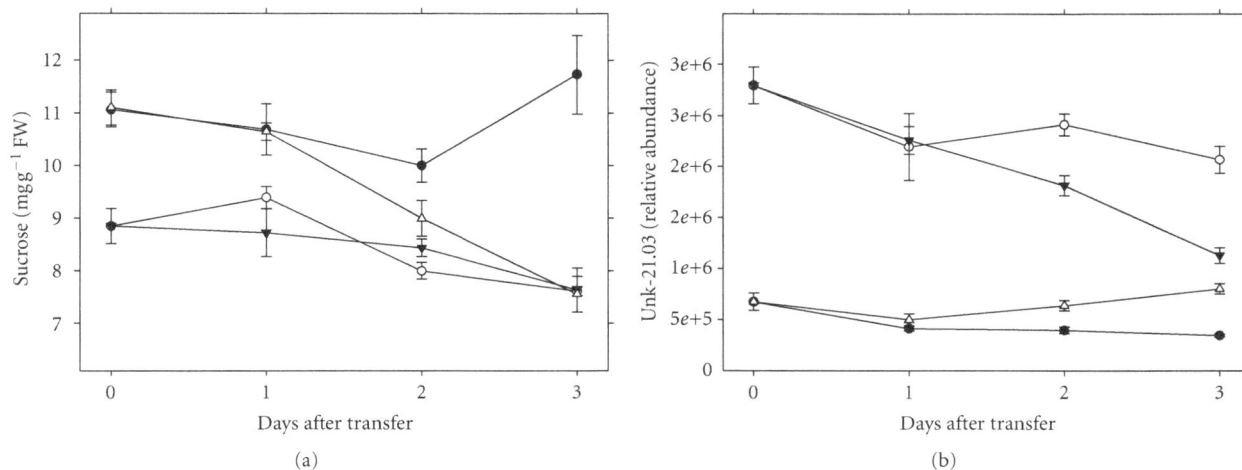

FIGURE 5: Effects of reversing the ambient and elevated CO_2 treatments on metabolite levels in soybean apices. Sucrose (a) and Unk-21.03 (b) were measured in soybean apices using plants that were grown for a total of 24 d in the ambient (∘) and elevated (•) CO_2 treatments. Reciprocal transfers were performed 21 DAS by switching one-half the plants from ambient to elevated CO_2 (△) and vice versa (▲). Values are means ± SE ($n = 8$).

the synthesis of Unk-21.03 was dependent on the activity of phospho(enol)pyruvate carboxylase much more rapid changes in tissue concentrations would be expected in response to fluctuating atmospheric CO_2 concentrations. Responses of sucrose levels in soybean apices were similarly delayed by a reciprocal shift in CO_2 levels. These findings suggested that Unk-21.03 may have been synthesized elsewhere on the plant and transported to the shoot apex.

The identification of Unk-21.03 is incomplete. However, the mass spectrum indicates that Unk-21.03 is probably an amine. A prominent ion with $m/z = 306$ is also present in aspartic acid and serine. These two aminoacids share the root structure $RCH_2CH(NH_2)COOH$. The R group in the serine molecule is a hydroxyl, and this is replaced in aspartic acid by a carboxyl moiety. To test this idea further, a concentrated extract of Unk-21.03 from soybean roots was separated by ascending thin layer chromatography, and the band associated with Unk-21.03 was detected with a spray containing bromphenol blue. A blue band was observed confirming that Unk-21.03 contained a carboxyl group. Since the ion attributed to $m/z = 306$ is not commonly found in small molecules from plant extracts, it is probable that Unk-21.03 contains the root structure described above. If this assumption is correct, Unk-21.03 was likely derived from serine or aspartate. It is interesting that Unk-21.03 decreased in soybean apices in response to CO_2 enrichment, whereas the possible precursors of Unk-21.03, serine and aspartate, were unchanged (see Figure 1). A possible explanation for these observations was that these compounds were synthesized elsewhere on the plant and imported into the apical tissue.

There is abundant evidence from higher plant species that multiple compounds involved in primary metabolism were altered in leaves and other plant tissues by CO_2 enrichment [7, 8, 10]. The finding that Unk-21.03 levels in soybean apices were decreased up to 80% by CO_2

enrichment is potentially significant. Identifying Unk-21.03 and determining how CO_2 enrichment affected levels of this compound in sink tissues could be an important step in understanding plant growth responses to CO_2 enrichment.

Abbreviations

DAS: Days after sowing
PPFD: Photosynthetic photon flux density.

Acknowledgments

The author thanks Steve Emche and Robert Erdman for assistance with gas chromatography and metabolite analysis. J. Y. Barnaby provided helpful comments on the paper.

References

[1] R. A. Houghton, "Revised estimates of the annual net flux of carbon to the atmosphere from changes in land use and land management 1850–2000," *Tellus B*, vol. 55, no. 2, pp. 378–390, 2003.

[2] G. Bowes, "Facing the inevitable: plants and increasing atmospheric CO_2," *Annual Review of Plant Physiology and Plant Molecular Biology*, vol. 44, no. 1, pp. 309–332, 1993.

[3] B. A. Kimball, J. R. Mauney, F. S. Nakayama, and S. B. Idso, "Effects of increasing atmospheric CO_2 on vegetation," *Vegetatio*, vol. 104-105, no. 1, pp. 65–75, 1993.

[4] L. H. Ziska, J. A. Bunce, and F. Caulfield, "Intraspecific variation in seed yield of soybean (*Glycine max*) in response to increased atmospheric carbon dioxide," *Australian Journal of Plant Physiology*, vol. 25, no. 7, pp. 801–807, 1998.

[5] E. A. Ainsworth, P. A. Davey, C. J. Bernacchi et al., "A meta-analysis of elevated [CO_2] effects on soybean (*Glycine max*) physiology, growth and yield," *Global Change Biology*, vol. 8, no. 8, pp. 695–709, 2002.

[6] G. Y. Nie, S. P. Long, R. L. Garcia et al., "Effects of free-air CO_2 enrichment on the development of the photosynthetic apparatus in wheat, as indicated by changes in leaf proteins," *Plant, Cell & Environment*, vol. 18, no. 8, pp. 855–864, 1995.

[7] R. C. Sicher and J. A. Bunce, "Relationship of photosynthetic acclimation to changes of Rubisco activity in field-grown winter wheat and barley during growth in elevated carbon dioxide," *Photosynthesis Research*, vol. 52, no. 1, pp. 27–38, 1997.

[8] M. Geiger, V. Haake, F. Ludewig, U. Sonnewald, and M. Stitt, "The nitrate and ammonium nitrate supply have a major influence on the response of photosynthesis, carbon metabolism, nitrogen metabolism and growth to elevated carbon dioxide in tobacco," *Plant, Cell & Environment*, vol. 22, no. 10, pp. 1177–1199, 1999.

[9] W. J. Arp, "Effects of source-sink relations on photosynthetic acclimation to elevated CO_2," *Plant, Cell & Environment*, vol. 14, no. 8, pp. 869–875, 1991.

[10] R. C. Sicher, "Interactive effects of inorganic phosphate nutrition and carbon dioxide enrichment on assimilate partitioning in barley roots," *Physiologia Plantarum*, vol. 123, no. 2, pp. 219–226, 2005.

[11] J. M. G. Thomas, K. J. Boote, L. H. Allen Jr., M. Gallo-Meagher, and J. M. Davis, "Elevated temperature and carbon dioxide effects on soybean seed composition and transcript abundance," *Crop Science*, vol. 43, no. 4, pp. 1548–1557, 2003.

[12] R. Sicher, J. Bunce, and B. Matthews, "Differing responses to carbon dioxide enrichment by a dwarf and a normal-sized soybean cultivar may depend on sink capacity," *Canadian Journal of Plant Science*, vol. 90, no. 3, pp. 257–264, 2010.

[13] L. H. Ziska, R. Palowsky, and D. R. Reed, "A quantitative and qualitative assessment of mung bean (*Vigna mungo* (L.) Wilczek) seed in response to elevated atmospheric carbon dioxide: potential changes in fatty acid composition," *Journal of the Science of Food and Agriculture*, vol. 87, no. 5, pp. 920–923, 2007.

[14] S. Y. Wang, J. A. Bunce, and J. L. Maas, "Elevated carbon dioxide increases contents of antioxidant compounds in field-grown strawberries," *Journal of Agricultural and Food Chemistry*, vol. 51, no. 15, pp. 4315–4320, 2003.

[15] E. A. Kinsman, C. Lewis, M. S. Davies et al., "Elevated Co_2 stimulates cells to divide in grass meristems: a differential effect in two natural populations of *Dactylis glomerata*," *Plant, Cell & Environment*, vol. 20, no. 10, pp. 1309–1316, 1997.

[16] R. C. Sicher, "Daily changes of amino acids in soybean leaflets are modified by CO_2 enrichment," *International Journal of Plant Biology*, vol. 1, no. 18, pp. 89–893, 2010.

[17] U. Roessner, C. Wagner, J. Kopka, R. N. Trethewey, and L. Willmitzer, "Simultaneous analysis of metabolites in potato tuber by gas chromatography-mass spectrometry," *Plant Journal*, vol. 23, no. 1, pp. 131–142, 2000.

[18] M. Stitt, "Rising CO_2 levels and their potential significance for carbon flow in photosynthetic cells," *Plant, Cell & Environment*, vol. 14, no. 8, pp. 741–762, 1991.

[19] R. Zrenner, K. Schüler, and U. Sonnewald, "Soluble acid invertase determines the hexose-to-sucrose ratio in cold-stored potato tubers," *Planta*, vol. 198, no. 2, pp. 246–252, 1996.

19

Nitrogen Fertilizer and Growth Regulator Impacts on Tuber Deformity, Rot, and Yield for Russet Potatoes

Mark S. Reiter,[1] Steven L. Rideout,[2] and Joshua H. Freeman[3]

[1] *Department of Crop and Soil Environmental Sciences, Eastern Shore Agricultural Research and Extension Center, Virginia Tech, Painter, VA 23420, USA*
[2] *Department of Plant Pathology, Physiology and Weed Science, Eastern Shore Agricultural Research and Extension Center, Virginia Tech, Painter, VA 23420, USA*
[3] *Department of Horticulture, Eastern Shore Agricultural Research and Extension Center, Virginia Tech, Painter, VA 23420, USA*

Correspondence should be addressed to Mark S. Reiter, mreiter@vt.edu

Academic Editor: H. Allen Torbert

Potatoes (*Solanum tuberosum*) are an important high-value commodity for producers in the Mid-Atlantic Region. Current production recommendations were based on white potatoes, and practices for Russet potatoes have not been researched in this region. The objective of this study was to test impacts of N rate (0, 67, 134, 201, and 268 kg N ha^{-1}), N application timing (100% applied with planter, 2-way split (30% with planter and 70% band applied approximately 30 days after planting at dragoff), and three-way split (30% with planter, 50% band applied prior to drag-off, and 20% band applied at first sight of bloom)), and additions of the growth regulator maleic hydrazide (MH-30). We tested "Goldrush" and "Norkotah" Russet potato varieties on marketability, total yield, tuber deformity, and tuber soft rot incidence for sandy loam soils in the Mid-Atlantic. Overall, year variations were significant with substantial rots (up to 86.5%) occurring in year 3. Maleic hydrazide and N application timing had little consistent effect on any tested parameter. Nitrogen rate and variety factors had the greatest impacts on deformity, tuber rots, and yields for Russet potatoes in the Mid-Atlantic Region with 134 kg N ha^{-1} producing the highest total yields in 2009 and 2010. If tuber rots can be controlled, both "Goldrush" and "Norkotah" are acceptable varieties under the Mid-Atlantic production practices.

1. Introduction

Potatoes are an important crop to Virginia and the rest of the Mid-Atlantic Region that includes Delaware and Maryland, USA. Annually, the Mid-Atlantic states produce 4049 hectares (ha) of potatoes with an average yield of 30,091 kg tubers ha^{-1} worth $9.97 million (5 year averages) [1]. Sandy loam soils in the Mid-Atlantic Region are favorable for potato production. However, a close proximity to sensitive water bodies, such as the Chesapeake Bay, means that fertilizer use efficiency and reduction of nutrient losses from production fields are more important than ever before.

Intensive fertilizer management is necessary in sandy loam soils in the Mid-Atlantic to ensure proper nutrient supplies to growing crops. Sandy loam soils generally have overall low organic matter, low cation exchange capacities, and low total nitrogen (N) in the upper horizon, which means

that little N is mineralized from soil organic N sources and N must be applied with fertilizer to match crop uptake needs [2]. For instance, Stanford and Smith [2] found that a Norfolk fine sandy loam had little initial N mineralized after 4 weeks of incubation (7.3 mg kg^{-1}) and this amount gradually fell for the next 30 weeks. Similarly, Van Veen and coworkers [3] found that any organic N applied to sandy loam soils was quickly mineralized into inorganic forms; therefore, soil supplies of N from crop to crop would be minimal in sandy loam crop production systems. Generally, only 1 to 3% of total organic N concentrations in the soil become available to a crop within the growing season in temperate regions of the world [4].

Nitrogen management is one of the most important aspects for potato production. Nitrogen fertilizer recommendations vary widely around the world. In loamy sand soils, Jamaati-E-Somarin and coworkers [5] found that

160 kg N ha^{-1} was sufficient for producing highest yields. Worthington et al. [6] found that 196 to 224 kg N ha^{-1} was necessary for the highest yields in Florida fine sands, with sidedress N applications being important for supplying necessary N fertilizer after leaching rain events. Nitrogen fertilizer recommendations for white potatoes in Virginia vary based on yield potential. For instance, producers yielding approximately 22,000 kg tubers ha^{-1} need 140 to 168 kg N ha^{-1} while higher producing yield goals require more N. For high yielding systems; growers are recommended to multiply yield goals in kg ha^{-1} by 0.006 to find recommended kg N ha^{-1} necessary [7]. Currently, producers in Virginia do not have separate guidelines for Russet potato varieties even though Russet varieties are more prone to deformity and other quality issues than white potatoes. Tuber deformity and secondary growth are historically correlated with moisture; however, other environmental factors that initiate and decrease growth, such as N fertility, may also be a cause [8].

Nitrogen application timing is one of the most important management techniques that producers can use to increase their fertilizer N use efficiency. For sandy loam and loamy sand soils, Virginia Cooperative Extension recommends two or three N fertilizer application splits to reduce potential N losses due to over irrigation or excessive rainfall. Fertilizer should be split between at planting, at dragoff (approximately 30 days after planting when potato plants are beginning to emerge, are bedded during cultivation, and the bed height is reduced), and immediately prior to bloom [7, 9]. Split N applications are recommended to increase overall yield and fertilizer use efficiency [10, 11]. However, N applications too late in the growing season can significantly delay maturity and decrease tuber quality [12].

Nutrient availability is not only important for overall yield in potato production, but is also important for disease management. Mackenzie [13] studied the relationship between N rate and potato yield in association with potato early blight in silt clay loam soils of Pennsylvania. Mackenzie [13] demonstrated that increasing N rates from 133 to 160 kg N ha^{-1} decreased overall rates of potato early blight infection. Research with potassium fertilizer also demonstrated that fertilizer rate and source impacted disease incidence. Panique and coworkers [14] found that potassium sulfate increased overall yields, but potassium chloride fertilizers decreased *Rhizoctonia solani* incidence. Therefore, fertilizer applications can significantly impact both foliar and tuber disease; however, the role of N fertilizer management in the Mid-Atlantic on tuber rots such as *Erwinia carotovora* ss. *carotovora* and *Pythium* sp. is not known.

Growth regulators have been researched for decades to help producers manage tuber sugar content, maturity, and sprouting after harvest and during storage [15, 16]. However, results are mixed depending on the factors studied. For instance, Yada et al. [17] found that foliar application of maleic hydrazide (MH-30) at a rate of 3.39 kg ha^{-1} had no significant effect on potato yield or sugar content, but did reduce sprout growth after harvest. Other research by Rex [18] found that foliar applications of chlormequat chloride, ethephon, and a combination of the two products reduced

tuber size, increased tuber deformity, and reduced overall yield. Work by Caldiz and coworkers [19] found that MH-30 was safe to use on potato foliage did not cause any phytotoxicity symptoms, and did increase yield in several varieties. However, no yield impact was found in any variety, but in all cases tuber sprouting was reduced. An analysis by Davis and Groskopp [20] found that overall tuber yield was reduced from MH-30 treatments; however, this reduction was mainly from lower numbers of undersized tubers. Overall, growth regulators have demonstrated positive effects on yield and sprouting, but impacts on tuber rot is not known.

Previous studies did not focus on Russet potatoes and most producers in the Mid-Atlantic currently use recommendations for white potatoes. The objectives of this study included finding: (1) impacts of N rate and N application timing and (2) impact of maleic hydrazide growth regulator on tuber yield, deformity, and rot on sandy loam soils in the Mid-Atlantic for Russet potatoes.

2. Materials and Methods

The trial was conducted on a Bojac sandy loam soil (Coarse-loamy, mixed, semiactive, and thermic Typic Hapludult) (65% sand, 25% silt, and 10% clay; 0.75% organic matter) at Virginia Tech's Eastern Shore Agricultural Research and Extension Center near Painter, VA, USA (37.5845808N, −75.8210421W) [21]. Prior to treatment establishment, soil was sampled per replication to a 15 cm depth, dried, ground, and analyzed using the Dumas method for total N and total carbon (C) analysis using a Vario EL cube (elementar Americas, Mt. Laurel, NJ, USA) [22]. Treatments included a 2 × 4 × 3 factorial combination of treatments of variety × N rate × N application timing. Ammonium nitrate (340 g N kg^{-1}) was applied with total N rates of 67, 134, 201, and 268 kg N ha^{-1} at various application timing intervals. Nitrogen treatments were applied either in a single treatment (100% applied with planter), 2-way split (30% with planter and 70% band applied prior to dragoff (approximately 30 days after planting when potato plants are beginning to emerge and are bedded during cultivation, and the bed height is reduced)), or with the high yielding white potato timing methodology of a three-way split (30% with planter, 50% band applied prior to dragoff, and 20% band applied at first sight of bloom). A 0-N control was also included. At-planting fertilizer treatments were spread evenly across the treatment area and incorporated using a field cultivator; dragoff treatments were surface applied and incorporated with bed shaping, while early bloom treatments were surface band applied. "Goldrush" and "Norkotah" Russet cultivars were seeded into conventionally-tilled land in early April following incorporation of fertility treatments. The growth regulator, maleic hydrazide, was applied at the manufacturer's recommended rate of 3.36 kg active ingredient (ai) ha^{-1} (MH-30; 0.18 kg ai L^{-1}). The MH-30 was broadcast foliar-applied using a CO_2-pressurized sprayer calibrated to deliver 280 L ha^{-1} at three weeks past full bloom in late June. The MH-30 was applied to both cultivars and plots that had combinations of all three N application timing treatments, and total N rates of 0, 134, and 268 kg N ha^{-1}. Comparable plots with identical variety, N rate, and N application timing treatments were maintained with and without MH-30 treatments, so direct comparisons of MH-30 impacts could be observed. Plots were two 7.62 m rows spaced 0.9 m apart and separated by a guard row, which did not receive any N fertilizer or growth regulator treatments. Treatments were replicated four times in a randomized complete block design. During the growing season, recommended practices for disease, weed, and insect control were followed as outlined by Wilson and coworkers [9]. The second row of each plot was mechanically harvested upon maturity in late July. Harvested tubers were sorted in the following categories: deformed (misshapen tubers), rotten, marketable (total of all tubers not rotten = misshapen + size B + size A + chef), and total (marketable + rotten tubers). Means were separated using Fisher's protected least significant difference (LSD) test at α = 0.10 that was established *a priori*. The 0-N control plots were not included in the factorial combination PROC ANOVA analysis as only 1 control plot per replication was included per variety; therefore, the design was not balanced for all combinations of treatments if 0-N treatments were included.

3. Results and Discussion

Total N attributed to potato plants from sandy loam soil organic matter mineralization is expected to be low in the Mid-Atlantic due to relatively low total C concentrations (Table 1). Ambient soil total N concentrations ranged from 0.42 to 0.49 g N kg^{-1} (Table 1), which would equate to 864 to 1,008 kg total soil ha^{-1} at an average bulk density of 1.35 g kg^{-1} (2,057,400 kg soil ha^{-1} at 15.24 cm depth) for a Bojac sandy loam on the Eastern Shore of Virginia [23]. Due to these low soil total N concentrations, we would expect only 26 to 30 kg N ha^{-1} to be available to potato tubers in temperate conditions from ambient soil sources via mineralization [4]. Therefore, substantial inorganic N fertilizer sources are necessary for optimal production and are the overall focus of a potato grower's fertilizer program.

In each year, *Erwinia carotovora* ss. *carotovora* and *Pythium* sp. occurred naturally in these trials causing tuber rots in the field. Data were analyzed as a year × treatment effect to test differences of treatments across years. In all cases, year × treatment was significant (P = 0.0063, <0.0001, <0.0001, and <0.0001 for percentage of deformed tubers, total yield, marketable yield, and percentage of rots, resp.). Large weather variations are likely responsible for varying yearly effects of N fertilizer, variety, and growth regulator treatments. Total yields were the highest in 2010, followed by 2009, and the lowest in 2011, which was inversely proportion to rots (2011 > 2009 > 2010). Therefore, each year will be discussed separately.

3.1. 2009 Growing Season. Maleic hydrazide had no significant effect on deformed tubers (55.8% of nonrotten tubers), rotten tubers (34.3% of total yield), marketable yield (13517 kg ha^{-1}), or total yield (20040 kg ha^{-1}) (P = 0.6380,

TABLE 1: Soil total nitrogen, carbon, and C : N ratio for a sandy loam soil in the Mid-Atlantic by year.

Year	Total nitrogen	Total carbon	C : N ratio
	g kg^{-1}		g C g N^{-1}
2009	0.49 a†	5.26 a	10.72 b
2010	0.46 ab	4.88 a	10.52 b
2011	0.42 b	4.87 a	11.47 a
LSD$_{0.10}$	0.04	0.54	0.40
Pr > F	0.0716	0.3500	0.0081

† Means within each dependent variable with the same letter are not significantly different ($P \geq 0.10$; Fisher's Protected LSD) and can be compared within column.

0.8236, 0.3860, and 0.2909, resp.) when compared to plots with identical N application timing and total N rate treatments that did not receive MH-30. These yield results are similar to work by Yada et al. [17] and Caldiz and coworkers [19] where no yield advantage was seen on several Russet potato varieties when MH-30 was used.

Generally, only the N rate and cultivar main effects were significant in 2009. When averaged across N application timing and cultivar, the percentage of deformed tubers and total yield increased as N rate increased (Table 2). The 201 kg ha^{-1} N rate had more deformed tubers than the lower N rate of 67 kg N ha^{-1} (58.3 versus 45.8% of nonrotten tubers, resp.; LSD$_{0.10}$ = 9.5%). For total yield, at least 134 kg N ha^{-1} was necessary to achieve commercially acceptable yields (19267 kg ha^{-1}), which is similar to N rates currently recommended in Virginia for white potato production [7]. However, the Virginia agronomic efficiency is significantly lower than efficiencies for Russet potatoes grown in Oregon and Washington states. Lauer [24] found that Russet Burbank returned 223 kg of tubers per kg N fertilizer applied, while yields in our study returned 144 kg tubers per kg N fertilizer. For the N rate main effect, marketable yield was not significant and averaged 13061 kg tubers ha^{-1} (P = 0.4200) (Table 2). For the cultivar main effect, "Goldrush" had significantly higher yields than "Norkotah" albeit higher percentage of deformed tubers, averaged across N rate and N application timing (Table 3). An N rate × N application timing × variety cultivar interaction was significant for rotten tubers (P = 0.0399, Table 4). Wide variation was seen in this experiment regarding tuber rots, which resulted in a relatively large LSD$_{0.10}$ (15.9% rotten tubers as a percentage of total yield). Generally, treatments that one would expect to have higher N use efficiency (more N splits) and higher N rates had more tuber rots. For example, "Norkotah" 268 kg N ha^{-1} with 3-splits (41.9% rots) compared to "Norkotah" 67 kg N ha^{-1} with all N applied at planting (24.5% rots) or "Norkotah" 0-N fertilizer treatments (21.8% tuber rots) (Table 4).

3.2. 2010 Growing Season. Maleic hydrazide had no impact on deformity (16.8% deformed tubers as a percentage of nonrotten tubers) or rot (23.4% of total yield) in 2010. An N rate × N application timing × variety × MH-30 interaction in 2010 for total yield was significant (P = 0.0474) and is illustrated in Table 5. Generally, "Norkotah" yielded the

TABLE 2: Nitrogen rate main effect in 2009 for deformed tubers, rotten tubers as a percentage of total yield, marketable tuber yield, and total tuber yield for Russet potatoes grown on sandy loam soils in the Mid-Atlantic, averaged across Russet cultivars and N application timing.

Nitrogen rate kg ha^{-1}	Deformed tubers Percentage of tubers	Rotten tubers Percentage of total yield	Marketable yield kg ha^{-1}	Total yield
0[†]	30.0	22.2	12728	15978
67	45.8 b[‡]	33.2 a	12248 a	17360 b
134	54.2 ab	34.0 a	14884 a	21515 a
201	58.3 a	35.5 a	12862 a	19267 ab
268	61.5 a	35.7 a	12248 a	21507 a
LSD$_{0.10}$	9.5	6.5	3053	3208
Pr > F	0.0445	0.8967	0.4200	0.0995

[†] No-fertilizer control plots were not included in Analysis of Variance and are included for informational purposes only.
[‡] Means within each dependent variable with the same letter are not significantly different ($P \geq 0.10$, Fisher's Protected LSD) and can be compared within column.

TABLE 3: Russet potato main effect in 2009 for deformed tubers, rotten tubers as a percentage of total yield, marketable tuber yield, and total tuber yield for Russet potatoes grown on sandy loam soils in the Mid-Atlantic, averaged across N rate and N application timing.

Variety	Deformed tubers Percentage of tubers	Rotten tubers Percentage of total yield	Marketable yield kg ha^{-1}	Total yield
Goldrush	66.7 a[†]	34.5 a	16894 a	24549 a
Norkotah	43.2 b	34.6 a	10343 b	15276 b
LSD$_{0.10}$	6.7	4.6	1928	2025
Pr > F	< 0.0001	0.9726	<0.0001	<0.0001

[†] Means within each dependent variable with the same letter are not significantly different ($P \geq 0.10$, Fisher's Protected LSD) and can be compared within column.

TABLE 4: Nitrogen rate × N application timing × variety interaction in 2009 for rotten tubers as a percentage of total yield for Russet potatoes grown on sandy loam soils in the Mid-Atlantic.

Nitrogen rate	Goldrush All at planting	2-splits	3-splits	Norkotah All at planting	2-splits	3-splits
kg ha^{-1}			% of total yield			
0[†]		11.8			21.8	
67	47.2 ab[‡]	29.3 cdef	34.6 bcdef	24.5 ef	29.6 cdef	33.7 bcdef
134	31.2 cdef	31.9 bcdef	35.4 abcdef	35.3 abcdef	26.7 def	43.4 abc
201	23.2 f	50.8 a	33.0 bcdef	39.8 abcde	29.7 cdef	36.6 abcdef
268	36.4 abcdef	26.3 def	35.2 abcdef	31.6 bcdef	42.9 abc	41.9 abcd
LSD$_{0.10}$			15.9			
Pr > F			0.0399			

[†] No-fertilizer control plots were not included in Analysis of Variance and are included for informational purposes only.
[‡] Means with the same letter are not significantly different ($P \geq 0.10$, Fisher's Protected LSD) and any mean can be compared within the table.

TABLE 5: Nitrogen rate × N application timing × variety × MH-30 interaction in 2010 for marketable yield for Russet potatoes grown on sandy loam soils in the Mid-Atlantic.

Nitrogen rate and MH-30	Goldrush All at planting	2-splits	3-splits	Norkotah All at planting	2-splits	3-splits
kg ha^{-1}			kg ha^{-1}			
134 no MH-30	9148 j[†]	13325 hij	19556 efgh	28348 abcd	32261 abc	29262 abcd
134 with MH-30	17187 fghi	17035 ghi	12237 hij	28337 abcd	32789 ab	31356 abc
268 no MH-30	24800 cdef	22615 defg	22513 defg	28601 abcd	27707 abcd	28277 abcd
268 with MH-30	17696 fghi	10815 ij	26833 cde	27239 abcde	34883 a	32250 abc
LSD$_{0.10}$			7724			
Pr > F			0.0474			

[†] Means with the same letter are not significantly different ($P \geq 0.10$, Fisher's Protected LSD) and any mean can be compared within the table.

TABLE 6: Nitrogen rate × N application timing × MH-30 interaction in 2010 for total yield for Russet potatoes grown on sandy loam soils in the Mid-Atlantic.

| Nitrogen rate | No MH-30 | | | MH-30 | | |
	All at planting	2-splits	3-splits	All at planting	2-splits	3-splits
kg ha^{-1}				kg ha^{-1}		
134	15185 c[†]	19144 bc	19378 bc	19200 bc	20211 abc	17756 c
268	24216 ab	20719 abc	20231 abc	18241 c	18204 c	25256 a
LSD$_{0.10}$			5766			
Pr > F			0.0588			

[†] Means with the same letter are not significantly different ($P \geq 0.10$, Fisher's Protected LSD) and any mean can be compared within the table.

TABLE 7: Nitrogen rate × variety interaction in 2010 for rotten tubers as a percentage of total yield, marketable tuber yield, and total tuber yield for Russet potatoes grown on sandy loam soils in the Mid-Atlantic, averaged across N application timing.

| Nitrogen rate | Rotten tubers | | Marketable yield | | Total yield | |
	Goldrush	Norkotah	Goldrush	Norkotah	Goldrush	Norkotah
kg ha^{-1}	Percentage of total yield		kg ha^{-1}		kg ha^{-1}	
0[†]	18.8	14.0	8507	7735	9839	8792
67	43.0 a[‡]	12.2 bc	9239 c	21266 b	13593 d	23777 bc
134	48.2 a	8.6 c	8396 c	27409 a	14009 d	29957 a
201	21.7 b	11.6 bc	21100 b	26972 a	26121 abc	30360 a
268	20.4 b	13.2 bc	19000 b	2444 c	23309 c	28195 ab
LSD$_{0.10}$	10.6		4744		4812	
Pr > F	0.0009		0.0033		0.0184	

[†] No-fertilizer control plots were not included in Analysis of Variance and are included for informational purposes only.
[‡] Means within each dependent variable with the same letter are not significantly different ($P \geq 0.10$, Fisher's Protected LSD).

TABLE 8: Growth regulator (MH-30) × variety interaction in 2011 for rotten tubers as a percentage of total yield and marketable tuber yield for Russet potatoes grown on sandy loam soils in the Mid-Atlantic, averaged across Nrates and N application timing.

| Russet cultivar | Rotten tubers | | Marketable yield | |
	With MH-30	No MH-30	With MH-30	No MH-30
	Percentage of total yield		kg ha^{-1}	
Goldrush	52.3 b[†]	39.2 c	7157 b	9423 a
Norkotah	86.4 a	86.5 a	2815 c	2573 c
LSD$_{0.10}$	6.5		1705	
Pr > F	0.0177		0.0873	

[†] Means within each dependent variable with the same letter are not significantly different ($P \geq 0.10$, Fisher's Protected LSD).

TABLE 9: Nitrogen application timing × variety interaction in 2011 for rotten tubers as a percentage of total yield, marketable tuber yield, and total tuber yield for Russet potatoes grown on sandy loam soils in the Mid-Atlantic, averaged across N rates.

| Nitrogen application timing | Rotten tubers | | Marketable yield | | Total yield | |
	Goldrush	Norkotah	Goldrush	Norkotah	Goldrush	Norkotah
	Percentage of total yield		kg ha^{-1}			
No nitrogen[†]	35.5	82.9	7871	1773	11603	9400
At planting	34.8 c[‡]	85.2 a	7874 b	2811 c	11947 c	15698 ab
2-split	42.6 c	86.4 a	10396 a	2696 c	17354 a	16200 ab
3-split	48.0 bc	79.7 ab	8303 b	3970 c	15729 ab	14284 bc
LSD$_{0.10}$	32.1		1835		2710	
Pr > F	0.0131		0.0823		0.0458	

[†] No-fertilizer control plots were not included in Analysis of Variance and are included for informational purposes only.
[‡] Means within each dependent variable with the same letter are not significantly different ($P \geq 0.10$, Fisher's Protected LSD).

higher than "Goldrush" with only isolated effects of N application timing and MH-30 application. An N rate × N application timing × MH-30 interaction, averaged across variety, was significant for marketable yields in 2010 ($P = 0.0588$) (Table 6). Similar to total yield, significant effects were isolated but at high N rates with MH-30 the 3-split application timing yielded highest.

An N rate × variety interaction was significant in 2010 for rotten tubers, marketable yield, and total yield (Table 7). Overall, "Goldrush" tubers had more rot incidence than "Norkotah" tubers at 67 and 134 kg N ha^{-1}. Tuber rot generally mirrored marketable yield with 134 and 201 kg N ha^{-1} providing the highest marketable yields for "Norkotah," while "Goldrush" tuber yield was lower at similar N rates (Table 7). Compared to white potato recommendations [4], a yield of 29957 kg tuber ha^{-1} would suggest a need of 180 kg N ha^{-1} based on the yield × 0.006 factor for "Norkotah," while only 134 kg N ha^{-1} was necessary in both 2009 and 2010.

3.3. 2011 Growing Season. The 2011 growing season resulted in low yields compared to 2009 and 2010. Generally, yields were only 25 to 50% of yield expectations for potatoes in Virginia. A wet Spring coupled with excessive heat and drought likely contributed to low yields. In 2011, there was an MH-30 × variety interaction for rotten tubers and marketable yield (Table 8). For rotten tubers, "Norkotah" had no impact if MH-30 was included; however, "Goldrush" with MH-30 had more rotten tubers than treatments with no MH-30 (52.3 versus 39.2, resp., LSD$_{0.10}$ = 6.5) (Table 8). Marketable yield followed an inversely proportional trend to rotten tubers with "Goldrush" with no MH-30 treatments having the highest marketable yields (9423 kg tubers ha^{-1}) (Table 8). A main effect was significant for deformed tubers with MH-30 treatments having 38.2% of nonrotten tubers being deformed with 27.6% being irregular if no MH-30 was used (LSD$_{0.10}$ = 6.3%), averaged across variety, N rate, and N application timing.

Nitrogen rate generally had no significant impacts on deformity, rot, or yield in 2011. However, N application timing × variety interactions, averaged over N rate, indicated that "Norkotah" tubers rotted twice as much as "Goldrush"; which resulted in significantly higher marketable yields for "Goldrush" (Table 9). A two-way N split (at planting and dragoff) was sufficient for providing the highest marketable and total tuber yields for "Goldrush" (10396 and 17354 kg ha^{-1}, resp.). A 3-split application timing decreased yield for "Goldrush," possibly due to reduced tuber formation and delayed maturity due to excessive N late in season as demonstrated by Ojala and coworkers [12]. Interestingly, total yield indicated that "Norkotah" generally yielded similar to "Goldrush" for total yield, so any management to reduce tuber rots would be beneficial.

4. Conclusions

In conclusion, N rate and variety factors had the greatest impacts on deformity, tuber rots, and yields for Russet potatoes in the Mid-Atlantic Region. Generally, findings indicate that 134 kg N ha^{-1} were adequate for producing the highest yields. If tuber rots can be controlled, both "Goldrush" and "Norkotah" are acceptable varieties under the Mid-Atlantic production practices. Neither maleic hydrazide nor application timing had a consistent impact on tuber rot, deformity, or yield.

Abbreviations

ai:	Active ingredient
C:	Carbon
ha:	Hectare
kg:	Kilogram
LSD:	Least significant difference
MH-30:	Maleic hydrazide
m:	Meter
N:	Nitrogen.

Acknowledgments

The authors gratefully acknowledge staff stationed at the Virginia Tech Eastern Shore Agricultural Research and Extension Center for assistance with plot establishment, maintenance, and data collection. They also thank the Virginia Irish Potato Board for funding this paper.

References

[1] United States Department of Agriculture—National Agricultural Statistics Service, *Quick Stats. U.S. & All States Data—Crops*, USDA-NASS, Washington, DC, USA, 2012, http://www.nass.usda.gov/Quickstats/.

[2] G. Stanford and S. J. Smith, "Nitrogen mineralization potentials of soils," *Soil Science Society of America Journal*, vol. 36, pp. 465–472, 1972.

[3] J. A. Van Veen, J. N. Ladd, J. K. Martin, and M. Amato, "Turnover of carbon, nitrogen and phosphorus through the microbial biomass in soils incubated with 14-C-, 15N- and 32P-labelled bacterial cells," *Soil Biology and Biochemistry*, vol. 19, no. 5, pp. 559–565, 1987.

[4] J. F. Power and J. W. Doran, "Nitrogen use in organic farming," in *Nitrogen in Crop Production*, R. D. Hauck, Ed., chapter 40, pp. 585–598, American Society of Agronomy, Crop Science Society of America, and Soil Science Society of America, Madison, Wis, USA, 1984.

[5] S. Jamaati-E-Somarin, A. Tobeh, K. Hashemimajd, M. Hassanzadeh, M. Saeidi, and R. Zabihi-E-Mahmoodabad, "Effects of nitrogen fertilizer and plant density on N-P-K uptake by potato tuber," *Indian Journal of Horticulture*, vol. 67, pp. 329–333, 2010.

[6] C. M. Worthington, K. M. Portier, J. M. White et al., "Potato (*Solanum tuberosum* L.) yield and internal heat necrosis incidence under controlled-release and soluble nitrogen sources and leaching irrigation events," *American Journal of Potato Research*, vol. 84, no. 5, pp. 403–417, 2007.

[7] M. S. Reiter, S. B. Phillips, J. G. Warren, and R. O. Maguire, *Nitrogen Management for White Potato Production*, Virginia Cooperative Extension and Virginia Agricultural Experiment Station, Blacksburg, Va, USA, 2009.

[8] J. E. Kraus, "Influence of certain factors on second growth on Russet Burbank potatoes," *American Potato Journal*, vol. 22, no. 5, pp. 134–142, 1945.

[9] H. P. Wilson, T. P. Kuhar, S. L. Rideout et al., *Commercial Vegetable Production Recommendations*, Virginia Cooperative Extension and Virginia Agricultural Experiment Station, Blacksburg, Va, USA, 2011.

[10] S. Roberts, W. H. Weaver, and J. P. Phelps, "Effect of rate and time of fertilization on nitrogen and yield of Russet Burbank potatoes under center pivot irrigation," *American Potato Journal*, vol. 59, no. 2, pp. 77–86, 1982.

[11] D. T. Westermann and G. E. Kleinkopf, "Nitrogen requirements of potatoes," *Agronomy Journal*, vol. 77, pp. 616–621, 1985.

[12] J. C. Ojala, J. C. Stark, and G. E. Kleinkopf, "Influence of irrigation and nitrogen management on potato yield and quality," *American Potato Journal*, vol. 67, no. 1, pp. 29–43, 1990.

[13] D. R. Mackenzie, "Association of potato early blight, nitrogen fertilizer rate, and potato yield," *Plant Disease*, vol. 65, pp. 575–577, 1981.

[14] E. Panique, K. A. Kelling, E. E. Schulte, D. E. Hero, W. R. Stevenson, and R. V. James, "Potassium rate and source effects on potato yield, quality, and disease interaction," *American Potato Journal*, vol. 74, no. 6, pp. 379–398, 1997.

[15] P. C. Marth and E. S. Schultz, "A new sprout inhibitor for potato tubers," *American Potato Journal*, vol. 29, no. 11, pp. 268–272, 1952.

[16] S. N. Rao and S. H. Wittwer, "Further investigations on the use of maleic hydrazide as a sprout inhibitor for potatoes," *American Potato Journal*, vol. 32, no. 2, pp. 51–59, 1955.

[17] R. Yada, R. Coffin, M. Keenan, M. Fitts, C. Dufault, and G. Tai, "The effect of maleic hydrazide (potassium salt) on potato yield, sugar content and chip color of kennebec and norchip cultivars," *American Journal of Potato Research*, vol. 68, pp. 705–709, 1991.

[18] B. L. Rex, "Effect of two plant growth regulators on the yield and quality of russet burbank potatoes," *Potato Research*, vol. 35, no. 3, pp. 227–233, 1992.

[19] D. O. Caldiz, L. V. Fernández, and M. H. Inchausti, "Maleic hydrazide effects on tuber yield, sprouting characteristics, and french fry processing quality in various potato (*Solanum tuberosum* L.) Cultivars grown under argentinian conditions," *American Journal of Potato Research*, vol. 78, no. 2, pp. 119–128, 2001.

[20] J. R. Davis and M. D. Groskopp, "Yield and quality of Russet Burbank potato as influenced by interactions of Rhizoctonia, maleic hydrazide, and PCNB," *American Potato Journal*, vol. 58, no. 5, pp. 227–237, 1981.

[21] United States Department of Agriculture—Natural Resources Conservation Servic, *Bojac Series*, USDA-NRCS, Washington, DC, USA, 2012, https://soilseries.sc.egov.usda.gov/OSD_Docs/B/BOJAC.html.

[22] J. M. Bremner, "Nitrogen—total," in *Methods of Soil Analysis: Part 3—Chemical mMethods*, J. M. Bigham, Ed., hapter 37, pp. 1085–1122, Soil Science Society of America, Madison, Wis, USA, 1996.

[23] United States Department of Agriculture—Natural Resources Conservation Service, *Map—Bulk Density, Bojac Series*, United States Department of Agriculture—Natural Resources Conservation Service, Washington, DC, USA, 2012, http://websoilsurvey.nrcs.usda.gov/app/WebSoilSurvey.aspx.

[24] D. A. Lauer, "Russet burbank yield response to sprinkler-applied nitrogen-fertilizer," *American Potato Journal*, vol. 63, pp. 61–69, 1986.

20

The Effect of Freezing Temperatures on *Microdochium majus* and *M. nivale* Seedling Blight of Winter Wheat

Ian M. Haigh[1] and Martin C. Hare[2]

[1] *Field Trials Department, Charles River, Tranent, Edinburgh, EH33 2NE, UK*
[2] *Crop and Enviromental Sciences, Harper Adams University College, Edgmond, Shropshire, TF10 8NB, UK*

Correspondence should be addressed to Ian M. Haigh, ian.haigh@crl.com

Academic Editor: Paul C. Struik

Exposure to pre-emergent freezing temperatures significantly delayed the rate of seedling emergence ($P < 0.05$) from an infected and a non-infected winter wheat cv. Equinox seed lot, but significant effects for timing of freezing and duration of freezing on final emergence were only seen for the *Microdochium*-infested seed lot. Freezing temperatures of $-5°C$ at post-emergence caused most disease on emerged seedlings. Duration of freezing (12 hours or 24 hours) had little effect on disease index but exposure to pre-emergent freezing for 24 hours significantly delayed rate of seedling emergence and reduced final emergence from the infected seed lot. In plate experiments, the calculated base temperature for growth of *M. nivale* and *M. majus* was $-6.3°C$ and $-2.2°C$, respectively. These are the first set of experiments to demonstrate the effects of pre-emergent and post-emergent freezing on the severity of *Microdochium* seedling blight.

1. Introduction

Microdochium nivale (Fr.) Samuels and Hallett (teleomorph *Monographella nivalis* (Schaffnit) E. Müller) and *Microchium majus* (Wollenw.) Glynn and S.G. Edwards (teleomorph *Monographella)* can cause seedling blight of cereals in the UK. *Microdochium nivale* var. *majus* and *M. nivale* var. *nivale* were reclassified as species by Glynn et al. [1]. Before this, mention of *M. nivale* refers to both subspecies unless stated. *Microdochium spp.* may be soil or seed borne; however, seed-borne inoculum is considered to be the predominant cause of seedling blight in the UK [2]. *Microdochium* seedling blight can cause death of cereal plants at the pre-emergent and post-emergence stages of development and surviving seedlings exhibit brown lesions on the coleoptile and roots [3]. Seedling death can result in significant yield losses when surviving plants cannot compensate for large reductions in establishment [4]. In addition, *M. majus* and *M. nivale* inoculum from coleoptile and root lesions has been demonstrated to be able to cause foot rot disease and stem colonisation in glasshouse experiments [5].

Microdochium seedling blight is more severe at cold temperatures and low soil moisture contents [6]. Hare et al. [7] described a strong correlation between the rate of seedling emergence from a wheat seed lot naturally infected with 72% *M. nivale*-infection and final emergence over a range of temperatures and soil moisture contents. In many situations, winter wheat seedlings are likely to be exposed to air temperatures below 0°C. However, the only published work at near freezing temperatures is that by Bateman [8] who reported that maintaining newly emerged wheat seedlings from *M. nivale*-infected seeds at 0-1°C for several weeks increased disease severity.

Despite the lack of evidence for the effect of temperatures below 0°C on the severity of seedling blight from naturally infected wheat seeds, freezing has been observed to increase both the incidence and severity of *Microdochium* seedling blight on oats and barley from surface-inoculated seeds and soil-borne *M. nivale* inoculum. Rawlinson and Colhoun [9] described 4-hour freezing ($-6°C$) on 4 occasions at weekly intervals beginning 1 month after planting increased the incidence of isolation of *M. nivale* from oat seedling mesocotyls and roots grown from untreated seeds in *M. nivale*-infected soil.

When surface inoculated barley seeds with conidia of *M. nivale* (1×10^6 conidia mL^{-1}) were frozen 10 days after planting at $-2°C$ for 48 hours, coleoptiles lesion index increased to 100% compared to 5% on seedlings exposed to 2°C, and 39% on seedlings maintained at 10°C [10]. It is possible that temperatures below 0°C stop winter wheat seedling growth [11] giving *M. majus* and *M. nivale* increased opportunity for infection. However, there is a lack of information for the effects of temperatures below 0°C on the development of seedling blight from seed-borne *Microdochium spp.* in wheat and on *M. majus* and *M. nivale* growth.

A series of controlled environment experiments were designed to test the following hypotheses: (i) timing, duration, and severity of freezing does not affect seedling blight from seeds naturally infected with *M. majus* and *M. nivale*; (ii) timing, duration, and severity of freezing does not affect seedling emergence from non-infected seeds; (iii) *in vitro* growth of *M. majus* and *M. nivale* does not occur below 0°C.

2. Materials and Methods

2.1. In Vitro Growth of Microdochium majus and M. nivale. Five *M. majus* and 5 *M. nivale* isolates from the Harper Adams culture collection were cultured on potato dextrose agar (PDA) at 15°C for 8 days. Plugs of 5 mm diameter from the edges of actively growing colonies were transferred to Petri dishes containing 20 mL wheat flour agar (5% (w/w) winter wheat cv. Equinox flour; 2% (w/w) No. 1 agar (Oxoid Ltd, Basingtoke, UK)). Four dishes of each isolate were incubated in darkness at 5, 10, 15, and 20°C. Fungus colony diameters were measured in 2 directions at 90° angles at 2 day intervals and fungus growth rates ($mm \, day^{-1}$) calculated. Base temperatures for growth of *M. majus* and *M. nivale* were calculated by extrapolation following simple regression of the growth rate of each isolate. Data was analysed using *t*-test.

2.2. Effect of Freezing on the Rate of Seedling Emergence, Final Seedling Emergence, and Severity of Microdochium Seedling Blight. Two winter wheat seed lots cv. Equinox (88% *Microdochium* infection; 95% germination potential (infected seed lot) and 0% *Microdochium* infection; 98% germination potential (non-infected seed lot)) were used in this experiment. Due to a lack of incubator space, experiments for each seed lot were conducted separately. Experiments were conducted testing exposure to temperatures of 0°C or −5°C for 12 hours or 24 hours. Each seed lot was surface-sterilised by immersion in 10% NaOCl solution (1% available chlorine) for 3 minutes, rinsed 3 times in sterile distilled water, placed on sterile filter paper, and dried in a flow of sterile air. The severity of *Microdochium spp.* infection was determined by plating 200 surface-sterilised seeds of each seed lot onto PDA amended with $130 \, \mu g \, mL^{-1}$ streptomycin sulfate (Sigma-Aldrich Company Ltd., Dorset, UK) and $25 \, \mu g \, mL^{-1}$ Bavistin DF (carbendazim 50% w/w; BASF, Bury St. Edmunds, UK).

The germination potential of each seed lot was assessed by the tetrazolium biochemical test [12]. PCR analysis [13] confirmed both *M. majus* and *M. nivale* to be present in the infected seed lot and not present in the non-infected seed lot.

John Innes No. 2 compost was passed through a 5 mm sieve and autoclaved (121°C; 1.08 bar) for 1 hour on 3 consecutive days and adjusted to 40% w/w soil water content. For each seed lot, 100 seeds were planted crease-down 20 mm deep in 45 seed trays. Trays were watered every 3 days to maintain constant 40% w/w soil water content. Trays were placed in an incubator set at 12 hours light (11°C) and 12 hours darkness (7°C) according to a fully randomised design and re-randomised daily. Freezing (12 hours or 24 hours at 0°C or −5°C) was applied 7 days (pre-emergent) or 28 days (post-emergent) after planting to 5 trays in a separate incubator. After freezing, trays were returned to their original incubator. Seedlings not exposed to freezing were used as controls.

Rate of seedling emergence (seedlings $days^{-1}$) was calculated from daily plant counts [14]. Final emergence and disease severity were measured at GS 12. A disease index on emerged seedlings was calculated (1), where *a* is the number of seedlings with category 0 symptoms (no symptoms), *b* is the number of seedlings with category 1 symptoms (≤2 lesions on coleoptile), *c* is the number with category 2 symptoms (>2 lesions on coleoptile), *d* is the number with category 3 symptoms (total necrosis of coleoptile), *e* is the number with category 4 symptoms (total necrosis of coleoptile and deformed seedling growth) and *N* is the number of seedlings assessed [15],

$$\frac{(a \times 0) + (b \times 1) + (c \times 2) + (d \times 3) + (e \times 4)}{N \times 4} \times 100. \quad (1)$$

Diseased cotyledons were surface sterilised and plated onto PDA amended with $130 \, \mu g \, mL^{-1}$ streptomycin sulfate to confirm *Microdochium spp.* were the causal agents of disease.

Each experiment was repeated twice and the data combined prior to analysis. Analysis of each seed lot was conducted separately. A factorial analysis of variance was conducted with rate of seedling emergence, final emergence and disease index as variables, and exposure to freezing, timing, severity and duration of freezing as factors using Genstat 5.0 (Rothamsted Experimental Station, Hertfordshire, UK). Significant probabilities are given as $P < 0.05$. Data for the infected and non-infected seed lots were square-root transformed prior to analysis to ensure normality. Disease index values for the infected seed lot could not be transformed to a normal distribution, therefore standard error values are presented.

3. Results

3.1. In Vitro Growth of Microdochium majus and M. nivale. There were no significant differences between growth rates of the 5 *M. nivale* and 5 *M. majus* isolates so data was pooled for each species. The calculated base temperature for growth of *M. nivale* was significantly lower than *M. majus*. The growth rate for *M. majus* was significantly faster than *M. nivale* (Figure 1).

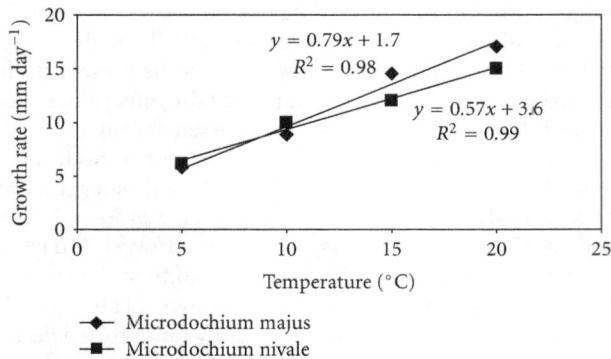

FIGURE 1: *In vitro* growth of 5 *Microdochium majus* and 5 *Microdochium nivale* isolates on winter wheat cv. Equinox flour agar.

3.2. Effect of Freezing on the Rate of Seedling Emergence and Final Seedling Emergence.

For the infected and non-infected seed lots, timing and duration of freezing had a significant effect on rate of seedling emergence. Only pre-emergent freezing and exposure to freezing for 24 hours signifi- cantly delayed the rate of seedling emergence for the infected and non-infected seed lots (Table 1). The timing of freezing∗ duration of freezing interaction only significantly affected rate of seedling emergence from the non-infected seed lot. Exposure to pre-emergent freezing for 24 hours significantly delayed rate of seedling emergence (Table 1).

For the infected seed lot, exposure to freezing, duration and timing of freezing, and the timing of freezing∗duration of freezing and freezing temperature∗timing of freezing interactions had a significant effect on final seedling emer-gence. Exposure to freezing significantly reduced final seed-ling emergence compared to non-frozen seedlings. Only ex-posure to pre-emergent freezing for 24 hours and pre-emergent freezing to −5°C significantly reduced final seed-ling emergence (Table 2). Timing, duration, and severity of freezing had no significant effect on final seedling emergence from the non-infected seed lot (data not shown).

3.3. Effect of Freezing on the Severity of Microdochium Seedling Blight.

Disease symptoms did not occur on seedlings grown from the non-infected seed lot. Isolations from diseased coleoptiles of seedlings grown from the infected seed lot con-firmed *Microdochium spp.* were the causal agents of disease. Seedlings exposed to 0°C had significantly less disease than non-frozen seedlings. Seedlings exposed to −5°C generally had significantly more disease than seedlings frozen to 0°C, but only exposure to −5°C for 24 hours post-emergence significantly increased disease above non-frozen seedlings (Figure 2). Timing and duration of freezing had no signif-icant effect on disease index when seedlings were exposed to 0°C or −5°C. Post-emergent freezing caused more severe seedling blight than pre-emergent freezing.

TABLE 1: Effect of timing (a) and duration (b) of freezing on rate of seedling emergence (seedlings day^{-1}) from a winter wheat cv. Equinox seed lot naturally infected (88%) with *Microdochium spp.* and a non-infected seed lot, and the timing of freezing∗duration of freezing interaction (c) on rate of seedling emergence (seedlings day^{-1}) from a non-infected winter wheat cv. Equinox seed lot.

(a)

Timing of freezing	Infected seedlot	Non-infected seedlot
None	0.055	0.253
Pre-emergent	0.046	0.221
Post-emergent	0.058	0.267
LSD ($P < 0.05$)	0.0267	0.0246
SEM	0.0120	0.0111
DF	79	80
CV (%)	16.6	7.1

Data are back transformed, statistical analysis performed on square root transformed data.

(b)

Duration of freezing	Infected seedlot	Non-infected seedlot
None	0.055	0.253
12 hours	0.056	0.258
24 hours	0.048	0.229
LSD ($P < 0.05$)	0.0267	0.0246
SEM	0.0120	0.0111
DF	79	80
CV (%)	16.6	7.1

Data are back transformed, statistical analysis performed on square root transformed data.

(c)

Timing of freezing	Duration of freezing		LSD	SED	DF	CV (%)
	12 hours	24 hours				
Pre-emergent	0.250	0.193	0.0269	0.0111	80	7.1
Post-emergent	0.267	0.267				

Rate of emergence of non-frozen seedlings 0.253 seedlings day^{-1}.
Data are back transformed, statistical analysis performed on square root transformed data.

4. Discussion

This is the first study to demonstrate that exposure to zero and sub-zero temperatures can affect the severity of *Microdochium* seedling blight on winter wheat from natu-rally infected seeds. For the infected and non-infected seed lots, freezing for 24 hours was required to significantly delay rate of seedling emergence. Freezing may increase the opportunity for seedling infection as *M. majus* and *M. nivale* could continue to grow *in vitro* at temperatures below the minimum air temperatures for growth of winter wheat seedlings [11]. Lowest emergence from the infected seed lot was caused by pre-emergent freezing (−5°C) for 24 hours 7 days after planting. This is in line with the results obtained

TABLE 2: Effect of exposure to freezing temperatures (a), timing and duration of freezing (b) and the timing of freezing∗duration of freezing and freezing temperature∗timing of freezing interactions (c) on final emergence (%) of seedlings from a winter wheat cv. Equinox seed lot naturally infected (88%) with *Microdochium spp.*

(a)

Exposure to freezing	Final emergence (%)
Yes	71
No	56
LSD ($P < 0.05$)	0.9
SEM	0.4
DF	79
CV (%)	17.8

Data are back transformed, statistical analysis performed on square root transformed data.

(b)

Timing of freezing	Final emergence (%)	Duration of freezing	Final emergence (%)
None	71	None	71
Pre-emergent	46	12 hours	64
Post-emergent	67	24 hours	48
LSD ($P < 0.05$)	1.0		1.0
SEM	0.4		0.4
DF	79		79
CV (%)	17.8		17.8

Data are back transformed, statistical analysis performed on square root transformed data.

(c)

	Timing of freezing					
Duration of freezing	Pre-emergent	Post-emergent	LSD	SED	DF	CV (%)
12 hours	65	60	0.9	0.4	79	17.8
24 hours	34	69				
Freezing temperature	Pre-emergent	Post-emergent				
0°C	52	62	0.9	0.4	79	17.8
−5°C	40	72				

Final emergence of non-frozen seedlings 71%.
Data are back transformed, statistical analysis performed on square root transformed data.

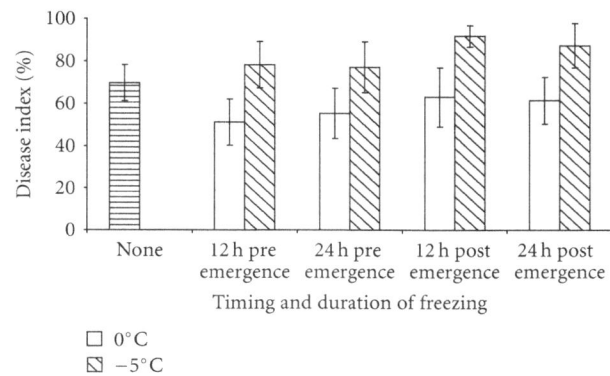

FIGURE 2: Disease index on seedlings grown from a winter wheat cv. Equinox seed lot naturally infected (88%) with *Microdochium spp.* exposed to different freezing temperatures at pre- and post-emergence. Vertical bars represent standard error.

severity has been reported for soil-borne *M. nivale* infecting ryegrass [18] and oats [9] and *Fusarium avenaceum* infecting barley from artificial soil inoculation [19].

Throughout this investigation no attempt was made to distinguish between *M. majus* and *M. nivale*. *Microdochium majus* had a faster *in vitro* growth rate than *M. nivale* which could confer a competitive advantage upon it but Glynn et al. [20] in *in vivo* experiments found no differences in pathogenicity. The effect of freezing temperatures on seed- lings growing in a range of soil moisture conditions is a further avenue for research. The results of this investigation may be used to more accurately target the use of fungicide seed treatments for the control of *Microdochium* seedling blight to planting conditions where seedling blight is likely to occur.

Acknowledgments

The authors acknowledge Harper Adams University College and Chemtura Europe Ltd. for funding this paper.

References

[1] N. C. Glynn, M. C. Hare, D. W. Parry, and S. G. Edwards, "Phylogenetic analysis of EF-1 alpha gene sequences from isolates of *Microdochium nivale* leads to elevation of varieties *majus* and *nivale* to species status," *Mycological Research*, vol. 109, no. 8, pp. 872–880, 2005.

[2] D. W. Parry, T. R. Pettitt, P. Jenkinson, and A. K. Lees, "The cereal *Fusarium* complex," in *Ecology of Plant Pathogens*, P. Blakeman and B. Williamson, Eds., pp. 301–320, CAB International, Wallingford, UK, 1993.

[3] C. S. Millar and J. Colhoun, "*Fusarium* diseases of cereals—IV. Observations on *Fusarium nivale* on wheat," *Transactions of the British Mycological Society*, vol. 52, pp. 57–66, 1969.

[4] J. Humphreys, B. M. Cooke, and T. Storey, "Effects of seed-borne *Microdochium nivale* on establishment and grain yield of winter-sown wheat," *Plant Varieties and Seeds*, vol. 8, no. 2, pp. 107–117, 1995.

[5] I. M. Haigh, P. Jenkinson, and M. C. Hare, "The effect of a mixture of seed-borne *Microdochium nivale* var. *majus* and

by Perry [10] when emergence was lowest from barley seeds surface-inoculated with *M. nivale* and exposed to −2°C for 48 hours 10 days after planting.

Post-emergent freezing (0°C and −5°C) increased the disease index compared to pre-emergent freezing. This is possibly because pre-emergent freezing resulted in heavily diseased seedlings not emerging. These results suggest that freezing increases the severity of *Microdochium* seedling blight rather than directly damaging the winter wheat seedlings as in line with previous observations [16, 17] no damage was seen on the coleoptiles of frozen seedlings from the non-infected seed lot. A similar trend for zero and sub-zero post-emergent temperatures increasing seedling blight

Microdochium nivale var. *nivale* infection on *Fusarium* seedling blight severity and subsequent stem colonisation and growth of winter wheat in pot experiments," *European Journal of Plant Pathology*, vol. 124, no. 1, pp. 65–73, 2009.

[6] C. S. Millar and J. Colhoun, "*Fusarium* diseases of cereals—VI. Epidemiology of *Fusarium nivale* on wheat," *Transactions of the British Mycological Society*, vol. 52, pp. 195–204, 1969.

[7] M. C. Hare, D. W. Parry, and R. A. Noon, "Towards the prediction of *Fusarium* seedling blight of wheat," in *A Vital Role for Fungicides in Cereal Production*, Hewitt et al., Ed., pp. 211–220, Bios Scientific, Oxford, UK, 1995.

[8] G. L. Bateman, "Control of seed-borne *Fusarium nivale* on wheat and barley by organomercury seed treatment," *Annals of Applied Biology*, vol. 83, pp. 245–250, 1976.

[9] C. J. Rawlinson and J. Colhoun, "The occurrence of *Fusarium nivale* in soil," *Plant Pathology*, vol. 18, pp. 41–45, 1969.

[10] D. A. Perry, "Pathogenicity of *Monographella nivalis* to spring barley," *Transactions of the British Mycological Society*, vol. 86, pp. 287–293, 1986.

[11] J. R. Porter and M. Gawith, "Temperatures and the growth and development of wheat: a review," *European Journal of Agronomy*, vol. 10, no. 1, pp. 23–36, 1999.

[12] "International rules for seed testing," *Seed Science & Technology*, vol. 13, pp. 464–480, 1985.

[13] N. C. Glynn, R. Ray, S. G. Edwards et al., "Quantitative *Fusarium* spp. and *Microdochium* spp. PCR assays to evaluate seed treatments for the control of *Fusarium* seedling blight of wheat," *Journal of Applied Microbiology*, vol. 102, no. 6, pp. 1645–1653, 2007.

[14] E. M. Khah, R. H. Ellis, and E. H. Roberts, "Effects of laboratory germination, soil temperature and moisture content on the emergence of spring wheat," *Journal of Agriculture and Science*, vol. 107, pp. 431–438, 1986.

[15] I. M. Haigh, *The effects of temperature and soil water on Fusarium seedling blight of winter wheat and its effective control by fungicide seed treatments*, Ph.D. thesis, Harper Adams University College, 2003.

[16] M. K. Pomeroy, C. J. Andrews, K. P. Stanley, and I. Y. Gao, "Physiological and metabolic responses of winter wheat to prolonged freezing stress," *Plant Physiology*, vol. 78, pp. 207–210, 1985.

[17] C. W. Windt and P. R. Van Hasselt, "Development of frost tolerance in winter wheat as modulated by differential root and shoot temperature," *Plant Biology*, vol. 1, no. 5, pp. 573–580, 1999.

[18] S. J. I. Holmes and A. G. Channon, "Glasshouse studies on the effect of low temperature on infection of perennial ryegrass seedlings by *Fusarium nivale*," *Annals of Applied Biology*, vol. 79, pp. 43–48, 1975.

[19] M. N. Smith and C. R. Olien, "Pathological factors affecting survival of winter barley following controlled freeze tests," *Phytopathology*, vol. 68, pp. 773–777, 1978.

[20] N. C. Glynn, M. C. Hare, and S. G. Edwards, "Fungicide seed treatment efficacy against *Microdochium nivale* and *M. majus* in vitro and in vivo," *Pest Management Science*, vol. 64, no. 8, pp. 793–799, 2008.

21

Enhanced Soil Solarization against *Fusarium oxysporum* f. sp. *lycopersici* in the Uplands

Radwan M. Barakat and Mohammad I. AL-Masri

Plant Protection Research Center, Hebron University, P.O. Box 40, Hebron, Palestine

Correspondence should be addressed to Radwan M. Barakat, radwanb@hebron.edu

Academic Editor: David Clay

Soil solarization tests against *Fusarium oxysporum* f. sp. *lycopersici*, the causal agent of tomato Fusarium wilt, were conducted for seven weeks through July and August 2008 and 2009 in the climatic conditions of Al-Aroub Agricultural Experimental Station, located in the southern mountains of the West Bank, Palestine. Double polyethylene (DPE) sheets, regular polyethylene (PE) sheets, and virtually impermeable films (VIF) were compared to examine their effects on soil temperature, disease severity, and plant growth. Results showed that in comparison to the control, PE, DPE, and VIF treatments increased the mean maximum soil temperatures by 10.2, 14.1, and 8.8°C, respectively, in 2008 and by 10.2, 12.6, and 8.3°C respectively, in 2009. The longest length of time recorded for temperature above 45°C under DPE sheets were 220 hours in 2008 and 218 hours in 2009. The treatments reduced the pathogen population by 86% and the disease by 43% under the DPE treatment in 2009 and to a lesser extent by the other treatments. Increases of up to 94% in fresh plant weight and up to 60% in plant dry weight were evident under the same treatment. The treatments also increased soil organic matter, both nitrogen forms, and major cations.

1. Introduction

Soil solarization is a natural hydrothermal process of disinfesting soil of plant pests and pathogens that is accomplished through passive solar heating. Solarization is commercially practiced mainly in areas which are characterized by high summer air temperatures such as the Mediterranean, deserts, and tropical areas and is affected by several factors, including solar irradiation intensity, air temperature, plastic color and type, soil moisture, soil properties, and other factors [1, 2]. The most well-known function of solarization is reduction of soilborne inoculum of plant pathogens including fungi, bacteria, and nematodes by direct thermal inactivation which is achieved at soil temperatures ranging from 40°C to more than 60°C [3]. In addition, soil solarization increases the release of soluble nutrients (inorganic N forms, extractable P, and K, available cations, and dissolved organic matter) due to soil heating, and consequently results in improved plant growth and yield increases [3–8].

Solarization's major drawbacks are its dependence on climate and its ineffectiveness in controlling heat tolerant soilborne pathogens such as *Fusarium oxysporum* f. sp. *lycopersici*, the causal agent of tomato wilt. Many efforts are being made to improve solarization efficiency in controlling soilborne pathogens including integration with a biological control agent, lower dosages of chemical fungicides, organic amendments (composts, plant residues, and green and animal manures), and physical methods (plastic mulch type, and double-layer mulch) [2, 9]. Using a double layer of polyethylene sheets makes soil solarization more feasible in areas with cooler climates and increases soil temperatures 2–5°C more than a single layer [10].

Tomato wilt disease caused by *Fusarium oxysporum* f. sp. *lycopersici* is a serious disease which causes heavy crop losses worldwide. Several management options have been suggested to control the disease, including soil solarization [11, 12]. In Palestinian agriculture, Fusarium wilt is a serious disease of greenhouses and open field crops. Various fungicides and soil fumigants are being used to control the disease, but because of the concern regarding the toxicity and cost of these compounds, there are strong efforts to reduce the amounts applied to soil and to use more environmentally friendly and cost-effective control options including soil solarization.

The present study aimed to improve the efficiency of soil solarization by using double layers of regular polyethylene sheets compared with regular polyethylene, and virtually impermeable films that contain ethylene vinyl alcohol against *Fusarium oxysporum* f. sp. *lycopersici* populations, and Fusarium wilt severity on tomato planted in the Southern Palestinian Uplands.

2. Materials and Methods

2.1. Soil Preparation and Treatments. Two soil solarization field experiments were conducted for seven weeks from July 6 to August 22, 2008 and from July 1 to August 17, 2009 in Al-Aroub Agricultural Experimental Station of the Faculty of Agriculture, Hebron University, Hebron-Palestinian Authority. The solarized soil was classified as clay soil (28% sand, 13% silt, and 59% clay; pH_{H_2O} 7.3; $EC_{1:2.5}$ (25°C) 0.4 ms cm^{-1}; 22% CaCO$_3$; 2.1% organic matter; 5.2 mg kg^{-1} NH$_4^+$; 28 mg kg^{-1} NO$_3^-$; 20.1 mg kg^{-1} P; 2203 mg kg^{-1} Ca^{+2}; 399 mg kg^{-1} Mg^{+2}; 195 mg kg^{-1} K$^+$; 74 mg kg^{-1} Na$^+$; 30 mg kg^{-1} Fe^{+3}). The soil was deeply plowed (30 cm) two weeks before starting the experiment and rotovated before mulching. Experimental plots were then irrigated to be 60% water-filled pore space two days before the start of the solarization period. The experimental design was a randomized complete block design with three replicates (plots, 3 × 4 m) for each treatment. Four treatments were involved: nonsolarized soil (CK), solarized soil using regular polyethylene (PE) sheets (50 µm thick), solarized soil using double polyethylene sheets separated by 2 cm (DPE), and solarized soil using virtually impermeable films that contain ethylene vinyl alcohol (VIF). Two sets of inoculum bags of *Fusarium oxysporum* f. sp. *lycopersici* were incorporated at 20 and 30 cm depths. The treated plots were mulched with PE, DPE, and VIF sheets, while the control plots were left uncovered. After solarization, three soil subsamples were randomly collected from the middle of each plot to a depth of 20 cm. After removing the top 2-3 cm of soil, the subsamples were combined into one composite sample. One set of inoculum bags was sampled 3 weeks after solarization, and the second set was sampled at the end of the solarization period.

2.2. Preparation of F. oxysporum Inoculum. The isolate of *Fusarium oxysporum* f. sp. *lycopersici* used in the experiment was isolated from diseased tomato stem. Diseased stem was sectioned into 3-4 mm pieces, surface-sterilized by immersion in 1% sodium hypochlorite solution for 4 min, and rinsed three times with sterile-distilled water. Two thin pieces of diseased samples were placed in 90 mm Petri dish containing selective peptone pentachloronitrobenzene (PCNB) agar medium [13]. The peptone PCNB agar medium ingredients were 15 g Difco peptone; 1 g KH$_2$PO$_4$; 0.5 g MgSO$_4$·7H$_2$O; 20 g agar; 1 g pentachloronitrobenzene (PCNB, 75% WP); 1 mL Lactic acid; 0.5 g Chloramphenicol; 1 L distilled water. The ingredients were mixed and dissolved, and the pH was adjusted to 4.5, and the medium were autoclaved.

Seven-day-old growing fungal hyphae were further subcultured on potato dextrose agar (PDA) medium amended with 300 mg L^{-1} chloramphenicol. A single-conidium culture was prepared and subcultured, and one of the growing colonies was used to inoculate further Petri dishes. Petri dishes were then incubated for 40 days in the growth chamber at 25°C, with 12 hours photoperiods. The dried paste made of the fungal growth in the growing media was used to prepare the chlamydospore inoculum. Forty days were enough for most of the mycelial cells to develop into chlamydospores. The chlamydospore inoculum was ground and mixed in dry sandy soil and propagules measured as CFU g^{-1} using the dilute plate technique: 2.5 g of previously prepared soil inoculum were placed in 23 mL distilled water (1 : 10), and 0.2 mL of the suspension were spread on each of the six Petri dishes containing 15 mL of selective peptone-PCNB agar medium prepared earlier. The Petri dishes were then incubated at 25°C under darkness for three days and under natural room light for 4 days. The numbers of *Fusarium* colonies were counted and the mean inoculum concentration was calibrated to 1260 CFU per 1 g of the dry inoculum-sandy soil mixture. Twenty grams of prepared soil inoculum were placed in each muslin bag. Small muslin bags containing the inoculum were closed with plastic silks and incorporated in experimental plots at a depth of 20 and 30 cm (as mentioned earlier) in both seasons experiments [14].

2.3. Estimation of F. oxysporum Population after Solarization. The population of *F. oxysporum* in the muslin bags buried earlier in solarized and nonsolarized plots in both seasons were assessed after 3 weeks of solarization and at the end of the solarization period. The pathogen population in the muslin bags was measured as CFU g^{-1} by using the dilute plate techniques on selective peptone-PCNB agar medium [13]. Soil dilutions were prepared by taking 2.5 g of soil in 23 mL of sterilized distilled water (1 : 10); 0.2 mL of the suspension was spread on each Petri dish. Petri dishes were incubated at 25°C under darkness for three days and under natural room light for 4 days. The number of propagules grown was counted and calculated as CFU per gram soil. The experimental design was completely randomized with five replicates (Petri dishes) seeded with dilutions of soil sampled earlier at 20 and 30 cm depths.

2.4. Disease Severity. Fusarium wilt (%) of tomato plants growing in solarized and nonsolarized soils was evaluated after 3 weeks of solarization and at the end of the solarization period (7 weeks). Under each experimental plot, three soil subsamples were randomly collected to a 20 cm depth and mixed thoroughly to make one composite sample. Seventeen grams of the inoculum bags were added to the respective soil sample taken from each plot for each sampling period. Each 500 g composite soil sample amended with the inoculated soil from bags from each experimental plot was divided into 5 small planting pots, each containing 100 g of soil. In each pot, a plastic grid was placed in the bottom to keep the soil in and a 20 mL of autoclaved perlite were added to permit excess water to drain. Tomato seeds (3–5) were then seeded in each pot and a layer of autoclaved perlite was added to cover the top. After emergence, the number of seedlings was reduced to two per pot. Plants were then incubated in a growth chamber at 25°C, with 15 hours photoperiods. Plants were irrigated regularly with deionized water. The number of wilted plants was recorded weekly from week 3 to week 10 after sowing. The accumulated number of wilted plants was documented and the percentage of wilt was calculated. A completely randomized design was used with five replicates. The same parameters were measured in the same manner for the second solarization experiment.

2.5. Plant's Growth Evaluation. Three soil subsamples were randomly collected from the upper 20 cm of each experimental plot. After removing the top 2-3 cm of soil, the subsamples were mixed thoroughly to make one composite sample. Each composite soil sample (1 kg) was incorporated in 5 pots (replicates). Tomato seeds (3–5) were then seeded in each pot. After emergence, the number of seedlings was reduced to 1 per pot. Plants were then incubated under greenhouse conditions at 25°C for two months. Plants were irrigated daily and the plant's fresh and dry weights were evaluated at the end of the experiment. A completely randomized design was used with five replicates (pots).

2.6. Soil Temperature. The soil temperature was recorded by HOBO data loggers (Onset Computer Corporation, Bourne, USA) during the two solarization periods. The loggers were set to take a reading every 40 minutes during the solarization period at the depth of 15 cm in the middle of all experimental plots. The loggers were removed at the end of the period, and the data downloaded using the BOXCar version 3.7 software (Onset Computer Corporation, Bourne, USA).

2.7. Soil Analysis. Composite soil samples (1000 g) were collected from the center of experimental plots at the end of the experiment. Soil samples were oven dried at 105°C for 24 h. Dry soil samples were then sieved (2 mm) and the fine soil was used for chemical analysis (pH_{H_2O}, $EC_{1:2.5}$, $CaCO_3$, organic matter, NH_4^+, NO_3^-, P, Ca^{+2}, Mg^{+2}, K^+, Na^+, and Fe^{+3}). The soil pH and EC were evaluated in water extracts (1 : 2.5, w/v) by pH meter (pH meter 3305, Jenway, UK) and conductivity meter (4010 conductivity meter, Jenway, UK). Calcium carbonate was evaluated using a calcimeter (Calcimeter, Eijkelkamp, Germany). The organic matter was evaluated by acidic wet oxidation with potassium dichromate according to the Walkley-Black wet combustion method [15]. The exchangeable ammonium and nitrate were evaluated according to the methods described by Keeney and Nelson [16]. Available phosphorus was measured by using the molybdate ascorbic acid method [17]. Exchangeable calcium, magnesium, sodium, and potassium were evaluated by the neutral ammonium acetate (pH = 7) method. Air-dry soil samples were ground and sieved using a 20-mesh sieve; 2.5 g of soil were placed in 125 mL Erlenmeyer flasks and 25 mL of 1 N ammonium acetate, pH 7 were then added. Flasks were shaken for 15 min, and the solutions

were filtrated and analyzed by flame atomic absorption (A Analyst 100, Perkin Elmer). The concentrations of cations were calculated using standard curves as mg kg^{-1} [18].

2.8. Statistical Analysis. The data were statistically analyzed using one-way repeated analysis of variance (ANOVA). Fishers LSD test ($P \leq 0.05$) was used for mean's separation (Sigma Stat 2.0 program, SPSS Inc., USA).

3. Results

3.1. Soil Temperature. Soil temperature was greatly increased in solarized soil treatments compared with the control (Table 1 and Figure 1). The means of maximum soil temperatures (°C) recorded during the solarization period were 32.6, 42.8, 46.7, and 41.4 during 2008 and 35.3, 45.5, 47.9, and 43.6 during 2009, under the control, PE, DPE, and VIF treatments, respectively. The means of maximum soil temperatures increased by 10.2, 14.1, and 8.8 during 2008 and by 10.2, 12.6, and 8.3 during 2009 under the same treatments, respectively, compared with the unsolarized control treatment. The double layer treatment (DPE) increased the mean of maximum temperature by 3.9 and 2.4°C during 2008 and 2009, respectively, compared to the one layer regular polyethylene sheet treatment (PE) and by 5.3 and 4.3°C during the 2008 and 2009, respectively, compared to the VIF sheet treatment. The number of hours recorded for the sublethal temperature class (45–50°C) were 220 h (19.1%) and 218 h (18.9%); 17 h (1.4%) and 28 h (2.4%); 5 h (0.4%) and 36 h (3.1%) under the DPE, PE, and VIF sheet treatments during 2008 and 2009 seasons, respectively, compared to the total solarization time. The absolute maximum soil temperatures measured during the solarization periods were 50.1°C and 49.3°C recorded under the DPE sheets in the summers of 2009 and 2008, respectively.

3.2. Pathogen Population and Disease Severity. After seven weeks of solarization, the population of *F. oxysporum* (CFU) in soil was reduced significantly at the depth of 20 cm by 44%, 53%, and 43% in 2008 and by 61%, 86%, and 60% in 2009 when PE, DPE, and VIF solarization sheets were used, respectively, compared with the control (Table 2). The highest reduction of pathogen population at both soil depths was obtained under DPE sheets during 2009 after 7 weeks of solarization.

Three weeks of soil solarization was not significantly effective in reducing Fusarium wilt (%) in either year. However, the disease was reduced significantly by 32%, 39%, and 25% during 2008 and by 30%, 43%, and 30% during 2009 when PE, DPE, and VIF sheets were used, respectively, compared with the control after seven weeks of solarization. The double polyethylene sheet recorded the highest disease reduction compared to control after seven weeks of solarization (Table 3).

3.3. Plant's Growth. Soil solarization significantly increased fresh and dry weights of tomato plants in both seasons. Fresh

TABLE 1: Number of hours for different temperature classes recorded under solarization treatments (control (CK), regular polyethylene (PE), double regular polyethylene (DPE), and virtually impermeable films (VIF)) during July 6-August 22, 2008 and July 1-August 17, 2009 in Al-Aroub Agricultural Research Station, South of the West Bank.

Temperature class	Number of hours (2008)				Number of hours (2009)			
	CK	VIF	PE	DPE	CK	VIF	PE	DPE
<30°C	730	47	67	41	670	200	103	22
30–35°C	422	401	430	328	383	384	457	408
35–40°C	0	410	362	296	99	310	289	262
40–45°C	0	289	276	267	0	222	275	242
45–50°C	0	5	17	220	0	36	28	218
Total (hours)	1152	1152	1152	1152	1152	1152	1152	1152
Min (°C)	20.9	26.1	23.1	23.7	20.5	23.2	24	23.6
Max (°C)	33.1	43.9	45.8	49.3	36.6	49.6	45.8	50.1
Average of Max.	32.6	41.4	42.8	46.7	35.3	43.6	45.5	47.9

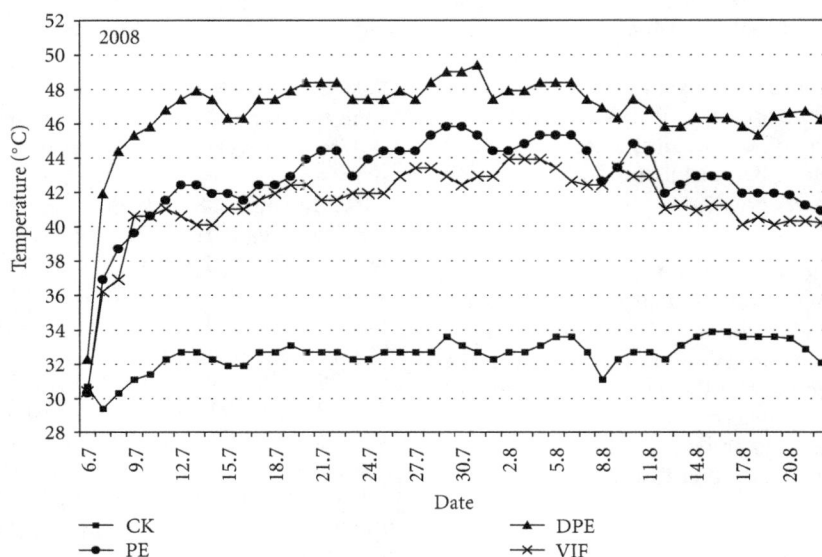

(a)

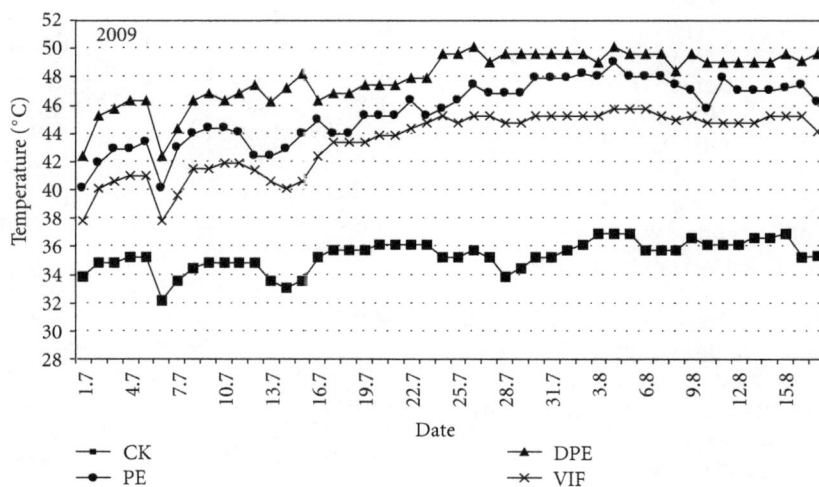

(b)

FIGURE 1: Maximum temperatures recorded under solarization treatments, regular polyethylene (PE), double regular polyethylene (DPE), virtually impermeable films (VIF), and nonsolarized soil (CK) from July 6-August 22, 2008 and July 1-August 17, 2009 in Al-Aroub Agricultural Research Station, South of the West Bank.

TABLE 2: Effect of solarization with regular polyethylene (PE), double regular polyethylene (DPE), and virtually impermeable films (VIF) on population of *Fusarium oxysporum* f. sp. *lycopersici* in soil.

Treatments	2008				2009			
	Solarization period (weeks)							
	3		7		3		7	
	20 cm	30 cm	20 cm	30 cm	20 cm	30 cm	20 cm	30 cm
CK	1233* ab	1563 ab	1242 a	1270 a	1132 a	1249 a	1176 a	1209 a
PE	923 b	1067 b	694 b	628 bc	766 ab	877 ab	458 bc	804 b
DPE	950 b	940 b	582 bc	528 bc	642 ab	603 b	161 d	174 d
VIF	1150 b	1050 bc	708 b	644 bc	804 ab	1078 ab	470 cd	679 c
	LSD = 534				LSD = 514			

*Values are means of *F. oxysporum* f. sp. *lycopersici* (CFU/g soil) of 3 soil sub-samples, each planted on five Petri dishes.

TABLE 3: Effect of solarization with regular polyethylene (PE), double regular polyethylene (DPE), and virtually impermeable films (VIF) on Fusarium wilt (%) caused by *Fusarium oxysporum* f. sp. *lycopersici*. (LSD = 11).

Treatments	2008		2009	
	Solarization period (Week)			
	3	7	3	7
CK	58 a	65 a	56 a	63 a
PE	46 a	33 c	51 a	33 bc
DPE	46 a	26 cd	40 ab	20 d
VIF	50 a	40 b	46 ab	33 bc

weights (%) increased by 37, 94, and 56, while dry weights (%) increased by 38, 60, and 57 under PE, DPE, and VIF treatments, respectively, compared to control in 2008. In 2009, plant fresh weights (%) increased by 46, 77, and 61, while dry weights increased by 46, 76, and 60 in PE, DPE, and VIF treatments, respectively, compared to the control (Table 4). The highest increase in plant growth was recorded under the DPE treatment and significant variation between solarization treatments was not observed.

3.4. Soil Chemical Properties. The solarization treatments significantly affected the soil pH, EC, organic matter, available ammonium, nitrate, calcium, magnesium, and potassium. Calcium carbonate, phosphorus, and iron were not affected (Table 5). The solarization treatment PE, DPE, and VIF slightly increased the pH by 4.1%, 5.5%, and 4.1% during 2008 and by 8.2%, 6.8%, and 4.1% during 2009, respectively, compared to the control. In addition, the solarization treatments increased the EC values by 50%, 200%, and 75% during 2008 and by 125%, 225%, and 250% during 2009, respectively, compared to the control.

The PE, DPE, and VIF solarization treatments stimulated the decomposition of organic matter and reduced the percentage of organic matter in soil by 19, 42, 33 percentage during 2008 and by 32, 45, 45 percentages during 2009 respectively, compared to the control. In addition, the available ammonium increased significantly by 2-3-fold and the nitrate by 1-2-fold during the solarization treatment in both seasons.

At the same treatments, available calcium (%) increased by 65, 67, and 76% during 2008 and by 65, 69, and 62% during 2009 respectively, compared to the control.

Available magnesium was increased by 12, 13, and 14%, respectively, during 2008 and by 10 and 14% under DPE and VIF, respectively; the PE treatment has not affected Mg^{+2}. Available potassium in the treated soils, increased by 100, 112, and 125% during 2008 and by 103, 120, and 111%, during 2009, respectively, under the three solarization treatments.

4. Discussion

Soil solarization is considered a relatively mild heating treatment for disinfesting soils where the population of soilborne pathogens including *Fusarium oxysporum* f. sp. *lycopersici* is reduced. The results showed significant reduction in pathogen population at the soil depths of 20 and 30 cm after seven weeks of two solarization periods under all treatments compared with the control; the highest reduction was observed under DPE treatment mainly during the 2009 solarization. In addition, tomato wilt was significantly (LSD = 11, $P \leq 0.05$) reduced after seven weeks of solarization, but was not affected after three weeks using all types of solarization sheets in both seasons. Both pathogen population and disease reduction were negatively correlated to soil temperatures recorded under solarization treatments. The mean maximum temperature increase over the control was in the range of 8.8–14.1°C in 2008 and 8.3–12.6°C in 2009. In addition, the number of hours recorded for temperatures above 45°C under VIF, PE, and DPE was 5, 17, and 220 hours for the 2008 season and 36, 28, and 218 hours, for the 2009 season, respectively. Similar results were obtained by Tamietti and Valentino, [12] in which significant reduction in Fusarium wilt of melon (*Fusarium oxysporum*

TABLE 4: Effect of solarization with regular polyethylene (PE), double regular polyethylene (DPE), and virtually impermeable films (VIF) on tomato plant fresh and dry weights (g plant^{-1}).

| Treatments | 2008 | | 2009 | |
| | Weight (g plant^{-1}) | | | |
	Fresh	Dry	Fresh	Dry
CK	88.1* c	11.6 b	97.6 b	13.2 b
PE	121.2 b	16 a	143 a	19.3 a
DPE	171.7 a	18.6 a	172.7 a	23.3 a
VIF	138.2 a	18.2 a	157.1 a	21.2 a
LSD ($P \leq 0.05$)	37.2	4.8	45.3	6.1

*Data are means of fifteen replicates; means followed by the same letter in colums are not significantly different according Fisher LSD test at ($P \leq 0.05$).

TABLE 5: Effect of solarization with regular polyethylene (PE), double regular polyethylene (DPE), and virtually impermeable films (VIF) on chemical properties of soil.

| Chemical propreties | 2008 | | | | | 2009 | | | | |
| | Treatments | | | | | Treatments | | | | |
	CK	PE	DPE	VIF	LSD	CK	PE	DPE	VIF	LSD
pH$_{H_2O}$	7.3* b	7.6 a	7.7 a	7.6 a	0.21	7.3 c	7.9 a	7.8 ab	7.6 b	0.26
EC (ms)	0.4 c	0.8 b	1.2 a	0.7 b	0.25	0.4 c	0.9 b	1.3 a	1.4 a	0.14
CaCO$_3$ (%)	22 NS***	23 NS	23 NS	22.3 NS		22 NS	23 NS	23 NS	22.3 NS	
O.M (%)	2.1 a	1.7 ac	1.2 b	1.4 bc	0.38	2.2 a	1.5 b	1.2 b	1.2 b	0.42
NH$_4^+$	5.2 b	13.1 a	14 a	14.4 a	4.41	5.3 b	10.7 a	12.2 a	12.1 a	4.24
NO$_3^-$	28 b	48 a	48 a	49 a	9.88	27 b	42 a	46 a	47 a	11.2
P	20.1 NS	20.7 NS	24.3 NS	23.1 NS		20 NS	20 NS	21 NS	21 NS	
Ca^{+2}	2203 b	3642 a	3694 a	3890 a	396	2136 b	3542 a	3627 a	3470 a	667
Mg^{+2}	399 b	449 a	450 a	455 a	44.5	399 b	415 b	439 b	455 a	30.6
K$^+$	195 b	390 a	413 a	439 a	109	191 b	388 a	421 a	403 a	80.9
Na$^+$	74 b	146 b	311 a	162 b	100	78 b	160 b	280 a	154 b	117
Fe	30 NS	32 NS	38 NS	36 NS		32 NS	30 NS	36 NS	33 NS	

*Data are means of three replicates; means followed by the same letters in rows are not significantly different according Fisher LSD ($P < 0.05$).
**The concentrations of NH$_4^+$, NO$_3^-$, P, Ca^{+2}, Mg^{+2}, K$^+$, Na$^+$ and Fe are in mg kg^{-1}.
***NS: not significant.

f. sp. *melonis*) was negatively correlated with the number of hours of soil temperatures above 40°C; the number of hours recorded for the temperatures above 40°C was 234 hours at the depth of 25 cm. The reduction in pathogen population may be due to increased temperature killing the pathogen directly or due to increased temperatures weakening the pathogen with sublethal heat, rendering it in capable of inducing crop damage [10].

Furthermore, it was evident that solarization stimulated fresh and dry weights of tomato plants in both seasons. Fresh weights increased by 37–94% and dry weights by 38–60% in 2008 and 2009, respectively, compared to control. The highest increase was recorded under the DPE treatment. The stimulation of tomato plant growth may be related to the decomposition of organic matter and the increase of available ammonium and nitrate and available calcium, magnesium, and potassium in solarized soils. Increases in mineral nutrients such as ammonium and nitrate can be attributed to the decomposition of organic components of soil during treatment, while other minerals, such as calcium, magnesium, and potassium may have been virtually cooked of the mineral soil particles undergoing solarization Similar

increased growth responses (IGRs) of several plant systems grown under solarized soils have been observed in other studies [3–8, 19]. In addition, Stapleton and DeVay [20] reported that the concentrations of NH$_4$ and NO$_3$ in the top 15 cm of solarized soil were increased and the concentrations of other soluble mineral nutrients, including calcium, magnesium, phosphorus, potassium, and others increased, but less consistently. IGR can be attributed to a number of mechanisms, including the increase in nutrient levels in soil solution, stimulation of beneficial microorganisms, and the control of minor pathogens [3, 4].

Solarization treatments slightly increased soil pH during both seasons, compared with nonsolarized soils. The light increases in soil pH may be due to an increase in the concentrations of some mineral nutrients such as calcium, magnesium, and potassium which are released from the mineral soil particles undergoing solarization and have alkaline buffering action. On the other hand, some studies reported that solarization can decrease pH [3, 19]. In addition, solarization significantly increased the EC of soil extract, probably due to the increase in nutrient minerals after solarization.

In conclusion, double layers of polyethylene sheets enhanced the positive effect of solarization in terms of disease reduction and plant growth improvement, which was only possible by a better elevation of soil temperature during the solarization period. This enhancement will definitely open doors for a wide use of the technique in regions characterized with lower average temperatures than those traditionally used for solarization over the years.

References

[1] J. Katan and J. E. De Vay, *Soil Solarization*, CRC Press, Boca Raton, Fla, USA, 1991.

[2] J. J. Stapleton, "Soil solarization in various agricultural production systems," *Crop Protection*, vol. 19, no. 8–10, pp. 837–841, 2000.

[3] A. Gelsomino, L. Badalucco, L. Landi, and G. Cacco, "Soil carbon, nitrogen and phosphorus dynamics as affected by solarization alone or combined with organic amendment," *Plant and Soil*, vol. 279, no. 1-2, pp. 307–325, 2006.

[4] Y. Chen, J. Katan, A. Gamliel, T. Aviad, and M. Schnitzer, "Involvement of soluble organic matter in increased plant growth in solarized soils," *Biology and Fertility of Soils*, vol. 32, no. 1, pp. 28–34, 2000.

[5] J. M. Gruenzwig, H. D. Rabinowitch, and J. Katan, "Physiological and devlopmental aspects of increased plant growth in solarized soils," *Annual Applied Biology*, vol. 122, pp. 579–591, 1993.

[6] M. I. Salerno, G. A. Lori, D. O. Giménez, J. E. Giménez, and J. Beltrano, "Use of soil solarization to improve growth of eucalyptus forest nursery seedlings in Argentina," *New Forests*, vol. 20, no. 3, pp. 235–248, 2000.

[7] J. J. Stapleton, J. Quick, and J. E. Devay, "Soil solarization: effects on soil properties, crop fertilization and plant growth," *Soil Biology and Biochemistry*, vol. 17, no. 3, pp. 369–373, 1985.

[8] C. Stevens, V. A. Khan, J. E. Brown et al., "Soil solarization: the effects of organic amendments on growth and fertility of cole crops," in *Proceedings of the 23rd National Agricultural Plastics Congress American Society of Plasticulture*, vol. 23, pp. 272–280, 1991.

[9] C. Stevens, V. A. Khan, R. Rodriguez-Kabana et al., "Integration of soil solarization with chemical, biological and cultural control for the management of soilborne diseases of vegetables," *Plant and Soil*, vol. 253, no. 2, pp. 493–506, 2003.

[10] J. E. De Vay and J. Katan, "Mechanisms of pathogen control in solarized soils," in *Soil Solarization*, J. Katan and J. E. De Vay, Eds., pp. 87–102, CRC Press, Boca Raton, Fla, USA, 1991.

[11] N. Ioannou, C. A. Poullis, and J. B. Heale, "Fusarium wilt of watermelon in Cyprus and its management with soil solarization combined with fumigation or ammonium fertilizers," *OEPP Bulletin*, vol. 30, pp. 223–230, 2000.

[12] G. Tamietti and D. Valentino, "Soil solarization as an ecological method for the control of Fusarium wilt of melon in Italy," *Crop Protection*, vol. 25, no. 4, pp. 389–397, 2006.

[13] P. E. Nelson, T. A. Tousson, and W. F. O. Marasas, *Fusarium Species: An Illustrated Manual for Identification*, Pennsylvania State University Press, University Park, Pa, USA, 1983.

[14] L. W. Burgess, B. A. Summerell, S. Bullock, K. P. Gott, and D. Backhouse, Eds., *Laboratory Manual for Fusarium Research*, University of Sydney, Australia, 1994.

[15] K. H. Tan, *Soil Sampling, Preparation, and Analysis*, Marcel Dekker, New York, NY, USA, 1995.

[16] D. R. Keeney and D. W. Nelson, "Nitrogen-Inorganic forms," in *Methods of Soil Analysis, Part 2-Chemical and Microbiological Properties*, A. L. Page, R. H. Miller, and D. R. Keeney, Eds., pp. 643–698, American Society of Agronomy, Soil Science Society of America, Madison, Wis, USA, 2nd edition, 1982.

[17] S. R. Olsen and L. E. Sommers, "Phosphorus," in *Methods of Soil Analysis, Part 2-Chemical and Microbiological Properties*, A. L. Page, R. H. Miller, and D. R. Keeney, Eds., pp. 403–430, American Society of Agronomy/Soil Science Society of America, Madison, Wis, USA, 2nd edition, 1982.

[18] Perkin Elmer, *Atomic Absorption Spectroscopy Analytical Methods*, Perkin Elmer, Norwalk, Conn, USA, 1996.

[19] J. M. Grünzweig, J. Katan, Y. Ben-Tal, and H. D. Rabinowitch, "The role of mineral nutrients in the increased growth response of tomato plants in solarized soil," *Plant and Soil*, vol. 206, no. 1, pp. 21–27, 1998.

[20] J. J. Stapleton and J. E. DeVay, "Soil solarization: a natural mechanism of integrated pest management," in *Novel Approaches to Integrated Pest Management*, R. Reuveni, Ed., pp. 309–322, Lewis Publishers, Boca Raton, Fla, USA, 1995.

22

Peanut Cultivar Response to Flumioxazin Applied Preemergence and Imazapic Applied Postemergence

W. J. Grichar,[1] P. A. Dotray,[2] and M. R. Baring[3]

[1] Texas A&M AgriLife Research, 3507 Highway 59E, Beeville, TX 78102, USA
[2] Texas A&M AgriLife Research and Texas A&M AgriLife Extension Service, Texas Tech University, Lubbock, TX 79403, USA
[3] Texas A&M AgriLife Research, Department of Soil and Crop Sciences, College Station 77843, USA

Correspondence should be addressed to W. J. Grichar; w-grichar@tamu.edu

Academic Editor: Kassim Al-Khatib

Field studies were conducted during 2009 and 2010 in Texas at Yoakum and Lamesa to determine peanut cultivar response to flumioxazin applied preemergence (0.053, 0.107, and 0.214 kg ai ha^{-1}) and imazapic applied postemergence (0.035, 0.071, and 0.141 kg ai ha^{-1}). At Yoakum, two cultivars (Tamrun OL01, Tamrun OL07) were evaluated while at Lamesa, four cultivars (FlavorRunner 458, Tamrun OL01, Tamrun OL02, and Tamrun OL07) were evaluated. In 2009, no stunting was noted at Yoakum with any herbicide regardless of cultivar. At Lamesa, FlavorRunner 458 and Tamrun OL01 were stunted at least 6% with the 0.21 kg ha^{-1} rate of flumioxazin and 6 to 17% with the 0.07 and 0.14 kg ha^{-1} rate of imazapic. Tamrun OL02 was stunted by all rates of flumioxazin (5%) and imazapic (5 to 18%) while Tamrun OL07 was stunted by all rates of flumioxazin (6 to 12%) and imazapic (7 to 15%) with the exception of flumioxazin at 0.05 kg ha^{-1}. Flumioxazin did not have an effect on yield while all imazapic rates reduced yields when compared with the non-treated control. In 2010 at Yoakum, little (<2%) or no herbicide stunting was noted on any cultivar and only imazapic at 0.14 kg ha^{-1} caused significant stunting (7%). No yield differences were noted between herbicides regardless of cultivar. At Lamesa, all cultivars were affected (6 to 9% stunting) by herbicide treatments. No peanut stunting was noted with flumioxazin at 0.05 kg ha^{-1} while imazapic at 0.04 kg ha^{-1} and flumioxazin at 0.11 kg ha^{-1} resulted in 4 and 6% stunting, respectively. Flumioxazin at 0.21 kg ha^{-1} and imazapic at 0.07 kg ha^{-1} resulted in 12% stunting and imazapic at 0.14 kg ha^{-1} stunted peanut 19%. Both Tamrun OL01 and Tamrun OL07 produced lower yields (≤6369 kg ha^{-1}) than FlavorRunner 458 (7252 kg ha^{-1}). Tamrun OL02 yields were intermediate (6889 kg ha^{-1}). Peanut yields from herbicide treatments were not different from the non-treated control.

1. Introduction

Flumioxazin is an N-phenyl phthalamide soil-applied herbicide that received a federal label in the US for use in peanut in 2001 [1]. Flumioxazin inhibits the enzyme protoporphyrinogen oxidase [2–4]. In Georgia, flumioxazin applied preemergence (PRE) was shown to control morningglory species (*Ipomoea* spp.), prickly sida (*Sida spinosa* L.), and Florida beggarweed (*Desmodium tortuosum* (Sw.) DC) [5] while in Texas, pitted morningglory (*Ipomoea lacunose* L.) was controlled greater than 75% [6]. Also, flumioxazin is used in the Virginia-Carolina area for early-season suppression of Palmer amaranth (*Amaranthus palmeri* S. Wats.) and other broadleaf weeds (D. Jordan, personal communication).

Imazapic is similar to imazethapyr and controls all the weeds controlled by imazethapyr [7–13]. In addition, imazapic provides control and suppression of Florida beggarweed and sicklepod (*Senna obtusifolia* (L.) Irwin & Barneby), which are not adequately controlled by imazethapyr [12]. Imazethapyr provides consistent control of many broadleaf and sedge species (*Cyperus* spp.) if applied within 10 d after emergence, but imazapic has a longer effectiveness period when applied postemergence (POST) [9, 11, 14, 15]. Imazapic also is effective control of rhizome and seedling johnsongrass (*Sorghum halepense* (L.) Pers.), Texas panicum (*Panicum texanum* Buckl.), large crabgrass (*Digitaria sanguinalis* (L.) Scop.), southern crabgrass (*Digitaria ciliaris* (Retz.) Koel.), and broadleaf signalgrass (*Brachiaria platyphylla* (Griseb.) Nash) [9].

Flumioxazin has been reported to cause peanut injury while imazapic has only caused minor stunting [16]. Grichar et al. [1] reported that flumioxazin plus metolachlor combinations, under cool, wet conditions resulted in peanut stunting which was evident throughout the growing season. They attributed this to increased uptake of flumioxazin and metolachlor with the heavy rainfall and the slowed metabolism of these herbicides as a result of cool temperatures [2]. Askew et al. [17] reported that flumioxazin at 0.07 and 0.11 kg ai ha^{-1} injured peanut 45 and 62%, respectively, when rated 2 weeks after peanut planting. Peanut stunting of greater than 60% was followed by as much as 35% leaflet discoloration, which was characterized as necrotic spots on foliage. Scott et al. [18] reported that flumioxazin treated peanuts were injured 10% when rated 3 weeks after planting. However, injury was transient and was not apparent 6 weeks after planting. Flumioxazin enters plants mainly by shoot and root uptake, and plant injury can be avoided via rapid metabolism [2, 3].

Peanut injury from imazapic was observed in North Carolina [19] and from soil applications in south Texas [12] but no injury was observed in other studies [6, 8, 12, 15, 20]. No reductions in peanut grade or yield following imazapic treatments have been observed in several studies [19, 21–23]. Dotray et al. [23] concluded that peanut injury observed in season does not appear to have any influence on grade or yield.

Many peanut cultivars now used in the southwest production area have not been evaluated with respect to response to flumioxazin or imazapic. Therefore, the objective of this research was to evaluate the effect of flumioxazin applied preemergence or imazapic applied postemergence on peanut growth and yield of four runner market-type peanut cultivars.

2. Materials and Methods

2.1. Research Sites. Peanut tolerance studies, under weed-free conditions, were conducted in south Texas near Yoakum and in the Texas High Plains near Lamesa with runner market-types during the 2009 and 2010 growing seasons. Soils at Yoakum were a Denhawken sandy clay loam (fine, smectitic, hyperthermic, Vertic Haplustepts, 1.6% organic matter, pH 7.6) while soils at Lamesa were an Amarillo fine sandy loam (fine-loamy, mixed, superactive, thermic Aridic Paleustalf, 0.4% organic matter, pH 7.8). Planting dates were July 1, 2009 and May 27, 2010 in south Texas and April 30, 2009 and April 28, 2010 at the High Plains location.

2.2. Peanut Cultivars and Herbicide Treatments. Treatments consisted of a factorial arrangement of two runner market-type peanut varieties in south Texas, Tamrun OL01 [24] and Tamrun OL07 [25] or four varieties at the High Plains location FlavorRunner 458 (a runner-type released by Mycogen Co.) [26], Tamrun OL01, Tamrun OL02 [27], and Tamrun OL07 with six herbicide treatments (flumioxazin at 0.053, 0.107, and 0.214 kg ha^{-1} and imazapic at 0.035, 0.071, and 0.141 kg ha^{-1}). This represents 0.5, 1.0, and 2 times the US labeled rate of either herbicide [28, 29].

A non-treated control was included in each study at both locations. At least 12.5 mm of overhead irrigation was applied to activate the herbicide at Lamesa after peanuts were planted and flumioxazin was applied. Each study was replicated three times. Imazapic treatments include Induce (Helena Chemical Company, Collierville, TN, USA) at 0.25% v/v at the south Texas location or Agridex (Helena Chemical Company, Memphis, TN, USA) at 2.3 L ha^{-1} at the High Plains location. Imazapic POST applications were made approximately five weeks after planting at Lamesa and four weeks after planting at Yoakum.

2.3. Plot Description and Rainfall. Individual plot size was 1.9 by 7.6 m at Yoakum and 2.1 by 9.5 m at Lamesa. Rainfall for Lamesa can be best described as average for 2009 and above average for 2010 (Table 1). In 2009, rainfall for April and May was below average but above average for June and July. The rest of the growing season experienced below average rainfall with less than 24 mm of rainfall for either August, September, or October. In 2010, the growing season was considered wet with above normal rainfall for April, July, and September, with slightly below average rainfall for May and June; however, no rain was received in August and little rain in October. Seasonal rainfall at Yoakum was above normal in both years (Table 1). Rainfall in 2009 was below normal for May through August with near normal rainfall for September and above normal for October. Peanuts were dug in October but due to the exceptional high rainfall throughout the month, peanuts could not be combined. In 2010, rainfall was slightly above normal for May and August and above normal rainfall for July and September. No rainfall was received in October.

Supplemental irrigation was supplied as needed at both locations. Traditional production practices were used to maximize peanut growth, development, and yield. All plots received a dinitroaniline herbicide applied preplant incorporated and were cultivated and hand-weeded throughout the growing season to maintain weed-free conditions. Clethodim at 0.18 kg ai ha^{-1} was applied POST to control annual grass escapes at the south Texas location. No insecticides were needed at any location in any year.

2.4. Herbicide Application and Peanut Response. Herbicides were applied using water as a carrier with a CO_2-pressurized backpack sprayer using TeeJet 11002 DG flat fan nozzles (Tee-Jet Spraying Systems Co., Wheaton, IL, USA) that delivered 190 L ha^{-1} at 180 kPa at Yoakum or Turbo TeeJet 110015 flat fan nozzles that delivered 140 L ha^{-1} at 207 kPa at Lamesa. Peanut stunting was rated approximately 60 d after peanut planting with the runner peanuts cultivars in south Texas and the High Plains area. Peanut stunting was based on a scale of 0 to 100 (0 = no peanut stunting) to 100 (peanut death). Peanut yield was determined by digging the pods based on maturity of non-treated control plots, air-drying in the field for 6 to 10 d, and harvesting individual plots with a combine. Yield samples were cleaned and adjusted to 10% moisture. For grades, a 200 g pod sample from each plot was obtained

TABLE 1: Rainfall amounts at Lamesa and Yoakum during the 2009 and 2010 growing seasons.

Month	Lamesa			Yoakum		
	2009	2010	60 yr avg	2009	2010	30 yr avg
	mm					
April	12.7	88.9	35.8	222.5	36.1	81.0
May	35.1	35.1	58.4	16.3	118.4	110.2
June	158.0	72.4	77.2	3.8	95.0	105.9
July	125.5	199.4	48.5	5.3	200.7	71.6
August	12.7	0	48.5	42.7	89.4	75.4
September	23.1	134.6	63.8	114.1	223.3	102.6
October	22.4	7.6	49.0	352.6	0	83.8
Total	389.5	538.0	381.2	757.3	762.9	630.5

and grades determined following procedures described by the Federal-State Inspection Service [30].

2.5. Data Analysis. Data for percentage of peanut injury and stunting were transformed to the arcsine square root prior to analysis; however, nontransformed means are presented because arscine transformation did not affect interpretation of this data. Data were subjected to ANOVA and analyzed using SAS PROC MIXED with locations and years designated as random effects in the model [31]. Treatment means were separated using Fisher's Protected LSD at $P \leq 0.05$. The nontreated control was used for peanut yield and grade calculation comparison and a visual comparison for peanut injury and was only included in yield and grade analysis.

3. Results

Peanut was planted at Yoakum in 2009 and peanut growth response was observed and recorded. However, extremely wet conditions persisted for approximately 4 weeks after the peanuts were dug which prevented entry into the field in time to harvest in a timely manner; therefore, yield data was collected only in 2010. Yield data was collected at Lamesa in both years.

3.1. Peanut Stunting. No peanut stunting was observed at Yoakum in 2009 (data not shown). There was a herbicide treatment by peanut cultivar interaction at Lamesa in 2009 (Table 2) while at Lamesa and Yoakum in 2010, peanut cultivar and herbicide treatment were significant (Table 3). In 2009 at Lamesa, flumioxazin at 0.053 kg ha^{-1} caused stunting only on Tamrun OL02 (5%) while flumioxazin at 0.107 kg ha^{-1} caused stunting on Tamrun OL02 and Tamrun OL07 (5 to 6%) and flumioxazin at 0.214 kg ha^{-1} caused at least 5% stunting on all four peanut cultivars (Table 2). Also, imazapic at 0.035 kg ha^{-1} caused at least 5% stunting on Tamrun OL02 and Tamrun OL07 while imazapic at 0.07 kg ha^{-1} (6 to 10%) and 0.141 kg ha^{-1} (13 to 18%) stunted all four peanut cultivars.

In 2010 herbicide and rate did have an effect on peanut growth at both locations. At Yoakum, Tamrun OL01 and

Tamrun OL07 responded similarly to both herbicides while at Lamesa all peanut cultivars resulted in stunting from the application of flumioxazin and imazapic (Table 3). Stunting was greatest with Tamrun OL02 (9%) while FlavorRunner 458 showed the least stunting (6%). At Yoakum, stunting was greatest (7%) with imazapic at 0.141 kg ha^{-1} with all other rates of imazapic and all flumioxazin rates showing 3% or less stunting (Table 3). At Lamesa, all rates of flumioxazin, with the exception of flumioxazin at 0.053 kg ha^{-1}, resulted in at least 6% peanut stunting while all rates of imazapic resulted in at least 4% peanut stunting. With both herbicides, peanut stunting increased as the herbicide rate increased.

Previous research has reported a reduction in peanut canopy development in some instances when flumioxazin has been used followed by cool, wet conditions [1, 32]. Vencill [33] found that if peanut germinated in above normal concentrations of flumioxazin due to soils with little subsoil moisture receiving water, injury occurred. Conversely, the use of imazapic has also been mentioned as having an effect on peanut canopy [23]. However, earlier work with Runner, Spanish, and Virginia peanut market types reported, regardless of application timing (ground cracking to 56 d after ground cracking), that imazapic at 0.07 kg ha^{-1} did not affect canopy height or width [20, 23].

3.2. Peanut Yield. At Lamesa, there was a herbicide and rate effect in 2009 while in 2010 there was a cultivar response but no herbicide and rate effect. In 2010 at Yoakum, there was no peanut cultivar or herbicide and rate response (Table 4).

In 2009 at Lamesa, no difference in peanut yield was noted between the non-treated control and any flumioxazin rate while all imazapic rates reduced yield when compared with the non-treated control (Table 4). In 2010 at Yoakum, no peanut cultivar or herbicide and rate effect were noted and at Lamesa there was only a peanut cultivar response. Both Tamrun OL01 and Tamrun OL07 resulted in lower yields than FlavorRunner 458 (Table 4).

4. Discussion

Other studies have reported that flumioxazin had no influence on the yield of various peanut cultivars [17, 32] although cool, wet conditions may result in some peanut stunting [1, 17, 34]. Main et al. [32] reported that yields of Georgia Green, C-99R, or MDR-98 were not influenced by flumioxazin applied at the US labeled rate. In another study, Wilcut et al. [34] showed that flumioxazin did not affect the yield of eight Virginia-type cultivars.

Imazapic has had no effect on peanut cultivar yield in other studies [20, 34]. Also, in weed efficacy studies, no reductions in peanut grade or yield have been observed with imazapic [11, 19, 21, 22]. However, the reduced yields with imazapic observed at Lamesa in 2009 may be attributed to several factors. Most peanut soils in south Texas have a pH of 6.5 to 7.5 and organic matter contents less than 1%. Therefore, in these soils, imidazolinone herbicides are readily available for microbial degradation [35]. However, in the Texas High Plains, the pH may range from 7.0 to 8.5 resulting in reduced

TABLE 2: Response of peanut cultivars to flumioxazin and imazapic in 2009 at Lamesa when rated 4 weeks after postemergence application.

Treatment	Rate	Application timing[a]	Stunting			
			FL 458[b]	OL01	OL02	OL07
	Kg ai ha^{-1}		%			
Untreated	—	—	0	0	0	0
Flumioxazin	0.053	PRE	0	0	5	0
Flumioxazin	0.107	PRE	2	2	5	6
Flumioxazin	0.214	PRE	7	6	5	12
Imazapic	0.035	POST	0	3	5	7
Imazapic	0.071	POST	6	10	7	6
Imazapic	0.141	POST	17	13	18	15
LSD (0.05)				4		

[a]Application timing: PRE: preemergence; POST: postemergence.
[b]Peanut cultivars: FL 458: FlavorRunner 458; OL01: Tamrun OL01; OL02: Tamrun OL02; OL07: Tamrun OL07.

TABLE 3: Response of peanut cultivars to flumioxazin and imazapic in 2010 at Yoakum and Lamesa when rated 4 wks after postemergence application.

	Yoakum	Lamesa
	Stunting (%)	
Cultivar		
FlavorRunner 458	—	6
Tamrun OL01	2	8
Tamrun OL02	—	9
Tamrun OL07	1	8
LSD (0.05)	NS	1
Herbicide and rate		
non-treated	0	0
Flumioxazin 0.053 kg ha^{-1}	1	2
Flumioxazin 0.107 kg ha^{-1}	1	6
Flumioxazin 0.214 kg ha^{-1}	3	12
Imazapic 0.035 kg ha^{-1}	2	4
Imazapic 0.071 kg ha^{-1}	2	12
Imazapic 0.141 kg ha^{-1}	7	19
LSD (0.05)	4	3

TABLE 4: Effect of peanut cultivar and herbicide on yield.

Treatment	2009	2010	
	Lamesa	Yoakum	Lamesa
	Kg ha^{-1}		
Cultivar			
FlavorRunner 458	3694	—	7252
Tamrun OL01	3430	2977	6352
Tamrun OL02	3392	—	6889
Tamrun OL07	3110	2993	6369
LSD (0.05)	NS	NS	761
Herbicide and rate			
non-treated	3660	3110	6710
Flumioxazin 0.053 kg ha^{-1}	3300	3110	6950
Flumioxazin 0.107 kg ha^{-1}	3600	3160	6630
Flumioxazin 0.214 kg ha^{-1}	3660	3100	6700
Imazapic 0.035 kg ha^{-1}	3260	3120	6750
Imazapic 0.071 kg ha^{-1}	3250	3130	6470
Imazapic 0.141 kg ha^{-1}	3120	2960	6810
LSD (0.05)	370	NS	NS

microbial degradation. With soils low in organic matter and near neutral pH, little of the imidazolinone herbicide should be adsorbed on soil particles [35]. Therefore, under certain conditions, more imazapic may have been adsorbed by the peanut cultivars themselves (author's personal opinion).

Conflict of Interests

The authors declare that they have no conflict of interests.

Acknowledgments

The authors wish to thank the Texas Peanut Producers Board for helping to fund this research. They also thank Kevin Brewer, Dwayne Drozd, Lyndell Gilbert, and Bill Klesel for their technical assistance.

References

[1] W. J. Grichar, B. A. Besler, P. A. Dotray, W. C. Johnson III, and E. P. Prostko, "Interaction of flumioxazin with dimethenamid or metolachlor in peanut (*Arachis hypogaea* L.)," *Peanut Science*, vol. 31, no. 1, pp. 12–16, 2004.

[2] R. M. Yoshida, R. Sakaki, R. Sato et al., "S-53482-a new N-phenyl phthalimide herbicide," in *Proceedings Brighton Crop Protection Conference*, vol. 1, pp. 69–75, Weeds, 1991.

[3] R. J. Anderson, A. E. Norris, and F. D. Hess, "Synthetic organic chemicals that act through the prophyrin pathway," in *Porphyric Pesticides: Chemistry, Toxicity, and Pharmaceutical Applications*, S. O. Duke and C. A. Rebeiz, Eds., ACS Symposium Series 559,

pp. 18–33, American Chemical Society, Washington, DC, USA, 1994.

[4] S. A. Senseman, *Herbicide Handbook*, Weed Science Society of America, Lawrence, Kan, USA, 9th edition, 2007.

[5] J. W. Wilcut, "Summary of flumioxazin performance in southeastern peanuts," in *Proceedings of the 50th Annual Meeting of the Southern Weed Science Society (SWSS '97)*, vol. 50, pp. 1–7, Southern Weed Science Society, 1997.

[6] W. J. Grichar and A. E. Colburn, "Flumioxazin for weed control in Texas peanuts (*Arachis hypogaea* L.)," *Peanut Science*, vol. 23, no. 1, pp. 30–36, 1996.

[7] P. R. Nester and W. J. Grichar, "Cadre combinations for broadleaf weeds control in peanut," in *Proceedings of the Annual Meeting of the Southern Weed Science Society (SWSS '93)*, vol. 46, 1993.

[8] W. J. Grichar, A. E. Colburn, and P. R. Nester, "Weed control in Texas peanut with Cadre," in *Proceedings of the American Peanut Research and Education Society*, vol. 26, p. 70, 1994.

[9] J. W. Wilcut, E. F. Eastin, J. S. Richburg III, W. K. Vercill, F. R. Wells, and G. Wiley, "Imidazolinone systems for southern weed management in resistant corn," *Weed Science Society America*, vol. 33, article 5, 1993.

[10] J. W. Wilcut, J. S. Richburg III, G. Wiley, F. R. Walls Jr., S. R. Jones, and M. J. Iverson, "Imidazolinone herbicide systems for peanut (*Arachis hypogaea* L.)," *Peanut Science*, vol. 21, no. 1, pp. 23–28, 1994.

[11] J. W. Wilcut, A. C. York, W. J. Grichar, and G. R. Wehtje, "The biology and management of weeds in peanut (*Arachis hypogaea*)," in *Advances in Peanut Science*, H. E. Pattee and H. T. Stalker, Eds., pp. 207–244, American Peanut Research Education Society, Stillwater, Okla, USA, 1995.

[12] W. J. Grichar and P. R. Nester, "Nutsedge (*Cyperus* spp.) control in peanut (*Arachis hypogaea*) with AC 263,222 and imazethapyr," *Weed Technology*, vol. 11, no. 4, pp. 714–719, 1997.

[13] T. L. Grey, D. C. Bridges, E. F. Eastin et al., "Residual weed control for peanut (*Arachis hypogaea*) with imazapic, diclosulam, flumioxazin, and sulfentrazone in Alabama, Georgia, and Florida: a multi-state and year summary," in *Proceedings of the Meeting of the American Peanut Research and Education Society*, vol. 33, p. 19, Oklahoma, Okla, USA, 2001.

[14] J. S. Richburg III, J. W. Wilcut, and G. R. Wehtje, "Toxicity of AC 263,222 to purple (*Cyperus rotundus*) and yellow nutsedge (*C. esculentus*)," *Weed Science*, vol. 42, no. 3, pp. 398–402, 1994.

[15] J. S. Richburg III, J. W. Wilcut, D. L. Colvin, and G. R. Wiley, "Weed management in southeastern peanut (*Arachis hypogaea*) with AC 263,222," *Weed Technology*, vol. 10, no. 1, pp. 145–152, 1996.

[16] J. T. Ducar, S. B. Clewis, J. W. Wilcut et al., "Weed management using reduced rate combinations of diclosulam, flumioxazin, and imazapic in peanut," *Weed Technology*, vol. 23, no. 2, pp. 236–242, 2009.

[17] S. D. Askew, J. W. Wilcut, and J. R. Cranmer, "Weed management in peanut (*Arachis hypogaea*) with flumioxazin preemergence," *Weed Technology*, vol. 13, no. 3, pp. 594–598, 1999.

[18] G. H. Scott, S. D. Askew, and J. W. Wilcut, "Economic evaluation of diclosulam and flumioxazin systems in peanut (*Arachis hypogaea*)," *Weed Technology*, vol. 15, no. 2, pp. 360–364, 2001.

[19] J. W. Wilcut, J. S. Richburg III, G. L. Wiley, and F. R. Walls Jr., "Postemergence AC 263,222 systems for weed control in peanut (*Arachis hypogaea*)," *Weed Science*, vol. 44, no. 3, pp. 615–621, 1996.

[20] J. S. Richburg III, J. W. Wilcut, A. K. Culbreath, and C. K. Kvien, "Response of eight peanut (*Arachis hypogaea* L.) cultivars to the herbicide AC 263,222," *Peanut Science*, vol. 22, no. 1, pp. 76–80, 1995.

[21] W. J. Grichar, "Control of palmer amaranth (*Amaranthus palmeri*) in peanut (*Arachis hypogaea*) with postemergence herbicides," *Weed Technology*, vol. 11, no. 4, pp. 739–743, 1997.

[22] T. M. Webster, J. W. Wilcut, and H. D. Coble, "Influence of AC 263,222 rate and application method on weed management in peanut (*Arachis hypogaea*)," *Weed Technology*, vol. 11, no. 3, pp. 520–526, 1997.

[23] P. A. Dotray, T. A. Baughman, J. W. Keeling, W. J. Grichar, and R. G. Lemon, "Effect of imazapic application timing on texas peanut (*Arachis hypogaea*)," *Weed Technology*, vol. 15, no. 1, pp. 26–29, 2001.

[24] C. E. Simpson, M. R. Baring, A. M. Schubert et al., "Registration of 'Tamrun OL01' peanut," *Crop Science*, vol. 43, article 2298, 2003.

[25] M. R. Baring, C. E. Simpson, M. D. Burow et al., "Registration of 'Tamrun OL07' peanut," *Crop Science*, vol. 46, no. 6, pp. 2721–2722, 2006.

[26] J. Beasley and J. Baldwin, "Peanut cultivars and descriptions," March 2011, http://www.caes.uga.edu/commodities/fieldcrops/peanuts/production/cultivardescription.html.

[27] C. E. Simpson, M. R. Baring, A. M. Schubert, M. C. Black, H. A. Melouk, and Y. Lopez, "Registration of 'Tamrun OL 02' peanut," *Crop Science*, vol. 46, pp. 1813–1814, 2006.

[28] Anonymous, *Label and MSDS for Valor Herbicide*, EPA Registration Number 59639-99, Valent USA Corporation, Walnut Creek, Calif, USA, 2011.

[29] Anonymous, *Imazapic Herbicide Label*, EPA Registration Number 241-381, BASF, Research Triangle Park, NC, USA, 2011.

[30] USDA, *Farmers Stock Peanut Inspection Instructions*, Agricultural Marketing Service, Washington, DC, USA, 2008.

[31] Statistical Analysis Systems, *SAS User's Guide, version 9.1*, SAS Institute, Cary, NC, USA, 2005.

[32] C. L. Main, J. T. Ducar, E. B. Whitty, and G. E. MacDonald, "Response of three runner-type peanut cultivars to flumioxazin," *Weed Technology*, vol. 17, no. 1, pp. 89–93, 2003.

[33] W. K. Vencill, "Flumioxazin injury to peanut," in *Proceedings of the Annual Meeting of the Southern Weed Science Society (SWSS '02)*, vol. 55, pp. 195–196, 2002.

[34] J. W. Wilcut, S. D. Askew, W. A. Bailey, J. F. Spears, and T. G. Isleib, "Virginia market-type peanut (*Arachis hypogaea*) cultivar tolerance and yield response to flumioxazin preemergence," *Weed Technology*, vol. 15, no. 1, pp. 137–140, 2001.

[35] G. Mangels, "Behavior of the imidazolinone herbicide in soil—a review of the literature," in *The Imidazolinone Herbicides*, D. L. Shaner and S. L. O'Connor, Eds., pp. 191–209, CRC Press, Boca Raton, Fla, USA, 1991.

23

Changes in Protein, Nonnutritional Factors, and Antioxidant Capacity during Germination of *L. campestris* Seeds

C. Jiménez Martínez,[1] **A. Cardador Martínez,**[2] **A. L. Martinez Ayala,**[3] **M. Muzquiz,**[4] **M. Martin Pedrosa,**[4] **and G. Dávila-Ortiz**[1]

[1] *Departamento de Graduados e Investigación en Alimentos, Escuela Nacional de Ciencias Biológicas,*
 Instituto Politécnico Nacional, Prol de Carpio y Plan de Ayala, 11340 México, DF, Mexico
[2] *Biotecnología Alimentaria, Instituto Tecnológico y de Estudios Superiores de Monterrey, Campus Querétaro,*
 Avenida Epigmenio González 500, Fraccionamiento San Pablo, 76130 Santiago de Querétaro, QRO, Mexico
[3] *Centro de Investigación en Biotecnología Aplicada, IPN, Carretera Estatal Tecuexcomac-Tepetitla Km 1.5,*
 90700 Tepetitla de Lardizábal, TLAX, Mexico
[4] *Departamento de Tecnología de Alimentos, INIA, Apdo 8111, 28080 Madrid, Spain*

Correspondence should be addressed to C. Jiménez Martínez, crisjm_99@yahoo.com and G. Dávila-Ortiz, gdavilao@yahoo.com

Academic Editor: Antonio M. De Ron

The changes in SDS-PAGE proteins patterns, oligosaccharides and phenolic compounds of *L. campestris* seeds, were evaluated during nine germination days. SDS-PAGE pattern showed 12 bands in the original protein seeds, while in the samples after 1–9 germination days, the proteins located in the range of 28–49 and 49–80 kDa indicated an important reduction, and there was an increase in bands about 27 kDa. On the other hand, oligosaccharides showed more than 50% of decrease in its total concentration after 4 germination days; nevertheless after the fifth day, the oligosaccharides concentration increases and rises more than 30% of the original concentration. Phenolic compounds increased their concentration since the first germination day reaching until 450% more than the original seed level. The obtained results are related with liberation or increase of phenolic compounds with antioxidant properties, allowing us to suggest that the germination would be used to produce legume foods for human consumption with better nutraceutical properties.

1. Introduction

Legume seeds are important staple foods, particularly in developing countries, due to their relatively low cost, long conservation time, and high nutritional value; among these meals it is *Lupinus* seeds and their derivatives. This legume is one of the richest sources of vegetable protein, and although the protein content and amino acid profile vary between species, the intraspecies variability is low. In 2009, the FAOST reported that the area harvested was 662712 Ha, and *L. albus* and *L. angustifolius* were the most widely used. About 100 wild species have been reported throughout México [1]. These wild lupins have not been exploited at a commercial level. For this reason, in the present work we consider them as potential providers of vegetable proteins for human consumption. *Lupinus campestris* seed, like other

Lupinus species, has high protein content (44%) [1, 2]. Lupin seeds offer some advantages in comparison with soy bean, since it contains only small amounts of trypsin inhibitors, tannins, phytates, saponins, α-galactosides, and so forth [3, 4]. However, a limitation for the wider use of lupins has been their high content of quinolizidine alkaloids [5, 6] as well as condensed tannins [7, 8]. Consequently, it is desirable to develop transformation processes which could improve the nutritional quality of legumes and also provide new derived products for the consumers. Germination is considered a potentially beneficial process for legume seed transformation which may decrease undesirable components such as alkaloids and phytates [9], and during germination, some grade of transformation of alkaloids to other more bioactive compounds, such as esters, occurs [7]. Cuadra et al. [3] and De Cortes-Sánchez et al. [7] found a slight increase in alkaloids during germination of *L. albus, L. angustifolius*, and *L. campestris*, and no α-pyridone alkaloids, such as the highly toxic anagyrine and cytosine, were detected in any of these species. Germination also increases nutrients such as vitamin C [10] and increase protein digestibility [11], consequently improving nutritional quality. Additional advantages of germination are reduction in cooking time and improvement of the product sensorial attributes [11]. Germination has been shown to decrease the level of α-galactosides of different legume seeds including soybean, black bean, and lupin seed, with the corresponding decrease in carbohydrates available for fermentation in the large human intestine. The content of trypsin inhibitors and phytates is also decreased, but considerable amounts of these factors are still present after germination [6]. On the other hand, it is widely accepted that antioxidant activity of food is related to high phenolic content. Phenolic compounds are capable of removing free radicals, chelating metal catalysts, activating antioxidant enzymes, and inhibiting oxidases [12]. Legume seeds are a rich source of many substances with antioxidant properties, including plant phenolics. *Lupinus* is a potential source of bioactive components with antioxidant activities. Although the interest in *Lupinus* species as a valuable component of functional food is increasing and has let to investigate on the determination of antioxidant activity in *Lupinus* seeds and its products, the information is scarce [13, 14]. The objective of this work was to evaluate the original content of proteins, oligosaccharides, and phenolic compounds, the antioxidant capacity in *Lupinus campestris* seed, and the changes of these parameters during the germination process.

2. Material and Methods

2.1. Samples and Germination Process.

L. campestris seeds (wild type) were collected along 50 km of the Oaxtepec-Xochimilco highway in the Morelos State, México.

Germination process was performed as described by De Cortes-Sánchez et al. [7]. Briefly, 800 *Lupinus campestris* seeds were used for the germination assay distributed in 10 trays, with 80 seeds each one. The seeds were spread on a moist sheet of filter paper (Albet 1516, 42–52 cm) and covered with another sheet of moist filter paper. They were put into a germination chamber under environmentally controlled conditions: 20°C, 8 h of light per day exposure, and watering of the seeds during germination keeps the paper always wet. Samples (80 seeds/tray) were taken at 0 (control), 1, 2, 3, 4, 5, 6, 7, 8, and 9 germination days. The germination process was repeated twice, and the germination capacity was evaluated by germination percentage and seed weights. Samples for analysis were constituted by germinated and moist seeds, discarding those that did not show any water absorption during the process. The germinated seeds were freeze-dried, milled, and passed through a sieve of 0.5 mm. The germinated flour was stored in darkness in a desiccator at 4°C until analysis.

2.2. Gel Electrophoresis.

Denaturing gel electrophoresis (SDS-PAGE) was carried out according to the method of Schagger and von Jagow [15] using 10% polyacrylamide gels in the presence of 1% SDS; the proteins (1 μg) were loaded with or without β-mercaptoethanol. Standards used were phosphorylase B (97.4 kDa), bovine serum albumin (66 kDa), ovalbumin (45 kDa), carbonic anhydrase (31 kDa), soybean trypsin inhibitor (20.1 kDa), and lysozyme (14.4 kDa).

2.3. Extraction and Quantification of Carbohydrates (CH).

The method of Muzquiz et al. [16] was used for CH extraction. 0.1 g of grounded seeds was homogenized with aqueous ethanol solution (50% v/v, 5 mL) for 1 min at 4°C. Then the mixture was centrifuged for 5 min (2100 ×g) at 4°C, and the supernatant was recovered. The procedure was repeated twice, and the combined supernatants were concentrated under vacuum at 35°C. The concentrated supernatant was dissolved in deionized water (1 mL) and passed through a Waters minicolumn (Waters C-18 at 500 mg/cc) with Supelco vacuum system.

Samples (20 μL) were analyzed using a Beckman HPLC chromatograph f156 with refraction index detector. A Spherisob-5-NH₂ column (250 × 4.6 mm id) was used with acetonitrile: water (65:35, v/v) as the mobile phase at a flow rate of 1 mL min⁻¹. Individual sugars were quantified by comparison with standards of sucrose, raffinose, stachyose, and verbascose. Calibration curves were prepared for all these sugars, and a linear response was obtained for the range of 0–5 mg/mL with a determination coefficient (r^2) > 0.99.

2.4. Extraction and Quantification of Phenolic Compounds (PC).

1 g of sample was extracted with 10 mL methanol previous to phenolic determination. Total phenols content was estimated by using the Folin-Ciocalteu colorimetric method [17]. Briefly, the 0.02 mL of the extracts was oxidized with 0.1 mL of 0.5 N Folin-Ciocalteu reagent, and then the reaction was neutralized with 0.3 mL sodium carbonate solution (20%). The absorbance values were obtained by the resulting blue color measured at 760 nm with a Beckman spectrophotometer (California, USA) model DU-65 after incubation for 2 h at 25°C. Quantification was done on the basis of a standard curve of gallic acid. Results were expressed as mg of gallic acid equivalent per 1 g of dry weight.

2.5. TLC Analysis of Phenolic Compounds.

TLC was performed on TLC sheets coated with 0.25 mm layers of silica gel 60 F254 (E. Merck, number 5554). Two mobile phases were used: ethyl acetate-formic acid-ethanol (65:15:20, v/v/v) and 1-butanol-acetic acid-water (7:0.5:2.5, v/v/v), upper phase. The chromatograms were evaluated in UV light at 360 nm before spraying them with 10% sulphuric acid [18].

2.6. HPLC Analysis of Phenolic Compounds.

HPLC analysis was performed on an Agilent Technologies 1200 series liquid chromatograph (G1311A quaternary pump, UV-VIS DAD G1315D detector, ALS G1329A injector, G1322A Deggaser, and TCC G1316A thermostat column), equipped with a Zorbax Eclipse XDB-C18 column (150 × 4.6 mm, 5 mm particle size) (Agilent Technologies, USA), and thermostated at 30°C. A gradient elution was used to separate the extracted phenolics. Solvent (A) was 5.0% formic acid in water, and solvent (B) was acetonitrile. Elution was performed at a solvent flow rate of 1.0 mL/min. The gradient profile of the system was 0% solvent B at the initial stage, 0% solvent B at 3 min, 30% solvent B at 5 min, 60% solvent B at 20 min, 100% solvent B at 25 min, and 0% solvent B at 30 to 35 min.

The eluted phenolic compounds were monitored at 280 nm. Quantitative levels were determined by comparing with a catechin standard curve. Phenolic concentration was expressed as mg catechin equivalent per gram of dry sample.

2.7. Free Radical DPPH Scavenging Capacity. 2,2-diphenyl-1-picrylhydrazyl (DPPH) is a free radical used for assessing antioxidant activity. Reduction of DPPH by an antioxidant or by a radical species results in a loss of absorbance at 515 nm. PC extracts were adjusted at a concentration of 0.24 mg gallic acid equivalent/mL prior to antioxidant capacity evaluation. Determination of antioxidant capacity, previously adapted for microplates [19], was performed as follows: 0.02 mL of extract (500 μM gallic acid equivalent) or standard (gallic acid, 500 μM) was added to a 96-well flat-bottom plates containing 0.2 mL of DPPH solution (125 μM DPPH in 80% methanol). Samples were prepared in triplicate. The plate was covered, left in the dark at room temperature, and read after 90 min in a visible-UV microplate reader (680 XR Microplate Reader, Bio-Rad Laboratories, Inc) using a 520 nm filter. Data are expressed as a percentage of DPPH· discoloration [20].

2.8. Statistical Analysis. All analyses were carried out in triplicate, and the report data are the average of the results and the standard error in each case.

3. Results

3.1. Germination. In Figure 1 it is shown the germination capacity expressed as percent of germinated seeds. This germination percentage increases from day 1 to day 4, and after that, no significant increase in germination is observed.

The gain in weight is observed in Figure 1, an increase in weight can be observed since the first day, this weight augmentation was due mainly to the water that has shrunk, and the germination percentage was 5% only. The total increase in weight was three times plus from the initial weight of the seed. By the second day, the germination and the weight have increased to 27% and 14 g/80 seeds (Figure 1), and additionally root has left the head. The greatest increment in the number of germinated seeds (from 27% to 82%) is observed between the second and third day. At the fourth day, 98% of the seeds showed development of the stem and root. A maximum germination percentage of 100% was obtained, which shows the good viability of *L. campestris*. These results agree with De Cuadra et al. [3] who reported a high degree of germination (up to 100%) for two *Lupinus* species.

3.2. Electrophoretic Analysis. Figure 2 shows the electrophoretic profile of *L. campestris* seeds subjected to different germination times. As it is observed the seed without any germination time showed greater amount of protein bands located between 20 and 75 kDa. As the germination time

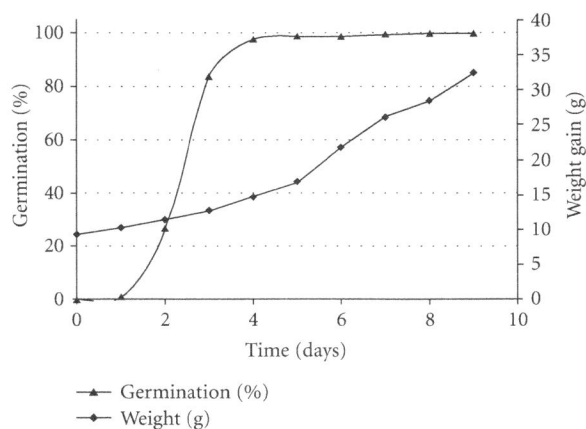

FIGURE 1: Germination (%) and weight gain (g) of the *L. campestris* seeds. The results represent the average of three independent experiments ± S. E.

advances, the proteins located in the range of 28–49 and 49–75 kDa almost disappeared after nine germination days of Lupinus seed, and there was an increase in bands about 27 kDa. These results confirm previous findings about storage proteins, which are hydrolysed and mobilised after germination [21, 22]. This behavior lets us to suggest that the principal storage protein molecules, the globulins 7 s, and 11 s constituted by three and six subunits, respectively, were hydrolyzed in lower molecular weight compounds which has a best digestibility and consequently a better biological value.

3.3. Changes of Oligosaccharides in L. campestris Seeds during the Germination. The *L. campestris* germinated seeds were also evaluated as for the variation of present oligosaccharides. The obtained results are showed in Table 1. The concentration of total oligosaccharides in the seed without germinating was of 90.26 mg/g; this concentration was diminished near 15% in the first day, and then 25, 46, and 58% in the period were comprised since the second to the fourth germination day. Then oligosaccharide concentration increased its value from the fifth to the ninth day, reaching 30% above than the original content. The composition of oligosaccharides varies during the germination process. In the seed without treatment, the sucrose was present with an initial content of 21.45 mg/g increasing to 55.36 mg/g at five days of germination. Since the sixth day, sucrose diminished until reaching 14.84 mg/g of seed at the nine day of germination. This increase in sucrose concentration can be due to hydrolysis of oligosaccharides by the α-galactosidase enzyme, which selectively acts on the galactosides such as raffinose, stachyose, and verbascose releasing sucrose [23]. Muzquiz et al. [16] has reported a similar behavior in other species of *Lupinus*. After the fourth day of germination, the oligosaccharides proportion has increased substantially, mainly in the stachyose percentage, which is almost twice of the originally presented. Even though there is a substantial reduction of these carbohydrates, they are not totally eliminated, since it has been informed for other species of *Lupinus* whose diminution is bigger than 80–100% after four days

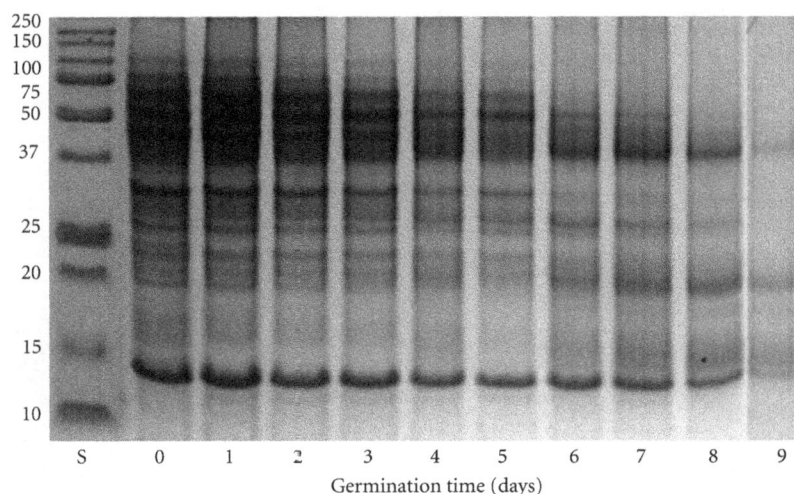

FIGURE 2: Sodium dodecyl sulfate polyacrylamide gel electrophoresis (SDS-PAGE) of present proteins in the *L. campestris* germinated seed. S = standards (kDa).

TABLE 1: Behavior of carbohydrates of the *L. campestris* germinated seed (mg/g of seed).

Time (days)	Sucrose	Raffinose	Stachyose	Verbascose	Total oligosaccharides
0	21.45 ± 0.76	13.65 ± 0.89	57.16 ± 0.95	19.45 ± 0.59	90.26 ± 0.73
1	26.49 ± 3.86	8.51 ± 0.51	54.35 ± 3.38	11.48 ± 2.14	74.34 ± 1.17
2	32.21 ± 1.80	7.92 ± 0.51	49.10 ± 0.51	10.94 ± 0.55	67.96 ± 0.02
3	34.38 ± 1.37	6.49 ± 0.68	36.75 ± 1.06	5.53 ± 1.06	48.77 ± 0.03
4	46.14 ± 0.34	4.03 ± 1.03	29.99 ± 0.13	6.78 ± 0.12	40.80 ± 0.07
5	55.36 ± 5.15	7.08 ± 1.85	47.95 ± 0.69	8.44 ± 0.69	63.47 ± 0.65
6	35.09 ± 1.65	8.18 ± 2.23	65.17 ± 0.12	9.30 ± 0.12	82.65 ± 0.32
7	27.85 ± 1.45	10.19 ± 1.01	84.75 ± 0.57	9.64 ± 0.58	104.58 ± 0.13
8	22.95 ± 1.57	11.38 ± 0.10	93.09 ± 0.24	10.82 ± 0.24	115.29 ± 0.02
9	14.84 ± 3.76	11.34 ± 4.23	98.17 ± 2.53	11.93 ± 0.23	121.44 ± 0.30

*The values represent the average of two separated germinations with extractions made by triplicate ± S.E.

of germination [3, 24]. This difference can be due to the germination conditions in which the seeds were carried out. With the obtained results it is observed that the germination diminishes the concentration of oligosaccharides, being the lower value in the fourth day.

3.4. Phenolic Compounds in Seed of L. campestris during the Germination.
Total phenolic compounds concentrations in the germinated *L. campestris* seed during nine germination days are presented in Figure 3. Control seed, without germination, presented 5.27 mg gallic acid equivalent per g of seed. This value remained almost constant during the days one and two. After that, the concentration of total phenolics increased gradually reaching twice the original value. However, phenolic content was in the range reported for other legumes such as yellow pea, green pea, lentils, common beans, and soybean [25] and similar to the content in other *Lupinus* species [26]. Contrary to the behavior of the oligosaccharides, the phenolic compounds increase as the time of germination occurs. The behavior shown for *Lupinus* germinated seeds differs from the observed by Muzquiz et al. [27] who indicated a reduction of 76% in phenolics from lentil but is similar to

Cajanus seed [28] which showed a fivefold increment in total phenolic content during a period of five germination days.

3.5. Phenolic Compounds by HPLC.
Although individual phenolics remain unidentified, they were quantified on the basis of a catechin calibration curve. In Table 2, the changes in composition and quantity of phenolics as determined by HPLC are shown.

There are two main groups of peaks. The first one which is presented since cero day increases at seven and eight days and decrease a little in the ninth day. This group is formed by peaks marked as peak 1 to 6 and peak number 10. The other group of peaks appears between the third and fourth day of germination, increases its concentration at the same days the other group does, and also diminished at nine day. This group is formed by peaks number 7–9 and 11–15 (Table 2). As the time of germination occurs, the complexity and quantity of total phenolics increase, being the germinated seeds on seven to nine day more complex in composition than no germinated seeds or those on the first germination days. Also, seeds in the seventh day are the richest in total phenolic composition.

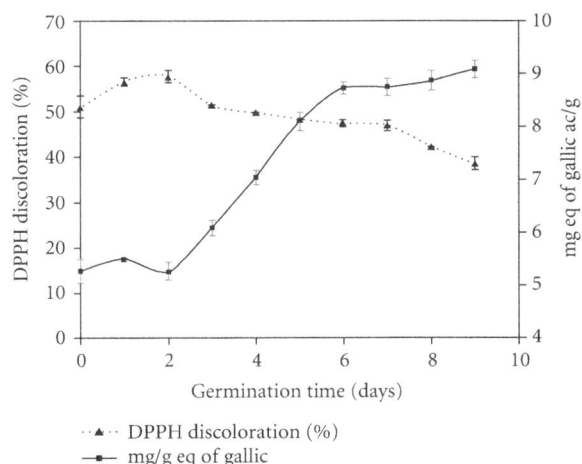

FIGURE 3: Phenolic compound variation and antioxidant capacity in *L. campestris* germinated seed.

Concentration of total phenolics by HPLC is lower than the obtained by the Folin-Ciocalteu method. Considering the heterogeneity of natural phenols and the possibility of interference from other readily oxidized substances such as ascorbic acid and mono- and disaccharides, this disagreement between methods is comprehensible [17, 29].

3.6. Antioxidant Capacity. All the extracts showed antioxidant capacity against DPPH-free radical, as measured by the decrease in absorbance at 520 nm. During the seeds germination, it was observed a light increase in antioxidant capacity nongemiated seeds until the second germination day (51–58%), followed by a continuing depression in antioxidant capacity until the ninth germination day (38%) (Figure 3). The initial antioxidant capacity (51%) is similar to that reported for other legumes [30]. An enhancement in antioxidant capacity by germination has been reported for *Lupinus albus* [14] and *Lupinus angustifolius* seeds [13] as measured in aqueous extracts. The *L. campestris* methanolic extracts showed a different behavior, which could be attributed to the kind of compounds that could be solubilized by methanol, and since water could solubilize other antioxidants such as vitamins. Fernandez-Orozco et al. [13] suggested that polyphenols extractability is better in buffer phosphate than in methanol. Correlations between antioxidant capacity toward DPPH-free radical and total polyphenols have been observed in beans [30] and in *L. angustifolius* germinated seeds [13]. In this study, polyphenol concentration did not correlate with antioxidant activity; while polyphenols increase as germination progress, antioxidant capacity decreases. It is interesting to note that polyphenol concentration was adjusted to 0.24 mg/mL in all samples, previously to antioxidant capacity determination. These results suggest again that it is composition but not concentration of polyphenols in the extracts, and possibly the presence of other antioxidants, which makes a difference in antioxidant capacity behavior. In order to confirm that composition affects antioxidant activity, the extracts were

analyzed by TLC. The best profile was obtained with ethyl acetate-formic acid-ethanol (65 : 15 : 20, v/v/v), which is shown in Figures 4(a) and 4(b). 360 nm UV light shows that there is a spot with an Rf value of 0.375; although this yellowish fluorescent spot is in all samples, its relative intensity is bigger at the last germination days. There is another spot (Rf 0.875) that is present in all germination days (Figure 4(a)). On the other hand, the extracts would contain flavonoids and phenolic acids due to the yellow and the blue fluorescent bands under 360 nm UV light [18]. The Figure 4(b) shows the TLC plate revealed with sulphuric acid. There is a group of three spots at the medium of the plate in all extracts (Rf values = 0.424, 0.515, and 0.606); however it has higher intensity around five–seven days, this intensity suggests higher concentration of phenolics, as all the samples were applied in the same volume (Figure 4(b)). Another group of phenolics is observed in 0.031, 0.156, and 0.219 Rf values. The behavior of this second group differs from the previous one; the spots can be visualized in the seed without germination, and at one and two days, later the group is disappeared in the next two days, increased its intensity in fifth day, and once again, decreased in the last germination days. Changes in composition of the phenolic extract were confirmed by HPLC analysis, as described previously (Table 2) germination process increase, the complexity of the phenolic extract (Figure 4). There is a group of compounds around 25 min that should appear as a consequence of germination. According to the HPLC analysis these compounds must be lower polar, suggesting that their antioxidant activity could be less than more polar compounds.

The antioxidant activity of phenolic compounds is affected by their chemical structure. Structure-activity relationships have been used as a theoretical method for predicting antioxidant activity. Polymeric polyphenols are more potent antioxidants than simple monomeric phenolics: Hagerman et al. [31] demonstrated the higher antioxidant ability of condensed and hydrolyzable tannins at quenching peroxyl radicals over simple phenols; Yamaguchi et al. [32] observed that the higher the polymerization degree of flavanols, the stronger the superoxide-scavenging activity. A similar effect was reported for the capacity to inhibit the O_2^- radical, which increased with the degree of procyanidin polymerization [33].

The antioxidant activity also depends on the type and polarity of the extracting solvent, the isolation procedures, purity of active compounds, the test system, and substrate to be protected by the antioxidant [34].

4. Conclusion

The germination is a simple technological process of easy application and low cost. This process allows to the protein modification, obtaining peptides of low molecular weight and improving the nutritional quality. The oligosaccharides ones show diminution in the third germination day, nevertheless tend to increase as of the fourth day of this one process. On the contrary, the phenolic compounds concentration increases from the first day. With this, we can

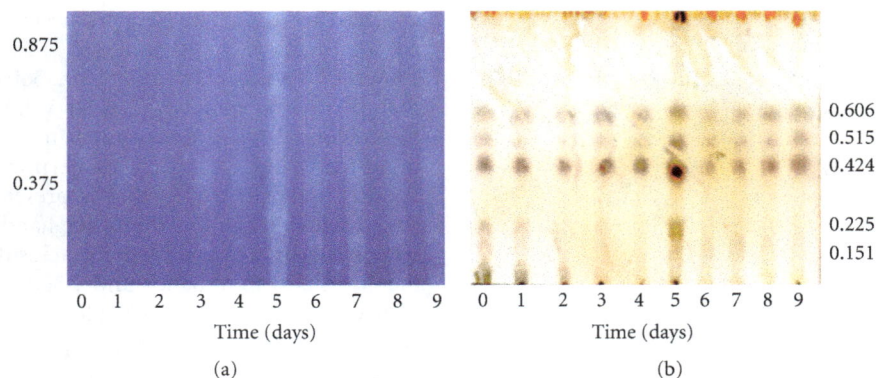

FIGURE 4: TLC analysis of phenolic compounds of *L. campestris* germinated seed eluted with a mixture of ethyl acetate-formic acid-ethanol (65 : 15 : 20, v/v/v) and revealed (a) UV light at 360 nm and (b) 10% sulphuric acid.

TABLE 2: Phenolic compounds in *L. campestris* germinated seeds by HPLC.

Peak number	Retention time (min)	Phenolic concentration, days of germination[1]									
		0	1	2	3	4	5	6	7	8	9
1	1.590	145.8	47.3	105.8	100.4	325.0	235.1	444.7	568.9	573.6	452.1
2	1.654	124.4	ND	76.4	118.4	238.9	252.1	347.0	499.2	498.0	408.9
3	1.764	362.8	ND	ND	ND	233.2	ND	231.0	306.8	310.7	272.5
4	7.454	524.7	133.6	149.7	108.3	321.6	137.2	357.7	474.2	482.8	431.3
5	8.941	665.0	ND	243.7	209.4	427.0	332.9	551.5	982.4	837.3	763.1
6	9.169	593.5	215.1	165.0	119.4	380.6	273.0	511.8	931.7	829.9	746.9
7	11.277	ND	ND	ND	ND	ND	111.4	ND	454.8	393.5	ND
8	12.038	ND	ND	ND	ND	ND	272.1	440.4	1270.5	1159.9	853.0
9	12.178	ND	ND	ND	ND	ND	162.2	257.3	696.4	785.6	561.1
10	12.626	395.7	64.5	98.8	ND	ND	134.7	288.5	451.1	400.7	341.3
11	14.455	ND	ND	ND	ND	ND	ND	214.6	354.0	308.4	265.1
12	15.953	ND	ND	ND	99.5	451.2	231.0	701.8	1157.9	668.6	452.8
13	25.195	ND	ND	ND	ND	212.4	87.0	318.9	262.2	253.2	209.4
14	25.361	ND	ND	ND	ND	85.4	ND	62.6	55.9	98.5	90.3
15	25.522	ND	ND	ND	ND	137.8	54.4	255.1	199.0	164.1	128.5
Total concentration		2811.9	460.5	839.5	755.4	2812.9	2282.9	4983.0	8665.1	7764.6	5976.3

[1] Phenolic concentration expressed as μg catechin equivalent per g of dry sample.
ND: not determined, under the detection limit.

conclude that it is necessary to control the time of germination to obtain an optimal concentration of nonnutritional factors to the third day.

Acknowledgments

The authors thank the Instituto Politécnico Nacional (IPN) and Consejo Nacional de Ciencia y Tecnología (CONACyT) through 33995 project for financial support.

References

[1] M. A. Ruiz and A. Sotelo, "Chemical composition, nutritive value, and toxicology evaluation of Mexican wild lupinst," *Journal of Agricultural and Food Chemistry*, vol. 49, no. 11, pp. 5336–5339, 2001.

[2] A. Sujak, A. Kotlarz, and W. Strobel, "Compositional and nutritional evaluation of several lupin seeds," *Food Chemistry*, vol. 98, no. 4, pp. 711–719, 2006.

[3] C. De la Cuadra, M. Muzquiz, C. Burbano et al., "Alkaloid, alpha-galactoside and phytic acid changes in germinating lupin seeds," *Journal of the Science of Food and Agriculture*, vol. 66, no. 3, pp. 357–364, 1994.

[4] M. A. Ruiz-López, P. M. García-López, H. Castañeda-Vazquez et al., "Chemical composition and antinutrient content of three lupinus species from jalisco, Mexico," *Journal of Food Composition and Analysis*, vol. 13, no. 3, pp. 193–199, 2000.

[5] D. Resta, G. Boschin, A. D'Agostina, and A. Arnoldi, "Evaluation of total quinolizidine alkaloids content in lupin flours, lupin-based ingredients, and foods," *Molecular Nutrition and Food Research*, vol. 52, no. 4, pp. 490–495, 2008.

[6] L. C. Trugo, L. A. Ramos, N. M. F. Trugo, and M. C. P. Souza, "Oligosaccharide composition and trypsin inhibitor activity of

P. vulgaris and the effect of germination on the α-galactoside composition and fermentation in the human colon," *Food Chemistry*, vol. 36, no. 1, pp. 53–61, 1990.

[7] M. De Cortes Sánchez, P. Altares, M. M. Pedrosa et al., "Alkaloid variation during germination in different lupin species," *Food Chemistry*, vol. 90, no. 3, pp. 347–355, 2005.

[8] C. Jiménez-Martínez, H. Hernández-Sánchez, G. Alvárez-Manilla, N. Robledo-Quintos, J. Martínez-Herrera, and G. Dávila-Ortiz, "Effect of aqueous and alkaline thermal treatments on chemical composition and oligosaccharide, alkaloid and tannin contents of *Lupinus campestris* seeds," *Journal of the Science of Food and Agriculture*, vol. 81, pp. 421–428, 2001.

[9] M. Muzquiz, M. Pedrosa, C. Cuadrado, G. Ayet, C. Burbano, and A. Brenes, "Variation of alkaloids, alkaloid esters, phytic acid, and phytase activity in germinated seed of Lupinus albus and *L. luteus*," in *Recent Advances of Research in Antinutritional Factors in Legume Seeds and Rapeseed*, A. M. Jansman, G. Hill, J. Huisman, and A. van der Poel, Eds., vol. 93, pp. 387–339, Wageningen Pers, Wageningen, The Netherlands, 1998.

[10] C. H. Riddoch, C. F. Mills, and G. G. Duthie, "An evaluation of germinating beans as a source of vitamin C in refugee foods," *European Journal of Clinical Nutrition*, vol. 52, no. 2, pp. 115–118, 1998.

[11] Y.-H. Kuo, P. Rozan, F. Lambein, J. Frias, and C. Vidal-Valverde, "Effects of different germination conditions on the contents of free protein and non-protein amino acids of commercial legumes," *Food Chemistry*, vol. 86, no. 4, pp. 537–545, 2004.

[12] K. E. Heim, A. R. Tagliaferro, and D. J. Bobilya, "Flavonoid antioxidants: chemistry, metabolism and structure-activity relationships," *Journal of Nutritional Biochemistry*, vol. 13, no. 10, pp. 572–584, 2002.

[13] R. Fernandez-Orozco, M. K. Piskula, H. Zielinski, H. Kozlowska, J. Frias, and C. Vidal-Valverde, "Germination as a process to improve the antioxidant capacity of *Lupinus angustifolius* L. var. Zapaton," *European Food Research and Technology*, vol. 223, no. 4, pp. 495–502, 2006.

[14] J. Frias, M. L. Miranda, R. Doblado, and C. Vidal-Valverde, "Effect of germination and fermentation on the antioxidant vitamin content and antioxidant capacity of *Lupinus albus* L. var. Multolupa," *Food Chemistry*, vol. 92, no. 2, pp. 211–220, 2005.

[15] H. Schagger and G. von Jagow, "Tricine-sodium dodecyl sulfate-polyacrylamide gel electrophoresis for the separation of proteins in the range from 1 to 100 kDa," *Analytical Biochemistry*, vol. 166, no. 2, pp. 368–379, 1987.

[16] M. Muzquiz, C. Rey, C. Cuadrado, and G. R. Fenwick, "Effect of germination on the oligosaccharide content of lupin species," *Journal of Chromatography*, vol. 607, no. 2, pp. 349–352, 1992.

[17] V. L. Singleton, R. Orthofer, and R. M. Lamuela-Raventós, "Analysis of total phenols and other oxidation substrates and antioxidants by means of folin-ciocalteu reagent," *Methods in Enzymology*, vol. 299, pp. 152–178, 1998.

[18] Ž. Maleš and M. Medić-Šarić, "Optimization of TLC analysis of flavonoids and phenolic acids of *Helleborus atrorubens* Waldst. et Kit," *Journal of Pharmaceutical and Biomedical Analysis*, vol. 24, no. 3, pp. 353–359, 2001.

[19] L. R. Fukumoto and G. Mazza, "Assessing antioxidant and prooxidant activities of phenolic compounds," *Journal of Agricultural and Food Chemistry*, vol. 48, no. 8, pp. 3597–3604, 2000.

[20] S. Burda and W. Oleszek, "Antioxidant and antiradical activities of flavonoids," *Journal of Agricultural and Food Chemistry*, vol. 49, no. 6, pp. 2774–2779, 2001.

[21] P. Gulewicz, C. Martínez-Villaluenga, J. Frias, D. Ciesiołka, K. Gulewicz, and C. Vidal-Valverde, "Effect of germination on the protein fraction composition of different lupin seeds," *Food Chemistry*, vol. 107, no. 2, pp. 830–844, 2008.

[22] G. Urbano, P. Aranda, A. Vílchez et al., "Effects of germination on the composition and nutritive value of proteins in *Pisum sativum*, L," *Food Chemistry*, vol. 93, no. 4, pp. 671–679, 2005.

[23] S. Jood, U. Mehta, R. Singh, and C. M. Bhat, "Effect of processing on flatus-producing factors in legumes," *Journal of Agricultural and Food Chemistry*, vol. 33, no. 2, pp. 268–271, 1985.

[24] H. A. Oboh, M. Muzquiz, C. Burbano et al., "Effect of soaking, cooking and germination on the oligosaccharide content of selected Nigerian legume seeds," *Plant Foods for Human Nutrition*, vol. 55, no. 2, pp. 97–110, 2000.

[25] B. J. Xu and S. K. C. Chang, "A comparative study on phenolic profiles and antioxidant activities of legumes as affected by extraction solvents," *Journal of Food Science*, vol. 72, no. 2, pp. S159–S166, 2007.

[26] E. Tsaliki, V. Lagouri, and G. Doxastakis, "Evaluation of the antioxidant activity of lupin seed flour and derivatives (*Lupinus albus* ssp. Graecus)," *Food Chemistry*, vol. 65, no. 1, pp. 71–75, 1999.

[27] M. Muzquiz, C. Cuadrado, G. Ayet, L. Robredo, M. Pedrosa, and C. Burbano, "Changes in non-nutrient compounds during germination," in *Effeccts of Antinutritional Value of Legume Diets*, S. Bardocz, E. Gelencsér, and A. Pusztai, Eds., vol. 1, pp. 124–129, Budapest, Hungary, 1996.

[28] R. A. Oloyo, "Chemical and nutritional quality changes in germinating seeds of *Cajanus cajan* L.," *Food Chemistry*, vol. 85, no. 4, pp. 497–502, 2004.

[29] P. Stratil, B. Klejdus, and V. Kubáň, "Determination of total content of phenolic compounds and their antioxidant activity in vegetables—evaluation of spectrophotometric methods," *Journal of Agricultural and Food Chemistry*, vol. 54, no. 3, pp. 607–616, 2006.

[30] N. E. Rocha-Guzmán, A. Herzog, R. F. González-Laredo, F. J. Ibarra-Pérez, G. Zambrano-Galván, and J. A. Gallegos-Infante, "Antioxidant and antimutagenic activity of phenolic compounds in three different colour groups of common bean cultivars (*Phaseolus vulgaris*)," *Food Chemistry*, vol. 103, no. 2, pp. 521–527, 2007.

[31] A. E. Hagerman, K. M. Riedl, G. A. Jones et al., "High molecular weight plant polyphenolics (Tannins) as biological antioxidants," *Journal of Agricultural and Food Chemistry*, vol. 46, no. 5, pp. 1887–1892, 1998.

[32] F. Yamaguchi, Y. Yoshimura, H. Nakazawa, and T. Ariga, "Free radical scavenging activity of grape seed extract and antioxidants by electron spin resonance spectrometry in an H_2O_2/NaOH/DMSO system," *Journal of Agricultural and Food Chemistry*, vol. 47, no. 7, pp. 2544–2548, 1999.

[33] N. S. C. de Gaulejac, C. Provost, and N. Vivas, "Comparative study of polyphenol scavenging activities assessed by different methods," *Journal of Agricultural and Food Chemistry*, vol. 47, no. 2, pp. 425–431, 1999.

[34] A. S. Meyer, M. Heinonen, and E. N. Frankel, "Antioxidant interactions of catechin, cyanidin, caffeic acid, quercetin, and ellagic acid on human LDL oxidation," *Food Chemistry*, vol. 61, no. 1-2, pp. 71–75, 1998.

24

Epidemiology of the Diseases of Wheat under Different Strategies of Supplementary Irrigation

Roberto P. Marano,[1] Roxana L. Maumary,[2] Laura N. Fernandez,[2] and Luis M. Rista[2]

[1] Department of Environmental Science, National University of Litoral, Kreder 2805, 3080 Esperanza, Argentina
[2] Department of Vegetal Production, National University of Litoral, Kreder 2805, 3080 Esperanza, Argentina

Correspondence should be addressed to Roberto P. Marano, rmarano@fca.unl.edu.ar

Academic Editor: María Rosa Simón

Wheat (*Triticum aestivum* L.) is one of the most important and highly productive crops grown under supplementary irrigation in the central region of Santa Fe. However, its production is limited by the presence of diseases in the main stages for yield definition. The objective of this work was to assess wheat health in response to different supplementary irrigation strategies under greenhouse and field conditions. The field experiment included three treatments: dry (D), controlled deficit irrigation (CDI), and total irrigation (TI) using the central pivot method. Disease incidence from stem elongation and severity in flag leaf and the leaf below the flag leaf were measured. Leaf area index (LAI), harvest index, air biomass, and yield components were determined. In greenhouse the treatments were TI and CDI, with evaluations similar to the field. The major leaf diseases observed were tan spot, leaf rust, and septoria leaf blotch. Significant differences in disease burden, LAI and yield components were observed in the different treatments. Under greenhouse conditions, only tan spot was observed. The results of this study indicated that the application of supplemental irrigation in wheat improved the yield, without increasing the incidence and severity of foliar diseases.

1. Introduction

The amount of water available for crops is defined by the balance between precipitation and evapotranspiration [1]. Wheat (*Triticum aestivum* L.) cultivated in the central region of Santa Fe is subjected to periods of water deficit that can significantly decrease yields [2].

Because most farmers are focused on grain yield potential, irrigation technology has become an important tool both to maximize production [3] and to reduce the interannual variability of yields [4]. Furthermore, wheat is one of the most important agricultural crops that are treated with supplementary irrigation in humid/subhumid regions [2]. Wheat is also important in crop rotation schemes, because its stubble has beneficial effects on soil structure and for diversify production [5].

Central pivot irrigation is the dominant technique used in this region, but it is not clear whether this technology affects disease susceptibility. This method wets the foliage, thus reducing its temperature while increasing the relative humidity and the length of time during which the leaves remain wet; both of them can promote foliar diseases.

Foliar diseases are the main biotic restrictions that reduce wheat yield in Argentina [6]. Photosynthesis, respiration, the translocation of water and nutrients, and reproduction are affected by pathogens. Any interference in these vital processes prevents the plant from taking advantage of the environmental factors necessary for their growth and development [7], resulting in decreased yield potential. This can be measured through the total amount of biomass generated and the proportion of it which is allocated to reproductive organs [6]. In wheat, the period from the beginning of stem elongation to flowering, during which the stalk and spike grow together and compete intensely, is crucial to determine the number of grains per unit area [6], the variable that is most closely associated with crop yield. Maintaining an adequate area of healthy and functional leaves during this period is essential to achieve higher rates of photosynthesis, allowing greater availability and partitioning of photoassimilates towards the ears and therefore a larger number of grains [6].

However, the negative effects of foliar diseases on wheat yield and quality have increased in Argentina over the last several years due to, among other things, the expansion of no-till, the dissemination of susceptible genotypes, and the use of infected seed [8]. Therefore, there has been an increase either in the prevalence of known foliar diseases like in the threat of the emergence of new diseases, according to Perelló and Moreno [8].

The major foliar fungal diseases caused by necrotrophic pathogens in Argentina have historically been tan spot (DTR) and septoria leaf blotch (SLB); the latter is caused by *Septoria tritici* Rob. ex Desm., teleomorph *Mycosphaerella graminicola* (Fuckel) J. Schröt. in Cohn. Together with some other pathogenic fungi (mainly *Bipolaris sorokiniana* (Sacc.) Schoem., teleomorph *Cochliobolus sativus* (Ito & Kuribayashi) Drechsler ex Dastur and *Alternaria* spp.), tan spot and septoria leaf blotch form a leaf spot disease complex in Argentina [9].

According to Fernández and Corro Molas [10], the most common and severe wheat diseases in Argentina are leaf rust (LR) (*Puccinia triticina* Eriks), SLB, DTR [*Pyrenophora tritici-repentis* (Died.) Drechs, anamorph *Drechslera tritici-repentis* (Died.) Shoemaker], and white blow or fusarium head blight (FHB) (*Fusarium graminearum*). Massaro et al. [11] conducted an experimental survey of wheat foliar diseases in the southern region of Santa Fe over seven consecutive years (2000 to 2006) and concluded that the most prevalent diseases were DTR and LR (71% and 86%, resp.); SLB was observed less frequently, in only one of the seven years studied (14%). Work carried out in the Santa Fe center since 2003 has shown that FHB is the most important disease of the year, with an erratic appearance that is highly dependent on the environmental conditions at the time of flowering [12].

Serrago et al., cited by Simón et al. [9], indicated that a complex of diseases formed by DTR, SLB, and LR reduced grain yield by 1020 kg ha^{-1} on average. Other authors have reported SLB yield losses of 2–50% [13–18] and as high as 75% [19]. In Argentina, yield losses from 21 to 37% [20, 21] and from 20 to 50% [21, 22] in high yielding cultivars have been found.

Additionally, in Argentina, the losses caused by the DTR can reach values as high as 14% in grain yield, as well as an 8 to 11% reduction in thousand grain weight and between 1.2 and 4.5% in hectoliter weight [23]. Globally, yield losses were reported up to 40% [24].

Wheat cultivars that are susceptible to LR regularly suffer yield reductions of 5–15% or greater, depending on the stage of crop development [25]. Reductions of 10–30% have also been reported [26, 27].

Seed quality is also essential; the health status of a seed lots is the main criterion for seed quality, together with purity, energy, and germinative power [28].

Few studies have investigated wheat diseases grown under supplementary central pivot irrigation in Argentina. Work carried out in southern Alberta (Canada) showed that wheat foliar diseases increased in the presence of sprinkler irrigation [29]. Crops cultivated under irrigation tend to be denser, and this modification of the microclimate influences not only the contraction of diseases but also the sporulation of pathogens and later spore dispersal [29]. The wetting of infested crops promotes the sporulation of pathogens, especially when the crop foliage is dense and the subhumid conditions produced by irrigation are prolonged. Pathogenic spores can be dispersed directly by irrigation water droplets or indirectly through the hydration of specialized fruiting bodies such as perithecia [29]. In southern Santa Fe, Andriani et al. [30] reported that wheat under central pivot irrigation developed a powdery mildew (*Blumeria graminis* f. sp. *tritici*) every year. In general, lower yields are closely related to the presence of diseases that affect the entire cycle of crop.

The concepts outlined above highlight the importance of obtaining local information about health problems in cultivated wheat and their possible effects on grain production, that is, comparing the yield maximization achieved through supplementary irrigation with the potential negative effects of irrigation on the evolution of diseases.

The objective of this work was to assess the relation between the health of a wheat crop (grown in the greenhouse or field) and the water management conditions used in the eastern/central region of Santa Fe.

2. Materials and Methods

2.1. General Procedures. The experiment was carried out over two successive growing seasons (2009-2010) in the "Miraflores" area (latitude 32°10′14″ S, longitude 60°59′57″ W), located in the eastern/central region of the Santa Fe province, with 800 ha under central pivot irrigation with water from the Coronda River. The system that they have has an intake in the river, which drives through channels, partly excavated and partly on an embankment, with four pumping stations. The central pivot covers an area of 32 ha (six towers, 325 m) with average irrigation flow and depth of 125 m^3 h^{-1} and 8 mm day^{-1}, respectively. The applied drops are between 1 and 2 mm, and the passage time on the leaves varies from a few minutes (extreme towers) to a few hours (central towers), depending on the applied depth.

The climate analysis considered historical information for the central region (Oliveros and Santa Fe), including rainfall, temperatures, pressure vapor, wind, radiation, and evaporation.

The soil is a Typic Argiudolls, which is suitable for agriculture (class I, INTA, 1992). Surface composite samples of soil (0–0.2 m) were extracted for chemical analysis (pH, total nitrogen, organic matter, phosphorous, sulfur) in order to calculate the fertilizer doses required.

2.2. Treatments. The treatments were as follows: D (rainfed, no irrigation) crops located outside the circle; TI, with irrigation managed according to the maximum expected yield and maximum demand for water; CDI, with irrigation managed strategically according to the water deficit. Three plots (replicates of 100 m^2 each) in each treatment area were selected for evaluation.

The Cronox cultivar was used for all treatments. Cronox is a short-intermediate cycle plant with moderate susceptibility to DTR and LR, and moderate-to-low suscepti-bility to SLB, according to the information provided by their respective breeder. Seeding was carried out on June 10 with a density of 150 kg ha^{-1} seed, resulting in a density of 453 plants m^{-2}. Fertilizer was applied based on a prior analysis of the soil: 150 kg ha^{-1} urea (broadcast applied), 70 kg ha^{-1} of diammonium phosphate, and 50 kg ha^{-1} calcium sulfate, and the harvest was on November 12. In the 2010 season, Cronox was sowed on June 23 but at a higher density (160 kg ha^{-1}), 409 plants m^{-2}. Plants were fertilized with 120 kg ha^{-1} urea (broadcast applied), 100 kg ha^{-1} of ammonium phosphate, and 80 kg ha^{-1} calcium sulfate, and the harvest was on November 14. Management practices, which were usually carried out by the farmers, included the preventive treatment of seeds with an antifungal agent (25% carbendazim + 25% tiram).

2.3. Blotter Test.

Seed samples with and without treatment (4 samples of 100 seeds each) were obtained and incubated to measure germination energy (GE) and germinative power (GP). Incubation was carried out using the top of paper method according to Peretti [31]. Seed health was measured in terms of pathogen burden as determined by "Blotter test" or, when it was necessary to isolate specific pathogens, through selective culture [32].

Incubation was carried out at 21 ± 1°C, a relative humid-ity of 80%, 12 h light, and 12 h of darkness [33] for four to ten days [31]. The protocol published by Peretti [31] was not used because the incubation temperature was inappropriate. The GE count was conducted after four days of incubation, and the final count to determine GP was conducted after eight days. Pathogen load was determined by visually observ-ing incubated seeds for fungal colonies with a magnifying binocular, both from above and from below [31]. A stereo-scopic microscope was used to diagnose fungal structures in specially made preparations.

2.4. Foliar Disease Incidence, Severity, and Biomass Deter-mination.

The Zadoks scale was used to monitor crop phenology [34]. Disease monitoring was conducted in the field, from stem elongation onwards because during tillering, new leaves quickly appear and there is a reduction in the intensity of disease [35]. Monitoring consisted of two weekly visits to evaluate LR and once per week to evaluate foliar spots caused by Drechslera tritici-repentis, Septoria tritici, or Bipolaris sorokiniana. The severity and incidence of these diseases were quantified as the percentage of affected leaf area on the flag leaf (FL) and the leaf below the flag leaf (FL-1).

Fusarium head blight (FHB) results from the develop-ment of a complex of pathogenic fungi. Fusarium consists of five main species (Fusarium graminearum, Fusarium cul-morum, Fusarium avenaceum, Fusarium poae, and Fusarium triticum), with several strains per species. The most common of these species, which causes FHB, is F. graminearum [36]. To quantify FHB, we measured the percent incidence (sick spikes/assessed spikes × 100). We also monitored the cereal disease Gaeumannomyces graminis, which has become important in the wheat region within the last several years due to the increase of inoculum in soil [37], and a powdery

mildew (Blumeria graminis f. sp. tritici) disease that was previously observed to develop upon irrigation [30]. Batch sampling was conducted by randomly selecting 50 tillers taken in a zigzag path of the sampling area. The Cobb scale was applied to evaluate the severity of LR and foliar diseases on FL and FL-1, and the Stack and McMullen scale was used to evaluate FHB [38, 39]. All scores were expressed as percentages [40]. The incidence (percentage of infected plants) and percentage of sick leaves (with respect to the total number of leaves) were calculated by separating green leaves and expanded bearers with symptoms from those that were healthy. Leaves that exhibited at least one lesion or leaf spots >2 mm [35] were considered to be infected with rust sheet. Because it is difficult to differentiate lesions caused by DTR and SLB, the accurate diagnosis was made based on the sign: Drechslera has long conidiophores and conidia, and those of septoria, pycnidia, and conidia are shorter, as observed at 40x with an optical microscope [41].

Distrain software was also used to estimate the severity of several diseases, including LR, powdery mildew, SLB, striated rust, stem rust, and DTR [42].

To estimate the total aboveground biomass (TAB), samples were taken from plants at three timepoints, Z 3.1, Z 6.5, and Z 9.2, according to Zadoks et al. [34]. Twenty stems were extracted from each of the first two samples, and leaf area index (LAI) was measured with a LIQUOR LI team index 3000.A instrument. The stem and leaf components were separated and dried to a constant weight at 65°C.

2.5. Yield and Yield Components.

Yield was determined from two samples extracted at random from physiologically mature plants along one linear meter per plot. In the laboratory, plants and stems were counted for each sample, and subsamples (20 stems) were separated by components (stalk and spike); the number of spikelets per spike and fertile and infertile spikelets was counted. Each component and the rest of the sample were dried separately at 65°C to a constant weight. Each sub-sample of spikes was threshed manually, and the resulting grains were subsequently weighed and counted.

2.6. Statistical Analysis.

The experiment was conducted in a random block design with three replicates, and analysis of variance (ANOVA) was used to evaluate the severity, impact, LAI, and yield parameters using the program INFOSTAT/ professional-version 2009 [43]. Homogeneity of variance was tested by comparing the error mean squares for all dependent variables and Shapiro-Wilks modified [43]. Means were compared by Tukey ($P \leq 0.05$). The data on severity per-centage and incidence percentage were arcsine square root transformed for analysis.

2.7. Greenhouse Experiment.

In addition to epidemiological studies in the field, we evaluated plants grown in a green-house in order to compare the health and yield of this cultivar under different irrigation conditions.

The same variety of wheat was used (Cronox) with a sow-ing date of May 31, 2010, in furrows of 0.3 m and separated by 0.2 m. Greenhouse plants received either TI (irrigated

TABLE 1: Incidence (%) and genera of pathogens identified through blotter tests of seeds treated with fungicide or untreated in 2009 and 2010.

Treatment	Year	*Alternaria* spp.	*Aspergillus* spp.	*Penicillium* spp.	*Drechslera tritici-repentis*	*Bipolaris sorokiniana*	*Rhizopus* spp.
Seeds treated	2009	0	0	0	0	0	0
	2010	1.25	0	10.5	0	0	0.25
Seeds untreated	2009	22	5	0	2.5	0.5	0
	2010	26.5	0	28	0	6.25	11.25

at 100% field capacity) or CDI (75% of field capacity) treatments, but it was not feasible to use D (rainfed, no irrigation). Cultivation occurred normally with a density of 47 pl per treatment, equivalent to 400 plants m^{-2}. The treatments began with an initial moisture equivalent to field capacity.

The plants were kept in a greenhouse with a temperature of 22°C and a photoperiod of 16:8 (light and dark) [44] with high relative humidity (100% for the first four hours, followed by 80%) [45] and grown in plastic containers with a capacity of 84 dm^3 (approximately 0.6 m long × 0.4 m wide × 0.35 m deep). Each container was divided into two equal parts such that each pot contained both treatments. The soil was textured silt/clay which had been conditioned by grinding and sifting (2 mm mesh). The bulk density was 1200 kg m^{-3}, and P, N, and S fertilization was conducted according to a soil analysis.

Interval irrigation was initiated when 75% of the available water had been depleted. A 20 mm fixed dose was used, representing the estimate of useful water in the container, and a pressurized sprayer (Giber) was used to simulate sprinkler irrigation.

Given that wheat stubble constitutes a natural reservoir of many fungi that cause necrotrophic "leaf spots," such as *Drechslera tritici-repentis*, *Septoria tritici*, and *Bipolaris sorokiniana* [46], plants were inoculated using a non-quantitative method through a recreation of the stubble on the surface of an infected crop [47]. The plants remained in the greenhouse until harvest (October 19).

Nondestructive methods (i.e., weekly observation through manual magnifiers) were used to evaluate disease from the beginning of tillering to the filling of grains. LAI was estimated by subsampling 10 plants per treatment and was repeated at three phenological timepoints: Z 2.3, Z 4, and Z 7. We used a non-destructive method to measure the length and maximum width of the sheet and subsequently multiplied this product by a correction coefficient previously obtained with LAI (LIQUOR LI 3000.A) measurement equipment.

Yield was determined using the same methodologies that were used in the growing field. The trial was conducted in a randomized block design with four replicates, and severity, impact, LAI, and yield parameters were evaluated using analysis of variance (ANOVA).

3. Results

3.1. Seed Analysis and Blotter Test. The GE and GP values obtained for the seeds from the 2009 season were 100% and

99%, respectively, for untreated seeds and 99.5% and 97.5%, respectively, for treated seeds; in 2010, these values were 98.75%, 98.25%, 99.5%, and 99%, respectively. According to Peretti [31], all of these values are within the acceptable ranges for regulated wheat seed.

In the untreated seeds from 2009, the incidence of microorganisms was 30.5%, predominantly "black point" grains caused by *Alternaria* spp. and, to a lesser degree, *Aspergillus* spp., *Drechslera tritici-repentis*, and *Bipolaris sorokiniana*. In contrast, the incidence of microorganisms in treated seeds was 0%.

The conditions of high humidity and high temperatures that occurred towards the end of the growing season in 2009, coupled with poor storage conditions, increased the incidence of the pathogens that cause discoloration and deterioration of seeds. This result was verified in the analyses performed on seeds that were stored by the farmer and used for seeding in the 2010 season, which contained *Alternaria* spp., *Bipolaris sorokiniana*, *Penicillium* spp., and *Rhizopus* spp. The overall incidence of pathogens was 72% for untreated seeds and 12% for treated seeds. A higher incidence of pathogens (*Penicillium* spp., *Rhizopus* spp. and *Alternaria* spp.) was detected during storage (Table 1), while the presence of *Penicillium* spp. in the seeds treated by the farmer would indicate an incorrect dose of fungicide. However, GE and GP were not affected by this pathogen.

Exposure to fungi in the field and during storage affects germination, seedling stand, grain size and weight, and industrial quality. In the case of wheat, these fungi are associated with the grain spotting known as "black scutellum," or "blackpoint." This pathology is characterized by a black or brown coloration in the area of the embryo, which could also be extended to the surrounding area and the groove [33].

3.2. Field Trials (2009 Season). A total of 310 mm effective rainfall was received in 2009, which was greater than the historical average (Figure 1). Because of this heavy rain, only two irrigation treatments totaling 64 mm were applied during the growing season, and both the TI and CDI treatment received the same amount of water. The first irrigation was administered on August 15 during phenological state Z 2.3, and the second was administered on September 5 during phenological state Z 3.1.5.

The daily average air temperatures were lower than 16°C in June, July, and September, as well as in the first ten days of August and the second ten days of October. The lowest temperature was recorded on July 14 (−8°C). Three consecutive days with temperatures greater than 21°C were

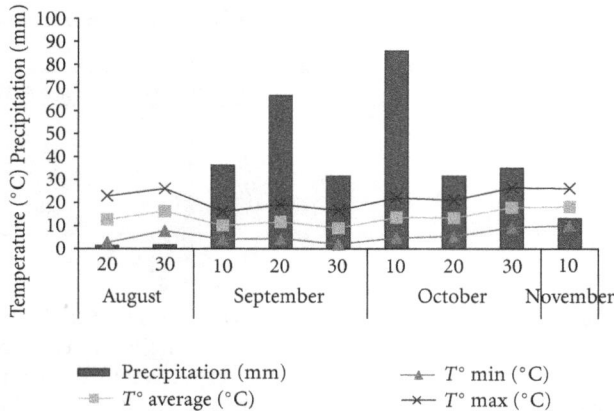

FIGURE 1: Average, minimum, and maximum air temperature values (°C) and precipitation during the 2009 growing season.

TABLE 2: Incidence (%) of foliar disease onset for total irrigation (TI), irrigation with controlled deficit (CDI) and dry (D) wheat at various phenological timepoints.

Treatment	Incidence (%)							
	10/09/2009		16/09/2009		30/09/2009		15/10/2009	
	Z 3.1		Z 3.9		Z 5.6		Z 7.05	
TI	17.1	A	19.6	A	14.2	A	27	A
CDI	18.8	A	13.5	A	15.7	A	16.8	A
D	51.3	B	31.3	B	46.7	B	59.5	B

Different letters indicate significant differences according to Tukey ($P \leq 0.05$).

recorded during August and October (last ten days) and two days in November (second ten days).

The ambient relative humidity remained above 60%, and wet leaves were still observed after 15 hours on two consecutive days during the second ten days of July and the first ten days of September.

The three most frequent leaf pathologies, LR, DTR, and SLB, were identified in all treatments.

The incidence of foliar diseases was higher in the D treatment than in the other treatments, although the severity remained below 1% in all treatments. The average disease incidence (percentage of sick leaves with respect to the total number of leaves), both in general and at different phenological stages, was significantly different (Table 2) between the D treatment and the two remaining treatments (CDI and TI).

This pattern was likely observed because the nonirrigated wheat did not achieve total coverage of furrows, even at advanced stages of development (Z 6, anthesis), suggesting that at the furrow minor coverage allowed the foliar disease to colonize the upper strata of the crop. This supposition is consistent with the LAI results, which were significantly different between D and the irrigation treatments at Z 4 (3.95 versus 4.99 and 5.56 for CDI and TI, resp.). In contrast, the length of leaf wetting caused by sprinkler irrigation was not significantly different than that from the normal wet period due to ambient humidity during the crop cycle.

The individual development of each foliar disease present during the crop cycle was analyzed. In general, epidemics of SLB is caused by a combination of favorable climatic conditions (usually characterized by long periods of light rain and moderate temperatures), certain cultivation practices, the availability of inoculum and the presence of susceptible varieties [48]. A relatively low intensity of foliar disease was observed during this growing season, and diseases were not identified in a uniform manner across treatments or sampling dates. The highest SLB incidence was just 10.5%, observed in samples from the D treatment analyzed on October 15. Injuries to FL-1 that corresponded with SLB

were observed in a total of two leaves (of 50 analyzed) in the CDI treatment, but one of these exhibited a severity of less than 1%.

DTR was the most frequently observed disease throughout the analysis period, with an average incidence value of 20.07%. DTR also made up 31.58% of leaf injuries, together with LR. These injuries were observed on both FL and FL-1. The fungus survives in the stubble and, under humid conditions and adequate rainfall, releases spores that infect the lower leaves. From there, the disease advances to higher leaves by rain splashing or air circulation [49], which are conditions that occurred in the D treatment because the furrow was not completely covered.

There were significant differences between the D treatment and the irrigation treatments (CDI and TI), with the exception of the sampling on September 16, in which the differences were not significant (Table 3). This finding can be attributed to a dilution of the disease by an increase in leaf area; DTR was present in basal leaves initially, but these leaves had dried up at more advanced phenological stages.

The average incidence of LR was 11.52% over two sampling dates. LR was first identified in Z 3.9 (September 16) and was more frequent in the D treatment (11.73% versus 1.2 and 0.5 for CDI and TI, resp.). During the next week (September 22), an application of 18.7% trifloxystrobin (strobilurin) and 8% cyproconazole (triazole) was made to control the disease. This application remained active up to 60 days, which allowed a reduction in the number of active pustules of *Puccinia triticina* and the control of this disease. However, a second LR infection cycle followed. This likely happened because the urediniospores are relatively long lived and can survive in the field without being deposited on host plants for periods of several weeks [26]; this is why very early treatment, insufficient wetting of the basal leaves, or favorable environmental conditions may allow reinfection by this polycyclic pathogen.

On September 30 (the sampling that was conducted before the new LR attack), the conditions in the experimental area were highly favorable for pathogen development. According to INTA Gálvez, in the first ten days of October, the maximum, minimum, and average temperatures were 22.1°C, 4.7°C, and 13.5°C, respectively. Rainfall of 101 mm accumulated in just 15 days (for comparison, the historical average for October is 105 mm), and several days were misty

TABLE 3: Incidence (%) of tan spot (DTR) for total irrigation (TI), irrigation with controlled deficit (CDI), and dry (D) treatments at different sampling dates in 2009.

Treatment	Incidence (%) DTR							
	10/09/2009		16/09/2009		30/09/2009		15/10/2009	
	Z 3.1		Z 3.9		Z 5.6		Z 7.05	
TI	16.7	A	19.1	A	14.2	A	11	A
CDI	18.8	A	12.3	A	15.7	A	3.3	A
D	49.73	B	20.1	A	46.7	B	29.5	B

Different letters indicate significant differences according to Tukey ($P \leq 0.05$).

TABLE 4: Incidence (%) of (LR) for total irrigation (TI), irrigation with controlled deficit (CDI), and dry (D) treatments at different sampling dates in 2009.

Treatment	Incidence (%) LR			
	16/09/2009		15/10/2009	
	Z 3.9		Z 7.05	
TI	0.5	A	10.2	A
CDI	1.2	A	7.7	A
D	11.7	B	48.2	B

Different letters indicate significant differences according to Tukey ($P \leq 0.05$).

and foggy, which resulted in water accumulation on the leaves.

At Z 7.05 (October 15), this disease was identified in all three treatments. D exhibited the highest incidence (48.2%), while CDI and TI exhibited incidences of 7.67 and 10.17%, respectively. LR infection was found in FL-1 (38% in D, 6% in CDI and 6% in TI), but with severity levels of less than 1% in all treatments. Some FL was also infected, but only in D, with an incidence of 4%. Significant differences ($P \leq 0.05$) were found between D and the two irrigation treatments (Table 4). These differences likely occurred because D, as a result of not adequately covering the furrows, allowed a greater remobilization of spores by wind and rain and consequently higher levels of infection, peaking in FL and FL-1.

In addition to all of the observed foliar diseases of fungal origin, large, dry, grayish-green lesions corresponding to bacterial blight caused by *Pseudomonas syringae* were observed in FL at the last sampling date (October 15). This "leaf blight" is favored by relatively cool temperatures (14 to 23°C) and high relative humidity, which are conditions that were present in the experimental field.

Finally, at physiological maturity (November 12), spikes were analyzed using wet chamber method. The presence of stained glumes caused by the saprotroph fungus *Alternaria* spp. was detected, resulting in 100% incidence in D, 50% in TI, and 46% in CDI treatments. The presence of this fungus was also observed through blotter analyses, as described in the previous section. No frost damage or *Fusarium graminearum, Gaeumannomyces graminis,* and *Erysiphe graminis* were observed, but insect damage was present.

Although foliar diseases were common throughout the growing season, high yields were obtained in all treatments,

TABLE 5: 1000 grains weight, biomass of harvested grain (BHG) and index harvest (HI) measured during 2009 for total irrigation (TI), irrigation with controlled deficit (CDI) and dry (D) treatments.

Treatment	1000 grains weight (g)		BHG (kg ha^{-1})		HI	
TI	33.57	A	8057	A	0.5	A
CDI	33.95	A	8128	A	0.48	A
D	30.88	B	6919	B	0.42	B

Different letters indicate significant differences according to Tukey ($P \leq 0.05$).

as evaluated by the number of spikes. The only significant differences observed between D and irrigation (DIC and TI) were in the weight of 1000 grains (Table 5).

The critical period for the main component of wheat yield (grains m^{-2}) ranges from 20 to 30 days before and 10 days after flowering. This is therefore the period during which leaf health is the most crucial for the plant to take advantage of incident radiation to maximize the growth and viability of the grains. Serious losses can also occur when the flag leaf is infected prior to anthesis. However, even the most prevalent diseases never exceeded 4% incidence or 1% severity in FL, so those were considered unlikely to have caused yield loss, regardless of the time of occurrence. Furthermore, crop health was generally very good, and yield differences between treatments were attributed to other causes (e.g., water availability differential, LAI achieved in each treatment).

3.3. Field Trials (2010 Season). During the wheat growing season, from implantation until the harvest, a total of 184 mm effective rainfall was received, well below the normal rainfall for the area of study. Due to the lack of rainfall, four irrigations were conducted, with a net sheet total of 180 mm. The first irrigation consisted of 40 mm conducted on August 7 (Z 2.2) with a blade, the second was 50 mm on September 28 (Z 3.9), the third was 40 mm on October 5 (Z 5.5), and the final irrigation was 50 mm on October 20 (Z 7).

The average daily temperature was below 16°C during the last third of June and during July, August, September, and October. In the first days of November, the daily average temperature exceeded 22°C (Figure 2). The lowest minimum temperature was recorded in the month of August at −7.7°C, and the highest maximum temperature was observed at the end of the growing season at 36.2°C. The average humidity remained above 57% over the whole growing season, and wet leaves were observed after 15 hours during the second ten days of July, the third ten days of August, and the first and third ten days of September.

Similar to the results from 2009, all three basic foliar diseases (LR, DTR and, to a lesser extent, SLB) were observed. Disease was significantly more prevalent in the D treatment than in either irrigation treatment ($P < 0,05$). Disease severity reached 15% on FL-1 and 10% on FL in the D treatment but only 5% and 1% on FL-1 in the CDI and TI treatment, respectively.

The first sampling was carried out in Z 2.2 (August 16), at which point some development of DTR could be observed

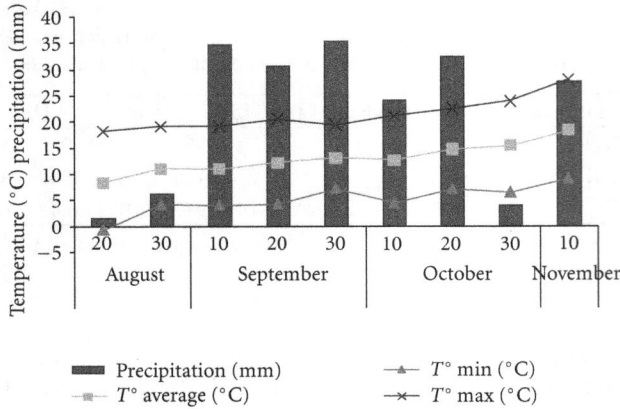

FIGURE 2: Average, minimum, and maximum air temperature (°C) and precipitation (mm) during the 2010 growing season.

TABLE 6: Leaf area index (LAI) at two sampling points for total irrigation (TI), irrigation with controlled deficit (CDI), and dry (D) treatments.

Treatment	LAI			
	16/09/2010		13/10/2010	
TI	6.39	A	7.95	A
CDI	6.38	A	6.15	AB
D	3.17	B	5.52	B

Different letters indicate significant differences according to Tukey ($P \leq 0.05$).

on the basal leaves in all three treatments. Later (September 16), the plants that had been irrigated and those that had not been irrigated exhibited different phenological states (Z 3.3 for D and Z 3.1 for TI and CDI). At this time, significant differences were observed between D and the two remaining treatments, both in terms of the variable incidence of diseases and in LAI (Table 6). Both DTR and, to a lesser degree, SLB were present. These diseases reached the upper strata in the D treatment but were restricted to the basal leaves in the irrigated treatments.

At the following sampling at Z 5 (September 30), only DTR was identified. This disease did remain confined to the lower strata in the irrigated treatments, in contrast to what was occurring upland, where DTR colonized the upper strata of the crop in the D treatment. This corresponded to an increased incidence of DTR: 52.7% in D compared to 23.7% and 20.1% for CDI and TI, respectively.

In the following sample, which was collected at Z 6.5 for D and Z 6.2 for TI and CDI (October 13), LR was observed in addition to DTR. Significant differences between irrigated and rainfed treatments were observed (Table 7). As suggested for the previous year, the differences in disease behavior could be due to the fact that plants in D did not totally cover the furrow, which is consistent with the measured LAI values (Table 6).

SLB infection levels were low due to the low rainfall and limited hours of wet leaves, which did not allow SLB establishment and dispersal. The registered incidence values

TABLE 7: Incidence (%) of foliar disease onset for individual phenological states under total irrigation (TI), irrigation with controlled deficit (CDI), and dry (D) treatments.

Treatment	Incidence (%)					
	16/09/2010		30/09/2010		13/10/2010	
	Z 3.2		Z 5		Z 6.5	
TI	10.3	A	20.1	A	50.2	A
CDI	11.7	A	23.7	A	48.2	A
D	21.6	B	52.7	B	74.3	B

Different letters indicate significant differences according to Tukey ($P \leq 0.05$).

TABLE 8: Incidence (%) of tan spot (DTR) in various phenological states for total irrigation (TI), irrigation with controlled deficit (CDI), and dry (D) treatments.

Treatment	Incidence of DTR (%)					
	16/09/2010		30/09/2010		13/10/2010	
	Z 3.2		Z 5		Z 6.5	
TI	6.8	A	20.1	A	46.77	A
CDI	9.3	A	23.7	A	45.2	A
D	20.4	B	52.7	B	71.2	B

Different letters indicate significant differences according to Tukey ($P \leq 0.05$).

were 1.22% in D, 3.14% in CDI, and 3.44% at TI. *Septoria tritici*, the causal organism of SLB, requires temperatures of 20 to 25°C [50] and water on leaves for 35 hours followed by 48 hours of high relative humidity, which favored heavy infection [51]. These conditions did not occur until November, and the disease was identified only in the first sampling.

DTR was present from the tillering stage to the end of the growing season. The stay of wheat straw at the soil surface, associated with moderately conducive weather conditions, favored the emergence and constant development of DTR throughout the entire crop cycle, with a variable but consistently increasing incidence according to phenological state. Significant differences were observed between D and the irrigated treatments (Table 8). DTR was observed more frequently on FL and FL-1 in D than in the irrigated treatments. For FL, the incidence of DTR was 24% in D versus 2% in CDI and 12% in TI; for FL-1, the incidence peaked at 86% in D versus 26% in CDI and 30% in TI.

The spread and infection of *Drechslera tritici-repentis* can occur under a wide range of environmental conditions; in general, temperatures between 10 and 30°C and 6- to 48-hour humid periods are sufficient [52–55]. Therefore, tan or DTR appears every year, in contrast to other diseases, such as FHB, which are strongly dependent on environmental conditions [41].

The onset of LR was significantly delayed in 2010 relative to 2009 and was first registered only at the beginning of flowering. According to INTA Galvez, the maximum, minimum, and average temperatures during the second ten days of September were 27.5°C, 8.1°C, and 16.8°C, respectively. A total of 56.4 mm of rainfall was recorded in

TABLE 9: 1000 grains weight, biomass of harvested grain (BHG) and harvest index (HI) measured during 2010 for total irrigation (TI), irrigation with controlled deficit (CDI) and dry (D) treatments.

Treatment	1000 grains weight (g)		BHG (kg ha^{-1})		HI	
TI	30.3	B	8898	B	0.49	A
CDI	27.2	BA	7820	BA	0.47	A
D	25.7	A	6899	A	0.48	A

Different letters indicate significant differences according to Tukey ($P \leq 0.05$).

the last two weeks of September and the first ten days of October, and leaves were wet for up to 17 consecutive hours for several days in the last third of September.

Statistical analysis showed significant differences in LR incidence between D and irrigation treatments (37.07% versus 8.73% in CDI and 9.5% in TI).

The disease reached FL-1 with an incidence of 40% in D, 2% in CDI, and 10% in TI. The severity reached levels of 15% in D but was less than 1% with only one to two pustules per leaf in the irrigation treatments. LR was observed in FL only in the D treatment, with an incidence of 6% and a maximum severity of 10%.

Finally, on November 11, samples were extracted to analyze the crop yield. Very good results were obtained in all treatments, although a significant difference ($P < 0.01$) in the weight of 1000 grains was observed between D and TI. The differences between CDI and D or TI were not significant (Table 9).

In terms of the health of the spikes and grains, *Fusarium* and *Alternaria* spp. were not identified because there were no environmental conditions that favor their appearance. The grain is susceptible to infection by *Alternaria* during filling or ripening stage, particularly in the states called milky, pasty ([56–58]). The sporulation of *Alternaria* range is between 0°C and 35°C, with optimum at 27°C, but is inhibited below 15°C or above 33°C [59]. Moschini et al. [57] determined that the severity of this disease in Argentina is favored by warmer temperatures, frequent rainfall, and days with relative humidities higher to 62%, in the grain development period spanning about 30 days after heading, but these conditions did not appear in the 2010 season.

Additionally, *Gaeumannomyces graminis* and *Erysiphe graminis* were not detected. On the other hand, agronomic frost did not generate the grains yield decrease, because it occurred in noncritical states for the crop.

3.4. Greenhouse Trials. The first irrigation was conducted during Z 2.2 (4 July) for both treatments. Over the entire growing season, TI received 340 mm, while CDI received 240 mm. The initial inoculum from the straw, accompanied by droplets of water from the first irrigation, led to the development of DTR and SLB.

The first symptoms were observed during full tillering (Z 2.2, 16 July), although differences became significant after September 14, when the TI treatment was in Z 7 and the CDI treatment was in Z 6. At this point, yield components had already been defined (Table 10).

DTR infection reached both FL-1 and FL. The maximum incidence in FL-1, observed at Z 7.0, was 65% in TI and 45% in CDI, and severity values reached 10% in both treatments. The incidence of DTR in FL reached 70% for TI and 25% for CDI, with a maximum severity of 5% for both treatments.

It should be noted that lower levels of incidence and severity were reported in CDI in the greenhouse trials than under field conditions.

LAI values were similar between treatments (Table 11), but significant differences ($P \leq 0.05$) in the weight of 1000 grains and BHG were observed between the two treatments (Table 12).

4. Discussion

The genera of fungi identified in this analysis correspond to those recognized by Can Xing et al. [60] and Ramirez et al. [61]. These results highlight the importance of using cured seed for seeding. The treatment of seeds with fungicide both eradicates the inoculum so that they do not constitute a primary or initial source of infection as well as protects the seed and seedlings from fungal infection in the soil, which indirectly leads to increased germination and ensures the implementation of cultivation [28].

During the two agricultural cycles evaluated, DTR and LR were the dominant foliar diseases. The cultivated plants remained healthy until advanced stages of development, and the severity of both foliar diseases was low in all of the treatments tested. In the 2009 season, 100% of plants in all treatments exhibited some degree of infection, although the severity was very low (less than 1%). Similarly, in the 2010 season, 100% of the experimental plants exhibited some degree of infection, again with relatively low severity (less than 15% in D, below 5% in CDI and 1% in TI).

Plants that received irrigation treatments exhibited lower levels of foliar diseases in both years. These results conflict with those of a previous study [29] conducted in southern Alberta, which suggested sprinkler irrigation to generate crops that are denser, thus modifying the local microclimate and creating optimal conditions for the development of diseases. However, these authors also suggested that irrigation influences the development of diseases not only through its impact on infection conditions but also through the sporulation of pathogens and later spore dispersal.

The lower disease burden of irrigated plants, observed during both years, may be attributed to the fact that better nourished plants (i.e., plants with greater water accessibility) are generally more tolerant of or less affected by foliar diseases. The work of Annone et al. [62] and that of Formento et al. [49] have shown that nitrogen fertilization at the right time may reduce the development of diseases such as DTR and increase the green tissue remaining in many leafy cultivars. The incidence of DTR in D was lower in 2010 than in the wet year 2009, despite the drier environmental conditions and thus limited water availability. In contrast, Annone and García [63] assert that any measure that directly or indirectly reduces the likelihood of secondary inoculum displacement, among plants both lower and higher levels of culture, reduces the final level of symptoms. Therefore,

TABLE 10: Incidence (%) of foliar disease onset at different times of measurement for total irrigation (TI) and irrigation with controlled deficit (CDI) treatments.

Treatments	Incidence of foliar disease (%)									
	13/08/2010		24/08/2010		02/09/2010		14/09/2010		28/09/2010	
	Z 3.1		Z 4		Z 6.5		Z 7		Z 7.5	
CDI	9.12	A	32.12	A	30.08	A	17.08	A	36.67	A
TI	9.63	A	31.38	A	33.21	A	57.38	B	70	B

Different letters indicate significant differences according to Tukey ($P \leq 0.05$).

TABLE 11: Leaf area index (LAI) measured during different phenological states under total irrigation (TI) and irrigation with controlled deficit (CDI) treatments.

Treatments	LAI							
	13/07/2010		24/08/2010		02/09/2010		14/09/2010	
	Z 2.3		Z 4		Z 6		Z 7	
CDI	6.57	A	6.16	A	5.63	A	3.56	A
TI	7.14	A	6.56	A	5.56	A	3.86	A

Different letters indicate significant differences according to Tukey ($P \leq 0.05$).

TABLE 12: 1000 grains weight, biomass of harvested grain (BHG) and harvest index (HI) measured for total irrigation (TI) and irrigation with controlled deficit (CDI) treatments.

Treatments	1000 grains weight (g)		BHG (kg ha^{-1})		HI	
TI	31.75	A	7328	A	0.39	A
CDI	28.59	B	4898	B	0.34	A

Different letters indicate significant differences according to Tukey ($P \leq 0.05$).

some management practices to obtain the highest possible density, such as the adjustment of seeding density based on grain weight, balanced fertilization to produce a compact cultivation structure, and the use of the lowest possible distance between lines and an appropriate cultivar for the desired sowing date, mitigate the development of "leaf spots." This is consistent with the higher incidence of DTR identified in D, which exhibited incomplete furrow coverage and low LAI, and therefore a less dense cultivation structure, which allowed higher levels of infection, even of FL and FL-1. The tests performed by Perello et al. [18] show that the disease incidence increased with the plant age and the severity increased with the growth stage when the evaluation was performed at 14 days compared to 7 days after inoculation. This coincides with the higher incidence found in more advanced stages of the crop at different treatments.

SLB was minimal (trace levels) in both years and was observed more frequently in D plants than in irrigated plants, especially during the more humid 2009 season. These results can be attributed to the density of plants generated in each treatment; as discussed above, plants in the D treatment did not fully cover the grooves, unlike the plants under irrigation, thus allowing the disease to develop further. This

finding is consistent with the work of Massaro et al. [64], who emphasized that growing crops at an optimal density, without large spaces between plants, can reduce the epidemic development of "SLB of the road" through secondary infections from the sites of primary infection (basal leaves) into the upper leaves. In contrast, Klatt and Torres [48] have noted that tall varieties of wheat tend to be less affected by SLB than short or semidwarf varieties. In general, this is due to a morphological resistance as a result of the increased distance between the leaves, which tends to impede the upward progress of the pathogen through the splashing of raindrops. In semiannual wheat cultivars, the leaves are closer to each other and the foliage tends to be denser, facilitating the upward spread of disease.

The results of our greenhouse experiments should not override those obtained in the field; significant differences in the parameters severity and incidence for both irrigation systems have not been verified.

Finally, significant differences in productivity were observed between irrigation and rainfed treatments. These differences were due to the application of water during the stem elongation stage (Z 3.0), which allowed the survival of more tillers and therefore more spikes than in the D treatment [65].

5. Conclusions

Based on tests carried out over two consecutive years, supplementary sprinkler irrigation of cultivated wheat at opportune moments, even in small quantities, increases grain weight and thus yield without increasing the incidence of foliar disease. Two fundamental principles should be considered for the correct management of wheat diseases: (1) the initial health of the crop should be optimized by using seeds with a low pathogen load and (2) appropriate monitoring should be conducted to properly quantify the diseases present in the field.

References

[1] P. E. Abbate, J. L. Dardanelli, M. G. Cantarero, M. Maturano, R. J. M. Melchiori, and E. E. Suero, "Climatic and water availability effects on water-use efficiency in wheat," *Crop Science*, vol. 44, no. 2, pp. 474–483, 2004.

[2] G. Camussi and R. P. Marano, "Wheat response to supplemental irrigation in the central region of Santa Fe," *Revista FAVE Sección Ciencias Agrarias*, vol. 7, no. 1, pp. 7–21, 1990.

[3] N. Formento, "Productive and health behaviour of wheat cultivars in zero tillage, rainfed and irrigation, season 1999/2000," http://www.planetasoja.com.ar/index.php?sec=10&tra=29376&tit=29379.

[4] A. Salinas, E. Martellotto, P. Salas, J. Giubergia, S. Lingua, and E. Lovera, "Economic performance in supplementary irrigation production system with supplementary irrigation in continuous zero tillage," INTA Manfredi, 2003, http://anterior.inta.gob.ar/f/?url=http://anterior.inta.gob.ar/manfredi/info/documentos/docsuelos/rieresecono.htm.

[5] M. D'Onofrio, "Liquid fertilization in wheat under irrigation. SolUAN application (tillering) combined with FoliarSOL U (anthesis)," Communications- Valle Inferior, pp. 12–15, 2007, http://anterior.inta.gob.ar/f/?url=http://anterior.inta.gob.ar/valleinferior/info/r60/03.pdf.

[6] R. Carretero, R. Serrago, and D. Miralles, "Foliar diseases of wheat. An eco-physiological perspective," 2nd Meeting Wheat in the Central Region, pp. 17–23, 2007, http://www.bccba.com.ar/bcc/images/00001773_Actas%20de%20la%20Jornada%20vf.pdf.

[7] M. Carmona and E. Reis, *Potential Evaluation System for Yield in Wheat. Its Usefulness for Application of Fungicides for Economic Control of Diseases*, Imprenta Rago, Buenos Aires, Argentina, 1st edition, 2001.

[8] A. E. Perelló and M. Moreno, "Survey of foliar diseases of wheat and identification of causal agents," in *Proceedings of the 6th Congreso Nacional de Trigo. IV Simposio Nacional de Cultivos de siembra otoño-invernal*, p. 257, INTA, Bahia Blanca, Argentina, 2004.

[9] M. R. Simón, F. M. Ayala, S. I. Golik et al., "Integrated foliar disease management to prevent yield loss in Argentinian wheat production," *Agronomy Journal*, vol. 103, no. 5, 2011.

[10] J. Fernández and A. Corro Molas, "Disease management in wheat," 2003, http://anterior.inta.gob.ar/f/?url=http://anterior.inta.gob.ar/anguil/info/pdfs/boletines/bol76/cap7.pdf.

[11] R. Massaro, J. Castellarín, and J. Andriani, "Foliar diseases of wheat: frequency recorded during 6 years in southern Santa Fe and its relationship to climatic variables," pp. 55–58, 2007, http://www.elsitioagricola.com/articulos/massaro/enfermedadesFoliaresDelTrigo.pdf.

[12] M. Sillon, E. Weder, G. Gianinetto, and J. Albrech, "Phytosanitary analysis of the 2006 season and possible future scenarios," 2nd Meeting Wheat in the Central Region, pp. 25–30, 2007, http://www.bccba.com.ar/bcc/images/00001773_Actas%20de%20la%20Jornada%20vf.pdf.

[13] R. M. Hosford, "A form of Pyrenphora tricosthoma pathogenic to wheat and other grasses," *Phytopathology*, vol. 61, pp. 28–32, 1971.

[14] R. M. Hosford and R. H. Busch, "Losses in wheat caused by *Pyrenophora trichostoma* and *Leptosphaeria avenaria* f. sp. Triticea," *Phytopathology*, vol. 64, pp. 184–187, 1974.

[15] R. G. Rees, G. J. Platz, and R. J. Meyer, "Yield losses in wheat from yellow spot comparison of estimates derived from single tillers and plots," *Australian Journal of Agricultural Research*, vol. 33, pp. 899–908, 1982.

[16] R. G. Rees and G. J. Platz, "Effects of yellow spot on wheat: comparison of epidemic at different stages of crop development," *Australian Journal of Agricultural Research*, vol. 34, pp. 39–46, 1983.

[17] A. Tekauz and R. G. Platford, "Tan spot of wheat," *Manitoba Agronomy Proceedings*, pp. 60–65, 1982.

[18] A. E. Perelló, V. Moreno, M. R. Simón, and M. Sisterna, "Tan spot of wheat (*Triticum aestivum* L.) infection at different stages of crop development and inoculum type," *Crop Protection*, vol. 22, no. 1, pp. 157–169, 2003.

[19] R. G. Rees, R. J. Mayer, and G. J. Platz, "Yield losses in wheat from yellow spot: a disease-loss relationship derivated from single tillers," *Australian Journal of Agricultural Research*, vol. 32, pp. 851–859, 1981.

[20] G. Kraan and J. E. Nisi, "Septoria of wheat in Argentina. Status of the crop against the disease," in *Proceedings of the Septoria Tritici Workshop, CIMMYT*, L. Gilchrist et al., Ed., pp. 1–8, Mexico City, DF, Mexico, September 1993.

[21] M. R. Simón, A. E. Perelló, C. A. Cordo, and P. C. Struik, "Influence of Septoria tritici on yield, yield components, and test weight of wheat under two nitrogen fertilization conditions," *Crop Science*, vol. 42, no. 6, pp. 1974–1981, 2002.

[22] J. Annone, A. Calzolari, O. Polidoro, and H. Conta, "Effect of tan spot caused by Septoria tritici on yield," Informe 122, INTA EEA Pergamino, Pergamino, Argentina, 1991.

[23] E. Alberione, C. Bainotti, J. Fraschina et al., "Evaluation of seed treatment products for the control of yellow spots (*Drechslera tritici repentis*) in wheat," pp. 59–63, 2010, http://agrolluvia.com/wp-content/uploads/2011/05/EVALUACION-DE-PRODUCTOS-CURASEMILLAS-PARA-EL-CONTROL-DE-MANCHA-AMARILLA-DRECHSLERA-TRITICI-REPENTI-EN-TRIGO.pdf.

[24] N. Formento, J. de Souza, and J. C. Velázquez, "Yield losses by yellow spot in wheat (*Pyrenophora tritici-repentis, anamorph: Drechslera tritici-repentis*). Preliminary results," 2007, http://anterior.inta.gob.ar/f/?url=http://anterior.inta.gob.ar/parana/info/documentos/produccion_vegetal/trigo/enfermedades/20323_071221_perd.htm.

[25] J. A. Kolmer, "Genetics of resistance to wheat leaf rust," *Annual Review of Phytopathology*, vol. 34, pp. 435–455, 1996.

[26] A. P. Roelfs, R. P. Singh, and E. E. Saari, *Rust Diseases of Wheat: Concepts and Methods of Disease Management*, CIMMYT, Mexico City, Mexico, 1992.

[27] S. G. Layva Mir, E. E. Rangel, E. Villaseñor Mir, and J. H. Espino, "Effect of leaf rust (*Puccinia triticina*. Eriks) on wheat yield (*Triticum aestivum* L.) in dry conditions," Revista Mexicana de Fitopatología, pp. 40–45, ISSN: 0185-3309, 2003.

[28] N. Formento, "Health and aging of wheat seeds," 2002, http://crea30agomlqn.wikispaces.com/file/view/Sanidad+y+Curado+de+las+Semillas+de+Trig1.pdf.

[29] T. K. Turkington, A. Kuzyk, R. Dunn et al., "Irrigation and plant disease management," 2004, http://www1.agric.gov.ab.ca/$department/deptdocs.nsf/all/ind10759/$file/irrigation_and_plant_disease_management.pdf?OpenElement.

[30] J. Andriani, N. Huguet, and C. Regis, "Evaluation of wheat cultivars with supplemental irrigation," in *To Improve Production 16, Wheat Season 2000/01*, INTA EEA Oliveros, Santa Fe, Argentina, 2001.

[31] A. Peretti, *Seed Testing Manual*, Hemisferio Sur, Buenos Aires, Argentina, 1994.

[32] R. Arango Perearnau, R. Craviotto, and C. Gallo, "New container for analysis of germination and seed health: use in pathology," in *Proceedings of the 3rd Congreso de soja del MERCOSUR, CD*, pp. 166–169, 2006.

[33] M. Sisterna and D. Minhot, "The genus Marielliottia (hyphomycetes, Ascomycetes): a new taxon mycoflora associated with the grain of wheat in Argentina," *Boletín Sociedad Argentina de Botánica*, vol. 41, pp. 177–182, 2006.

[34] J. C. Zadoks, T. T. Chang, and C. F. Konzak, "A decimal code for growth stages of cereals," *Weed Research*, vol. 14, pp. 415–421, 1974.

[35] M. Carmona, "Chemical control of foliar diseases in wheat," 2007, http://www.concienciarural.com.ar/agricultura/control-quimico-de-enfermedades-foliares-en-el-cultivo-de-trigo_a226.

[36] A. Champeil, T. Doré, and J. F. Fourbet, "Fusarium head blight: Epidemiological origin of the effects of cultural practices on head blight attacks and the production of mycotoxins by Fusarium in wheat grains," *Plant Science*, vol. 166, no. 6, pp. 1389–1415, 2004.

[37] N. Formento, "Dose and timing of fungicides to control foliar diseases in wheat," Año 2001, 2002, http://dl.dropbox.com/u/12187844/Aguas/P%C3%A1gina/bibliografia%20adicional/Formento%2037.pdf/.

[38] R. F. Peterson, A. B. Campbell, and A. E. Hannah, "A diagrammatic scale for estimating rust intensity of leaves and stem of cereals," *Canadian Journal of Research Section C*, vol. 26, pp. 496–500, 1948.

[39] R. W. Stack and M. P. McMullen, "A visual scale to estimate severity of fusarium head blight in wheat," Extensión Serv. North Dakota State University, USA, 1995.

[40] N. Formento, "Effectiveness of tebuconazole and epoxiconazole more Carbendazim in the control of foliar diseases and ear of wheat," 2003, http://agrolluvia.com/wp-content/uploads/2011/05/Eficacia-del-Metconazole-y-Epoxiconazole-+-Carbendazim-en-el-Control-de-Enfermedades-Foliares-y-de-la-Espiga-del-Trigo.pdf.

[41] M. Carmona, E. M. Reis, and P. Cortese, *Leaf Spot of Wheat. Diagnosis, Epidemiology and New Approaches to the Management*, BASF, Buenos Aires, Argentina, 1999.

[42] M. Carmona, E. M. Reis, and P. Cortese, *Wheat Rusts: Symptoms, Epidemiology and Control Strategies*, Imprenta Commiso Industria Gráfica, Buenos Aires, Argentina, 1st edition, 2000.

[43] INFOSTAT, Universidad Nacional de Córdoba, Editorial Brujas, Córdoba Argentina, 1st edition, 2009.

[44] C. Jobet, J. Zúñiga, H. Campos de Quiroz, P. Rathgeb, and G. Marín, "Doubled haploid plants (DH) generated by intergeneric rosses of wheat x maize," 2002, http://www.inia.cl/medios/biblioteca/serieactas/NR29064.pdf.

[45] S. Leyva Mir, J. Terrones Rodriguez, J. Herrera Espino, and H. Villaseñor Mir, "Analysis of inductive resistance components of a slow development of wheat rust (*Puccinia triticina Eriks.*)," 2007, http://redalyc.uaemex.mx/redalyc/pdf/302/30211241007.pdf.

[46] N. Formento and Z. Burne, "Effectiveness of fungicides in the control of tan spot (*Drechslera tritici-repentis*) in wheat," 2002, https://dl.dropbox.com/u/12187844/Aguas/P%C3%A1gina/bibliografia%20adicional/Formento%2046.pdf.

[47] Z. Eyal, A. L. Scharen, J. M. Prescott, and M. Van Ginkel, *Wheat Disease Septoria: Concepts and Methods Related to the Management of these Diseases*, CIMMYT, Mexico City, Mexico, 1987.

[48] A. R. Klatt and E. Torres, "An overview of diseases of wheat caused by Septoria, Programa de Trigo del CIMMYT," in *Proceedings of the Conferencia Regional Sobre la Septoria Leaf Blotch del Trigo*, M. M. Kohli and L. T. Van Beuningen, Eds., CIMMYT, Mexico City, México, 1990.

[49] N. Formento, J. C. Velázquez, and J. de Souza, "Epidemiology of diseases in wheat," *Boletín Fitopatológico. Cultivo de Trigo*, vol. 1, no. 2, 2009.

[50] M. Díaz de Ackermann, "Leaf spot in wheat caused by *Mycosphaerella graminicola* (Fuckel) schroeter perfect state of *Septoria tritici Rob. Es Desm*," in *Curso Manejo de Enfermedades del Trigo*, M. M. Kohli, J. Annone, and R. García, Eds., pp. 100–117, INTA-CIMMYT, Pergamino, Argentina, 1995.

[51] G. Shaner and R. Finney, "Wheater and epidemics of septoria leaf blotch of wheat," *Phytopathology*, vol. 66, pp. 781–785, 1976.

[52] C. R. Larez, R. M. Hosford, and T. P. Freeman, "Infection of wheat and oats by *Pyrenophora tritici-repentis* and initial characterization of resistance," *Phytopathology*, vol. 76, pp. 931–938, 1986.

[53] R. M. Hosford, C. R. Larez, and J. J. Hammond, "Interaction of wheat period and temperature on *Pyrenophora tritici-repentis* infection and development in wheat of differing resistance," *Phytopathology*, vol. 77, pp. 1021–1027, 1987.

[54] D. N. Sah, "Effects of leaf wetness duration and inoculums level on resistance of wheat genotypes to *Pyrenophora tritici-repentis*," *Journal of Phytopathology*, vol. 142, pp. 324–330, 1994.

[55] J. Annone, "The tan spot or yellow wheat," in *Curso Manejo de Enfermedades del Trigo*, M. M. Kohli, J. Annone, and R. García, Eds., pp. 118–134, INTA-CIMMYT, Pergamino, Argentina, 1995.

[56] R. L. Conner, "Influence of irrigation and precipitation on incidence of blackpoint in soft white spring wheat," *Canadian Journal of Plant Pathology*, vol. 11, pp. 388–392, 1989.

[57] R. C. Moschini, M. N. Sisterna, and M. A. Carmona, "Modelling of wheat black point incidence based on meteorological variables in the southern Argentinean Pampas region," *Australian Journal of Agricultural Research*, vol. 57, no. 11, pp. 1151–1156, 2006.

[58] M. Miravalles, V. Beaufort, and F. Möckel, "Relative susceptibility to blackpoint in durum wheat varieties of Argentina," *Phyton*, vol. 77, pp. 263–273, 2008.

[59] L. Carrillo, "Fungi in Foods and Feeds," in *Alternaria*, pp. 81–86, Universidad Nacional de Salta, Tartagal, Salta, Argentina, 2003.

[60] D. Can Xing, W. Xiao Ming, Z. Zhen Dong, and W. Xao Fei, "Testing of seedborne fungi in wheat germplasm conserved in the national crop genebank of China," *Agricultural Sciences in China*, vol. 6, no. 6, pp. 682–687, 2007.

[61] N. A. Ramirez, M. Mezzalama, A. Carballo, and A. Livera, "Effect of fungicides on seed physiological quality of wheat miller (*Triticum aestivum* L.) and their effectiveness in controlling *fusarium graminearum* Schwabe," *Revista Mexicana de Fitopatología*, vol. 24, no. 2, pp. 115–121, 2006.

[62] J. Annone, R. García, O. Polidoro, and A. Calzolari, "Response of cultivars of wheat under different tillage combinations of nitrogen, supplemental foliar fungicide treatment," Trigo en Siembra Directa, pp. 61–67, AAPRESID, 2003.

[63] J. Annone and R. García, "The main wheat leaf spot. Importance, epidemiology and strategies to reduce their impact on production," 2003, http://www.inta.gov.ar/ediciones/idia/cereales/trigo10.pdf.

[64] R. A. Massaro, A. Gargicevich, M. González et al., "Association between cultural variables and severity of wheat leaf diseases," 2005, http://www.agrositio.com/vertext/vertext.asp?id=44901&se=12.

[65] R. A. Fischer, "Wheat physiology: a review of recent developments," *Crop and Pasture Science*, vol. 62, no. 2, pp. 95–114, 2011.

25

Advances in Agronomic Management of Indian Mustard (*Brassica juncea* (L.) Czernj. Cosson): An Overview

Kapila Shekhawat, S. S. Rathore, O. P. Premi, B. K. Kandpal, and J. S. Chauhan

Directorate of Rapeseed-Mustard Research, Sewar, Rajasthan Bharatpur 321 303, India

Correspondence should be addressed to Kapila Shekhawat, drrathorekapila@gmail.com

Academic Editor: Sascha Rohn

India is the fourth largest oilseed economy in the world. Among the seven edible oilseeds cultivated in India, rapeseed-mustard contributes 28.6% in the total oilseeds production and ranks second after groundnut sharing 27.8% in the India's oilseed economy. The mustard growing areas in India are experiencing the vast diversity in the agro climatic conditions and different species of rapeseed-mustard are grown in some or other part of the country. Under marginal resource situation, cultivation of rapeseed-mustard becomes less remunerative to the farmers. This results in a big gap between requirement and production of mustard in India. Therefore site-specific nutrient management through soil-test recommendation based should be adopted to improve upon the existing yield levels obtained at farmers field. Effective management of natural resources, integrated approach to plant-water, nutrient and pest management and extension of rapeseed-mustard cultivation to newer areas under different cropping systems will play a key role in further increasing and stabilizing the productivity and production of rapeseed-mustard. The paper reviews the advances in proper land and seedbed preparation, optimum seed and sowing, planting technique, crop geometry, plant canopy, appropriate cropping system, integrated nutrient management and so forth to meet the ever growing demand of oil in the country and to realize the goal of production of 24 million tonnes of oilseed by 2020 AD through these advanced management techniques.

1. Introduction

Rapeseed-mustard is the third important oilseed crop in the world after soybean (*Glycine max*) and palm (*Elaeis guineensis* Jacq.) oil. Among the seven edible oilseed cultivated in India, rapeseed-mustard (*Brassica spp.*) contributes 28.6% in the total production of oilseeds. In India, it is the second most important edible oilseed after groundnut sharing 27.8% in the India's oilseed economy. The share of oilseeds is 14.1% out of the total cropped area in India, rapeseed-mustard accounts for 3% of it. The global production of rapeseed-mustard and its oil is around 38–42 and 12–14 mt, respectively. India contributes 28.3% and 19.8% in world acreage and production. India produces around 6.7 mt of rapeseed-mustard next to China (11-12 mt) and EU (10–13 mt) with significant contribution in world rapeseed-mustard industry. The rapeseed-mustard group broadly includes Indian mustard, yellow sarson, brown sarson, raya, and toria crops. Indian mustard (*Brassica juncea (L.)* Czernj. & Cosson) is predominantly cultivated in Rajasthan, UP, Haryana, Madhya Pradesh, and Gujarat. It is also grown under some nontraditional areas of South India including Karnataka, Tamil Nadu, and Andhra Pradesh. The crop can be raised well under both irrigated and rainfed conditions. Brown sarson (*B. rapa* ssp *sarson*) has 2 ecotypes lotni and toria. Yellow sarson (B. *rapa* var. *trilocularis*) is cultivated in Assam, Bihar, Orissa, and West Bengal as rabi crop. In Punjab, Haryana, UP, Himachal Pradesh, and Madhya Pradesh, it is grown mainly as a catch crop. Taramira (*Eruca sativa*) is grown in the drier parts of North-West India comprising the states of Rajasthan, Haryana, and UP. Gobhi sarson (*B. napus* L. ssp. *oleferia DC.* var *annua L.*) and karan rai (*Brassica carinata*) are the new emerging oilseed crops having limited area of cultivation. Gobhi sarson is a long duration crop confined to Haryana, Punjab, and Himachal Pradesh. It has good yield potential, wide adaptability and possesses high oil content of good quality. Karan rai yields well and shows better environment adoption and substantial resistance to pests and diseases. The country witnessed yellow revolution through a phenomenal

TABLE 1: Salient features of cultivated species of rapeseed-mustard (Cruciferous) group of crops.

SN	Common name	Botanical name	Days to maturity (days)	Yield potential, Kg/ha	Oil %
(1)	Indian mustard	*Brassica juncea*	105–160	1500–3000	38–42
(2)	Yellow mustard	*Brassica rapa* var. *yellow sarson*	120–155		41–47
(3)	Brown sarson	*Brassica campestris* syn. *B. rapa* var. *brown sarson*	100–235	900–2000	40–45
(4)	Black mustard	*Brassica nigra*	70–90	1000–1200	40-41
(5)	Karan rai	*Brassica carinata*	150–200		36–43
(6)	Toria	*Brassica rapa* var. *toria*	70–100	600–1800	36–44
(7)	Taramira	*Eruca sativa*	140–150	700–1400	34–38
(8)	Gobhi sarson	*Brassica napus*	145–180	1300–2700	37–45

increase in production and productivity from 2.68 MT and 650 kg/ha in 1985-86 to 6.96 MT and 1022 kg/ha in 1996-1997, respectively. In spite of these achievements, there exists a gap between production potential and actual realization. In India rapeseed-mustard is grown on an area of 5.53 Mha with production and productivity of 6.41 MT and 1157 Kg/ha, respectively [1].

Mustard is cultivated in mostly under temperate climates. It is also grown in certain tropical and subtropical regions as a cold weather crop. Indian mustard is reported to tolerate annual precipitation of 500 to 4200 mm, annual temperature of 6 to 27°C, and pH of 4.3 to 8.3. Rapeseed-mustard follows C_3 pathway for carbon assimilation. Therefore, it has efficient photosynthetic response at 15–20°C temperature. At this temperature the plant achieve maximum CO_2 exchange range which declines thereafter. *Rai* is mostly grown as a rainfed crop, moderately tolerant to soil acidity, preferring a pH from 5.5 to 6.8, thrives in areas with hot days and cool night and can fairly sustain drought. Mustard requires well-drained sandy loam soil. Rapeseed-mustard has a low water requirement (240–400 mm) which fits well in the rainfed cropping systems. Nearly 20% area under these crops is rainfed. A review is prepared on advances on agronomic practices for enhancing the rapeseed-mustard production in India. A review of the work done on the different aspects in India and abroad especially under advance agronomic practices is done in this paper.

2. Crop Adaptation and Distribution

The rapeseed-mustard group includes *brown sarson, raya,* and *toria* crops. Indian mustard (*Brassica juncea* (L.) Czernj. & Cosson) is predominantly cultivated in Rajasthan, UP, Haryana, Madhya Pradesh, and Gujarat. It is also grown under some nontraditional areas of South India including Karnataka, Tamil Nadu, and Andhra Pradesh. The crop can be raised well under both irrigated and rainfed conditions. Being more responsive to fertilizers, it gives better return under irrigated condition. Brown sarson (*B. rapa* ssp. sarson) has 2 ecotypes *lotni* and *toria*. Yellow sarson (*B. rapa* var. *trilocularis*) is cultivated in Assam, Bihar, Orissa, and West Bengal as *rabi* crop. In Punjab, Haryana, UP, Himachal Pradesh, and Madhya Pradesh, it is grown mainly as a catch crop. Taramira (*Eruca sativa*) is grown in the drier parts

of North-West India comprising the states of Rajasthan, Haryana and UP. *Gobhi sarson* (*B. napus l.* ssp. *oleferia DC.* Var. *annua* L.) and *karan rai* (*Brassica carinata*) are the new emerging oilseed crops having limited area of cultivation. *Gobhi sarson* is a long duration crop confined to Haryana, Punjab, and Himachal Pradesh. It is photo- and thermosensitive and makes little growth up to middle of February, but in the end of this month, plants make a quick growth. It has good yield potential, wide adaptability, and possesses high oil content of good quality. There are eight cultivated crops in rapeseed-mustard crop; the main characteristics features have been explained in Table 1.

Karan rai also yields well under a wide range of climate partly because it has a large number of primary and secondary racemes. It shows better environment adoption and substantial resistance to pests and diseases. Mustard is cultivated in most temperate climates. It is also grown in certain tropical and subtropical regions as a cold weather crop. Indian mustard is reported to tolerate annual precipitation of 500 to 4200 mm, annual temperature of 6 to 27°C, and pH of 4.3 to 8.3. Rai is mostly grown as a rainfed crop, moderately tolerant to soil acidity, preferring a pH from 5.5 to 6.8, thrives in areas with hot days and cool night, and fairly resistant to drought. Mustard requires good sandy loamy soil. The agro-climatic conditions of various locations under study have been explained in Table 2.

3. Varietals Development

Since, there is a vast variability in the climatic and edaphic conditions in the mustard growing areas of India, the selection of appropriate cultivars is important as it helps in increasing the productivity. Introduction of relatively short duration cultivar found favor with the environment where effective growing seasonal length is short. Improved varieties of mustard stabilize oil and seed yield through insulation of cultivars against major biotic and abiotic stresses enhance oil (low erucic acid) and seed meal (low glucosinolate) quality. The first Indian mustard hybrid, named "NRCHB-506," has been developed at Directorate of Rapeseed-Mustard Research, Bharatpur which can catapult the output of the country's key oil crop. The new hybrid is meant for cultivation in Rajasthan and Uttar Pradesh. Other high yielding varieties include "JM-1," "JM-3," and "Pusa Bold,"

TABLE 2: Agroclimatic conditions of various locations during mustard crop season.

| Location | Longitude | Latitude | Temp, °C | | Rain fall, mm | RH % | | Soil texture | Soil fertility, Kg/ha | | |
			Max	Min		Max	Min		N	P	K
Hisar	75°43′6″ E	29°9′11″ N	3.2	34.2	50–200	38	96	Sany loam	130	12	480
Pantnagar	79°24′36″ E	28°58′12″ N,	4.8	32.3	150–400	47	92	Clay loam	155	15	310
Dholi	85° 35′22″ E	26°0′2.2″ N	6.6	33.3	200–550	52	94	Clay loam	140	12.5	275
Ludhiana	75°18′ E	30°34′ N	3.5	32.0	30–120	45	95	Loamy sand	150	24	220
Bhubneshwar	85°50′ E	20°16′ N	14.8	34.8	180–250	38	94	Clay loam	130	19	175

TABLE 3: Varieties tolerant to various abiotic and biotic stresses of mustard (*Brassica juncea*).

SN	Specific abiotic/biotic stress	Tolerant verities
(1)	Rainfed	Aravali, Geeta, GM 1, PBR 97, PusaBahar, Pusa Bold, RH 781, RH 819, RGN 48, Shivani, TM 2, TM 4, Vaibhav, RB 50
(2)	Salinity tolerant	CS 52, CS 54, Narendra Rai (NDR8501)
(3)	Frost tolerant	RGN 13, RH 819, Swaranjyoti, RH 781, RGN 48
(4)	High temperature tolerant	Kanti, Pusa Agrani, RGN 13, Urvashi, NRCDR 02, Pusa mustard 25 (NPJ 112), Pusa mustard 27 (EJ 17)
(5)	White rust resistant	Basanti, JM 1, JM 2, NRCDR-2
(6)	*Alternaria* blight tolerant	Jawahar Mustard 3, Him Sarson 1 (ONK 1), Ashirwad (RK-01-03)

"NRCDR-2," "NRCDR 601." Their yield potentials vary from 16 to 25 q/ha. At IARI, an early-maturing and bold seeded mustard variety has been developed called "Mehak" (*B. juncea*). This improved variety is suitable for early sowing to replace *toria (B. rapa* var. *toria)* in Delhi and adjoining areas. *Gobhi sarson* has a good yield potential, wide adaptability and possesses high oil content of good quality. "Hyola" (PAC-401) is canola type hybrid rapeseed, developed in India by Advanta India Ltd, Holland-based multinational company. "Neelam" (HPN-3) and "Sheetal" (HPN-1) are the popular varieties of *gobhi sarson* [2]. Since inception of mustard research programme in India, number of tolerant varieties to various abiotic and biotic stresses of rapeseed-mustard has been developed (Table 3).

"Pusa Jaikisan" of *B. juncea* is the first variety though tissue culture. "TL-15," a toria variety has been recommended as summer crop for high altitude of Himachal Pradesh. In an attempt to incorporate resistance/tolerance to biotic and abiotic stresses in high yielding varieties, aphid tolerant strains like "RH-7846," "RH-7847," "RH-9020" and "RWAR-842," *Alternaria* blight moderately resistant variety "Saurabh"; white rust resistant variety, "Jawahar Mustard-1"; salt tolerant varieties "Narendra Rai" and "CS-52" frost tolerant "RH-781" and "RH-7361" varieties have been identified. "RH-781" is also drought tolerant and suitable for intercropping. For nontraditional areas, Indian mustard varieties "Rajat," "Pusa Jaikisan" and "Sej.2" have been recommended.

4. Land and Seedbed Preparation

A mustard seedbed should be firm, moist, and uniform which allows good seed-to-soil contact, even planting depth and quick moisture absorption leading to a uniform germination. Tillage affects both crop growth and grain yield. The various tillage systems are as follows: conventional tillage includes moldboard ploughing followed by disc harrowing; reduced tillage includes disc ploughing followed by disc harrowing and complete zero tillage in which crop is sown under uncultivated soil. Minimum tillage, with or without straw, enhances soil moisture conservation and moisture availability during crop growth. As a consequence, the root mass, yield components and seed yield increase [3]. Zero tillage is preferred in mustard as it conserves more moisture in the soil profile during early growth period. Subsequent release of conserved soil moisture regulates proper plant water status, soil temperature, lower soil mechanical resistance, leading to better root growth and higher grain yield of mustard [4]. Success with minimum or zero tillage requires even distribution of crop residues, as a well-designed crop rotation and evenly distributing residue will create a firm, moist and uniform seedbed.

Continuous zero tillage results in redistribution of extractable soil nutrients with greater concentration near the soil surface, compared with conventional tillage where mixing of soil, residues, fertilizers, and lime results in a relatively homogeneous soil to the depth of tillage [6]. With zero tillage having greater root density in the surface soil but lesser root density below a depth of 15 cm in the soil profile. Therefore, P and K uptake by crops grown under zero tillage is greater than those grown by conventional methods. But the plant growth and dry matter yields of mustard under zero tillage will be higher only if N fertilizers are applied in appropriate amount [7]. Under AICRP on RM at Dholi, Kanke, Bhubaneshwar, and Behrampur maximum seed yield of *toria* and mustard was obtained in line sowing under zero

TABLE 4: Seed yield (kg/ha) and oil content (%) of *toria* as influenced by different N levels in *utera* cropping system at Bhubaneshwar.

Cropping system	N levels (kg/ha)		
	0	40	80
Rice: yellow sarson (broadcast) in *utera* cropping (at dough stage of rice)	428 (33.3)	823 (40.3)	810 (37.6)
Rice: yellow sarson (broadcast) in *utera* cropping (sowing before harvest of rice)	530 (30.2)	729 (38.2)	642 (37.1)
Rice: yellow sarson (line sowing) under zero tillage in rice field	506 (34.4)	924 (41.5)	886 (39.6)
Rice: yellow sarson (line sowing) after land preparation in rice fields	388 (32.5)	846 (40.4)	820 (38.4)
Rice: yellow sarson (broadcast) after land preparation in rice fields	301 (28.2)	460 (37.6)	440 (35.5)

CD at 5% cropping system: 79 (0.7), N levels: 32 (0.4), Cropping system × N levels: 98 (1.0). Figures in the parenthesis denotes oil content (%).
Source: AICRP-RM, 2003 [5].

tillage practice which indicated that mustard can be grown well under zero tillage.

At Bhubaneshwar, line sowing of mustard under zero tillage after rice gave the maximum seed yield (933 kg/ha) and oil content (38.4%) (Table 4). The soil under zero tillage system contains higher amount of organic matter having more carbohydrate, amino acid and amino sugar that results in qualitative and quantitative improvement in soil and soil structure due to least soil disturbance. Energy output and input ratio are higher in zero tillage as compared to conventional tillage.

5. Seed and Sowing

Vigorous seedling growth, good root development, early stem elongation, rapid ground covering ability, and early flowering and radiation are important yield determining traits under low temperature and radiation regime. These traits can be successfully exploited in mustard if a good seed is grown at appropriate time along with maintaining an optimum plant population.

5.1. Seed Priming. Seed treatment is a useful practice for healthy plant growth. Seed priming through controlled hydration and dehydration enhances early germination of mustard seed in less time, even in compacted soil [8]. The soaking of mustard seeds in 0.025% aqueous pyridoxine hydrochloride solution for 4 hours improved germination. The combination of pyridoxine + $N_{60}P_{20}$ + $N_{15}P_5$ (top dressing) accelerated the crop performance by enhancing seed yield and oil yield by 15.8 and 13.5%, respectively, over the control [9]. The differential response of varieties for imbibition gives advantage to some of them to germinate early as compared to others. At Hisar, maximum rate of imbibition was reported in "NRCDR-2" (41.7%) and minimum in "NRCDR-509" (7.5%). Such drastic difference in rate of imbibition is important for identification of suitable varieties under abiotic stress conditions namely drought, frost, and temperature abnormalities.

5.2. Sowing Time. Sowing time is the most vital nonmonetary input to achieve target yields in mustard. Production efficiency of different genotypes greatly differs under different planting dates. Soil temperature and moisture influence the sowing time of rapeseed-mustard in various zones of the country. Sowing time influences phenological development of crop plants through temperature and heat unit. Sowing at optimum time gives higher yields due to suitable environment that prevails at all the growth stages. Though different varieties have a differential response to date of sowing, mustard sown on 14 and 21 October took significantly more days to 50% flowering (55 and 57) and maturity (154 and 156) compared to October 7 planting [10]. Delayed sowing resulted in poor growth, low yield, and oil content. The reduction in yield was maximum in "RH-30" and minimum in "Rajat" [11, 12].

Date of sowing influence the incidence of insect-pest and disease also. Sowing on October 21 resulted in least *Sclerotinia* incidence [13]. The maximum (20.5–25.4°C) and minimum (3.9–10.7°C) temperatures at the flowering stage of crops established through sowing on October 21 were negatively correlated with the development of *Sclerotina* stem rot. Mustard aphid (*Lipaphis erysimi (Kaltenbach)*) has been reported as one of the most devastating pests in realizing the potential productivity of Indian mustard. Normal sowing (1st week of November) also helps in reducing the risk of mustard aphid incidence.

5.3. Planting Technique. Sowing technique depends upon land resources, soil condition, and level of management and thus broadcast, line sowing, ridge and furrow method and broad bed and furrow method are common sowing techniques. At higher soil moisture regimes, broadcasting followed by light planking gives early emergence and growth. Under normal and conserved moisture regime, seed placement in moist horizon under line sowing becomes beneficial.

At Shillongani, broadcast method was found to be more successful. Significantly higher seed yield of *toria (Brassica rapa var. toria)* was harvested in broadcast sowing of *toria* over other practices. *Toria* broadcast at dough stage along with 80 kg N/ha gave the highest yield (AICRP-RM, 2006). At Bhubaneshwar, line sowing of yellow sarson after land preparation produced maximum seed yield (870 kg/ha) with 40 kg N/ha [14]. At Behrampore, 40% higher seed yield of *toria* was obtained when sown in line after land preparation in the rice-based cropping system over broadcast (AICRP on RM, 2006). *Paira* or *utera* is a method of *cropping* in which the sowing of next crop is done in the standing previous crop without any tillage operation. Mustard sowing under *paira/utera* in the rice field has shown its edge over

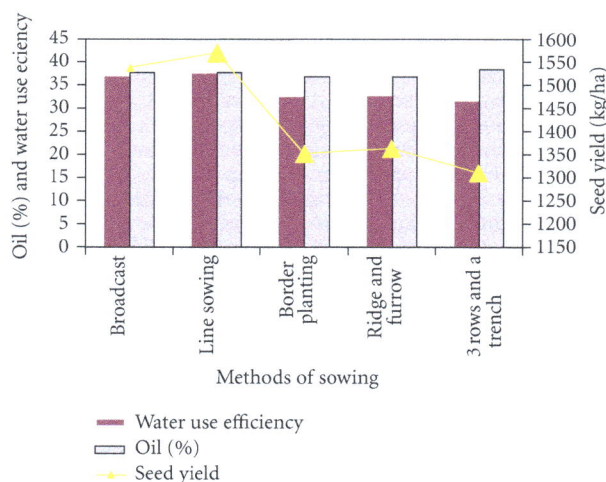

FIGURE 1: Seed yield, water use efficiency, Kg/ha-mm (WUE), and oil content of mustard (*Brassica juncea*) as influenced by various planting methods.

line sowing and broadcasting (Sowing of seeds by broad casting the seeds in the field) in eastern parts of India. At Dholi, mustard sown with *paira* cropping recorded significantly higher seed yield (1212 kg/ha) over line sown and broadcast method, while these 2 methods yielded at par. At Bhubaneswar, significantly higher yield (887 kg/ha) of mustard was recorded when sown as *utera* crop over line and broadcast sown crop [15].

Ridge and furrow sowing was superior to conventional flat sowing for growth parameters and yield of *Brassica juncea* [16]. Under saline condition, seed yield of canola in ridge sowing was higher by 45, 31, and 28% than broadcast, drill and furrow sowing methods, respectively [17]. The highest yield was associated with less saline environment at the ridges which allowed the seed to germinate and increase the yield. Transplanting of mustard has also been reported thereby saving time, and resources. Transplanting reduces days to maturity and results in higher seed yield. Ridge transplanting reduced water applied by 30% for each furrow as compared to 45 cm row spacing in flat method without any loss in seed yield. The corresponding increase in water use efficiency (WUE) was 27%. In bed planting, there was a 35% saving in water resulting in 32% increase in WUE (Figure 1).

5.4. Crop Geometry. The competitive ability of a rapeseed-mustard plant depends greatly upon the density of plants per unit area and soil fertility status. The optimum plant population density/unit area varies with the environment, the genotype, the seeding time, and the season. Uniform distribution of crop plants over an area results in efficient use of nutrients, moisture, and suppression of weeds leading to high yield. In wider row spacing, solar radiation falling within the rows gets wasted particularly during the early stages of crop growth whereas in closer row spacing upper part of the crop canopy may be well above the light saturation capacity but the lower leaves remain starved of light and contribute negatively towards yield.

Gobhi sarson (*Brassica napus*) being more vigourous, the days to maturity, plant height, branches, pod, seed weight per plant, seed index, seed yield, and oil content were higher at 60 cm row spacing [18]. An increase in rows up to 30 cm correspondingly prolonged maturity days followed by optimum 45 cm and wider rows 60 cm spacing. The plants receiving narrow row spacing increased vegetative growth. Due to shade and competition for nutrients and moisture the crop matures later by increasing developmental phases Taller plants were observed in the plots where crop was planted in rows of 60 cm apart followed by 45 cm and 30 cm row spacing due to sufficient space resulting in plants grown well and showed greater height [19] (Gupta, 1988). The regression coefficient indicated that each increase in row spacing up to 60 cm resulted in increased crop maturity by 0.54 days, plant height by 0.44 cm, number of branches would increase by 0.11, pods per plant by 1.96, seeds per pod by 0.04, seed weight per plant by 0.45, seed index by 0.152 g, oil content by 0.8% and increase in seed yield by 10.32 kg/ha. The recommended spacing for mustard is 30×10 and for hybrids it is 45×10. At Kumher, plant spacing 45×15 recorded significant higher seed yield over other spacing but was on a par with 45×10 cm. At Pantnagar, 30×15 recorded significantly higher seed yield which remained on a par with 45×10 and 45×15 cm plant spacing [20].

5.5. Plant Population and Inter-Plant Shading. The dense plant population reduces the yield due to reduction in the photosynthetically active leaf area caused by mutual shading. In an experiment on *Brassica juncea* (*Var. laxmi*) the reduction is more due to shading at 91–110 DAS over 71–90 DAS. The specific leaf weight (SLW), crop growth rate (CGR), and net assimilation rate (NAR) were more adversely affected by 50% shading at 71–90 DAS. Net assimilation ratio remained unaffected by 25% shading, while it reduced significantly by 50% shading at both the stages; the reduction was more with 50% due to shading at 91–110 DAS. On an average 50% shading at 91–110 DAS was more deleterious than 25% shading at 91–110 DAS, that is, at terminal seed development stage (Table 5).

6. Cropping System

Physiography, soils, geological formation, climate, cropping pattern, and development of irrigation and mineral resources greatly influence selection of variety and cropping system. Fallow mustard is popular sequence in major mustard growing areas but studies show that some of the crop result in better resource utilization and high remuneration if included in mustard-based cropping system.

6.1. Mustard Productivity under Various Crop Sequences. Under AICRP trials at Dholi, fallow-mustard sequence gave significantly higher seed yield which was on a par with blackgram-mustard sequence: urdbean-mustard at Morena; greengram-mustard, guar-mustard, and pearl-millet-mustard at S. K. Nagar and Hisar; maize-mustard

TABLE 5: Effect of shading on yield and growth parameters in Indian mustard at Hisar.

Treatment	Seed yield (kg/ha)	SLW (mg/cm^2)	CGR (g/m^2/day)	NAR (mg/m^2/day)
Control	571.6	8.3	11.3	0.93
25% shading at 71–90 DAS	546.0	7.0	10.8	0.93
25% shading at 91–110 DAS	490.9	7.9	9.4	0.87
50% shading at 71–90 DAS	527.0	6.4	9.9	0.95
50% shading at 91–110 DAS	380.0	7.0	8.3	0.83
CD at 5%	33.1	1.3	1.0	0.05

Source: AICRP-RM, 2004.

TABLE 6: Seed yield (kg/ha), mustard equivalent yield (MEY), and gross return (Rs./ha) as influenced by various intercropping combinations under rainfed conditions at Hisar.

Treatment	Main crop	Intercrop	MEY	Gross return (Rs./ha)
Pure mustard	2565	—	2565	29,497
Mustard + chickpea (1:5)	966	1035	1956	22,494
Mustard + fieldpea (1:5)	1002	189	1230	14,145
Mustard + linseed (1:5)	996	642	1721	19,791
Mustard + lentil (1:5)	1015	—	1015	11,672
Mustard without intercropping at same distance as in intercropping	1097	—	1092	12,668
CD at 5%	—	—	350	—

Source: AICRP-RM, 1997 [11].

at Kangra and Pantnagar revealed superiority to fallow-mustard. The productivity of the system also depends upon the fertility status and the nutrient supply. When mustard was grown after soybean or bajra, the response to S was observed up to 40 kg S/ha [21]. Productivity measured in terms of land equivalent ratio (LER) was higher for intercropping of chickpea and mustard in the 4:1 row ratio than for sowing of chickpea and mustard in sole stands [22].

6.2. Inclusion of Gobhi Sarson (Brassica Napus) under Various Cropping Sequences.
Gobhi sarson is comparatively recent introduction and hence needs identification of suitable cropping systems. Growing *gobhi sarson* and toria in alternate rows at 22.5 cm spacing is very remunerative. Maize-*gobhi sarson*, blackgram-gobhi sarson, rice-*gobhi sarson*, and soybean-*gobhi sarson* were identified remunerative cropping systems at Kangra [21].

6.3. Mustard-(Brassica Juncea) Based Cropping System under Rainfed Areas.
There are possibilities of increasing cropping intensity in monocropping mustard areas under rainfed condition. Green manuring or guar during rainy season enhance seed yield of succeeding mustard [12]. In addition to efficient resource use, intercropping imparts stability to productivity and reduces the risk of crop failure. Under irrigated conditions, at Bharatpur, the seed yield equivalent of mustard (*Brassica juncea*) was significantly higher where mustard was grown in combination with potato (1:3), mustard + wheat (1:5), mustard + barley (1:5) than pure mustard. At Hisar, intercropping *Brassica juncea* (variety RH-30) with rabi crops had revealed highest gross return (Rs. 29,498) when mustard was grown as a pure crop. The mustard seed equivalent was highest in mustard + chickpea

(1:5). Intercropping of mustard with chickpea, field pea, or linseed proved superior over their cultivation as a pure crop (Table 6).

7. Fertilizer Management

Adequate nutrient supply increases the seed and oil yields by improving the setting pattern of siliquae on branches, number of siliquae/plant, and other yield attributes [23]. Recommended dose of fertilizers (RDF) for different zones changes with climate, soil type, time, and type of cropping system followed.

7.1. Nitrogen and Phosphorus Fertilization.
Nitrogen use efficiency is greatly influenced by the rate, source, and method of fertilizer application. The rate of nitrogen depends upon the initial soil status, climate, topography, cropping system in practice, and crop. Crop under zero tillage is also more productive (695 kg/ha) with 80 kg N/ha [14]. Increase in the nitrogen level up to 60 kg N/ha consistently and significantly increased the number of primary branches, number of seeds per siliquae and 1000 seed weight [24]; however, increasing the nitrogen level up to 90 kg/ha increased the number of secondary branches per plant, number of siliquae per plant, and seed and straw yield with maximum cost benefit ratio of 3.03 [25]. Split application of total nitrogen in three equal doses one-each as basal, second after first irrigation and remaining one-third after second irrigation resulted in maximum increase in yield attributes and yield of *Brassica juncea* compared to application of total nitrogen in two split doses [26]. Top dressing of N fertilizers should be done immediately after first irrigation. Delaying of first irrigation, results in yield reduction of mustard crop. The application

TABLE 7: Effect of N and S levels (kg/ha) application on fatty acid composition and glucosinolate content in *Brassica juncea* cv. Varuna at Ludhiana.

N (Kg/ha)	S (Kg/ha)	Glucosinolate content (μmoles/g in defatted meal)	Palmitic acid	Stearic acid	Oleic acid	Linoleic acid	Linolenic acid	Eicosenoic acid	Erucic acid
75	0	64	2.61	1.17	11.78	14.99	6.48	50.91	11.80
75	20	72	2.88	1.31	10.15	14.53	5.14	52.75	12.28
100	0	52	2.58	1.58	13.16	15.31	7.01	49.55	10.57
100	20	42	2.91	1.65	11.94	15.06	6.13	49.63	12.18
125	0	52	3.01	1.33	12.19	16.17	5.91	47.71	12.26
125	20	42	4.42	1.31	16.12	16.55	6.57	44.77	9.55

Source: AICRP-RM, 2007 [14].

of nitrogen with presowing irrigation was superior to that of nitrogen application with last preparatory tillage. In case of nitrogen applied with pre-sowing irrigation single application of nitrogen was on a par with split application [27].

Application of phosphorus up to 60 kg/ha significantly enhanced dry matter/plant. Plant height, branches per plant and leaf chlorophyll content increased with up to 40 kg P/ha. The uptake of NPK and sulphur by both seed and stover increased significantly with successive increase in nitrogen levels up to 120 kg N/ha, sulphur levels up to 60 kg S/ha, and P_2O_5 level up to 60 kg P_2O_5/ha. Seed yield and yield attributes increased while oil content decreased with increasing level of nitrogen up to 120 kg/ha. Different levels of phosphorus increased seed yield, maximum being at 80 kg P/ha due to higher number of secondary branches/plant and consequently siliquae/plant. Oil content also increased with increase in levels of N, P_2O_5, and S. Activities of all nitrogen assimilating enzymes, namely; nitrate reductase, nitrite reductase, glutamine synthetase, and glutamate synthetase were found to be maximum at 100 kg N/ha.

7.2. Sulphur Fertilization.
7.2. Sulphur Fertilization. Among the oilseed crops, rapeseed-mustard has the highest requirement of sulphur [28]. Sulphur promotes oil synthesis. It is an important constituent of seed protein, amino acid, enzymes, glucosinolate and is needed for chlorophyll formation [29]. Sulphur increased the yield of mustard by 12 to 48% under irrigation, and by 17 to 124% under rainfed conditions [30]. In terms of agronomic efficiency, each kilogram of sulphur increases the yield of mustard by 7.7 kg [31].

Oil content in Canola-4 and Hyola-401 is 3% higher than the hybrid "PGSH-51" due to the effect of various doses of nitrogen and sulphur, while the oleic acid content in these hybrids is double that "PGSH-51." "PGSH-51" had erucic acid ranging from 23.2 in to 29.4%. At higher sulphur level there is 2-3% reduction in erucic acid content. However, lower level of nitrogen reduced erucic acid content by 3% with a concomitant increase in oleic acid (Table 7). Higher doses of sulphur along with low doses of nitrogen affect the chain elongation enzyme system thereby leading to reduction in erucic synthesis.

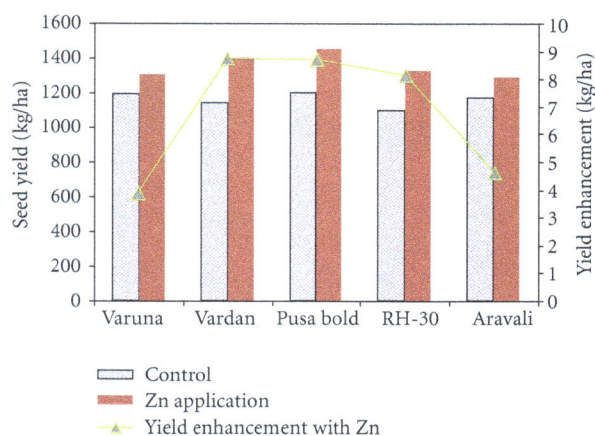

FIGURE 2: Influence of zinc application on seed yield of different cultivars of mustard.

A significant increase in yield was observed with increase in sulphur levels up to 40 kg S/ha in mustard-based cropping system. At Bawal, the highest seed yield of mustard was recorded in green gram-mustard cropping sequence while the lowest (2686 kg/ha) in pearl millet-mustard sequence. In rice-mustard sequence, the optimum seed yield of mustard was obtained at 40 kg S/ha at Behrampore and for blackgram-mustard at Dholi. Each successive increase in S level increased seed yield up to 20 kg S/ha at Dholi and Ludhiana, 40 kg S/ha at S. K. Nagar, and 60 kg S/ha at Behrampore and Morena conditions [32].

7.3. Micronutrients.
7.3. Micronutrients. Mustard, in general is very sensitive to micronutrient deficiency, specially zinc and boron. The increase in seed yield was 8.5% at 12.5 kg $ZnSO_4$/ha. The harvest index (HI) was significantly affected by Zn application, although seed yield showed diminishing return with additional $ZnSO_4$ doses (Table 8).

The response of various ideotype to the applied micronutrients varies considerably. The response of Indian mustard varieties, *viz.* 'Pusa Bold' and 'Vardan' to applied zinc was found higher (AICRP-RM, 2000) as compared to Varuna, RH- 30 and Aravali (Figure 2).

The concentration of Zn at flowering, pod formation stage, concentration and uptake of Zn in straw and grain

TABLE 8: Effect of Zn on yield and yield attributes of indian mustard.

$ZnSO_4$ (Kg/ha) levels	Seed yield (kg/ha)	Secondary branches/plant	Oil content (%)	Oil yield (kg/ha)	Protein (%)	Protein yield (kg/ha)	Harvest index (%)
0	1161	6.5	40.2	465.6	22.1	255.2	21.6
12.5	1260	8.1	39.9	501.1	22.5	281.9	22.4
25.0	1336	9.6	39.9	532.4	22.6	301.6	22.9
50	1414	12.4	39.9	570.0	22.5	318.6	22.2
CD at 5%	33	0.7	NS	22.8	NS	18.8	0.8

Source: AICRP-RM, 2000 [33].

at maturity and uptake of Zn in grain and straw at maturity of Indian mustard increased significantly with increase in Zn levels [34]. Similarly, the seed yield increased significantly (16–47%) with the application of boron. The average response to boron application ranged from 21 to 31%. The yield increase was due to 27% and 10% increase, respectively, in seeds/siliqua and 1000 seed weight, indicating the importance role in seed formation [35, 36].

7.4. Organic Sources of Nutrients. Bulky organic manures are applied to improve overall soil health and reduce evaporation losses of soil moisture. Depending upon the availability of raw material and land use conditions various organic sources, *namely,* clusterbean (green manure), *Sesbania* (green manure), mustard straw @ 3 t/ha and Vermicompost (2.5–7.5 t/ha) have been evaluated at Bharatpur. Green manure with *Sesbania* gave significantly higher mustard seed yield at Bharatpur and Bawal. *Sesbania* green manuring has shown higher mustard yield and improved soil environment (AICRP-RM, 2006).

Many biostimulants also encourage higher production. At Hisar, foliar spray of Bioforce (an organic formulation) 2 mL/L at the flowering and siliqua formation stage enhanced mustard seed yield (2059 kg/ha) [14].

7.5. Integrated Nutrient Management (INM). It is important to exploit the potential of organic manures, composts, crop residues, agricultural wastes, biofertilizers and their synergistic effect with chemical fertilizers for increasing balanced nutrient supply and their use efficiency for increasing productivity, sustainability of agriculture, and improving soil health and environmental safety. Balanced fertilization at right time by proper method increases nutrient use efficiency in mustard. Experiments have been conducted at different AICRP centres with the integrated use of organic manure, green manure, crop residue, and biofertilizers along with inorganic fertilizers. INM not only reduces the demand of inorganic fertilizers but also increases the efficiency of applied nutrients due to their favourable effect on physical, chemical and biological properties of soil. The introduction of leguminous crops in the rotational and intercropping sequence and use of bacterial and algal cultures play an important role in increasing the nutrient use efficiency [37].

7.5.1. Growth Promoter, BioFertiliser as a Component of INM. Biofertilizers are inoculants or preparation containing micro-organims that apply nutrients especially N and P. Two types of N-fixing microorganisms namely free living (*Azotobacter*) and associative symbiosis (*Azospirillum*) and two P supplying microorganisms, namely, phosphate solubilizing bacteria and vesicular arbuscular mycorrhiza (VAM) were extensively tested at various AICRP-RM centers. Inoculation of mustard seeds with efficient strains of *Azotobacter* and *Azospirillum* enhanced the seed yield up to 389 and 305 kg respectively with 40 Kg N/ha. The total NPK uptake was also higher with *Azotobacter* inoculation. The combined application of 10 t FYM + 90 : 45 : 45 NPK kg/ha with *Azotobacter* inoculation gave the highest B : C ratio of 1.51. At lower N levels, without inoculation, the seed yield decline was more as compared to inoculated treatment. Growth promoter's formulations like bioforce and biopower contain bio-amino acid, plant growth promoting terpenoid, siderophores, and attenuated bacteria fortified with BGA helped to increase water and nutrient absorption from the soil. Similarly, bioforce contains natural free amino acid, phytohormones, macro- and microelements and plant growth promoting terpenoid activated the cell division and stimulates plant growth, development, and photosynthate translocation. RDF (80 : 40 : 0) along with 25 kg Biopower/ha + spray of Bioforce (1 l in 500 litres of water) at 50% flowering and pod filling stage gave significant higher yield of mustard over other combinations [35, 36].

7.5.2. Effect of INM on Quality of Mustard Oil. At Kanpur, INM studies were evaluated in maize-mustard, bajra-mustard, and fallow mustard sequence. In maize-mustard sequence, 100/75% of RDF + 2 t FYM gave highest seed yield and quality of the oil (Table 9).

7.5.3. Integrated Nutrient Management (INM) and Nutrient Use Efficiency. INM improves the nutrient uptake by mustard and hence enhances the use efficiency of various nutrients from the soil. The incorporation of 25% nitrogen through FYM + 75% by chemical fertilizer + 100% sulphur significantly enhanced the uptake use efficiency and of nitrogen and sulphur in both seed and stover of crop followed by 100% NS and 50% N through FYM + 50% by chemical fertilizer + 100% S [38]. The highest mustard-equivalent yield, which includes converted yield of other crops in to mustard seed yield based on market price of the crops (24.88 q/ha), net monetary returns (Rs. 15,537/ha), B : C ratio (2.07), and agronomic efficiency (16.1) were

TABLE 9: Effect of INM on quality of mustard (Kanti-RK 9807) under maize-mustard sequence.

Treatment	Legends	Oil content (%)	Fatty Acid composition (%)					
			16:1	18:1	18:2	18:3	20:1	22:1
			Palmtic acid	Oleic acid	Linoleic acid	Linolenic acid	Eicosenoic acid	Erucic acid
RDF (120-40-40)	T1	40.4	2.8	18.4	10.1	10.6	4.3	52.7
T$_1$ + 10 t FYM/ha	T2	40.9	2.8	16.3	13.3	10.4	4.1	52.2
T$_2$ + 40 Kg S/ha	T3	40.4	2.9	18.0	14.4	12.2	3.2	48.6
T$_3$ + Zn SO4 25 kg/ha	T4	40.3	2.8	17.8	14.9	10.1	6.1	47.3
T$_4$ + B 1 kg/ha	T5	40.7	2.7	23.0	16.2	9.0	5.2	43.3
T$_1$ + Crop residue (Maize)	T6	40.1	2.7	20.0	14.3	9.2	4.4	48.6
75% RDF		40.4	2.6	17.8	15.1	7.9	6.3	49.7

Source: Modified from AICRP-RM, 2002 [21].

recorded with the application of 100% recommended N in the rainy season through FYM and 100% recommended NP in the winter season through inorganic fertilizers [39]. Agronomic efficiency is the response in terms of increase in mustard seed yield per unit use of nitrogen.

At Bharatpur and Jobner, 17.8 and 8.6% increase in seed yield was recorded with 50% RDF + 50% N through FYM and vermin-compost. Sole organic treated plot recorded 29.9% lesser seed yield over RDF at Jobner [32]. Amount of available phosphorus increased over initial value when organic manures and crop residues were incorporated. Organic carbon status builds up in organic source incorporated plots. The application of 10 t FYM/ha in addition to recommended dose of fertilizer (RDF) improved soil physical condition by improving aggregation, increased saturated hydraulic conductivity, and reducing bulk density and penetration resistance of the surface soil [40].

8. Water Management

Rapeseed-mustard crop is sensitive to water shortage. A substantial rapeseed-mustard area in Rajasthan (82.3%), Gujrat (98%), Haryana (75.6%), and Punjab (92.4%) is covered under irrigation. A positive effect of irrigating rapeseed-mustard at critical stages is observed. Water use efficiency was highest when irrigation was applied at 0.8 IW : CPE ratio and increased with increasing N rate [41, 42]. Number of irrigations is important for working out the most efficient water use by mustard. For mustard, two irrigations, one at flowering stage and at siliqua formation stage increased seed yield by 28% over the rainfed plots [43]. Increase in the amount of water increased leaf water potential, stomatal conductance, light absorption, leaf area index, seed yield, and evapotranspiration and decreased canopy temperature [44]. In similar study by Panda et al. [45], an average increase in seed yield with irrigation at the flowering and pod development stages and irrigation at the flowering stage over the control was 62.9% and 41.7%, respectively. However, for number of seeds per siliqua and oil content, single irrigation at 45 DAS remained parallel with two irrigations [46]. The water use efficiency was highest with one irrigation at 45 DAS. Crop receiving two irrigations at preflowering and pod-filling stages produce about 33 percent more seed than unirrigated crops [47]. Single irrigation given at vegetative

stage is found to be most critical, as irrigation at this stage produces the highest yield. When two irrigations are given, the irrigation at vegetative and pod formation stages is of maximum benefit. The irrigation at vegetative, flowering, and pod formation stages resulted in the highest yield, where three irrigations were given. Oil and protein yield were also significantly affected by number and stages of irrigation (Table 10).

Irrigation is very important for getting the optimum productivity potential of mustard, but equally important is the quality of irrigation water. If the quality of irrigation water is poor, it needs certain treatment and management before being utilized for crop production. The increasing levels of salinity of the irrigation water applied at presowing and flower initiation reduces the plant height, the branching pattern, and the pod formation [48]. Irrigation with saline water (12 and 16 dS/m) decreased the dry matter yield significantly when applied at pre-sowing or later. The saline irrigation at the pre-flowering stage or later reduced the grain yield by 50% and 70%, respectively.

As a result of saline water irrigation, the soil water infiltration was reduced up to 7%. The EC and exchangeable sodium percentage (ESP) were increased by 2.2 dSm^{-1} and 9.0, respectively. The yield of mustard crop could be further increased by better leveling the plots, reducing the level difference to less than 10 cm [49]. The ill effects of saline water can be overcome with proper N management. Nonsaline water can be substituted by applying N and saline water [50].

9. Weed Management

Weeds cause alarming decline in crop production ranging from 15–30% to a total failure in rapeseed-mustard yield. The critical period is 15–40 days. Weeds compete with crop plants for water, light, space, and nutrients. Therefore, timely and appropriate weed control greatly increases the crop yield and thus nutrient use efficiency. The common weeds of mustard are *Chenopodium album, C. murale, Cyperus rotundas, Cynodon dactylon, Melilotus alba, Asphodelus tenuifolius, Orobanche* spp. *and Anagallis arvensis*.

Farmers have adopted herbicides for weed control because the chemicals can increase the profit, weed control efficiency, production flexibility and reduce time and

TABLE 10: Influence of irrigation levels and stages on seed yield, oil yield and protein yield of Indian mustard.

Treatment	Seed yield (Kg/ha)	Oil yield (kg/ha)	Protein yield (kg/ha)
4 irrigations at V + F + P + S	2260	909	454
3 irrigations at V + F + P	2250	901	454
3 irrigations at V + F + S	2200	886	442
2 irrigations at V + P	2150	879	436
2 irrigations at V + F	2090	841	422
2 irrigations at F + P	2020	803	417
2 irrigations at P + S	1520	574	316
1 irrigation at V	1920	773	386
1 irrigation at F	1790	727	371
CD at 5%	480	144	94

Note: V: vegetative stage; F: flowering stage; P: pod formation; S: seed development.
Source: AICRP-RM, 1999 [15].

labour requirement for weed management. Hand weeding at 20DAS, fluchloralin preplant incorporation @ 0.75 kg/ha, wooden hand plough between the lines at 35 DAS on Indian mustard was found effective [51]. Polythene mulch was also found effective in controlling the weeds in mustard [52]. At Bawal, reductions in weed population and dry matter were obtained with fluchloralin supplemented with hand weeding at 30 and 60 DAS, which remained on a par with isoproturon and pendimethalin supplemented with hand weeding at 30 and 60 DAS. Weed-free plot recorded 39.9% higher seed yield over weedy check [32].

Broomrape (*Orobanche*) is a major devastating parasitic weed of mustard. Broomrape weed infestation caused 28.2% average reduction in Indian mustard yield. Among *Orobanche* spp., *O. aegyptiaca* is one of the most important parasitic weed causing severe yield and quality reducing factor in rapeseed-mustard. It is endemic in semiarid region and may reach epidemic proportions depending upon soil moisture and temperature. Preceding crop of cowpea, black gram, moth bean, sunn hemp, cluster bean, and sesame significantly reduced *Orobanche* menace in succeeding mustard crop while sorghum, pearl millet, chilies, and green gram did not influence broomrape infestation in mustard [53]. At Bharatpur, S. K. Nagar and Bawal directed spray of glyphoste (0.25–1.0%) and 2 drops of soybean oil per young shoot of *Orobanche* showed effective control and recorded 91.9% higher seed yield over infected sick plot.

Some cultural practices like mulching and hoeing are also helpful to curb some of the major weeds in mustard by providing a shield against sunlight, reducing the soil temperature and acting as a physical barrier for emergence of weeds. Maximum seed yield (2540 kg/ha) was obtained in the treatments where plots were kept weed-free followed by the treatment where mulching was done after hoeing (Table 11).

10. Response to Plant Growth Regulators

Plant growth regulators (PGR) involved in manipulating plant developments, enhancing yield and quality have been actualized in recent years. Indeterminate plant growth habit, shattering, or dehiscence of fruits and lodging are the

TABLE 11: Seed yield (kg/ha) and weed population/m^2 as influenced by different weed control practices.

Treatment	Seed yield	Weed population/m^2
Control	1620	57.0
Weed free (Khurpi)	2520	0.0
Hoeing at 25 DAS	2300	19.3
Mulching with bajra florets	1960	23.0
Fluchloralin @ 1 kg a.i./ha PPI	2000	23.0
Pendimethalin @ 1 kg a.i./ha PE	2050	22.1
Isoproturon @ 1 kg a.i./ha PE	1740	26.3
Hoeing at 25 DAS + mulching	2400	17.9
Fluchloralin @ 1 kg a.i./ha PPI + Hoeing	2210	20.3
Fluchloralin @ 1 kg a.i./ha PPI + Mulching	2100	22.5
Pendimethalin @ 1 kg a.i./ha PE + hoeing	2300	18.9
Pendimethalin @ 1 kg a.i./ha PE + mulching	1860	19.5
Isoproturon @ 1 kg a.i./ha PE + hoeing	1950	22.5
Isoproturon @ 1 kg a.i./ha PE + mulching	1910	22.9

Source: AICRP-RM, 2002 [21].

most significant and consistent limitations to maximum seed yields in *Brassica* spp. Considerable seed loss takes place, before or during harvest, due to shattering of fruits, which is correlated with hormonal imbalances and poorly developed lignified cells in the fruit wall. Further, lodging of the crop canopy adversely affects seed quality and yield due to decreased photosynthesis, increased disease severity, impaired rate of drying, and reduced harvest efficiency. Chemical plant growth regulators are being increasingly used as an aid to yield enhancement [54].

Brassinolide is the most bioactive form of the growth-promoting plant steroid termed as Brassinosteroids. Biologically active brassinosteroids show high growth-promoting as well as antistress activity besides other multiple effects on

TABLE 12: Seed yield (kg/ha) and net returns (Rs./ha) of mustard as influenced by foliar application of agrochemicals at different locations.

Treatment	S. K. Nagar		Sriganganagar	Ludhiana		
	Seed yield (kg/ha)	Net returns over control	Seed yield (kg/ha)	Oil content (%)	Oil yield (kg/ha)	Glucosinolate (μ mole/g defatted meal)
Control	1707	—	1604	34.7	375	130
Thiourea (0.1%)	2087	3226	1696	35.9	429	142
S @ 40 kg/ha	2249	6712	1799	35.2	405	149
S @ 40 kg/ha + Thiourea (0.1%)	2039	4070	1883	33.4	411	134
Urea (2%)	2019	5409	1845	34.7	396	124
$ZnSO_4$ (0.5%)	1921	4622	1667	33.2	372	126
Boric acid (0.1%)	1928	3418	1650	34.3	387	115
CD at 5%	150	—	158	—	—	—

Source: AICRP-RM, 2003 [5].

TABLE 13: Effect of low monetary agrotechniques on seed yield and oil content of mustard at Bharatpur during 1997-1998.

Treatments	Seed yield (kg/ha)	% increase over local practice	Oil content (%)	Oil yield (kg/ha)
Local Practice (T1)	1200	—	40.3	463
RP (No thinning and gypsum) (T2)	1371	14.2	40.3	525
RP + thinning at 15 & 25 DAS (T3)	1407	17.3	40.5	560
T_3 + N-S sowing (T4)	1376	14.7	40.7	560
T_3 + Removal of 4th row and 4th plant (T5)	1156	3.7	40.4	467
T_5 + 56.75% N as top dressing (T6)	1073	10.6	40.3	432
T_3 + I irrigation at 40–50 DAS (T7)	1232	2.7	40.6	500
T_1 + 200 kg gypsum/ha (T8)	1217	1.4	40.9	500
T_3 + removal of 4 older leaves (T9)	1343	11.9	40.5	544
RP + de-topping at bud-initiation stage (T10)	1464	22	40.7	596

Source: AICRP-RM, 1998 [12].

growth and development. As botanical juvenile hormones, they enhance the growth of young plant tissue and stimulate in submicromolar concentrations metabolic, differentiation and growth processes. Brassinosteroid caused accumulation of maximum total dry matter as compared to rest of the treatment at physiological maturity.

NPK accumulation and yield were maximal when spraying of GA3 was done at 40 DAS [55]. An increase in secondary and tertiary branching with consequent enhancement in seed yield through increased number of infloresence and siliquae per plant was observed with the application of Mixatalol (a mixture of long aliphatic alcohols varying in chain length from C_{24} to C_{32}) to Brassica plants as foliar spray [56]. The percentage of immature siliquae and shattering of siliquae decreased with this treatment. Mixtalol increased total dry matter of plants, partitioning coefficient, and harvest index. The contents of starch, protein, and oil were also higher in seeds from mixtalol treated plants.

The maximum plant height (169.1 cm), number of primary branches per plant (8.2), seed yield (2031 kg/ha), stover yield (5752 kg/ha), harvest index (26.1%), oil content (42%), and net returns (Rs. 20,471/ha) were recorded with thiourea (Shrama and Jain, 2003). At Bawal and Morena, highest seed yield (2060 kg/ha) was obtained with 40 kg S/ha + thiourea (0.1%). At Sriganganagar, significantly higher seed

yield (1883 kg/ha) was recorded on a par with 40 kg S/ha + thiourea (0.05%), urea (2%), H_2SO_4 (0.1%), and 40 kg S/ha. 40 kg S/ha + thiourea (0.1%) resulted into 17.67% higher seed yield over no spray. The highest oil content (35.9%) was recorded with thiourea 0.1% spray. Glucosinolate content ranged from 115 to 154 (μmole/g defatted meal) in different treatment (Table 12).

11. Impact of Low Monetary Agrotechniques on Mustard Productivity

Agricultural inputs like fertilizer, irrigation, insecticides, pesticides, and herbicides, and so forth, are very expensive. Some nonmonetary or low monetary inputs can enhance the yield considerably with a slight increase in the cost of cultivation. There are a number of low monetary agro techniques which enhance the mustard yield considerably (Table 13). For harvesting the maximum yield of rapeseed-mustard at a given situation, all the production technologies, like, soil amendments, thinning, nutrient supply, sowing direction, irrigation, plant protection, and so forth should be planned well in advance. At Bharatpur, highest seed yield (1464 kg/ha) was recorded with the application of recommended practice (RP) + thinning at 15 and 25 DAS + detopping at bud-initiation stage followed by RP + thinning at 15 and 25 DAS.

12. Future Line of Research

Rapeseed-mustard will continue to contribute considerably to the oilseed bowl of the country. A streamlined research programme for rapeseed-mustard should be focused on the below-mentioned points.

(i) Horizontal and vertical intensification in rapeseed-mustard production needs to be done for self-sufficiency in oilseed production. It is possible through varietal improvement and introduction of mustard in nontraditional areas.

(ii) An optimum agronomic package of practices for high yielding and insect, pest, and disease resistant varieties, along with the upcoming hybrids needs to be worked out.

(iii) Adoption of site-specific nutrient management (SSNM), precision agriculture, and conservation agriculture can bring more profits to the mustard growers.

(iv) An integrated weed management approach needs to be developed for problematic and parasitic weeds in mustard. *Orobanche* is becoming a serious constraint and for its management a holistic approach which includes GM techniques needs to be explored.

(v) Suitable crop models and simulation for various inputs like water and nutrients will be helpful to target the most productive and most potential mustard growing zones of India.

13. Conclusion

The tremendous increase in oilseed production is attributed to the development of high yielding varieties coupled with improved production technology, their widespread adoption and good support price. To meet the ever-growing demand of oil in the country, the gap is to be bridged through management techniques. The vertical growth in mustard production can be brought by exploiting the available genetic resources with breeding and biotechnological tools which will break the yield barriers. Horizontal growth in rapeseed-mustard can be brought in those rapeseed-mustard growing areas/districts of the country, wherever, the yield is lower than the national average. Production technologies for different agroecological cropping systems, crop growing situations like intercropping, salinity, rainfall, and so forth, under unutilized farm situations like rice-fallows, mustard to be followed after cotton, sugarcane, soyabean, and so forth, and mustard as a *paira* crop in rice with lathyrus, lentil or any other competing *rabi* crop in traditional and nontraditional areas, need to be worked out. It is estimated that at least 1 million hectares can be brought under cultivation, through adoption of such cropping systems.

Proper land preparation, proper time of sowing, selection of better quality seeds, and so forth are always neglected.

Fertilizer application is little or nonexistent leading to poor productivity. Whether little is spent on fertilizer input goes entirely on nitrogenous fertilizers. This results in a big gap between requirement and production of mustard in India. Therefore site-specific nutrient management through soil-test recommendation based should be adopted to improve upon the existing yield levels obtained at farmers field. Optimum crop geometry, balanced NPK fertilizers, intercultural operations, and inclusion of farmyard manure are the building blocks for achieving the utmost yield targets of rapeseed-mustard. Effective management of natural resources, integrated approach to plant-water, nutrient and pest management and extension of rapeseed-mustard cultivation to newer areas under different cropping systems will play a key role in further increasing and stabilizing the productivity and production of rapeseed-mustard to realize 24 million tonnes of oilseed by 2020 AD.

References

[1] India. Directorate of Economics and Statistics, *Agricultural Statistics at a Glance*, Department of Agricultural and cooperation. Ministary of Agriculture, Government of India, 2010.

[2] J. S. Chauhan, K. H. Singh, and A. Kumar, "Compendium of Rapeseed-mustard varieties notified in India," Directorate of Rapeseed-Mustard Research, Bharatpur, Rajasthan, pp. 7–13, 2006.

[3] M. A. Asoodari, A. R. Barzegar, and A. R. Eftekhar, "Effect of different tillage and rotation on crop performance," *International Journal of Agrcultural Biology*, vol. 3, no. 4, article 476, 2001.

[4] A. L. Rathore, A. R. Pal, and K. K. Sahu, "Tillage and mulching effects on water use, root growth and yield of rainfed mustard and chickpea grown after lowland rice," *Journal of Science of Food and Agriculture*, vol. 78, no. 2, pp. 149–161, 1999.

[5] AICRP-RM, Annual Progress Report of National Research Centre on Rapeseed-mustard.2002-2003, pp. 11–14, 2003.

[6] T. Nagra, R. E. Phillip, and J. E. Legett, "Diffusion and mass flow of nitrate-N into corn roots under field conditions," *Agronomy Journal*, vol. 68, pp. 67–72, 1976.

[7] R. L. Blevins, M. S. Smith, and G. W. Thomas, "Change in soil properties under no tillage," in *No Tillage Agriculture*, pp. 190–230, New York, NY, USA, 1984.

[8] S. Snapp, R. Price, and M. Morton, "Seed priming of winter annual cover crops improves germination and emergence," *Agronomy Journal*, vol. 100, no. 5, pp. 1506–1510, 2008.

[9] N. A. Khan and S.O. Aziz, "Response of mustard to seed treatment with pyridioxine and basal and foliar application of nitrogen and phosphorus," *Journal of Plant Nutrition*, vol. 16, no. 9, pp. 1651–1659, 1993.

[10] A. Kumar, B. Singh, Yashpal, and J. S. Yadava, "Effect of sowing time and crop geometry on tetralocular Indian mustard," *Indian Journal of Agricultural Sciences*, vol. 62, no. 4, pp. 258–262, 2001.

[11] AICRP-RM, Annual Progress Report of All India Coordinated Research Project on Rapeseed-Mustard, pp. 97–147, 1997.

[12] AICRP-RM, Annual Progress Report of National Research Centre on Rapeseed-mustard. 1997-98, pp. 8–18, 1998.

[13] R. Gupta, R. P. Avasthi, and S.J. Kolte, "Influence of sowing dates on the incidence of Sclerotinia stem rot of Rapeseed-mustard," *Annals of Plant Protection Sciences*, vol. 12, no. 1, pp. 223–224, 2004.

[14] AICRP-RM, Annual Progress Report of All India Coordinated Research Project on Rapeseed-Mustard, pp. A1–16, 2007.

[15] AICRP-RM, Annual Progress Report of All India Coordinated Research Project on Rapeseed-Mustard, pp. A1–44, 1999.

[16] G. M. Khan and S. K. Agarwal, "Influence of sowing methods, moisture Stress and nitrogen levels on growth, yield components and seed yield of mustard," *Indian Journal of Agricultural Science*, vol. 55, no. 5, pp. 324–327, 1985.

[17] M. J. Khan, R. A. Khattak, and M. A. Khan, "Influence of sowing methods on the production of canola grown in saline field," *Pakistan Journal of Biological Sciences*, vol. 3, no. 4, pp. 687–691, 2000.

[18] F. C. Oad, B. K. Solangi, M. A. Samo, A. A. Lakho, Hassan-Ul-Zia, and N. L. Oad, "Growth, yield and relationship of Rapeseed (*Brassica napus* L.) under different row spacing," *International Journal of Agriculture and Biology*, vol. 3, no. 4, pp. 475–476, 2001.

[19] H. P. Sierts and G. Geister, "Yield components stability in winter rape (*Brassica napus* L.) as a function of competition within the crop," in *Proceedings of the 7th International Rapeseed Congress*, p. 182, Poznan, Poland, May 1987.

[20] AICRP-RM, Annual Progress Report of All India Coordinated Research Project on Rapeseed-Mustard, pp. 97–144, 1996.

[21] AICRP-RM, Annual Progress Report of National Research Centre on Rapeseed-mustard. 2001-02, pp. 29–34, 2002.

[22] K. K. Singh and K. S. Rathi, "Dry matter production and productivity as influenced by staggered sowing of mustard intercropped at different row ratios with chickpea," *Journal of Agronomy and Crop Science*, vol. 189, no. 3, pp. 169–175, 2003.

[23] S. Chitale and M. C. Bhambri, "Response of Rapeseed-mustard to crop geometry, nutrient supply, farmyard manure and interculture—a review," *Ecology, Environment and Conservation*, vol. 7, no. 4, pp. 387–396, 2001.

[24] R. Sharma, K. S. Thakur, and P. Chopra, "Response of N and spacing on production of Ethopian mustard under mid-hill conditions of Himachal Pradesh," *Research on Crops*, vol. 8, no. 1, pp. 65–68, 2007.

[25] D. Sah, J. S. Bohra, and D. N. Shukla, "Effect of N, P, S on growth attributes and nutrient uptake of mustard," *Crop Research*, vol. 31, no. 1, pp. 234–236, 2006.

[26] M. L. Reager, S. K. Sharma, and R. S. Yadav, "Yield attributes, yield and nutrient uptake of Indian mustard (*Brassica juncea*) as influenced by N levels and its split application in arid Western Rajasthan," *Indian Journal of Agronomy*, vol. 51, no. 3, pp. 213–216, 2006.

[27] A. S. Sidhu and K. S. Sandhu, "Response of mustard to method of N application and timing of first irrigation," *Journal of Indian Society of Soil Science*, vol. 43, no. 3, pp. 331–334, 1995.

[28] H. L. S. Tandon, *S Research and Agricultural Production in India*, Fertilizer Development and Consultation Organization, New Delhi, India, 2nd edition, 1986.

[29] M. R. J. Holmes, *Nutrition of the Oilseed Rape Crops*, Applied science publishers, Essex, UK, 1980, In TSI/FAI/IFA Symposium.

[30] M. S. Aulakh and N. S. Pasricha, "S fertilization of oilseeds for yield and quality," in *Proceedings of the TSI-FAI symposium. S in agriculture-S-11/3*, 1988.

[31] J. C. Katyal, K. L. Sharma, and K. Srinivas, S in Indian agriculture. pp. KS-2/1-2/12, 1997.

[32] AICRP-RM, Annual Progress Report of All India Coordinated Research Project on Rapeseed-Mustard, pp. A1–22, 2007.

[33] AICRP-RM, Annual Progress Report of National Research Centre on Rapeseed-mustard. 1999-2000, pp. 24, 2000.

[34] M. Gupta and R. D. Kaushik, "Effect of saline irrigation water and Zn on the concentration and uptake of Zn by mustard," in *Proceedings of the 18th World Congress of Soil science*, Philadelphia, Pa, USA, July 2006.

[35] AICRP-RM, Annual Progress Report of All India Coordinated Research Project on Rapeseed-Mustard, pp. A1–28, 2005.

[36] AICRP-RM, Annual Progress Report of National Research Centre on Rapeseed-mustard. 2004-05, pp. 9–11, 2005.

[37] R. Prasad, S. N. Sharma, S. Singh, and R. Lakshaman, "Agronomic practices for increasing nutrient use efficiency and sustained crop production," in *Proceedings of the National Seminar on Resource Management for Sustainable Production*, New Delhi, India, February 1992.

[38] M. A. Bhat, Singh, Room, and D. Dash, "Effect of INM on uptake and use efficiency of N and S in Indian mustard on an inceptisol," *Crop Research*, vol. 30, pp. 23–25, 2005.

[39] B. S. Kumpawat, "Integrated nutrient management for maize-mustard cropping system," *Indian Journal of Agronomy*, vol. 49, pp. 4–7, 2004.

[40] K. M. Hati, A. K. Mishra, K. G. Mandal, P. K. Ghosh, and K. K. Bandopadhyay, "Irrigation and nutrient management effect on soil physical properties under soybean-mustard cropping system," *Agricultural Water Management*, vol. 85, no. 3, pp. 279–286, 2006.

[41] N. Pandey, R. S. Tripathi, and B. N. Mittra, "N, P and water management in greengram-rice-mustard cropping system," *Annals of Agricultural Reseaarch*, vol. 25, no. 2, pp. 298–302, 2004.

[42] S. S. Parihar, "Influence of N and irrigation schedule on yield, water use and economics of summer rice," *International Journal of Tropical Agriculture*, vol. 19, no. 1–4, pp. 157–162, 2001.

[43] R. K. Ghosh, P. Bandopadhyay, and N. Mukhopadhyay, "Performance of Rapeseed-mustard cultivars under various moisture regimes on the Gangetic Alluvial Plain of West Bengal," *Journal of Agronomy and Crop Sciences*, vol. 173, no. 1, pp. 5–10, 1994.

[44] S. K. Yadav, K. Chander, and D. P. Singh, "Response of late-sown mustard to irrigation and N," *The Journal of Agricultural Sciences*, vol. 123, pp. 219–224, 1994.

[45] B. B. Panda, S. K. Bandyopadhyay, and Y. S. Shivay, "Effect of irrigation level, sowing dates and varieties on yield attributes, yield, consumptive water use and water-use efficiency of Indian mustard," *Indian Journal of Agricultural Sciences*, vol. 74, no. 6, pp. 339–342, 2004.

[46] I. Piri, "Effect of irrigation on yield, quality and water-use-efficiency of Indian mustard," in *Proceedings of the 14th Australian Society of Agronomy Conference*, Adelaide, Australia, September 2008.

[47] Gangasaran and G. Giri, "Growth and yield of mustard as influenced by irrigation and plant population," *Annals of Agricultural Research*, vol. 7, no. 1, pp. 68–74, 1986.

[48] C. P. S. Chauhan and R. B. Singh, "Mustard performs well even with saline irrigation," *Indian Farming*, vol. 42, pp. 17–20, 2004.

[49] M. A. Kahlown, M. Akram, Z. A. Soomro, and W. D. Kemper, "Prospectus of growing barley and mustard with saline ground water irrigation in fine and coarse textured soils of Cholistan desert," *Irrigation and Drainage*, vol. 51, no. 4, pp. 328–338, 2008.

[50] D. K. Majumdar, "Effect of supplementary saline irrigation and applied N on the performance of dryland seeded Indian

mustard," *Experimental Agriculture*, vol. 31, pp. 423–428, 1995.

[51] J. Pandey and B. N. Mishra, "Effect of weed management practices in a rice-mustard-mungbean cropping system on weeds and yield of crops," *Annals of Agricultural Reaseach*, vol. 24, no. 4, pp. 36–39, 2003.

[52] B. R. Bazaya, D. Kachroo, and R. K. Jat, "Integrated weed management in mustard (*Brassica juncea*)," *Indian Jounal of Weed Science*, vol. 38, no. 1-2, pp. 16–19, 2006.

[53] S. Kumar, "Identification of trap crop for reducing broomrape infestation in the succeeding mustard," *Agronomy Digest*, vol. 2, pp. 99–101, 2002.

[54] M. Mobin, H. R. Ansari, and N. A. Khan, "Timing of GA3 application to indian mustard: DM distribution, growth analysis and nutrient uptake," *Journal of Agronomy*, vol. 6, no. 1, pp. 53–60, 2007.

[55] N. A. Khan, H. R. Ansari, and Samiullah, "Effect of gibberellic acid spray during ontogeny of mustard on growth, nutrient uptake and yield characteristics," *Journal of Agronomy and Crop Science*, vol. 181, no. 1, pp. 61–63, 1998.

[56] R. C. Setia, Richa, N. Setia, K. L. Ahuja, and C. P. Malik, "Effect of Mixtalol on growth, yield and yield components of Indian mustard (*Brassica juncea*)," *Plant Growth Regulation*, vol. 8, no. 2, pp. 185–192, 1989.

26

Cover Crop Biomass Harvest Influences Cotton Nitrogen Utilization and Productivity

F. Ducamp,[1] F. J. Arriaga,[2] K. S. Balkcom,[2] S. A. Prior,[2] E. van Santen,[1] and C. C. Mitchell[1]

[1] *Department of Agronomy and Soils, Auburn University, 202 Funchess Hall, Auburn, AL 36849, USA*
[2] *USDA-ARS National Soil Dynamics Laboratory, 411 S. Donahue Drive, Auburn, AL 36832, USA*

Correspondence should be addressed to F. J. Arriaga, francisco.arriaga@ars.usda.gov

Academic Editor: H. Allen Torbert

There is a potential in the southeastern US to harvest winter cover crops from cotton (*Gossypium hirsutum* L.) fields for biofuels or animal feed use, but this could impact yields and nitrogen (N) fertilizer response. An experiment was established to examine rye (*Secale cereale* L.) residue management (RM) and N rates on cotton productivity. Three RM treatments (no winter cover crop (NC), residue removed (REM) and residue retained (RET)) and four N rates for cotton were studied. Cotton population, leaf and plant N concentration, cotton biomass and N uptake at first square, and cotton biomass production between first square and cutout were higher for RET, followed by REM and NC. However, leaf N concentration at early bloom and N concentration in the cotton biomass between first square and cutout were higher for NC, followed by REM and RET. Seed cotton yield response to N interacted with year and RM, but yields were greater with RET followed by REM both years. These results indicate that a rye cover crop can be beneficial for cotton, especially during hot and dry years. Long-term studies would be required to completely understand the effect of rye residue harvest on cotton production under conservation tillage.

1. Introduction

Nitrogen is the most difficult nutrient to manage when growing cotton. About 5,445,749 ha of the cotton were planted in the USA in 2003 [1]. Applying optimum N rates is necessary to maximize economic yields and minimize the negative impacts that N overapplication can have on the crop and environment [2]. Higher N rates than required can result in excessive vegetative growth which increases the proportion of immature bolls, reduces lint quality and cotton yields, and increases disease and insect damage [3–6]. However, N deficiencies can reduce vegetative and reproductive growth and decrease yields [3]. Many parameters combine to determine the optimum N rates for cotton, such as soil type, location, N application method, tillage system, water availability, use of winter cover crops, and potential yield [7].

Conservation systems for cotton production in the southeastern US have increased in adoption to approximately 50% of the 2.9 million ha planted in this area [8]. The use of winter cover crops has been well documented as an effective method for improving soil chemical, biological, and physical properties [9, 10]. Among winter crop species, winter cereals like rye can have many benefits because they produce high amounts of biomass, are easy to establish and kill, and provide good ground cover during the winter [8, 11]. However, the high biomass grass cover crops can produce combined with their high C/N ratios and can lead to N immobilization, which can increase the N fertilizer demand for maximizing cotton yields [10, 12, 13]. Additionally, the probability of N immobilization increases when the N fertilizer is broadcast over a soil covered with grass residue [7].

Higher N fertilizer requirements for cotton following small grain cover crops were reported by Howard et al. [7], Varco et al. [14], and Mitchell [15]. Varco et al. [14] reported that 120 kg N ha^{-1} was required to achieve maximum cotton lint yields after a rye cover crop compared to 96 kg N ha^{-1} needed after winter fallow, but lint yields were greater after rye than winter fallow. Howard et al. [7] stated that for achieving similar yields, 101 and 67 kg N ha^{-1} were required for maximizing lint yields when cotton followed corn stover and native winter weed vegetation, respectively. However, it is expected that the long-term use of high biomass cover crops in conservation tillage systems will increase the soil organic carbon levels with a simultaneous increase of organic fractions of N in the soil, and once a new equilibrium is reached, N rates for crops could be reduced due to an increase of N provided through mineralization [16].

Recently, it has been proposed that winter cover crop biomass could be used as an alternative source of energy or for animal feed. Alternative uses for cover crop biomass would help farmers to increase revenue while diversifying market opportunities [17]. Cover crop biomass removal could cause significant changes in soil C and N dynamics and also impact crop yields and their response to N fertilization. Crop biomass removal can cause reductions in soil organic C levels with a subsequent deterioration of soil physical, chemical, and biological properties [18–23]. As a result of these changes in soil properties, reductions in crops yields are expected to occur [24, 25]. The impact of residue removal on soil properties and crop productivity has been well documented, but no research has been conducted emphasizing the potential impact of winter cover crop biomass removal on cotton yields and its response to N fertilization under conservation tillage.

We speculate that when rye residue is removed, N rates required for maximizing cotton production could be reduced because of the lower effect of N immobilization under conditions of low levels of residue with a high C/N ratio. Even though differences in soil properties in response to new management practices require some time to occur, we consider that short-term rye residue removal may produce enough changes in the soil environment to cause reductions in cotton yields. The objectives of this research were (i) to determine the effect of rye residue management on cotton growth parameters and yield, (ii) to quantify the impact of rye residue management and cotton response to N fertilization, and (iii) to determine if optimum N rates for cotton can be reduced under rye residue removal conditions.

2. Materials and Methods

A 2-year field experiment under supplemental irrigation was established in November 2005 at the Alabama Agricultural Experiment Station's E.V. Smith Research Center, Field Crops Unit (32° 25′ 19″N, 85° 53′ 7″W), near Shorter in central Alabama, USA. The soil was a Marvyn loamy sand (fine-loamy, kaolinitic, thermic Typic Kanhapludult). This region is characterized by a humid subtropical climate, with an average annual precipitation of about 1100 mm [8]. The experimental area was previously managed with conventional tillage. Three rye RM schemes and four N rates were evaluated for cotton production. Rye RMs were no cover (NC), residue removed (REM), and residue retained (RET). Each RM was evaluated with cotton N fertilization rates of 0, 50, 100, and 140 kg ha^{-1} applied at the first pinhead square stage. The RMs were the main plots (18 m long by 8 m wide) and N rates for cotton were the subplots (9 m long by 4 m wide).

2.1. Soil Management.
Before planting rye the first year, the entire area was deep-tilled with a noninversion, bent-leg

subsoiler to a depth of 46 cm to remove any soil compaction present, and leveled with a field cultivator. In early May each year the experimental area was tilled in-row (1 m between rows) with a narrow-shanked subsoiler to a depth of 40 cm. The in-row tillage was conducted using a tractor equipped with an automatic steering system with centimeter level precision. The NC treatment was kept free of weeds during winter by applying herbicide when required.

2.2. Crop Management.
Rye (cultivar "Elbon") was drilled at 100 kg ha^{-1}, in early November each year, using a no-till drill. Plots planted with rye received 40 and 30 kg N ha^{-1} as ammonium nitrate applied manually three weeks after planting and in late February, respectively. In the RET treatment, rye was rolled down at the early milk development stage [26] in late April each year and then sprayed with glyphosate (N-phosphonomethyl glycine) at a rate of 0.9 kg a.i. ha^{-1}. At the same time rye was terminated in the RET treatment, the aboveground rye biomass in the REM treatment was harvested with a small forage harvester to a height of 10 cm over the soil surface and removed from the plots.

The entire experimental area received an application of 21, 10, 42, and 6 kg ha^{-1} of nitrogen, phosphorus (P) as P_2O_5, potassium (K) as K_2O, and sulfur (S) as SO_4, respectively, each year by early May, based on the Alabama Cooperative Extension System soil test recommendations [27]. Cotton, cultivar DP 454 BG/RR (Delta Pine and Land Co., Scott, MS), was planted on May 19 and 18 in 2006 and 2007, respectively, using a four-row vacuum planter at a rate of 17 seeds m^{-1}. Row spacing was one meter. Herbicides, insecticides, defoliant, and boll opener applied to cotton were based on the Alabama Cooperative Extension System recommendations. The entire research area received supplemental irrigation of 70 and 160 mm during the 2006 and 2007 cotton seasons, respectively, using a linear-movement sprinkler irrigation system. Nitrogen treatments for cotton were applied manually as ammonium nitrate at the first pinhead square stage (37 days after planting (DAP)). Cotton was chemically defoliated and a boll opener was applied when 60–70% of the bolls in RET were opened. Before cotton harvest, one meter of each end of the plots was cut off with a rotary mower. After harvesting, cotton stalks were shredded with a rotary mower.

2.3. Data Collection.
Cotton population, leaf blade samples, and seed cotton yield were determined from the two middle rows of each subplot and cotton biomass from the two exterior rows of each subplot. Cotton population was determined by counting the number of plants from a 3 m length in each of the two middle rows of the subplots at 37 DAP. Ten upper-most fully developed blades of leafs were collected from recently matured leaves in the upper canopy of each subplot, at 37 and 65 DAP in 2006 and at 37 and 69 DAP in 2007. Total aboveground cotton biomass was determined at 37 and 92 DAP in 2006 and 2007 by randomly cutting eight plants per subplot. Leaf blade and whole plant samples were oven dried at 55°C until constant weight, finely ground to pass a 1 mm sieve, and analyzed for total N by dry combustion using a LECO TruSpec analyzer (LECO

Table 1: Analysis of variance for cotton population, leaf N concentration, plant N concentration, cotton biomass, and N uptake at first square as affected by year and rye residue management. P values within a row in bold are significant at $\alpha \leq 0.05$.

Source of variation	Cotton population	Leaf N concentration	Plant N concentration	Cotton biomass	N uptake
			$P > F$		
Year	**≤0.01**	**≤0.01**	**≤0.01**	0.64	0.15
RM[†]	**≤0.01**	**≤0.01**	**≤0.01**	**≤0.01**	**≤0.01**
Year∗RM	0.58	**0.02**	**≤0.01**	0.33	0.58

[†] Rye residue management.

Corp., St. Joseph, MI). Total cotton biomass was estimated using the dry weight per plant and cotton population. Plant N uptake at each sampling time was calculated based on the total biomass and plant N concentration. The cotton biomass and N uptake between first square and cutout were calculated by subtracting the amount at first square from the amount at cutout for each parameter. Seed cotton yield was determined at 139 and 125 DAP in 2006 and 2007, respectively, by harvesting a 14.6 m² (2 m wide by 7.3 m long) area from each subplot using a spindle picker.

2.4. Weather.
Daily average temperature data for both years were taken from an automated weather station located at the Experimental Station, beginning when cotton was planted and ending at the cutout stage of cotton development. Daily heats units (HUs) between planting and cutout were calculated as the difference between the average daily temperature and a base temperature of 15.6°C [28]. Rainfall and irrigation during each season were measured directly in the experimental area with a rain gauge connected to a data-logger.

2.5. Experimental Design and Statistical Analyses.
The experiment was arranged in a randomized complete block design (RCBD) with a split-plot restriction on the randomization and four replications. Rye RM was the main factor and N rates for cotton the subfactor. As N treatments were applied at the first pinhead stage of cotton development, data collected before this N application were analyzed using the MIXED procedure of SAS [29] only considering the RM effect (RCBD). The LSMEANS PDIFF option was used to establish mean differences between RM treatments. Data collected after applying N treatments to cotton were analyzed through covariance analysis using the MIXED procedure of SAS [29] considering N as covariate. Replication and its interactions were considered as random effects. Treatments and year were considered fixed effects. When a significant interaction including year occurred, data were presented separately for each year. When Year × RM × N or RM × N interactions were not significant, the LSMEANS PDIFF option was used to establish means differences between RM treatments. The covariance analysis was used to evaluate linear and quadratic effects of N rates on cotton parameters measured and to fit the best linear or quadratic regression model. Linear or quadratic effects were considered significant when $P \leq 0.15$ [30]. Treatment effects and differences of least squares means were considered significant when $P \leq 0.05$.

3. Results and Discussion

3.1. Climate Data.
Rainfall and irrigation during both years were different in amount and distribution. In 2006, rainfall and irrigation between one week before planting cotton and cutout were 247 and 70 mm, respectively. For the same period during 2007, they were 207 and 176 mm, respectively. Rainfall in 2006 and 2007 was 23 and 36% lower than the 10-year average. In 2006, rainfall was below the 10-year average until midseason, after which it was similar or greater. However, in 2007 rainfall was below the 10-year average early and late in the cotton season and it was not uniformly distributed, with 75% of rainfall occurring during the first 10 days of July, resulting in a higher amount of irrigation applied during 2007. The main difference in HU between years occurred at the end of the cotton season. For the last 20-day period before cutout, HUs in 2007 were 20% higher than that in 2006, indicating that higher temperatures occurred during this period in 2007 with respect to 2006.

3.2. Cotton Population.
Rye residue management had a significant effect on cotton population 37 DAP, across years (Table 1). Rye residue retained had a significantly higher population than NC ($P \leq 0.01$), but population for REM was not significantly different with respect to the other two treatments. Population for RET was 4 and 7% greater than REM and NC, respectively. Tillage operations were identical among RM, so differences in cotton populations can be attributed to differences in soil water content among treatments during the establishment period of the crop. Higher soil water content was measured in RET compared to REM and NC until 20–25 days after cotton planting in both years (data not shown) which probably contributed to better plant establishment.

Cotton population was also significantly different ($P \leq 0.01$) between years when averaged across RM. Higher cotton populations were observed in 2006 compared to 2007. In both years, the quality of the seed bed at planting and the soil water content between planting and the following two weeks were similar, indicating that other factors could be responsible for this difference between years. Accumulated HUs during the first 13 days after planting were 24% lower in 2007 compared to 2006, indicating that this period of 2007 was colder than 2006. These low temperatures could explain the population reduction in 2007, which probably contributed to slower plant growth, extending the period of time that young plants are susceptible to water deficit and

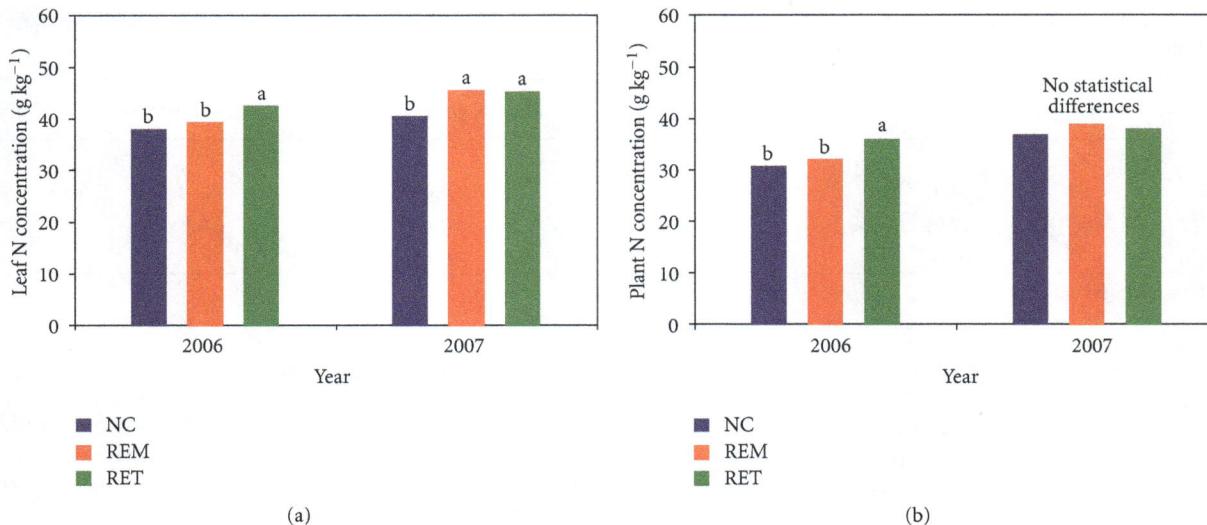

Figure 1: Effect of rye residue management and year on (a) leaf N concentration and (b) plant N concentration, at first square. Columns sharing the same letter for each parameter inside of year are not significantly different ($P \leq 0.05$). NC: no winter cover crop; REM: rye residue removed; RET: rye residue retained.

pest damage. Cotton populations in 2006 and 2007 were about 150,000 and 140,000 plants ha^{-1}, respectively, which were in a range considered high for cotton production, even though seed cotton yields can be stable for a wide range of plant densities [31]. However, this yield stability may be threatened if dry periods occur later in the growing season.

3.3. Leaf and Plant N Concentration at First Square.
There was a significant Year × RM interaction for leaf and plant N concentration at first square (Table 1). In 2006, RET had a significantly higher leaf N concentration than NC and REM ($P \leq 0.01$ and $P = 0.03$, resp.), and NC was not significantly different from REM (Figure 1(a)). In 2007, RET and REM had significantly higher leaf N concentration than NC ($P \leq 0.01$ and $P \leq 0.01$, resp.), but differences between these two treatments were not significant. Leaf N concentration values ranged between 38 and 43 g kg^{-1} in 2006 and between 40 and 46 g kg^{-1} in 2007. These values were lower than the 54 g kg^{-1} critical level reported by Bell et al. [32] at first pinhead square, for cotton in the southern USA. The N applied before planting and the N possibly provided through mineralization were not enough to increase leaf N concentration to this critical value. Nonetheless, Bell et al. [32] also stated that high cotton yields can still be achieved by crops having low leaf N at first pinhead square, if N deficiencies are corrected at this stage of development and leaf N concentrations at early bloom are in the sufficiency range. Further, leaf N concentration levels in our study were in the sufficiency range reported by Mills and Jones [33].

In 2006, RET had a significantly higher plant N concentration than REM and NC ($P \leq 0.01$ and $P \leq 0.01$, resp.), but REM was not significantly different from NC (Figure 1(b)). However, in 2007 differences among RM treatments were not significant and plant N concentration values were very similar for each treatment. Plant N concentration for RM

in 2006 followed a similar pattern to leaf N concentration, but their values were lower. This is expected because plant samples that include older tissues other than upper leaves are characterized by lower N concentrations. Further, there was a higher accumulation of HU during June 2007 compared to June 2006, which could help explain the greater leaf and plant N concentrations at first square. Additionally, the amount of soil mineral N at first square was 19% higher in 2007 compared to 2006 averaged across RM (data not shown), indicating that N availability during June was probably greater in 2007. However, soil mineral N amounts for all RM treatments in both years appeared to be sufficient, indicating that N availability was not a limiting factor for cotton growth at first square.

3.4. Cotton Biomass and N Uptake at First Square.
Rye RM had a significant influence on cotton biomass and N uptake at first pinhead square, averaged across years (Table 1). The RET treatment had significantly higher cotton biomass than NC and REM ($P \leq 0.01$ and $P \leq 0.01$, resp.; Figure 2). Rye residue removed had a cotton biomass 35% higher than NC, but this difference was not significant. Similar results were obtained for N uptake, with RET having values 96 and 166% higher than REM and NC ($P \leq 0.01$ and $P \leq 0.01$, resp.). No significant differences occurred between REM and NC, but N uptake was 39% higher for REM (Figure 2). Differences in N uptake among RM treatments can be partially explained by differences in cotton biomass and plant N concentration. When averaged across years, plant N concentration was 34, 35.7, and 37.1 g N kg^{-1} for NC, REM, and RET, respectively. Although these two growth parameters influenced N uptake in the same manner, cotton biomass could have had the highest impact on N uptake, because its variability between RM treatments was higher in proportion to plant N concentration.

TABLE 2: Analysis of variance for the effect of year, rye residue management, and N fertilization on leaf N concentration at early bloom, cotton biomass, cotton biomass N concentration and N uptake between first square and cutout, and seed cotton yield. P values within a row in bold are significant at $\alpha \leq 0.05$.

Effect	Leaf N concentration[†]	Cotton biomass[‡]	Cotton biomass N concentration[‡]	N uptake[‡]	Seed cotton yield
			$P > F$		
Year	≤**0.01**	**0.04**	≤**0.01**	≤**0.01**	≤**0.01**
RM[δ]	**0.03**	≤**0.01**	≤**0.01**	0.63	≤**0.01**
N	≤**0.01**	≤**0.01**	≤**0.01**	≤**0.01**	≤**0.01**
Year∗RM	0.23	0.95	0.30	0.79	≤**0.01**
Year∗N	0.57	**0.02**	0.15	≤**0.01**	≤**0.01**
RM∗N	0.76	0.07	0.82	**0.05**	**0.02**
Year∗RM∗N	0.16	0.99	0.45	0.83	≤**0.01**

[†] At early bloom.
[‡] Between first square and cutout.
[δ] RM: rye residue management.
N: nitrogen.

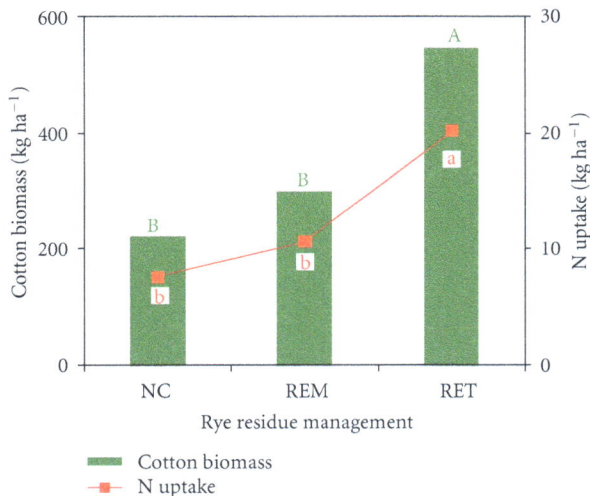

FIGURE 2: Effect of rye residue management on cotton biomass and N uptake at first square, averaged across years. Means followed by the same letter for each parameter are not significantly different ($P \leq 0.05$). NC: no winter cover crop; REM: rye residue removed; RET: rye residue retained.

These results show that in both years RET provided better conditions for cotton growth and N uptake early in the season. This could have been a consequence of greater N availability between planting and first square, as indicated by the residual levels of N at this time, and also because of the higher soil water content during this period of time. Although plant populations were greater with RET than REM and NC, plant biomass was also greater with RET at first square (130 and 77% greater relative to NC and REM, resp.). Further, the lack of a Year × RM interaction and a Year effect on cotton biomass and N uptake suggests that, at least until this stage of development, growth of the cotton crop was very similar in both seasons.

3.5. Leaf N Concentration at Early Bloom. Leaf N concentration at early bloom was significantly affected by RM, N rate, and year, but interactions were not significant (Table 2). No winter cover crop and REM had significantly higher leaf N concentration compared to RET ($P \leq 0.01$ and $P \leq 0.01$, resp.). It is possible that the rye residue immobilized some of the soil mineral N between first square and early bloom, decreasing the availability of N for cotton and reducing the N concentration in cotton tissues. Another possible explanation can be that higher cotton biomass production in RET could have caused a N dilution effect within cotton tissues. Cotton biomass was not measured at this growth stage, but plant heights at this sampling time averaged across years and N rates showed that plants in RET were taller than in REM and NC (data not shown), indicating a possible higher cotton biomass and potential N dilution effect. Similar results were reported by Fridgen and Varco [34] and Balkcom et al. [35], who found a dilution of leaf N when cotton biomass production was high.

Averaged across years and RM, leaf N concentration responded in a quadratic manner to fertilizer N rate, with a maximum leaf N concentration of $40\,g\,kg^{-1}$ observed at the highest N rate applied (Figure 3(a)). This value was very close to the sufficiency range of $43\,g\,kg^{-1}$ reported by Bell et al. [32]. However, our results do not agree with Fridgen and Varco [34], who reported a higher leaf N concentration using similar N rates as we did. Further, a maximum leaf N concentration was not achieved even with the highest N rate applied. This indicates that to reach the maximum leaf N concentration in the conditions of this experiment would have required a greater fertilizer N application than $140\,kg\,ha^{-1}$.

Year also significantly affected leaf N concentration at early bloom. The leaf N concentration was significantly lower ($P \leq 0.01$) in 2007, with a decrease of 22% with respect to 2006 (Figure 3(b)). Rainfall distribution and HU during 2007 could explain this trend between years. A more detailed analysis of the rainfall effect on cotton growth parameters will be provided when data of cotton biomass N concentration are presented hereinafter.

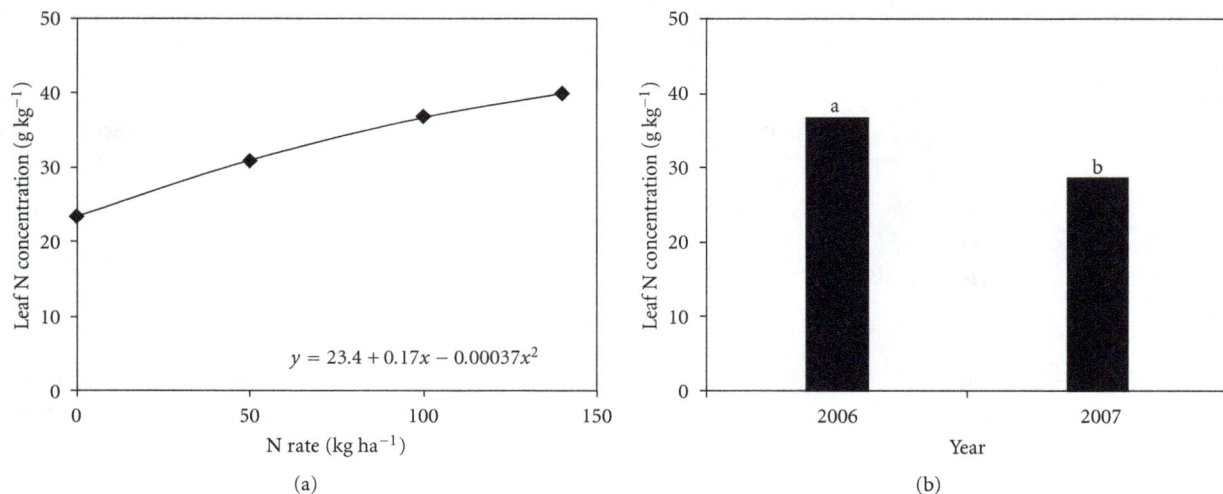

$$y = 23.4 + 0.17x - 0.00037x^2$$

(a)

(b)

FIGURE 3: Leaf N concentration at early bloom as affected by (a) N rate (averaged across years and rye residue management) and (b) year (averaged across rye residue management and N rates). Columns followed by different letters are significantly different ($P \leq 0.05$).

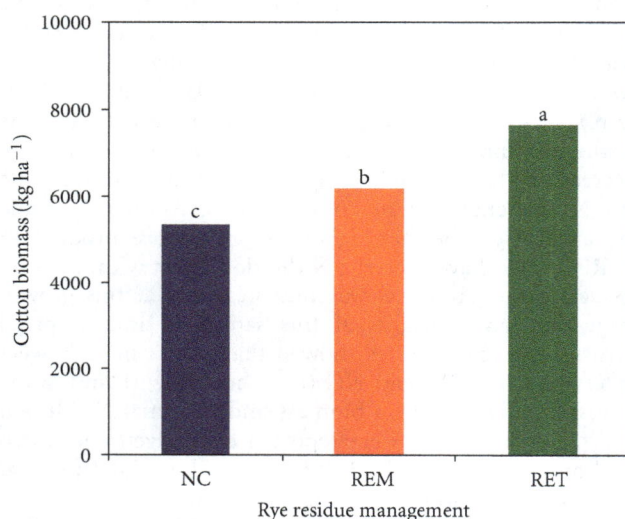

FIGURE 4: Effect of rye residue management on cotton biomass production between first square and cutout, averaged across years and N rates. Columns sharing the same letter are not significantly different ($P \leq 0.05$). NC: no winter cover crop; REM: rye residue removed; RET: rye residue retained.

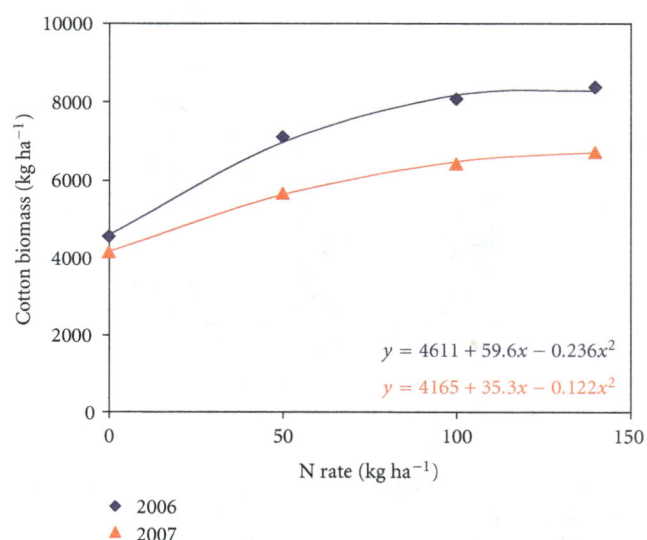

$$y = 4611 + 59.6x - 0.236x^2$$
$$y = 4165 + 35.3x - 0.122x^2$$

♦ 2006
▲ 2007

FIGURE 5: Cotton biomass production between first square and cutout as affected by year and N rate, averaged across rye residue management.

3.6. Cotton Biomass Production between First Square and Cutout. Rye RM had a significant effect on cotton biomass production between first square and cutout, averaged across years, and N rates (Table 2). Rye residue retained had significantly higher cotton biomass production than REM and NC (both of which were $P \leq 0.01$), and REM was significantly higher than NC ($P \leq 0.01$) (Figure 4). Cotton biomass production for RET was 24 and 43% higher than REM and NC, respectively, while REM was 16% higher than to NC. These results demonstrate that RET provided better conditions for cotton growth. Govaerts et al. [19] reported that keeping residue on the soil surface improves infiltration, increasing water available for plants.

The cotton biomass response to N produced a significant interaction with year, when averaged across RM treatments (Table 2). In both seasons, cotton biomass response to N was quadratic (Figure 5). The small increase between the 100 and 140 kg ha^{-1} N rates indicates that the N rate required to maximize cotton biomass would be similar to the highest rate used in this experiment. Cotton biomass in 2006 was similar to the one reported by Bassett et al. [36] for an N rate of 134 kg ha^{-1}, but it was extremely low compared to the findings of Boquet and Breitenbeck [2]. In spite of the similar trend between both seasons, cotton biomass was lower for all N rates in 2007 than that in 2006. The difference for the no N control was very low between years, with a decrease of

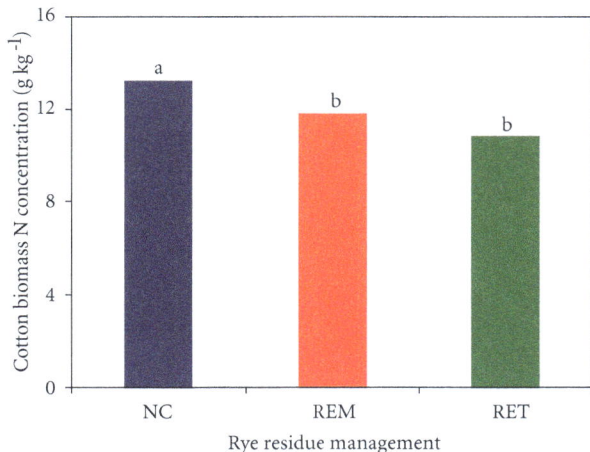

FIGURE 6: Effect of rye residue management on cotton biomass N concentration between first square and cutout, averaged across years and N rates. Columns sharing the same letter are not significantly different ($P \leq 0.05$). NC: no winter cover crop; REM: rye residue removed; RET: rye residue retained.

9% in 2007 compared to 2006. However, about 20% less biomass was produced in 2007 than that in 2006 when N was applied independent of the N rate used. This cotton biomass reduction could be explained by the lower rainfall and nonuniform distribution during 2007. Additionally, the last 20 days before cutout in 2007 were characterized by elevated temperatures as indicated by the higher accumulation of HU units relative to 2006. High temperatures and low rainfall in 2007 could have imposed a stress to the crop causing lower biomass production. This may have occurred even though irrigation was applied since low amounts of water were applied with each irrigation event (10 to 12 mm) and the high temperatures would have increased water loss due to evapotranspiration. These results agree with Balkcom et al. [35], who reported lower cotton biomass in hot and dry years regardless of irrigation.

3.7. Cotton Biomass N Concentration. The N concentration in cotton biomass accumulated between first square and cutout was significantly affected by RM, N, and year, but interactions were not significant (Table 2). No winter cover crop had 12 and 22% significantly higher cotton biomass N concentration compared to REM and RET, respectively ($P = 0.03$ and $P \leq 0.01$, resp.; Figure 6). Rye residue removed had a numerically higher cotton biomass N concentration compared to RET (9%), but this difference was not significant. As previously mentioned, N immobilization probably occurred in both growing seasons but at low levels. This would indicate that the reduction in cotton biomass N concentration in REM and RET relative to NC could be explained by the higher cotton biomass compared to NC (Figure 4) which may have contributed to a dilution of N in cotton tissues. Rye residue retained and REM accumulated 43 and 16% more biomass between first square and cutout than NC, respectively, but their increment in N uptake relative to

NC was only 18 and 5%, respectively. This result also suggests an occurrence of N dilution in the accumulated biomass. Gerik et al. [3] and Bell et al. [32] reported that under conditions of high availability of N, cotton plants increase vegetative growth very quickly which leads to a N dilution in the biomass produced and a subsequent drop in tissue N concentration.

Cotton biomass N concentration response to N rates was linear (averaged across years and RM), indicating that the highest N rate applied did not maximize N concentration in the biomass (Figure 7(a)). This trend was similar to that observed for leaf N concentration, even though that response to N was quadratic.

Cotton biomass N concentration was significantly influenced by year (Table 2). In 2007, there was a significant decrease ($P \leq 0.01$) of about 28% in cotton biomass N concentration compared to 2006 (Figure 7(b)). A similar pattern was also observed at early bloom for leaf N concentration and cotton biomass production. There was a simultaneous decrease in cotton biomass and N concentration, but the reduction in cotton biomass N concentration was greater with respect to cotton biomass (28 versus 18%, resp.), providing strong evidence that N dilution in plant tissues occurred. These results indicate that the 2007 crop was affected by N dynamics in the soil-plant system. The rainfall regime during 2007 may have played an important role in these findings. The high rainfall that occurred during the first 10 days of July (about 150 mm) was twice than the 70 mm of available water that the soil in the experimental area can retain to a depth of 50 cm. The excess rainfall above the soil water holding capacity could have leached part of the N fertilizer out of the root zone. These high rainfall events at the beginning of July in 2007 occurred only one week after the N fertilizer was applied to cotton.

3.8. Nitrogen Uptake between First Square and Cutout. Table 2 shows that there was a significant RM × N interaction for N uptake between first square and cutout. Nitrogen uptake response to N rates was linear for RET and REM, whereas for NC this relationship was best described by a quadratic model (Figure 8(a)). The response of N uptake per kg of N added up to the highest N rate was 0.49, 0.61, and 0.68 for NC, REM, and RET, respectively. Cotton plants in RET absorbed more N independent of the N rates applied. Even though RET had higher values of N uptake than REM and NC at all N rates, these differences were magnified with increasing rates of N fertilizer (Figure 8(a)). The highest N uptake for each RM treatment occurred at the highest N rate applied, where at this rate, N uptake for RET was 32 and 15% higher than NC and REM, respectively, and it was 15% higher for REM compared to NC. The linear relationship between N uptake and N rate for RET and REM indicates that the highest N rate applied was not enough to maximize N uptake under the conditions of this experiment. Conversely, NC had a quadratic relationship with a very low N uptake increment between the 100 and 140 kg ha^{-1} N rates, indicating that the N rate required for maximizing N uptake was very similar to the highest N rate we applied. Our results

(a)

(b)

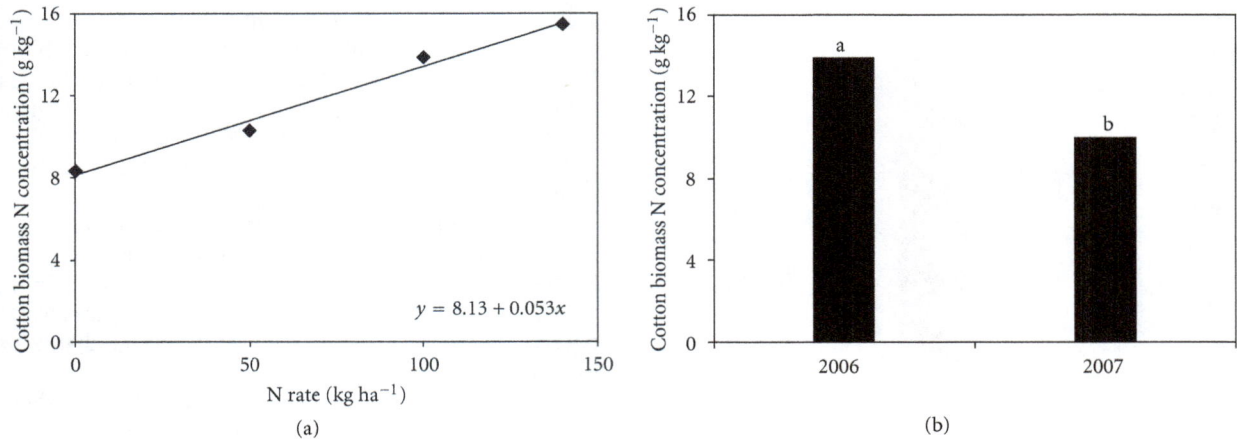

FIGURE 7: Cotton biomass N concentration between first square and cutout as affected by (a) N rate (averaged across years and rye residue management) and (b) year (averaged across rye residue management and N rates). Columns followed by the same letter are not significantly different ($P \leq 0.05$).

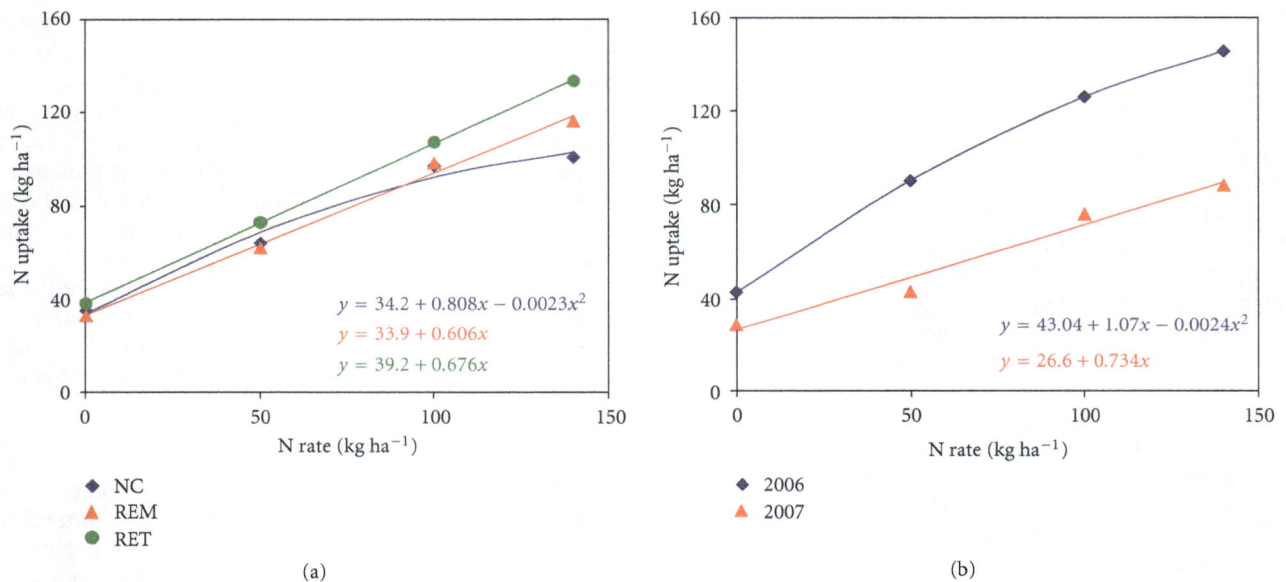

(a)

(b)

FIGURE 8: Cotton N uptake between first square and cutout as affected by (a) rye residue management and N rate (averaged across years) and (b) year and N rate (averaged across rye residue management).

for RET were similar to the findings of Basset et al. [36], who found a total N uptake of 142 kg ha^{-1} for irrigated cotton that received 134 kg N ha^{-1}. However, a study by Mullins and Burmester [37] revealed greater N taken up with an N rate of 112 kg ha^{-1}. The N uptake by cotton plants at the highest N rate (140 kg N ha^{-1}) represented 72, 83, and 95% of the N added, for NC, REM, and RET, respectively. This would indicate that RET provided better growing conditions for cotton that possibly improved the N use efficiency of the fertilizer applied.

The amount of N absorbed by a crop depends on its biomass production and its N tissue concentration. When averaged across years, the interaction RM $\times$ N was not significant for cotton biomass N concentration ($P = 0.82$)

and cotton biomass production ($P = 0.07$) between first square and cutout (Table 2). Even though no significant interaction existed, cotton biomass N concentration values were slightly higher for NC, followed by REM and RET, but cotton biomass was higher for RET, followed by RM and NC, at each N rate (Table 3). This tendency supports the findings of higher biomass production with RET compared to REM and NC (particularly for the 100 and 140 kg ha^{-1} N rates) and its greater N uptake. These results agree with Gastal and Lemaire [38], who stated that N taken up by crops is mainly affected by the crop growth rate.

There was a significant Year $\times$ N interaction for cotton N uptake between first square and cutout (Table 2). Nitrogen uptake response was quadratic in 2006 and linear in 2007

TABLE 3: Effects of rye residue management and N fertilization on cotton biomass production and cotton biomass N concentration between first square and cutout, averaged across years.

N rates	Cotton biomass			Cotton biomass N concentration		
	NC[†]	REM	RET	NC	REM	RET
kg ha^{-1}	kg ha^{-1}			g ha^{-1}		
0	3,620	4,190	5,281	9.8	7.9	7.3
50	5,430	6,060	7,684	11.4	10.1	9.4
100	6,371	7,034	8,365	15.0	13.9	12.7
140	5,954	7,468	9,258	16.7	15.4	14.3

[†]NC: no winter cover crop.
REM: rye residue removed.
RET: rye residue retained.

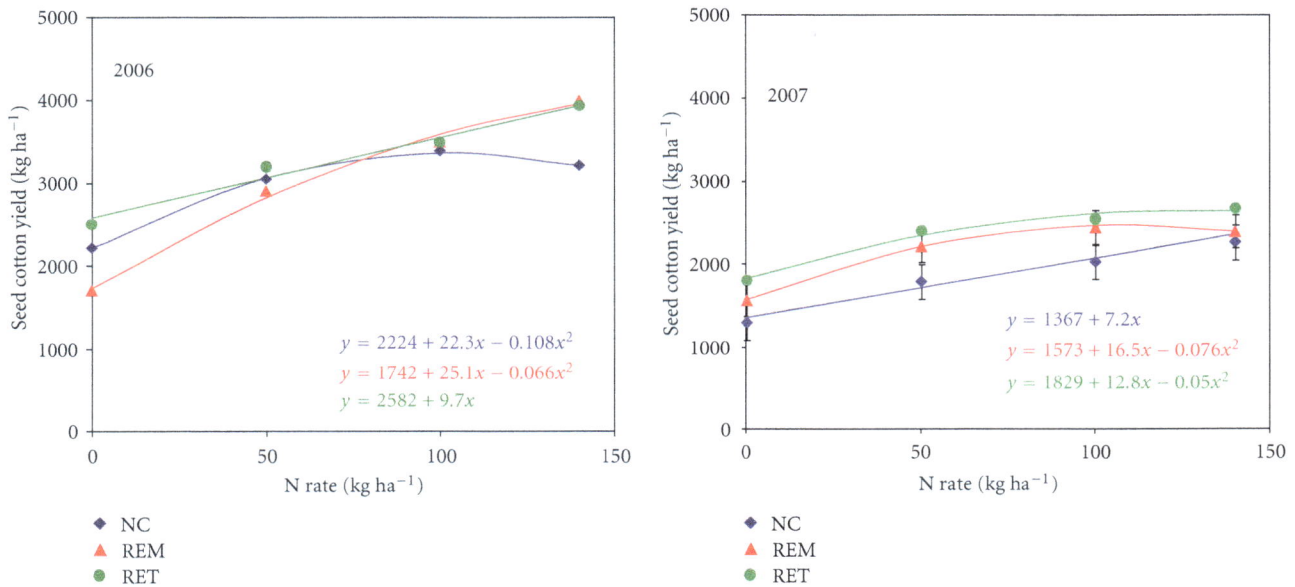

FIGURE 9: Seed cotton yield as affected by rye residue management and N rate, for the 2006 and 2007 seasons. NC: no winter cover crop; REM: rye residue removed; RET: rye residue retained.

(Figure 8(b)). In both years, the uptake response to N occurred up to the highest N rate applied, with 0.7 and 0.4 kg of N taken up per kg of N added in 2006 and 2007, respectively. Nitrogen uptake was lower for 2007 compared to 2006 for all N rates, but differences were greater when N was applied. Nitrogen uptake in 2007 was 33 and 39% lower for the no N control and for the 140 kg ha^{-1} N rate, respectively, relative to 2006 (Figure 8(b)). This observed reduction in N uptake can be attributed to the lower cotton biomass production in 2007 for all N rates and the reduction in cotton biomass N concentration measured for both years (Figures 5 and 7(b)).

3.9. Seed Cotton Yield.

A significant Year × RM × N interaction occurred for seed cotton yield (Table 2). In 2006, observed seed cotton yields ranged from 1,740 (REM, no N control) to 3,970 kg ha^{-1} (REM, 140 kg of N ha^{-1}). The seed cotton yield response to N for RET was linear, while

REM and NC were quadratic (Figure 9(a)). Highest observed and predicted yields corresponded to RET and REM with the application of 140 kg N ha^{-1}, with both treatments producing similar yields (Figure 9). Seed cotton yield response to N for RET and REM occurred up to the highest N rate applied without reaching a maximum. The fact that a plateau yield was not achieved in RET and REM indicates that the maximum yield potential with a cover crop was greater than that with no cover in 2006. Further, seed cotton yield in NC reached an estimated maximum at a N rate of 102 kg ha^{-1} and higher N rates caused a decrease in yields. The highest predicted seed cotton yield for RET and REM was about 17% higher than that for NC, showing that growing a cover was the best scenario for obtaining higher yields during 2006, whether or not the cover crop residue was removed or left on the soil surface. Our results for RET and REM are similar to the findings of Reiter et al. [12] who stated that conservation-tilled cotton on a Decatur silt loam responded up to 134 kg N ha^{-1}. Seed cotton yields observed in RET and

REM were also similar to results of Clawson et al. [39], who found cotton response to N up to 151 kg ha^{-1}. However, Wiatrak et al. [40] reported a linear increase in lint yields up to 200 kg N ha^{-1}, a rate considerably greater than that used in this experiment.

The seed cotton yield response to N during 2006 was 11.3, 15.9, and 9.7 kg of seed cotton per kg of N added for NC, REM, and RET at N rates of 102, 140, and 140 kg ha^{-1}, respectively. The highest yield increase with respect to the no N control corresponded to REM, followed by RET and NC (128, 53, and 45%, resp.), at the previously mentioned N rates. The lower response to N for RET can be somewhat explained by the greater seed cotton yield for the no N control relative to REM and NC. The REM treatment had the lowest seed cotton yield when no N was applied, with yield lower by 22 and 33% compared to NC and RET, respectively. Lower yields in REM compared to NC for the no N control were unexpected, since all cotton growth parameters for REM were at least similar or better than for NC when N was not applied. This yield decrease could be related to a factor or combination of factors directly affecting some of the yield components. However, the application of 50 and 100 kg of N ha^{-1} was enough to increase yields up to levels similar to NC and RET, and with 140 kg of N ha^{-1} the yield for REM was one of the highest. This observed trend for REM indicates that a severe N deficiency possibly occurred with this treatment when no N was added.

In 2007, seed cotton yields ranged from 1,295 kg ha^{-1} (NC, no N control) to 2,677 kg ha^{-1} (RET, 140 kg of N ha^{-1}). Rye residue retained and REM had a quadratic seed cotton yield response to N fertilization, while the yield increase for NC followed a linear trend (Figure 9). Rye residue retained had the highest predicted yield with 125 kg N ha^{-1}, followed by REM and NC at N rates of 106 and 140 kg N ha^{-1}, respectively (Figure 9). Boquet et al. [41] and Varco et al. [14] reported an optimum N rate of about 118 kg ha^{-1} for conservation tillage cotton, a value similar to the one we found for RET in 2007. Rye residue retained required 21 kg N ha^{-1} more than REM for maximizing yields but it had a higher yield. The highest estimated seed cotton yield for RET was 12 and 8% higher than that for NC and REM, respectively. Nitrogen rates above the optimum for REM tended to slightly decrease yields. A similar reduction in seed cotton yields occurred for NC in 2006 with N rates higher than 102 kg ha^{-1}. Cotton yield reductions with application of high N rates were reported by McConnell et al. [42] and Boquet et al. [43]. High N levels in soil can cause excessive vegetative growth, with a subsequent competition between vegetative and reproductive structures, which generally is detrimental to bolls and lint development, lint quality, and yield [4]. Regardless of its linear response to N, the decreasing yield with increasing N rate for NC indicates that the 140 kg N ha^{-1} we applied was near the optimum rate. In 2007, not only did NC have the lowest yield but also it required the highest N rate for achieving its highest seed cotton yield. Yields during 2007 were highly dependent on residue management. The best situation for achieving high seed cotton yields was to have a cover crop and keeping the residue on the soil surface.

The seed cotton yield response to N in 2007 was very similar among RM treatments, 7, 8, and 7 kg of seed cotton per kg of N for NC, REM, and RET (at N rates of 140, 106, and 140 kg ha^{-1}, resp.). No winter cover crop had the highest yield increase relative to the no N control, followed by REM and RET (75, 53, and 43%, resp.), for the previously indicated N rates. As in 2006, in 2007 RET had the lowest yield increment compared to the no N control, even though it had the highest estimated seed cotton yield. This pattern is explained by its greater seed cotton yield when no N was added.

In both years, RET had higher seed cotton yield than REM and NC in the no N control. This result was not expected because the presence of rye residue with a high C/N ratio on the soil surface has been commonly associated with the occurrence of N immobilization, which reduces levels of soil mineral N and decreases yields [10]. In situations with no N added this effect would have a greater negative impact on crop yields. However, results of cotton N uptake between first square and cutout in the no N control averaged across years (Figure 8(a)) followed a similar trend as seed cotton yield. The cotton N uptake in RET was 9 and 15% higher than that in NC and REM, respectively, for the no N control. These results indicate that under the conditions of our experiment N immobilization was not high enough to reduce seed cotton yields in RET.

4. Conclusions

Rye residue management treatments significantly influenced cotton growth parameters and seed cotton yield. In general, cotton population, leaf and plant N concentration, cotton biomass and N uptake at first square, and cotton biomass production between first square and cutout were higher for RET. However, leaf N concentration at early bloom and cotton biomass N concentration between first square and cutout were higher for NC. Leaf N concentration at early bloom, cotton biomass, and N concentration between first square and cutout increased with increasing N rates, when averaged across RM treatments. The highest N uptake was measured in RET, at the highest N rate. In 2006, the highest predicted seed cotton yield corresponded to RET and REM with the application of 140 kg N ha^{-1} (about 3,950 kg ha^{-1}). In 2007, RET had the highest predicted seed cotton yield with 125 kg N ha^{-1} (2,657 kg ha^{-1}) followed by REM with 106 kg N ha^{-1} (2,466 kg ha^{-1}). In both years, the lowest predicted yield was for NC. In 2006, the increase in cotton biomass for RET compared to REM did not necessarily result in an increase in seed cotton yields. However, a stronger association between cotton biomass production and seed cotton yields was observed in the hot and dry 2007 season. Even though RET had low leaf N concentration values at early bloom, it had high yields in both years, indicating that in our study leaf N concentration was not a good predictor of seed cotton yields. Results of this study show that short-term effects of rye residue removal can occur mainly in vegetative cotton parameters, but its effect on seed cotton yield and cotton response to N fertilization would depend more on the characteristics of the season. No rye residue removal effect would be expected in years with average temperatures and rainfall. However, during hot and dry years, rye residue removal may lead to a decrease in cotton yields. We anticipate that cotton N requirements under rye residue removed conditions would not be lower compared to residue retained. The year dependence of rye residue removal impact on seed cotton yields and cotton response to N fertilization suggests that long-term studies are required to strengthen conclusions concerning this management practice.

Abbreviations

RM: Rye residue management
NC: No winter cover crop
REM: Rye residue removed
RET: Rye residue retained
N: Nitrogen
P: Phosphorus
K: Potassium
S: Sulfur
DAP: Days after planting
HU: Heat units.

Acknowledgments

The authors would like to acknowledge the staff of the Field Crops Unit at the E. V. Smith Research Center, especially the late Mr. Bobby Durbin, for their support with this experiment. This paper contributes to the USDA-Agricultural Research Service cross-location Renewable Energy Assessment Project (REAP).

References

[1] Economic Research Service (ERS), "Cotton and Wool Yearbook," 2003, http://usda.mannlib.cornell.edu/usda/ers/89004/2003/table04uplandplantedacreage.xls.

[2] D. J. Boquet and G. A. Breitenbeck, "Nitrogen rate effect on partitioning of nitrogen anti dry matter by cotton," *Crop Science*, vol. 40, no. 6, pp. 1685–1693, 2000.

[3] T. J. Gerik, B. S. Jackson, C. O. Stockle, and W. D. Rosenthal, "Plant nitrogen status and boll load of cotton," *Agronomy Journal*, vol. 86, no. 3, pp. 514–518, 1994.

[4] J. S. McConnell, R. E. Glover, E. D. Vories, W. H. Baker, B. S. Frizzell, and F. M. Bourland, "Nitrogen fertilization and plant development of cotton as determined by nodes above white flower," *Journal of Plant Nutrition*, vol. 18, no. 5, pp. 1027–1036, 1995.

[5] A. S. Hodgson and D. A. MacLeod, "Seasonal and soil fertility effects on the response of waterlogged cotton to foliar-applied nitrogen fertilizer," *Agronomy Journal*, vol. 80, pp. 259–265, 1988.

[6] C. H. Harris and C. W. Smith, "Cotton production affected by row profile and nitrogen rates," *Agronomy Journal*, vol. 72, pp. 919–922, 1980.

[7] D. D. Howard, C. O. Gwathmey, M. E. Essington, R. K. Roberts, and M. D. Mullen, "Nitrogen fertilization of no-till cotton on loess-derived soils," *Agronomy Journal*, vol. 93, no. 1, pp. 157–163, 2001.

[8] H. H. Schomberg, R. G. McDaniel, E. Mallard, D. M. Endale, D. S. Fisher, and M. L. Cabrera, "Conservation tillage and cover crop influences on cotton production on a Southeastern U.S. Coastal Plain Soil," *Agronomy Journal*, vol. 98, no. 5, pp. 1247–1256, 2006.

[9] G. W. Langdale, R. L. Wilson, and R. R. Bruce, "Cropping frequencies to sustain long-term conservation tillage systems," *Soil Science Society of America Journal*, vol. 54, no. 1, pp. 193–198, 1990.

[10] S. M. Dabney, J. A. Delgado, and D. W. Reeves, "Using winter cover crops to improve soil and water quality," *Communications in Soil Science and Plant Analysis*, vol. 32, no. 7-8, pp. 1221–1250, 2001.

[11] S. M. Brown, T. Whitwell, J. T. Touchton, and C. H. Burmester, "Conservation tillage systems for cotton production (USA)," *Soil Science Society of America Journal*, vol. 49, no. 5, pp. 1256–1260, 1985.

[12] M. S. Reiter, D. W. Reeves, and C. H. Burmester, "Nitrogen management for cotton in a high-residue cover crop conservation tillage system," in *Making Conservation Tillage Conventional: Building a Future on 25 Years of Research, Proceedings of the 25th Southern Conservation Tillage Conference for Sustainable Agriculture*, E. van Santen, Ed., pp. 136–141, Auburn University, Auburn, Ala, USA, 2002.

[13] M. G. Wagger, "Time of dessication effects on plant composition and subsequent nitrogen release form several winter annual cover crops," *Agronomy Journal*, vol. 81, pp. 236–241, 1989.

[14] J. J. Varco, S. R. Spurlock, and O. R. Sanabria-Garro, "Profitability and nitrogen rate optimization associated with winter cover management in no-tillage cotton," *Journal of Production Agriculture*, vol. 12, no. 1, pp. 91–95, 1999.

[15] C. C. Mitchell, "Fertility," in *Conservation Tillage Cotton Production Guide—ANR-952*, C. D. Monks and M. G. Paterson, Eds., pp. 7–9, Agronomy and Soils Department, Auburn University, Auburn, Ala, USA, 1996.

[16] D. Dinnes, D. Jaynes, T. Kaspar, T. Colvin, C. A. Cambardella, and D. L. Karlen, "Plant-soil-microbe N relationships in high residue management systems," 2007, http://www.sdnotill.com/Newsletters/Relationships.pdf.

[17] G. Siri-Prieto, D. W. Reeves, and R. L. Raper, "Tillage requirements for integrating winter-annual grazing in cotton production: plant water status and productivity," *Soil Science Society of America Journal*, vol. 71, no. 1, pp. 197–205, 2007.

[18] C. E. Clapp, R. R. Allmaras, M. F. Layese, D. R. Linden, and R. H. Dowdy, "Soil organic carbon and 13C abundance as related to tillage, crop residue, and nitrogen fertilization under continuous corn management in Minnesota," *Soil and Tillage Research*, vol. 55, no. 3-4, pp. 127–142, 2000.

[19] B. Govaerts, M. Fuentes, M. Mezzalama et al., "Infiltration, soil moisture, root rot and nematode populations after 12 years of different tillage, residue and crop rotation managements," *Soil and Tillage Research*, vol. 94, no. 1, pp. 209–219, 2007.

[20] S. S. Malhi, R. Lemke, Z. H. Wang, and B. S. Chhabra, "Tillage, nitrogen and crop residue effects on crop yield, nutrient uptake, soil quality, and greenhouse gas emissions," *Soil and Tillage Research*, vol. 90, no. 1-2, pp. 171–183, 2006.

[21] W. W. Wilhelm, J. M. F. Johnson, J. L. Hatfield, W. B. Voorhees, and D. R. Linden, "Crop and soil productivity response to corn residue removal: a literature review," *Agronomy Journal*, vol. 96, no. 1, pp. 1–17, 2004.

[22] H. Tsuji, H. Yamamoto, K. Matsuo, and K. Usuki, "The effects of long-term conservation tillage, crop residues and P fertilizer

on soil conditions and responses of summer and winter crops on an Andosol in Japan," *Soil and Tillage Research*, vol. 89, no. 2, pp. 167–176, 2006.

[23] J. R. Salinas-Garcia, A. D. Báez-González, M. Tiscareño-López, and E. Rosales-Robles, "Residue removal and tillage interaction effects on soil properties under rain-fed corn production in Central Mexico," *Soil and Tillage Research*, vol. 59, no. 1-2, pp. 67–79, 2001.

[24] D. K. Cassel and M. G. Wagger, "Residue management for irrigated maize grain and silage production," *Soil and Tillage Research*, vol. 39, no. 1-2, pp. 101–114, 1996.

[25] J. W. Doran, W. W. Wilhelm, and J. F. Power, "Crop residue removal and soil productivity with no-till corn, sorghum, and soybean," *Soil Science Society of America Journal*, vol. 48, no. 3, pp. 640–645, 1984.

[26] J. C. Zadoks, T. T. Chang, and C. F. Konzak, "A decimal code for the growth of cereals," *Weed Research*, vol. 14, pp. 415–421, 1974.

[27] J. F. Adams and C. C. Mitchell, "Soil test nutrient recommendations for Alabama crops," 2000, http://www.ag.auburn.edu/agrn/croprecs/CropRecs/cc10.html.

[28] S. Peng, D. R. Krieg, and S. K. Hicks, "Cotton lint yield response to accumulated heat units and soil water supply," *Field Crops Research*, vol. 19, no. 4, pp. 253–262, 1989.

[29] R. C. Littell, G. A. Milliken, W. W. Stroup, R. D. Wolfinger, and O. Schabenberger, *SAS for Mixed Models*, SAS Institute, Cary, NC, USA, 2nd edition, 2006.

[30] C. C. Mitchell and S. Tu, "Long-term evaluation of poultry litter as a source of nitrogen for cotton and corn," *Agronomy Journal*, vol. 97, no. 2, pp. 399–407, 2005.

[31] C. W. Bednarz, D. C. Bridges, and S. M. Brown, "Analysis of cotton yield stability across population densities," *Agronomy Journal*, vol. 92, no. 1, pp. 128–135, 2000.

[32] P. F. Bell, D. J. Boquet, E. Millholln et al., "Relationship between leaf-blade nitrogen and relative seed cotton yields," *Crop Science*, vol. 43, pp. 1367–1374, 2003.

[33] H. A. Mills and J B. Jones, *Plant Analysis Handbook II: A Practical Sampling, Preparation, Analysis, and Interpretation Guide*, Micro-Macro, Athens, Ga, USA, 1996.

[34] J. L. Fridgen and J. J. Varco, "Dependency of cotton leaf nitrogen, chlorophyll, and reflectance on nitrogen and potassium availability," *Agronomy Journal*, vol. 96, no. 1, pp. 63–69, 2004.

[35] K. S. Balkcom, J. N. Shaw, D. W. Reeves, C. H. Burmester, and L. M. Curtis, "Irrigated cotton response to tillage systems in the Tennessee Valley," *Journal of Cotton Science*, vol. 11, no. 1, pp. 2–11, 2007.

[36] D. M. Bassett, W. D. Anderson, and C. H. E. Werkhoven, "Dry matter production and nutrient uptake in irrigated cotton (*Gossypium hirsutum*)," *Agronomy Journal*, vol. 62, pp. 299–303, 1970.

[37] G. L. Mullins and C. H. Burmester, "Dry matter, nitrogen, phosphorus, and potassium accumulation by four cotton varieties," *Agronomy Journal*, vol. 82, pp. 729–736, 1990.

[38] F. Gastal and G. Lemaire, "N uptake and distribution in crops: an agronomical and ecophysiological perspective," *Journal of Experimental Botany*, vol. 53, no. 370, pp. 789–799, 2002.

[39] E. L. Clawson, J. T. Cothren, and D. C. Blouin, "Nitrogen fertilization and yield of cotton in ultra-narrow and conventional row spacings," *Agronomy Journal*, vol. 98, no. 1, pp. 72–79, 2006.

[40] P. J. Wiatrak, D. L. Wright, J. J. Marois, W. Koziara, and J. A. Pudelko, "Tillage and nitrogen application impact on cotton following wheat," *Agronomy Journal*, vol. 97, no. 1, pp. 288–293, 2005.

[41] D. J. Boquet, R. L. Hutchinson, and G. A. Breitenbeck, "Long-term tillage, cover crop, and nitrogen rate effects on cotton: yield and fiber properties," *Agronomy Journal*, vol. 96, no. 5, pp. 1436–1442, 2004.

[42] J. S. McConnell, W. H. Baker, D. M. Miller, B. S. Frizzell, and J. J. Varvil, "Nitrogen fertilization of cotton cultivars of differing maturity," *Agronomy Journal*, vol. 85, no. 6, pp. 1151–1156, 1993.

[43] D. J. Boquet, E. B. Moser, and G. A. Breitenbeck, "Boll weight and within-plant yield distribution in field-grown cotton given different levels of nitrogen," *Agronomy Journal*, vol. 86, no. 1, pp. 20–26, 1994.

Permissions

The contributors of this book come from diverse backgrounds, making this book a truly international effort. This book will bring forth new frontiers with its revolutionizing research information and detailed analysis of the nascent developments around the world.

We would like to thank all the contributing authors for lending their expertise to make the book truly unique. They have played a crucial role in the development of this book. Without their invaluable contributions this book wouldn't have been possible. They have made vital efforts to compile up to date information on the varied aspects of this subject to make this book a valuable addition to the collection of many professionals and students.

This book was conceptualized with the vision of imparting up-to-date information and advanced data in this field. To ensure the same, a matchless editorial board was set up. Every individual on the board went through rigorous rounds of assessment to prove their worth. After which they invested a large part of their time researching and compiling the most relevant data for our readers. Conferences and sessions were held from time to time between the editorial board and the contributing authors to present the data in the most comprehensible form. The editorial team has worked tirelessly to provide valuable and valid information to help people across the globe.

Every chapter published in this book has been scrutinized by our experts. Their significance has been extensively debated. The topics covered herein carry significant findings which will fuel the growth of the discipline. They may even be implemented as practical applications or may be referred to as a beginning point for another development. Chapters in this book were first published by Hindawi Publishing Corporation; hereby published with permission under the Creative Commons Attribution License or equivalent.

The editorial board has been involved in producing this book since its inception. They have spent rigorous hours researching and exploring the diverse topics which have resulted in the successful publishing of this book. They have passed on their knowledge of decades through this book. To expedite this challenging task, the publisher supported the team at every step. A small team of assistant editors was also appointed to further simplify the editing procedure and attain best results for the readers.

Our editorial team has been hand-picked from every corner of the world. Their multi-ethnicity adds dynamic inputs to the discussions which result in innovative outcomes. These outcomes are then further discussed with the researchers and contributors who give their valuable feedback and opinion regarding the same. The feedback is then collaborated with the researches and they are edited in a comprehensive manner to aid the understanding of the subject.

Apart from the editorial board, the designing team has also invested a significant amount of their time in understanding the subject and creating the most relevant covers. They scrutinized every image to scout for the most suitable representation of the subject and create an appropriate cover for the book.

The publishing team has been involved in this book since its early stages. They were actively engaged in every process, be it collecting the data, connecting with the contributors or procuring relevant information. The team has been an ardent support to the editorial, designing and production team. Their endless efforts to recruit the best for this project, has resulted in the accomplishment of this book. They are a veteran in the field of academics and their pool of knowledge is as vast as their experience in printing. Their expertise and guidance has proved useful at every step. Their uncompromising quality standards have made this book an exceptional effort. Their encouragement from time to time has been an inspiration for everyone.

The publisher and the editorial board hope that this book will prove to be a valuable piece of knowledge for researchers, students, practitioners and scholars across the globe.

List of Contributors

Olivia E. Saunders, Craig G. Cogger and Andy I. Bary
Department of Crop and Soil Sciences, Washington State University, Pullman, WA 99164, USA

Ann-Marie Fortuna
Department of Crop and Soil Sciences, Washington State University, Pullman, WA 99164, USA
Department of Soil Science, North Dakota State University, 1301 12th Avenue North, Fargo, ND 58105, USA

Joe H. Harrison and Elizabeth Whitefield
Department of Animal Sciences, Washington State University, Puyallup, WA 98371, USA

Ann C. Kennedy
Land Management and Water Conservation Research Unit, USDA-ARS, Washington State University, Pullman, WA 99164, USA

Rodrick D. Lentz and Gary A. Lehrsch
North West Irrigation and Soils Research Laboratory, USDA-ARS, 3793 N 3600 W, Kimberly, ID, 83341 5076, USA

Katy Butchee
Department of Agriculture, Western Oklahoma State College, 2801 N. Main, Altus, OK 73521, USA

Daryl B. Arnall, Apurba Sutradhar, Chad Godsey, Hailin Zhang and Chad Penn
Department of Plant and Soil Sciences, Oklahoma State University, Stillwater, OK 74078, USA

Moncef Hammami
University of Carthage, High School of Agriculture of Mateur, 7030 Mateur, Tunisia

Khemaies Zayani
University of Carthage/High Institute of Environmental Sciences and Technologies of Borj Cedria, B.P. 1003, 2050 Hammam Lif, Tunisia

Hédi Ben Ali
Agricultural Investment Promotion Agency, 6000 Gab`es, Tunisia

Jonathan J. Halvorson and Javier M. Gonzalez
USDA-ARS Appalachian Farming Systems Research Center, 1224 Airport Road, Beaver, WV 25813-9423, USA

Ann E. Hagerman
Department of Chemistry and Biochemistry, Miami University, 160 Hughes Laboratories, 701 East High Street, Oxford, OH 45056, USA

Aman Chandi, David L. Jordan, Alan C. York and Susana R. Milla-Lewis
Department of Crop Science, North Carolina State University, P.O. Box 7620, Raleigh, NC 27695-7620, USA

James D. Burton
Department of Horticulture Science, North Carolina State University, P.O. Box 7609, Raleigh, NC 27695, USA

A. Stanley Culpepper
Department of Crop and Soil Sciences, University of Georgia, P.O. Box 478, Tifton, GA 31794, USA

Jared R. Whitaker
Department of Crop and Soil Sciences, University of Georgia-Southeast District, P.O. Box 8112, Statesboro, GA 30460, USA

Aman Chandi, Susana R.Milla-Lewis, David L. Jordan and Alan C. York
Department of Crop Science, North Carolina State University, Box 7620, Raleigh, NC 27695, USA

Darci Giacomini and Philip Westra
Department of Bio agricultural Sciences and Pest Management, Colorado State University, 1179 Campus Delivery, Fort Collins, CO 80523, USA

Christopher Preston
School of Agriculture, Food & Wine, University of Adelaide, Adelaide, SA 5005, Australia

James D. Burton
Department of Horticulture Science, North Carolina State University, Box 7609, Raleigh, NC 27695, USA

Jared R. Whitaker
Crop and Soil Sciences, University of Georgia Southeast District, P.O. Box 8112, Statesboro, GA 30460, USA

Kingdom Kwapata, Thang Nguyen and Mariam Sticklen
Department of Plant, Soil and Microbial Sciences, Michigan State University, East Lansing, MI 48824, USA

Zine El Abidine Fellahi and Abderrahmane Hannachi
National Institute of Agricultural Research, Setif Agricultural Research Unit, Setif 19000, Algeria

Hamenna Bouzerzour
Faculty of Life and Natural Sciences, Ecology and Plant Biology Department, University of Ferhat Abbas, Setif 1 19000, Algeria

Ammar Boutekrabt
Faculty of Agro-Veterinary and Biological Sciences, Agronomy Department, University of Saad Dahlab, Blida 09000, Algeria

Dexter B. Watts and H. A. Torbert
National Soil Dynamics Laboratory, USDA-ARS, 411 S. Donahue Drive, Auburn, AL 36832, USA

Katy E. Smith
Department of Math, Science, and Technology, University of Minnesota-Crookston, 2900 University Avenue, Crookston, MN 56716, USA

Shilpi Chawla
Department of Plant and Soil Science, Texas Tech University, Lubbock, TX 79409, USA

Jason E. Woodward
Department of Plant and Soil Science, Texas Tech University, Lubbock, TX 79409, USA
Texas AgriLIFE Extension Service, Texas A&M System, Lubbock, TX 79403, USA

Terry A. Wheeler
Texas AgriLIFE Research, Texas A&M System, Lubbock, TX 79403, USA

Carlos J. Fernandez and W. James Grichar
Texas AgriLife Research, Corpus Christi, TX 78406, USA

Dan D. Fromme
Texas AgriLife Extension Service, Corpus Christi, TX 78406, USA

E. Bonora and G. Costa
Department of Agricultural Sciences, 46 Via Fanin, 40127 Bologna, Italy

D. Stefanelli
Department of Primary Industries, Knoxfield Centre, Private Bag 15, 621 Burwood Highway, Ferntree Gully 3156, Australia

Gisela Jansen, Hans-Ulrich Jürgens, Edgar Schliephake and Frank Ordon
Institute for Resistance Research and Stress Tolerance, Julius Kuhn-Institut (JKI), Federal Research Centre for Cultivated Plants, 18190 Sanitz, Germany

W. James Grichar
Texas AgriLife Research, 3507 Hwy 59E, Beeville, TX 78102, USA

Peter A. Dotray
Texas Tech University, Texas AgriLife Research, and Texas AgriLife Extension Service, Lubbock, TX 79403, USA

Todd A. Baughman
Texas AgriLife Extension Service, Vernon, TX 76384, USA

Graham Brodie, Carmel Ryan and Carmel Lancaster
Melbourne School of Land and Environment, Dookie Campus, University of Melbourne, Nalinga Road, Dookie, VIC 3647, Australia

Chance W. Riggins and Patrick J. Tranel
Department of Crop Sciences, University of Illinois, 1201 West Gregory Drive, Urbana, IL 61801, USA

Richard Sicher
Crop Systems & Global Change Laboratory, Agricultural Research Service-USDA, Room 342, Building 001, BARC-west, 10300 Baltimore Avenue, Beltsville, MD 20705, USA

Mark S. Reiter
Department of Crop and Soil Environmental Sciences, Eastern Shore Agricultural Research and Extension Center, Virginia Tech, Painter, VA 23420, USA

Steven L. Rideout
Department of Plant Pathology, Physiology and Weed Science, Eastern Shore Agricultural Research and Extension Center, Virginia Tech, Painter, VA 23420, USA

Joshua H. Freeman
Department of Horticulture, Eastern Shore Agricultural Research and Extension Center, Virginia Tech, Painter, VA 23420, USA

Ian M. Haigh
Field Trials Department, Charles River, Tranent, Edinburgh, EH33 2NE, UK

Martin C. Hare
Crop and Enviromental Sciences, Harper Adams University College, Edgmond, Shropshire, TF10 8NB, UK

Radwan M. Barakat and Mohammad I. AL-Masri
Plant Protection Research Center, Hebron University, P.O. Box 40, Hebron, Palestine

W. J. Grichar
Texas A&M AgriLife Research, 3507 Highway 59E, Beeville, TX 78102, USA

P. A. Dotray
Texas A&M AgriLife Research and Texas A&M AgriLife Extension Service, Texas Tech University, Lubbock, TX 79403, USA

M. R. Baring
Texas A&M AgriLife Research, Department of Soil and Crop Sciences, College Station 77843, USA

C. Jiménez Martínez and G. Dávila-Ortiz
Departamento de Graduados e Investigaci´on en Alimentos, Escuela Nacional de Ciencias Biol´ogicas, Instituto Polit´ecnico Nacional, Prol de Carpio y Plan de Ayala, 11340 M´exico, DF, Mexico

A. Cardador Martínez
Biotecnolog´ıa Alimentaria, Instituto Tecnol´ogico y de Estudios Superiores de Monterrey, Campus Quer´etaro, Avenida Epigmenio Gonz´alez 500, Fraccionamiento San Pablo, 76130 Santiago de Quer´etaro, QRO, Mexico

A. L. Martinez Ayala
Centro de Investigaci´on en Biotecnolog´ıa Aplicada, IPN, Carretera Estatal Tecuexcomac-Tepetitla Km 1.5, 90700 Tepetitla de Lardiz´abal, TLAX, Mexico

M. Muzquiz and M. Martin Pedrosa
Departamento de Tecnolog´ıa de Alimentos, INIA, Apdo 8111, 28080 Madrid, Spain

Roberto P. Marano
Department of Environmental Science, National University of Litoral, Kreder 2805, 3080 Esperanza, Argentina

Roxana L. Maumary, Laura N. Fernandez and Luis M. Rista
Department of Vegetal Production, National University of Litoral, Kreder 2805, 3080 Esperanza, Argentina

Kapila Shekhawat, S. S. Rathore, O. P. Premi, B. K. Kandpal and J. S. Chauhan
Directorate of Rapeseed-Mustard Research, Sewar, Rajasthan Bharatpur 321 303, India

F. Ducamp, E. van Santen and C. C. Mitchell
Department of Agronomy and Soils, Auburn University, 202 Funchess Hall, Auburn, AL 36849, USA

F. J. Arriaga, K. S. Balkcom and S. A. Prior
USDA-ARS National Soil Dynamics Laboratory, 411 S. Donahue Drive, Auburn, AL 36832, USA

www.ingramcontent.com/pod-product-compliance
Lightning Source LLC
Chambersburg PA
CBHW080637200326
41458CB00013B/4656